Springer Series in Statistics

Perspectives in Statistics

Springer
New York
Berlin
Heidelberg
Barcelona
Budapest
Hong Kong
London
Milan
Paris
Santa Clara
Singapore
Tokyo

Springer Series in Statistics

Andersen/Borgan/Gill/Keiding: Statistical Models Based on Counting Processes.
Andrews/Herzberg: Data: A Collection of Problems from Many Fields for the Student and Research Worker.
Anscombe: Computing in Statistical Science through APL.
Berger: Statistical Decision Theory and Bayesian Analysis, 2nd edition.
Bolfarine/Zacks: Prediction Theory for Finite Populations.
Borg/Groenen: Modern Multidimensional Scaling: Theory and Applications
Brémaud: Point Processes and Queues: Martingale Dynamics.
Brockwell/Davis: Time Series: Theory and Methods, 2nd edition.
Daley/Vere-Jones: An Introduction to the Theory of Point Processes.
Dzhaparidze: Parameter Estimation and Hypothesis Testing in Spectral Analysis of Stationary Time Series.
Fahrmeir/Tutz: Multivariate Statistical Modelling Based on Generalized Linear Models.
Farrell: Multivariate Calculation.
Federer: Statistical Design and Analysis for Intercropping Experiments.
Fienberg/Hoaglin/Kruskal/Tanur (Eds.): A Statistical Model: Frederick Mosteller's Contributions to Statistics, Science and Public Policy.
Fisher/Sen: The Collected Works of Wassily Hoeffding.
Good: Permutation Tests: A Practical Guide to Resampling Methods for Testing Hypotheses.
Goodman/Kruskal: Measures of Association for Cross Classifications.
Gouriéroux: ARCH Models and Financial Applications.
Grandell: Aspects of Risk Theory.
Haberman: Advanced Statistics, Volume I: Description of Populations.
Hall: The Bootstrap and Edgeworth Expansion.
Härdle: Smoothing Techniques: With Implementation in S.
Hart: Nonparametric Smoothing and Lack-of-Fit Tests.
Hartigan: Bayes Theory.
Heyer: Theory of Statistical Experiments.
Huet/Bouvier/Gruet/Jolivet: Statistical Tools for Nonlinear Regression: A Practical Guide with S-PLUS Examples.
Jolliffe: Principal Component Analysis.
Kolen/Brennan: Test Equating: Methods and Practices.
Kotz/Johnson (Eds.): Breakthroughs in Statistics Volume I.
Kotz/Johnson (Eds.): Breakthroughs in Statistics Volume II.
Kotz/Johnson (Eds.): Breakthroughs in Statistics Volume III.
Kres: Statistical Tables for Multivariate Analysis.
Küchler/Sørensen: Exponential Families of Stochastic Processes.
Le Cam: Asymptotic Methods in Statistical Decision Theory.
Le Cam/Yang: Asymptotics in Statistics: Some Basic Concepts.
Longford: Models for Uncertainty in Educational Testing.
Manoukian: Modern Concepts and Theorems of Mathematical Statistics.
Miller, Jr.: Simultaneous Statistical Inference, 2nd edition.
Mosteller/Wallace: Applied Bayesian and Classical Inference: The Case of *The Federalist Papers.*

(continued after index)

Samuel Kotz Norman L. Johnson
Editors

Breakthroughs in Statistics
Volume III

Samuel Kotz
Department of Statistics
The George Washington University
Washington, DC 20052
USA

Norman L. Johnson
Department of Statistics
Phillips Hall
The University of North Carolina
at Chapel Hill
Chapel Hill, NC 27599
USA

Library of Congress Cataloging-in-Publication Data
Breakthroughs in statistics / Samuel Kotz, Norman L. Johnson, editors.
p. cm. — (Springer series in statistics. Perspectives in statistics)
Includes bibliographical references and index.
Contents: v. 2. Methodology — v. 1. Foundations and basic theory.
v. 2. Methodology and distribution.
ISBN 0-387-94039-1
1. Mathematical statistics. I. Kotz, Samuel. II. Johnson, Norman Lloyd. III. Series.
QA276.B68465 1993
519.5—dc20 93-3854

Printed on acid-free paper.

Production coordinated by Brian Howe and managed by Bill Imbornoni; manufacturing supervised by Jeffrey Taub.
Typeset by Asco Trade Typesetting Ltd., Hong Kong.
Printed and bound by Edwards Brothers, Inc., Ann Arbor, MI.
Printed in the United States of America.

9 8 7 6 5 4 3 2 1

ISBN 0-387-94988-7 Springer-Verlag New York Berlin Heidelberg (hardcover) SPIN 10557449
ISBN 0-387-94989-5 Springer-Verlag New York Berlin Heidelberg (softcover) SPIN 10557473

Preface to Volume III

The Preface to Volumes I and II of *Breakthroughs in Statistics* concludes:

> If it happens that our first sample proves insufficiently representative, we may be able to consider taking another sample (perhaps of similar size, *without replacement*).

Inevitably our first sample could not hope to be fully representative of influential publications during the past 100 years or so, over so broad a field. The possibility of our considering "taking another sample" owed its origin to the initiative of Dr. John Kimmel of Springer-Verlag. We welcomed the opportunity to supplement our original selections, and hope that readers will find these as interesting and informative as the first group.

The present collection includes one paper (Galton, 1889) earlier than any in the first two volumes, and several papers later than any in those volumes. Selection of the latter called for considerable care. In common with the literature in many fields of scientific endeavor, there has been a remarkable expansion in the amount of published material, on topics relevant to this volume, in recent years. This can be associated with several factors—increases in numbers, of workers in statistical sciences, and of journals devoted to their interests, and also speed of publication (in some cases). It has been said that output over the last 10 years is comparable to that over a 25-year "Golden Era"—of expansion of statistical theory and applications—in the middle fifties of the present century, and perhaps to that of the whole of the nineteenth century. It is quite possible that the *average* content of the more recent publications is lower than in those of earlier periods, in regard to groundbreaking contributions, but there is undoubtedly much of value for the future among them. Consequently, while we needed to avoid the possibility of being unduly influenced by passing fashions, we tried not to allow the profusion and variety of recent rapid developments to cause us to "not see the wood for the trees." This risk was, however, deemed to be worth taking, in the interests of representativeness. Our hope of "similar size" has not been fulfilled, but we feel that we have been able to achieve a worthwhile enrichment of the collection. The contents of this third volume are not classified into the two main categories—*Foundations and Basic Theory* (Vol. I) and *Methodology and Distribution* (Vol. II).

As before, we thank scientific colleagues from all over the world who provided us with their mature judgment on publications that might be considered for inclusion. We have not always acted on their suggestions, but have taken them all into account. We thank, especially, those who have written commentaries on the chosen papers. Their work constitutes an essential part of the collection, and indispensable assistance for readers.

We expect (and hope) for comments (constructive or otherwise) on our selection of items. This is our own responsibility, but the opinions of the commentators are, of course, their own.

Despite the present widespread use of electronic reproduction, the publishers have decided to retain traditional typeset methods (which have served humankind since the fifteenth century) for the present volume, which can thus be regarded as being among the "last of the Mohicans" in this respect.

Samuel Kotz
College Park, Maryland
November 1996

Norman L. Johnson
Chapel Hill, North Carolina
November 1996

Preface to Volumes I and II

> McCrimmon, having gotten Grierson's attention, continued: "A breakthrough, you say? If it's in economics, at least it can't be dangerous. Nothing like gene engineering, laser beams, sex hormones or international relations. That's where we don't want any breakthroughs." (Galbraith, J.K. (1990) *A Tenured Professor*, Houghton Mifflin: Boston.)

> To judge [*astronomy*] in this way [*a narrow utilitarian point of view*] demonstrates not only how poor we are, but also how small, narrow, and indolent our minds are; it shows a disposition always to calculate the payoff before the work, a cold heart and a lack of feeling for everything that is great and honors man. One can unfortunately not deny that such a mode of thinking is not uncommon in our age, and I am convinced that this is closely connected with the catastrophes which have befallen many countries in recent times; do not mistake me, I do not talk of the general lack of concern for science, but of the source from which all this has come, of the tendency to everywhere look out for one's advantage and to relate everything to one's physical well-being, of the indifference towards great ideas, of the aversion to any effort which derives from pure enthusiasm: I believe that such attitudes, if they prevail, can be decisive in catastrophes of the kind we have experienced. [Gauss, K.F.: *Astronomische Antrittsvorlesung* (cited from Buhler, W.K. (1981) *Gauss: A Biographical Study*, Springer-Verlag: New York)].

This collection of papers (reproduced in whole or in part) is an indirect outcome of our activities during the decade 1979–1988, in the course of compiling and editing the *Encyclopedia of Statistical Science* (nine volumes and a supplementary volume published by John Wiley and Sons, New York). It is also, and more directly, motivated by a more recent project, a systematic rereading and assessment of Presidential Addresses delivered to the Royal Statistical Society, the International Statistical Institute, and the American Statistical Association during the last 50 years.

Our studies revealed a growing, and already embarrassingly noticeable, diversification among the statistical sciences that borders on fragmentation. Although our belief in the unified nature of statistics remains unshaken, we must recognize certain dangers in this steadily increasing diversity accompanying the unprecedented penetration of statistical methodology into many branches of the social, life, and natural sciences, and engineering and other applied fields.

The initial character of statistics as the "science of state" and the attitudes summed up in the Royal Statistical Society's original motto (now abandoned) of *aliis exterendum* ("let others thresh")—reflecting the view that statisticians are concerned solely with the collection of data—have changed dramatically over the last 100 years and at an accelerated rate during the last 25 years.

To trace this remarkably vigorous development, it seemed logical (to us) to search for "growth points" or "breakthrough" publications that have initiated fundamental changes in the development of statistical methodology. It also seemed reasonable to hope that the consequences of such a search might result in our obtaining a clearer picture of likely future developments. The present collection of papers is an outcome of these thoughts.

In the selection of papers for inclusion, we have endeavored to identify papers that have had lasting effects, rather than search to establish priorities. However, there are introductions to each paper that do include references to important precursors, and also to successor papers elaborating on or extending the influence of the chosen papers.

We were fortunate to have available S.M. Stigler's brilliant analysis of the history of statistics up to the beginning of the twentieth century in his book, *The History of Statistics: The Measurement of Uncertainty* (Belknap Press, Cambridge, MA, 1986), which, together with Claire L. Parkinson's *Breakthroughs: A Chronology of Great Achievements in Science and Mathematics* 1200–1930 (G.K. Hall, Boston, MA, 1985), allowed us to pinpoint eleven major breakthroughs up to and including F. Galton's *Natural Inheritance*. These are, in chronological order, the following:

C. Huyghens (1657). *De Ratiociniis in Aleae Ludo* (Calculations in Games of Dice), in *Exercitationum Mathematicarum* (F. van Schooten, ed.). Elsevier, Leiden, pp. 517–534.
(The concept of *mathematical expectation* is introduced, as well as many examples of combinatorial calculations.)
J. Graunt (1662). *Natural and Political Observations Mentioned in a Following Index and Made upon the Bills of Mortality.* Martyn and Allestry, London.
(Introduced the idea that vital statistics are capable of scientific analysis.)
E. Halley (1693). An estimate of the degrees of mortality of mankind, drawn

from the curious "*Tables of the Births and Funerals* at the City of Breslaw; with an attempt to ascertain the price of *annuities* upon *lives*." *Philos. Trans. Roy. Soc. London*, **17**, 596–610, 654–656.
[Systematized the ideas in Graunt (1662).]

J. Arbuthnot (1711). An argument for Divine Providence, taken from the constant regularity observed in the births of both sexes. *Philos. Trans. Roy. Soc. London*, **27**, 186–190.
(This is regarded as the first use of a test of significance, although not described as such explicitly.)

J. Bernoulli (1713). *Ars Conjectandi* (The Art of Conjecture). Thurnisorium, Basel.
(Development of combinatorial methods and concepts of statistical inferences.)

A. De Moivre (1733, 1738, 1756). *The Doctrine of Chances*, 1st–3rd eds. Woodfall, London.
(In these three books, the normal curve is obtained as a limit and as an approximation to the binomial.)

T. Bayes (1763). Essay towards solving a problem in the doctrine of chances. *Philos. Trans. Roy. Soc. London*, **53**, 370–418.
(This paper has been the source of much work on inverse probability. Its influence has been very widespread and persistant, even among workers who insist on severe restrictions on its applicability.)

P.S. Laplace (1812). *Théorie Analytique des Probabilités*. Courcier, Paris.
(The originating inspiration for much work in probability theory and its applications during the nineteenth century. Elaboration of De Moivre's work on normal distributions.)

K.F. Gauss (1823). *Theoria Combinationis Observationum Erroribus Minimis Obnoxiae*. Dieterich, Gottingen.
(The method of least squares and associated analysis have developed from this book, which systematized the technique introduced by A.M. Legendre in 1805. Also, the use of "optimal principles" in choosing estimators.)

L.A.J. Quetelet (1846). *Lettres à S.A.R. le Duc Regnant de Saxe-Cobourg et Gotha, sur la Théorie des Probabilités, appliquée aux Sciences Morales et Politiques*. Hayez, Brussels. (English translation, Layton, London, 1849.)
(Observations on the stability of certain demographic indices provided empirical evidence for applications of probability theory.)

F. Galton (1889). *Natural Inheritance*. Macmillan, London.
[This book introduces the concepts of correlation and regression; also mixtures of normal distributions and the bivariate normal distribution. Its importance derives largely from the influence of Karl Pearson. In regard to correlation, an interesting precursor, by the same author, is "Co-relations and their measurement, chiefly from anthropometric data," *Proc. Roy. Soc. London*, **45**, 135–145 (1889).]

In our efforts to establish subsequent breakthroughs in our period of study (1890–1989), we approached some 50 eminent (in our subjective evaluation) statisticians, in various parts of the world, asking them if they would supply us with "at least five (a few extra beyond five is very acceptable) possibly suitable references ...". We also suggested that some "explanations of reasons for choice" would be helpful.

The response was very gratifying. The requests were sent out in June–July 1989; during July–August, we received over 30 replies, with up to 10 references each, the modal group being 8. There was remarkable near-unanimity recommending the selection of the earlier work of K. Pearson, "Student," R.A. Fisher, and J. Neyman and E.S. Pearson up to 1936. For the years following 1940, opinions became more diverse, although some contributions, such as A. Wald (1945), were cited by quite large numbers of respondents. After 1960, opinions became sharply divergent. The latest work cited by a substantial number of experts was B. Efron (1979). A number of replies cautioned us against crossing into the 1980s, since some time needs to elapse before it is feasible to make a sound assessment of the long-term influence of a paper. We have accepted this viewpoint as valid.

Originally, we had planned to include only 12 papers (in whole or in part). It soon became apparent, especially given the diversity of opinions regarding the last 50 years, that the field of statistical sciences is now far too rich and heterogeneous to be adequately represented by 12 papers over the last 90 years. In order to cover the field satisfactorily, it was decided that at least 30 references should be included. After some discussion, the publisher generously offered to undertake two volumes, which has made it possible to include 39 references! Assignment to the two volumes is on the basis of broad classification into "Foundations and Basic Theory" (Vol. I) and "Methodology and Distribution" (Vol. II). Inevitably, there were some papers that could reasonably have appeared in either volume. When there was doubt, we resolved it in such a way as to equalize the size of the two volumes, so far as possible. There are 19 Introductions in the first volume and 20 in the second. In addition, we have included Gertrude Cox's 1956 Presidential Address "Frontiers of Statistics" to the American Statistical Association in Vol. I, together with comments from a number of eminent statisticians indicating some lines on which statistical thinking and practice have developed in the succeeding years.

Even with the extension to two volumes, in order to keep the size of the books within reasonable limits, we found it necessary to reproduce only those parts of the papers that were relevant to our central theme of recording "breakthroughs," points from which subsequent growth can be traced. The necessary cutting caused us much "soul-searching," as did also the selection of papers for inclusion. We also restricted rather severely the lengths of the Introductions to individual items. We regret that practical requirements made it necessary to enforce these restrictions. We also regret another consequence of the need to reduce size—namely, our inability to follow much of the advice of our distinguished correspondents, even though it was most cogently advocated. In certain instances the choice was indeed difficult, and a decision was reached only after long discussions. At this point, we must admit that we have included two or three choices of our own that appeared only sparsely among the experts' suggestions.

The division between the two volumes is necessarily somewhat arbitrary. Some papers could equally appear in either. However, papers on fundamental concepts such as probability and mathematical foundation of statistical inference are clearly more Vol. I than Vol. II material (though not entirely so, because concepts can influence application).

There have been laudable and commendable efforts to put the foundations of statistical inference, and more especially probability theory on a sound footing, according to the viewpoint of mathematical self-consistency. Insofar as these may be regarded as attempts to reconcile abstract mathematical logic with phenomena observed in the real world—via interpretation (subjective or objective) of data—we feel that the aim may be too ambitious and even doomed to failure. We are in general agreement with the following remarks of the physicist H.R. Pagels:

> Centuries ago, when some people suspended their search for absolute truth and began instead to ask how things worked, modern science was born. Curiously, it was by abandoning the search for absolute truth that science began to make progress, opening the material universe to human exploration. It was only by being provisional and open to change, even radical change, that scientific knowledge began to evolve. And ironically, its vulnerability to change is the source of its strength. (From *Perfect Symmetry: The Search for the Beginning of Time*, Simon and Schuster, New York, 1985, p. 370.)

It is evident that this work represents the fruits of collaboration among many more individuals than the editors. Our special thanks go to the many distinguished statisticians who replied to our inquiries, in many cases responding to further "follow-up" letters requiring additional effort in providing more details that we felt were desirable; we also would like to thank those who have provided Introductions to the chosen papers. The latter are acknowledged at the appropriate places where their contributions occur.

We take this opportunity to express our gratitude to Dean R.T. Lamone of the College of Business and Professor B.L. Golden, Chairman of the Department of Management Science and Statistics at the University of Maryland at College Park, and to Professor S. Cambanis, Chairman of the Department of Statistics at the University of North Carolina at Chapel Hill, for their encouragement and the facilities they provided in support of our work on this project.

We are also grateful to the various persons and organizations who have given us reprint permission. They are acknowledged, together with source references, in the section "Sources and Acknowledgments."

We welcome constructive criticism from our readers. If it happens that our first sample proves insufficiently representative, we may be able to consider taking another sample (perhaps of similar size, *without replacement*).

Samuel Kotz
College Park, Maryland
November 1990

Norman L. Johnson
Chapel Hill, North Carolina
November 1990

Contents of Volume III

Contents of Volume I

Foundations and Basic Theory

Contents of Volume II

Methodology and Distribution

Contributors

FANCHER, R.E. Department of Psychology, York University, 4700 Keele Street, North York, Ontario M3J 1P3, Canada.

DIETZ, K. Institut für Medizinische Biometri, Universität Tübingen, Westbahnhofstrasse 55, Tübingen 72070, Germany.

COX, D.R. Nuffield College, Oxford OX1 1NF, United Kingdom.

SEN, P.K. Department of Biostatistics, School of Public Health, University of North Carolina, Chapel Hill, NC 27599-7400, USA.

GAIL, M.H. Chief, Biostatistics Branch, Division of Cancer Epidemiology and Genetics, National Cancer Institute, 6130 Executive Boulevard, EPN 431, Rockville, MD 20892, USA.

HUBER, P.J. Lehrstuhl Mathematik VII, Universität Bayreuth, Universitätstrasse 30, Gebaude NW 11, D-95440 Bayreuth, Germany.

CALDER M. AND DAVIS, R.A. Department of Statistics, Colorado State University, Fort Collins, CO 80523-0002, USA.

RONCHETTI, E.M. Department of Social Science, University of Geneve, COMIN, 1211 Geneva, Switzerland.

GOOD, I.J. Department of Statistics, Virginia Polytechnic Institute and State University, Blacksburg, VA 24061-0439, USA.

BERAN, R.J. Department of Statistics, University of California at Berkeley, Berkeley, CA 94720, USA.

LAI, T.L. Department of Statistics, Stanford University, Stanford, CA 94305, USA.

DEELY, J.J. Department of Mathematics, University of Canterbury, Christchurch, New Zealand.

SMITH, R.L. Department of Statistics, University of North Carolina, Chapel Hill, NC 27599-3260, USA.

PFANZAGL, J.F. Mathematisches Institut der Universität Köln, Weyertal 86-90, D-50931 Köln, Germany.

MCKEAGUE, I.W. Department of Statistics, Florida State University, Tallahassee, FL 32306, USA.

LETAC, G.G. Laboratoire de Statistique et Probabilités, URA CNRS D745, Université Paul Sabatier, 118 Route de Narbonne, 31062 Toulouse Cedex, France.

DICICCIO, T. Department of Statistics, 358 Ives Hall, Cornell University, Ithaca, NY 14853-3901, USA.

SIMPSON, D.G. Department of Statistics, 101 Illini Hall, University of Illinois, 725 South Wright Street, Champaign, IL 61820, USA.

DIGGLE, P.J. Department of Mathematics and Statistics, Lancaster University, Lancaster LA1 4YF, United Kingdom.

MAMMEN, E. Institut für Angewandte Mathematik, Im Neyenheimer Feld 294, Heidelberg 1 D-6900, Germany.

TITTERINGTON, D.M. Department of Statistics, University of Glasgow, Glasgow G12 8QQ, United Kingdom.

Sources and Acknowledgments

Galton, F. (1889). Co-relations and their measurement, chiefly from anthropometric data. *Proc. Roy. Soc. London*, **45**, 134–145. Reproduced by the kind permission of the Royal Society of London.

McKendrick, A.G. (1926). Applications of mathematics to medical problems. *Proc. Edinburgh Math. Soc.*, **44**, 98–130. Reproduced by the kind permission of the Edinburgh Mathematical Society.

Yates, F. and Cochran, W.G. (1938). The analysis of groups of experiments. *J. Agric. Sci.*, **28**, 556–580. Reproduced by the kind permission of the Cambridge University Press.

Rao, C.R. (1948). Large sample tests of statistical hypotheses concerning several parameters with application to problems of estimation. *Proc. Cambridge Philos. Soc.*, **44**, 50–57. Reproduced by the kind permission of the Cambridge Philosophical Society.

Cornfield, J. (1951). A method of estimating comparative rates from clinical data. Applications to cancer of the lung, breast and cervix. *J. National Cancer Inst.*, **11**, 1269–1275.

Metropolis, N., Rosenbluth, A.W., Rosenbluth, M.N., Teller, A.H., and Teller, E. (1953). Equation of state calculations by fast computing machines. *J. Chem. Phys.*, **21**, 1087–1092. Reproduced by the kind permission of the American Chemical Society.

Geman, S. and Geman, D. (1984). Stochastic relaxation, Gibbs distributions, and the Bayesian restoration of images. *IEEE Trans. Pattern Anal-Machine Intell.*, **6**, 721–741.

Whittle, P. (1953). The analysis of multiple stationary time series. *J. Roy. Statist. Soc., Ser. B*, **15**, 125–139. Reproduced by the kind permission of the Royal Statistical Society.

Daniels, H.E. (1954). Saddlepoint approximations in statistics. *Ann. Math. Statist.*, **25**, 631–650. Reproduced by the kind permission of the Institute of Mathematical Statistics.

Cooley, J.W. and Tukey, J.W. (1965). An algorithm for the machine calculation of complex Fourier series. *Math. Comput.*, **19**, 297–301. Reproduced by the kind permission of the American Mathematical Society.

Hájek, J. (1970). A characterization of limiting distributions of regular estimates. *Z. Wahrsch. Verw. Gebiete*, **14**, 323–330. Reproduced by the kind permission of Springer-Verlag, Berlin.

Hastings, W.K. (1970). Monte Carlo sampling methods using Markov chains and their applications. *Biometrika*, **57**, 97–109. Reproduced by the kind permission of the Biometrika Trustees.

Lindley, D.V. and Smith, A.F.M. (1972). Bayes estimates for the linear model. *J. Roy. Statist. Soc. Ser. B*, **34**, 1–18. Reproduced by the kind permission of the Royal Statistical Society.

Besag, J. (1974). Spatial interaction and the statistical analysis of lattice systems. *J. Roy. Statist. Soc. Ser. B*, **36**, 192–225. Reproduced by the kind permission of the Royal Statistical Society.

Levit, B.Ya. (1974). On optimality of some statistical estimates. *Proceedings of the Prague Symposium on Asymptotic Statistics*, **2**, 215–238.

Aalen, O. (1978). Nonparametric estimation of partial transition probabilities in multiple decrement models, *Ann. Statist.*, **6**, 534–545. Reproduced by the kind permission of the Institute of Mathematical Statistics.

Morris, C.N. (1982). Natural exponential families with quadratic variance functions. *Ann. Statist.*, **10**, 65–80. Reproduced by the kind permission of the Institute of Mathematical Statistics.

Barndorff-Nielsen, O.E. (1983). On a formula for the distribution of the maximum likelihood estimator. *Biometrika*, **70**, 343–365. Reproduced by the kind permission of the Biometrika Trustees.

Rousseeuw, P.J. (1984). Least median of squares regressions. *J. Amer. Statist. Assoc.*, **79**, 871–880. Reproduced by the kind permission of the American Statistical Association.

Liang, K.Y. and Zeger, S.L. (1986). Longitudinal data analysis using generalized linear models. *Biometrika*, **73**, 13–22. Reproduced by the kind permission of the Biometrika Trustees.

Hall, P. (1988). Theoretical comparison of bootstrap confidence intervals. *Ann. Statist.*, **16**, 927–953. Reproduced by the kind permission of the Institute of Mathematical Statistics.

Gelfand, A.E. and Smith, A.F.M. (1990). Sampling-based approaches to calculating marginal densities. *J. Amer. Statist. Assoc.*, **85**, 398–409. Reproduced by the kind permission of the American Statistical Association.

Introduction to Galton (1889) Co-Relations and Their Measurement, Chiefly from Anthropometric Data

Raymond E. Fancher
York University

The versatile Englishman Sir Francis Galton (1822–1911) contributed importantly to many different fields. His *Tropical South Africa* (1853) described his exploration and mapping of the present-day country of Namibia. In *Meteorographica* (1863) he presented the world's first weather maps, and announced the discovery of the "anticyclone" or high-pressure weather system. In "Hereditary Talent and Character" (1865) and *Hereditary Genius* (1972, originally published 1869) he introduced the linked ideas of "hereditary genius" and eugenics: the notion that human intellectual abilities are hereditarily determined to the same extent as physical attributes, and therefore that human evolution can potentially be accelerated or self-consciously guided through the adoption of selective mating practices. His "The History of Twins, as a Criterion of the Relative Powers of Nature and Nurture" (1875a) initiated modern behavior genetics with the twin-study method, and "A Theory of Heredity" (1875b) correctly anticipated major aspects of Weismann's germ–plasm theory of hereditary transmission. Galton's "Anthropometric Laboratory" (1882) constituted the first large-scale attempt at what today we call intelligence testing, and his *Finger Prints* (1892) developed the method for classification and analysis of fingerprints still used by law enforcement agencies today.

For Galton's biographer, Karl Pearson, however, his subject's single most important work was not one of those mentioned above, but rather the short article on "Co-relations and their measurement" which is reproduced in this volume:

> Galton's very modest paper from which a revolution in our scientific ideas has spread is in its permanent influence, perhaps, the most important of his writings. Formerly the quantitative scientist could only think in terms of causation, now he can think also in terms of correlation. This has not only

> enormously widened the field to which quantitative and therefore mathematical methods can be applied, but it has at the same time modified our philosophy of science and even of life itself. (Pearson, 1930, pp. 56–57.)

Pearson's judgment was perhaps not completely impartial on this matter, because he himself played the leading role in perfecting, elaborating, and promoting the statistic that was sketched in Galton's paper. Nevertheless, there can be no doubting the importance of correlational methods in modern statistics, and Galton's paper was crucial to their development.

Although short, Galton's paper drew on more than two decades of experimenting and tinkering with data from a variety of types of investigation. Perhaps the deepest root for his ultimate conception of correlation was a longstanding desire to give mathematical precision to descriptions of hereditary resemblances, a root still revealed in his 1888 paper by its designation of the two intercorrelated variables as the "subject" and its "relative." In his early work such as *Hereditary Genius* he had to content himself with very simple comparative statistics, such as the demonstration that approximately 10 percent of the intellectually eminent individuals who get listed in biographical dictionaries have at least one relative who is also eminent enough to be listed, while the proportion of eminent people within the general population was less than one in four thousand. Galton further showed that these eminent relations were more often close as opposed to remote; e.g., fathers, sons, and brothers substantially outnumbered grandfathers, grandsons, uncles, and nephews, who in turn were more frequent than "third-degree" relatives such as first cousins or great-grandfathers and great-grandsons.

Hereditary Genius also first revealed Galton's fascination with another key idea—namely the normal distribution and the possibility of describing variables in statistical (probable error) units derived from it. After presenting illustrative data collected by Quetelet to demonstrate the normal distributions of height and chest measurements in French soldiers, Galton went on to assert that differences in human "natural ability" are also normally distributed, and thus can be defined in terms of "grades ... separated by equal intervals," and defined by the probable error of the distribution (Galton, 1972, p. 75). Thus one person in four lies within the first grade above (or below) the population mean, a further one in six between the first and second grades, one in sixteen between the second and third, and so on through the seventh grade at the tails of the distribution, containing fewer than one person in a million. But while Galton here showed his inclination to classify and describe presumably hereditary ability in terms of statistically defined units or grades, he did not yet connect that idea with his attempts to assess degrees of hereditary resemblance.

A few years after *Hereditary Genius*, in an unpublished notebook that has been described by Hilts (1975, pp. 24–26), Galton constructed the first of his known "scatter plots" to illustrate the relationship between two variables. Using data collected from 92 scientists from Britain's Royal Society for his

1874 book *English Men of Science*, Galton plotted the relationship between their head sizes (measured by inches around the inside of their hats) and their heights. The plot suggested an essentially random relationship between the two variables, and was unaccompanied by further or detailed mathematical calculations.

Shortly after this, however, in a further quest for some index of hereditary resemblances, Galton produced a more sophisticated kind of plot. He and several of his friends had planted "parent" pea seeds in seven different size categories, and subsequently harvested and measured the seeds from the offspring plants. In a notebook Galton summarized the results in what Pearson has called "the first 'regression line' which I suppose ever to have been computed" (Pearson, 1930, p. 3)—a graph plotting the seven mean diameter sizes of the parent seeds on the abscissa, against the means of their offspring's diameters on the ordinate. The points approximated a line with a positive slope of one-third, indicating that the offspring seeds deviated from their population mean in the same direction as their parents, but only by an average of about a third as much; e.g., parental seeds with diameters 0.03 inch above or below the mean for all parents produced offspring averaging only about 0.01 inch above or below the overall offspring mean. When Galton published this graph (1877), he summarized the relationship with the equation "$R = 0.33$." He did not specifically state what the letter stood for but his text described the offspring seeds as manifesting a "reversion" to ancestral type. The term "regression" did not appear in this text, so it is unlikely that Galton intended that signification for the letter "r" until some time later.

In this graph and article, Galton recorded seed sizes in units of hundredths of an inch. In an independent notebook entry made at about the same time, however, he classified the sizes of another population of pea seeds in terms of "degrees" defined as probable error units above or below the mean (Pearson, 1924, p. 392; Hilts, 1973, p. 225). But as in 1869, he did not combine his insights and attempt a correlational or regression analysis of the parental and offspring seed measurements after they had been converted into the statistical units.

In 1883, Galton attempted to solve a new kind of correlational problem, this time involving two variables originally measured in qualitatively different units. Keenly interested in the predictive value of competitive examinations, he obtained the examination results achieved by some 250 candidates for the Indian Civil Service in the early 1860s, along with their salary levels in the Service two decades later. In still another unpublished study (described in Fancher, 1989), Galton attempted to calculate whether, and to what extent, the examination scores had predicted later occupational success as measured by salary levels. In his most interesting approach to the problem, Galton separately summed the raw examination scores achieved by the top 25, the middle 25, and the bottom 25 candidates in the group. Then he arbitrarily set the middle group's average at 1000 and adjusted the top and bottom average scores proportionately. This exercise revealed that the high scorers differed from the low by a ratio of 1336 to 838. Galton then repeated the analysis for the salary levels later achieved by these same three groups of men. Here the ratio worked out to 1154 for the high examination scorers versus 1028 for the low (who therefore were earning slightly *more* than the middle group, whose scores had arbitrarily been set at 1000). Galton summarized these results by writing, "Groups of marks vary from 1336 to 838 or 100 to 63; corresponding groups of pay 1154 to 1028 or 100 to 89. The lowest class men are getting 10 percent less pay after 20 years service though their marks were 40 percent less" (Fancher, 1989, p. 453). Despite its crude methodology, this unpublished study marked the first time Galton explicitly attempted to "standardize" the units of his two variables in a correlation-type problem.

He seems to have forgotten about that possibility, however, when he returned to the issue of hereditary size resemblances in 1885. At his Anthropometric Laboratory in London's South Kensington Museum, he collected height measurements for both parents and their adult children from several hundred families. After multiplying all females' height measurements by 1.08 to render their mean equal to the males', Galton plotted the heights of the children against those of their "mid-parents" (i.e., the father's and corrected mother's heights summed and divided by two). He found that the children's deviations from the average were about two-thirds those of their mid-parents, and now for the first time publicly used the term "regression" to describe the phenomenon; he entitled the paper describing his results, "Regression Towards Mediocrity in Hereditary Stature" (Galton, 1885). In this paper Galton also reported that regression could be described as going in two directions—that is, that he could take the *children's* heights as his base and plot their mid-parents' against them. When he did this, however, the regression fraction turned out to be only about one-third, a difference that puzzled and disturbed him.

By now, most of the pieces necessary to complete the puzzle of correlation were present although not yet assembled. Galton had empirically investigated a variety of relationships involving not only the same variable (seed diameter, height) in parents and offspring, but also "correlated parts" (head size and height) within the same individuals, and measures from the same individuals recorded in completely different units (examination scores and later salary levels). He had adopted the habit of constructing scatter plots of these relationships, had drawn regression lines to summarize them, and had recognized that regression could be described as going both ways between them. And in several other contexts he had constructed standardized measurement units based on the probable error.

Sometime in late 1888, Galton put these pieces together in a rather sudden insight which resulted in "Co-Relations and Their Measurement." Some of the details regarding the circumstances of his inspiration have been subject to debate (e.g., see Pearson, 1924, p. 393; Cowan, 1972, pp. 516–517;

Hilts, 1973, p. 233, n. 41), but the end result was clear. Galton now realized that the two regression fractions from his 1885 study had differed because the children's heights had shown greater variation than those of the constructed "mid-parents," and that the two fractions for this (or any other pair of correlated variables) could be rendered equal if the variables were measured not in absolute units but in terms of probable error units from their own distributions. Moreover, the slope of the resulting regression line would be directly interpretable as the long-sought, direct quantitative measure of the strength of the relationship, approaching 1.0 in cases of nearly perfect relationships and 0.0 for complete independence of the variables.

In many respects, Galton's invention of correlation was typical of his contributions. Beginning with his African expedition and continuing with his more intellectual explorations, Galton tended to break new ground and then leave it to others with more persistence and specialized expertise to fully map and develop the field. In statistics, he himself lacked extensive mathematical training and had to rely on others to augment his own intuitive and "amateur," if inspired efforts. For correlation, this assistance arrived via the biologist Walter Weldon, who read Galton's 1888 paper and used its method to calculate the correlations between pairs of organ sizes in several shrimp and crab species. More importantly, Weldon brought the new method to the attention of his friend and colleague at University College, London, Karl Pearson. A much more sophisticated and polished mathematician than either Galton or Weldon, Pearson in an 1895 lecture (published in 1896), proposed a "product moment" computing formula for r, which he showed presented "no practical difficulty in calculation," and which was based on the standard deviation rather than the probable error (Pearson, 1896). Pearson at this point referred to the resulting statistic as the "Galton function," but it soon became more widely known as the *Pearson* coefficient of correlation or more simply as "Pearson's r." Although Pearson and others would engage in extensive and heated polemics regarding some technical issues for the ensuing quarter-century, the modern form of correlation was now essentially established. For further details regarding these events see Cowles (1989, Chapter 10).

Pearson knew and acknowledged in 1895 that the French mathematician Auguste Bravais had anticipated the product-sum method and even used the term "correlation" in a much earlier paper (Bravais, 1846), although without positing a single symbol for the regression fraction. Bravais's mathematical concerns had been primarily theoretical, however, and there is little doubt that the practically minded Galton knew nothing about them as he pursued his own empirically driven and idiosyncratic studies. Accordingly, a fair summation of the priority issue is to credit Galton with anticipating the usefulness of a mathematical measure of correlation in a wide variety of situations, and of introducing in his 1889 paper a rough but practical method for calculating such a measure based on the probable error. Pearson in 1896, relying in part on Bravais's much earlier introduction of the product-sum, further developed the theoretical mathematical basis of correlation and introduced the modern computing formula. Pearson and others went on to develop these ideas much further still in the direction of negative, multiple, and partial correlation—but that is part of a different chapter in the history of statistics.

References

Bravais, A. (1846). Sur les probabilités des erreurs de situation d'un point. *Mem. Acad. Roy. Sci. Inst. France*, **9**, 255–332.

Cowan, R.S. (1972). Francis Galton's statistical ideas: The influence of eugenics. *Isis*, **63**, 509–528.

Cowles, M. (1989). *Statistics in Psychology: An Historical Perspective*. Lawrence Erlbaum, Hillsdale, NJ.

Fancher, R.E. (1989). Galton on examinations: An unpublished step in the invention of correlation. *Isis*, **80**, 446–455.

Galton, F. (1853). *Tropical South Africa*. Murray, London.

Galton, F. (1863). *Meteorographica, or Methods of Mapping the Weather*. Macmillan, London.

Galton, F. (1865). Hereditary talent and character. *Macmillan's Magazine*, **12**, 157–166, 318–327.

Galton, F. (1875a). The history of twins, as a criterion of the relative powers of nature and nurture. *Fraser's Magazine*, **12**, 566–576.

Galton, F. (1875b). A theory of heredity. *Contemp. Rev.*, **27**, 80–95.

Galton, F. (1877). Typical laws of heredity. *Proc. Roy. Inst.*, **8**, 282–301.

Galton, F. (1882). The anthropometric laboratory. *Fortnightly Review*, **183**, 332–338.

Galton, F. (1885). Regression towards mediocrity in hereditary stature. *J. Anthropol. Inst.*, **15**, 246–263.

Galton, F. (1889). Co-relations and their measurement, chiefly from anthropometric data. *Proc. Roy. Soc. London*, **45**, 135–145.

Galton, F. (1892). *Finger Prints*. Macmillan, London.

Galton, F. (1972). *Hereditary Genius*. Peter Smith, Gloucester, MA (originally published 1869).

Hilts, V.L. (1973). Statistics and social science. In R.N. Giere and R.S. Westfall, eds., *Foundations of Scientific Method: The Nineteenth Century*. University of Indiana Press, Bloomington, IN.

Hilts, V.L. (1975). A guide to Francis Galton's *English Men of Science*. *Trans. Amer. Philos. Soc.*, **65**(5), 1–85.

Pearson, K. (1896). Mathematical contributions to the theory of evolution III: Regression, heredity, and panmixia. *Philos. Trans. Roy. Soc. London Ser. A*, **187**, 253–318.

Pearson, K. (1914, 1924, 1930). *The Life, Letters and Labours of Francis Galton*, 3 vols. Cambridge University Press, Cambridge, UK.

Co-Relations and Their Measurement, Chiefly from Anthropometric Data

F. Galton

"Co-relation or correlation of structure" is a phrase much used in biology, and not least in that branch of it which refers to heredity, and the idea is even more frequently present than the phrase; but I am not aware of any previous attempt to define it clearly, to trace its mode of action in detail, or to show how to measure its degree.

Two variable organs are said to be co-related when the variation of the one is accompanied on the average by more or less variation of the other, and in the same direction. Thus the length of the arm is said to be co-related with that of the leg, because a person with a long arm has usually a long leg, and conversely. If the co-relation be close, then a person with a very long arm would usually have a very long leg; if it be moderately close, then the length of his leg would usually be only long, not very long; and if there were no co-relation at all then the length of his leg would on the average be mediocre. It is easy to see that co-relation must be the consequence of the variations of the two organs being partly due to common causes. If they were wholly due to common causes, the co-relation would be perfect, as is approximately the case with the symmetrically disposed parts of the body. If they were in no respect due to common causes, the co-relation would be *nil*. Between these two extremes are an endless number of intermediate cases, and it will be shown how the closeness of co-relation in any particular case admits of being expressed by a simple number.

To avoid the possibility of misconception, it is well to point out that the subject in hand has nothing whatever to do with the average proportions between the various limbs, in different races, which have been often discussed from early times up to the present day, both by artists and by anthropologists. The fact that the average ratio between the stature and the cubit is as 100 to 37, or thereabouts, does not give the slightest information about the nearness with which they vary together. It would be an altogether erroneous inference to suppose their average proportion to be maintained so that when the cubit was, say, one-twentieth longer than the average cubit, the stature might be expected to be one-twentieth greater than the average stature, and conversely. Such a supposition is easily shown to be contradicted both by fact and theory.

The relation between the cubit and the stature will be shown to be such that for every inch, centimetre, or other unit of absolute length that the cubit deviates from the mean length of cubits, the stature will on the average deviate from the mean length of statures to the amount of 2.5 units, and in the same direction. Conversely, for each unit of deviation of stature, the average deviation of the cubit will be 0.26 unit. These relations are not numerically reciprocal, but the exactness of the co-relation becomes established when we have transmuted the inches or other measurement of the cubit and of the stature into units dependent on their respective scales of variability. We thus cause a long cubit and an equally long stature, as compared to the general run of cubits and statures, to be designated by an identical scale-value. The particular unit that I shall employ is the value of the probable error of any single measure in its own group. In that of the cubit, the probable error is 0.56 inch = 1.42 cm; in the stature it is 1.75 inch = 4.44 cm. Therefore the measured lengths of the cubit in inches will be transmuted into terms of a new scale, in which each unit = 0.56 inch, and the measured lengths of the stature will be transmuted into terms of another new scale in which each unit is 1.75 inch. After this has been done, we shall find the deviation of the cubit as compared to the mean of the corresponding deviations of the stature, to be as 1 to 0.8. Conversely, the deviation of the stature as compared to the mean of the corresponding deviations of the cubit will also be as 1 to 0.8. Thus the existence of the co-relation is established, and its measure is found to be 0.8.

Now as to the evidence of all this. The data were obtained at my Anthropometric Laboratory at South Kensington. They are of 350 males of 21 years and upwards, but as a large proportion of them were students, and barely 21 years of age, they were not wholly full-grown; but neither that fact nor the small number of observations is prejudicial to the conclusions that will be reached. They were measured in various ways, partly for the purpose of this inquiry. It will be sufficient to give some of them as examples. The exact number of 350 is not preserved throughout, as injury to some limb or other reduced the available number by 1, 2, or 3 in different cases. After marshalling the measures of each limb in the order of their magnitudes, I noted the measures in each series that occupied respectively the positions of the first, second, and third quarterly divisions. Calling these measures in any one series, Q_1, M, and Q_3, I take M, which is the median or middlemost value, as that whence the deviations are to be measured, and $\frac{1}{2}\{Q_3 - Q_1\} = Q$, as the probable error of any single measure in the series. This is practically the same as saying that one-half of the deviations fall

Table I.

	M.		Q.	
	Inch	Centim.	Inch	Centim.
Head length	7.62	19.35	0.19	0.48
Head breadth	6.00	15.24	0.18	0.46
Stature	67.20	170.69	1.75	4.44
Left middle finger	4.54	11.53	0.15	0.38
Left cubit	18.05	45.70	0.56	1.42
Height of right knee	20.50	52.00	0.80	2.03

Note. The head length is its maximum length measured from the notch between and just below the eyebrows. The cubit is measured with the hand prone and without taking off the coat; it is the distance between the elbow of the bent left arm and the tip of the middle finger. The height of the knee is taken sitting when the knee is bent at right angles, less the measured thickness of the heel of the boot.

within the distance of $\pm$Q from the mean value, because the series run with fair symmetry. In this way I obtained the following values of M and Q, in which the second decimal must be taken as only roughly approximate. The M and Q of any particular series may be identified by a suffix, thus M_c, Q_c might stand for those of the cubit, and M_s, Q_s for those of the stature.

Tables were then constructed, each referring to a different pair of the above elements, like Tables II and III, which will suffice as examples of the whole of them. It will be understood that the Q value is a universal unit

Table II.

	Length of left cubit in inches, 318 adult males								
Stature in inches	Under 16.5	16.5 and under 17.0	17.0 and under 17.5	17.5 and under 18.0	18.0 and under 18.5	18.5 and under 19.0	19.0 and under 19.5	19.5 and above	Total cases
71 and above	—	—	—	1	3	4	15	7	30
70	—	—	—	1	5	13	11	—	30
69	—	1	1	2	25	15	6	—	50
68	—	1	3	7	14	7	4	2	48
67	—	1	7	15	28	8	2	—	61
66	—	1	7	18	15	6	—	—	48
65	—	4	10	12	8	2	—	—	36
64	—	5	11	2	3	—	—	—	21
Below 64	9	12	10	3	1	—	—	—	34
Totals	9	25	49	61	102	55	38	9	348

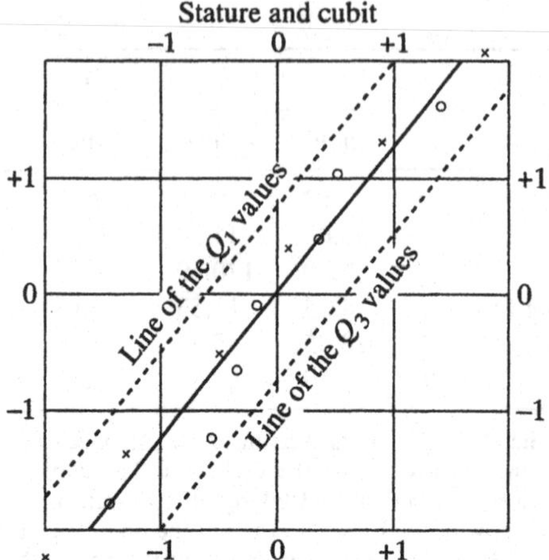

applicable to the most varied measurements, such as breathing capacity, strength, memory, keenness of eyesight, and enables them to be compared together on equal terms notwithstanding their intrinsic diversity. It does not only refer to measures of length, though partly for the sake of compactness, it is only those of length that will be here given as examples. It is unnecessary to extend the limits of Table II, as it includes every line and column in my MS. table that contains not less than twenty entries. None of the entries lying within the flanking lines and columns of Table II were used.

The measures were made and recorded to the nearest tenth of an inch. The heading of 70 inches of stature includes all records between 69.5 and 70.4 inches; that of 69 includes all between 68.5 and 69.4, and so on.

The values derived from Table II, and from other similar tables, are entered in Table III, where they occupy all the columns up to the three last, the first of which is headed "smoothed." These smoothed values were obtained by plotting the observed values, after transmuting them as above described into their respective Q units, upon a diagram such as is shown in the figure. The deviations of the "subject" are measured parallel to the axis of y in the figure, and those of the mean of the corresponding values of the "relative" are measured parallel to the axis of x. When the stature is taken as the subject, the median positions of the corresponding cubits, which are given in the successive lines of Table III, are marked with small circles. When the cubit is the subject, the mean positions of the corresponding statures are marked with crosses. The firm line in the figure is drawn to represent the general run of the small circles and crosses. It is here seen to be a straight line, and it was similarly found to be straight in every other figure drawn from the different pairs of co-related variables that I have as yet tried. But the inclination of the line to the vertical differs considerably in different cases. In the present one the inclination is such that a deviation of 1 on the part of the subject, whether it be stature or cubit, is accompanied by

Table III. Stature $M_s = 67.2$ inches; $Q_s = 1.75$ inch. Left cubit $M_c = 18.05$ inches; $Q_c = 0.56$ inch.

No. of cases	Stature	Deviation from M_s reckoned in: Inches	Deviation from M_s reckoned in: Units of Q_s	Mean of corresponding left cubits	Deviation from M_c reckoned in: Inches	Deviation from M_c reckoned in: Units of Q_c, Observed	Deviation from M_c reckoned in: Units of Q_c, Smoothed	Smoothed values multiplied by Q_c	Added to M_c
	inches			inches					
30	70.0	+2.8	+1.60	18.8	+0.8	+1.42	+1.30	+0.73	18.8
50	69.0	+1.8	+1.03	18.3	+0.3	+0.53	+0.84	+0.47	18.5
38	68.0	+0.8	+0.46	18.2	+0.2	+0.36	+0.38	+0.21	18.3
61	67.0	−0.2	−0.11	18.1	+0.1	+0.18	−0.08	−0.04	18.0
48	66.0	−1.2	−0.69	17.8	−0.2	−0.36	−0.54	−0.30	17.8
36	65.0	−2.2	−1.25	17.7	−0.3	−0.53	−1.00	−0.56	17.5
21	64.0	−3.2	−1.83	17.2	−0.8	−1.46	−1.46	−0.80	17.2

No. of cases	Left cubit	Deviation from M_s reckoned in: Inches	Deviation from M_s reckoned in: Units of Q_c	Mean of corresponding statures	Deviation from M_s reckoned in: Inches	Deviation from M_s reckoned in: Units of Q_s, Observed	Deviation from M_s reckoned in: Units of Q_s, Smoothed	Smoothed values multiplied by Q_s	Added to M_s
	inches			inches					
38	19.25	+1.20	+2.14	70.3	+3.1	+1.8	+1.70	+3.0	70.2
55	18.75	+0.70	+1.25	68.7	+1.5	+0.9	+1.00	+1.8	69.0
102	18.25	+0.20	+0.36	67.4	+0.2	+0.1	+0.28	+0.5	67.7
61	17.75	−0.30	−0.53	66.3	−0.9	−0.5	−0.43	−0.8	66.4
49	17.25	−0.80	−1.42	65.0	−2.2	−1.3	−1.15	−2.0	65.2
25	16.75	−1.30	−2.31	63.7	−3.5	−2.0	−1.85	−3.2	64.0

a mean deviation on the part of the relative, whether it be cubit or stature, of 0.8. This decimal fraction is consequently the measure of the closeness of the co-relation. We easily retransmute it into inches. If the stature be taken as the subject, then Q_s is associated with $Q_c \times 0.8$; that is, a deviation of 1.75 inches in the one with 0.56×0.8 of the other. This is the same as 1 inch of stature being associated with a mean length of cubit equal to 0.26 inch. Conversely, if the cubit be taken as the subject, then Q_c is associated with $Q_s \times 0.8$; that is, a deviation of 0.56 inch in the one with 1.75×0.8 of the other. This is the same as 1 inch of cubit being associated with a mean length of 2.5 inches of stature. If centimeter be read for inch the same holds true.

Six other tables are now given in a summary form, to show how well calculation on the above principle agrees with observation.

From Table IV the deductions given in Table V can be made; but they may be made directly from tables of the form of Table III, whence Table IV was itself derived.

When the deviations of the subject and those of the mean of the relatives are severally measured in units of their own Q, there is always a regression in the value of the latter. This is precisely analogous to what was observed in kinship, as I showed in my paper read before this Society on "Hereditary Stature" ('*Roy. Soc. Proc.*,' vol. 40, 1886, p. 42). The statures of kinsmen are co-related variables; thus, the stature of the father is correlated to that of the adult son, and the stature of the adult son to that of the father; the stature of the uncle to that of the adult nephew, and the stature of the adult nephew to that of the uncle, and so on; but the index of co-relation, which is what I there called "regression," is different in the different cases. In dealing with kinships there is usually no need to reduce the measures to units of Q, because the Q values are alike in all the kinsmen, being of the same value as that of the population at large. It however happened that the very first case that I analysed was different in this respect. It was the reciprocal relation between the statures of what I called the "mid-parent" and the son. The mid-parent is an ideal progenitor, whose stature is the average of that of the father on the one hand and of that of the mother on the other, after her stature had been transmuted into its male equivalent by the multiplication of the factor of 1.08. The Q of the mid-parental statures was found to be 1.2, that of the population dealt with was 1.7. Again, the mean deviation measured in inches of the statures of the sons was found to be two-thirds of the deviation of the mid-parents, while the mean deviation in inches of the mid-parent was one-third of the deviation of the sons. Here the regression, when calculated in Q units, is in the first case from $\frac{1}{1.2}$ to $\frac{2}{3} \times 1.7 = 1$ to 0.47, and in the second case from $\frac{1}{1.7}$ to $\frac{1}{3} \times \frac{1}{1.2} = 1$ to 0.44, which is practically the same.

The *rationale* of all this will be found discussed in the paper on "Hereditary Stature," to which reference has already been made, and in the appendix to it by Mr. J.D. Hamilton Dickson. The entries in any table, such

Table IV.

No. of cases	Length of head	Mean of corresponding statures: Observed	Calculated	No. of cases	Height	Mean of corresponding lengths of head: Observed	Calculated
32	7.90	68.5	68.1	26	70.5	7.72	7.75
41	7.80	67.2	67.8	30	69.5	7.70	7.72
46	7.70	67.6	67.5	50	68.5	7.65	7.68
52	7.60	66.7	67.2	49	67.5	7.65	7.64
58	7.50	66.8	66.8	56	66.5	7.57	7.60
34	7.40	66.0	66.5	43	65.5	7.57	7.69
26	7.30	66.7	66.2	31	64.5	7.54	7.65

No. of cases	Height	Mean of corresponding lengths of left middle finger: Observed	Calculated	No. of cases	Length of middle finger	Mean of corresponding statures: Observed	Calculated
30	70.5	4.71	4.74	23	4.80	70.2	69.4
50	69.5	4.55	4.68	49	4.70	68.1	68.5
37	68.5	4.57	4.62	62	4.60	68.0	67.7
62	67.5	4.58	4.56	63	4.50	67.3	66.9
48	66.5	4.50	4.50	57	4.40	66.0	66.1
37	65.5	4.47	4.44	35	4.30	65.7	65.3
20	64.5	4.33	4.38				

No. of cases	Left middle finger	Mean of corresponding lengths of left cubit: Observed	Calculated	No. of cases	Length of left cubit	Mean of corresponding length of left middle finger: Observed	Calculated
23	4.80	18.97	18.80	29	19.00	4.76	4.75
50	4.70	18.55	18.49	32	18.70	4.64	4.69
62	4.60	18.24	18.18	48	18.40	4.60	4.62
62	4.50	18.00	17.87	70	18.10	4.56	4.55
57	4.40	17.72	17.55	37	17.80	4.49	4.48
34	4.30	17.27	17.24	31	17.50	4.40	4.41
				28	17.20	4.37	4.34
				24	16.90	4.32	4.28

(*Continued*)

as Table II, may be looked upon as the values of the vertical ordinates to a surface of frequency, whose mathematical properties were discussed in the above-mentioned appendix, therefore I need not repeat them here. But there is always room for legitimate doubt whether conclusions based on the strict properties of the ideal law of error would be sufficiently correct to be serviceable in actual cases of co-relation between variables that conform

Table IV (*Continued*)

No. of cases	Length of head	Mean of corresponding breadths of head: Observed	Calculated	No. of cases	Breadth of head	Mean of corresponding lengths of head: Observed	Calculated
32	7.90	6.14	6.12	27	6.30	7.72	7.84
41	7.80	6.05	6.08	36	6.20	7.72	7.75
46	7.70	6.14	6.04	53	6.10	7.65	7.65
52	7.60	5.98	6.00	58	6.00	7.68	7.60
58	7.50	5.98	5.96	56	5.90	7.50	7.55
34	7.40	5.96	5.91	37	5.80	7.55	7.50
26	7.30	5.85	5.87	30	5.70	7.45	7.46

No. of cases	Stature	Mean of corresponding heights of knee: Observed	Calculated	No. of cases	Height of knee	Mean of corresponding statures: Observed	Calculated
30	70.0	21.7	21.7	23	22.2	70.5	70.6
50	69.0	21.1	21.3	32	21.7	69.8	69.6
38	68.0	20.7	20.9	50	21.2	68.7	68.6
61	67.0	20.5	20.5	68	20.7	67.3	67.7
49	66.0	20.2	20.1	74	20.2	66.2	66.7
36	65.0	19.7	19.7	41	19.7	65.5	65.7
				26	19.2	64.3	64.7

No. of cases	Left cubit	Mean of corresponding heights of knee: Observed	Calculated	No. of cases	Height of knee	Mean of corresponding left cubit: Observed	Calculated
29	19.0	21.5	21.6	23	22.25	18.98	18.97
32	18.7	21.4	21.2	30	21.75	18.68	18.70
48	18.4	20.8	20.9	52	21.25	18.38	18.44
70	17.1	20.7	20.6	69	20.75	18.15	18.17
37	17.8	20.4	20.2	70	20.25	17.75	17.90
31	17.5	20.0	19.9	41	19.75	17.55	17.63
28	17.2	19.8	19.6	27	19.25	17.02	17.36
23	16.9	19.3	19.2				

only approximately to that law. It is therefore exceedingly desirable to put the theoretical conclusions to frequent test, as has been done with these anthropometric data. The result is that anthropologists may now have much less hesitation than before, in availing themselves of the properties of the law of frequency of error.

I have given in Table V a column headed $\sqrt{(1-r^2)} = f$. The meaning of f is explained in the paper on "Hereditary Stature." It is the Q value of the distribution of any system of x values, as x_1, x_2, x_3, etc., round the mean of

Table V.

Subject	Relative	In units of Q		In units of ordinary measure	
		r	$\sqrt{(1-r^2)}=f$	As 1 to	f
Stature	Cubit	0.8	0.60	0.26	0.45
Cubit	Stature			2.5	1.4
Stature	Head length	0.35	0.93	0.38	1.63
Head length	Stature			3.2	0.17
Stature	Middle finger	0.7	0.72	0.06	0.10
Middle finger	Stature			0.2	1.26
Middle finger	Cubit	0.85	0.61	3.13	0.34
Cubit	Middle finger			0.21	0.09
Head length	Head breadth	0.45	0.89	0.43	0.16
Head breadth	Head length			0.48	0.17
Stature	Height of knee	0.9	0.44	0.41	0.35
Height of knee	Stature			0.20	0.77
Cubit	Height of knee	0.8	0.60	0.14	0.64
Height of knee	Cubit			0.56	0.45

all of them, which we may call X. The knowledge of f enables dotted lines to be drawn, as in the figure above, parallel to the line of M values, between which one half of the x observations, for each value of y, will be included. This value of f has much anthropological interest of its own, especially in connexion with M. Bertillon's system of anthropometric identification, to which I will not call attention now.

It is not necessary to extend the list of examples to show how to measure the degree in which one variable may be co-related with the combined effect of n other variables, whether these be themselves co-related or not. To do so, we begin by reducing each measure into others, each having the Q of its own system for a unit. We thus obtain a set of values that can be treated exactly in the same way as the measures of a single variable were treated in Tables II and onwards. Neither is it necessary to give examples of a method by which the degree may be measured, in which the variables in a series each member of which is the summed effect of n variables, may be modified by their partial co-relation. After transmuting the separate measures as above, and then summing them, we should find the probable error of any one of them to be $\sqrt{n}$ if the variables were perfectly independent, and n if they were rigidly and perfectly co-related. The observed value would be almost always somewhere intermediate between these extremes, and would give the information that is wanted.

To conclude, the prominent characteristics of any two co-related variables, so far at least as I have as yet tested them, are four in number. It is supposed that their respective measures have been first transmuted into others of which the unit is in each case equal to the probable error of a single measure in its own series. Let $y =$ the deviation of the subject, whichever of the two variables may be taken in that capacity; and let x_1, x_2, x_3, etc., be the corresponding deviations of the relative, and let the mean of these be X. Then we find:

(1) that $y = r\text{X}$ for all values of y;
(2) that r is the same, whichever of the two variables is taken for the subject;
(3) that r is always less than 1; and
(4) that r measures the closeness of co-relation.

Introduction to McKendrick (1926) Applications of Mathematics to Medical Problems

K. Dietz
University of Tübingen

The Author

The following biographical sketch is based on three sources:

(a) two obituaries by W.F. Harvey, McKendrick's old colleague in the Indian Medical Service (IMS), who succeeded him as Superintendent of the Royal College of Physicians Laboratory in Edinburgh (Harvey, 1943a, b); and
(b) a biographical article by Aitchison and Watson (1990) who consider McKendrick to be "one of the greatest contributors to our knowledge of the relationship of mathematics and medicine." (See also Gani (1997).)

Anderson Gray McKendrick was born on 8 September 1876 in Edinburgh. Soon after his birth, the family moved to Glasgow where his father had been appointed to the Chair of the Institutes of Medicine which was renamed the Chair in Physiology in 1893. At the age of 24, McKendrick graduated in medicine; during his studies he had spent 1 year in Jena where he lived with the Zeiss family. After his entry examination for the IMS in February 1901, he was invited by Ronald Ross to join him on the Fifth Liverpool Malarial Expedition to Sierra Leone. Ross received the Nobel Prize in 1902 for his discovery of the connection between the Anopheles mosquito and malaria during his time in the IMS, and can be considered as the founder of deterministic mathematical epidemic theory. This meeting turned out to be very influential in directing the professional interests of McKendrick. They returned home together on board the same ship. Ross stimulated McKendrick to familiarize himself with mathematical methods. A book by Mellor on *Higher Mathematics for Students of Chemistry and Physics* (London, 1902) became McKendrick's "mathematical bible," as

Aitchison and Watson (1990) put it. It was customary in the IMS to start with a military assignment, 18 months of which were spent by McKendrick in the Somaliland Campaign against the Mahdi.

His first civil medical post was as a surgeon at Nadia in Bengal where he also acted as superintendent of the district jail. In 1905 he was appointed to the Research Department of the Government of India at the Pasteur Institute at Kasauli in the Punjab. There he began to study rabies, a subject on which he ultimately became the recognized British authority. During a sick leave in Aberdeen which was planned for 2 years starting in March 1913 he wrote that he wanted "... to settle rigidly down to Mathematics...." Because of the start of the First World War he had to return to India earlier than planned. Until 1920 he was Director of the Pasteur Institute in Kasauli. For medical reasons he left the IMS and was appointed to the Royal College of Physicians Laboratory Edinburgh as superintendent until his retirement in 1941 2 years before his death on 30 May 1943.

Harvey's obituary (1943b) lists 58 publications covering a wide range of subjects. For example, in 1938 he published his *Eighth Analytical Report from Pasteur Institutes of the Results of Anti-Rabies Treatment* for the Health Organization of the League of Nations and (together with W.O. Kermack) a paper in the *Mathematical Gazette* on "Some properties of points arranged at random on a Möbius surface." It is quite likely that some of his publications contain gems still to be discovered.

The Paper

McKendrick's *Mathematical Applications to Medical Problems* were presented to the Edinburgh Mathematical Society on 15 January 1926 and published in the same year in the *Proceedings* of this Society. The key person to rescue it from complete oblivion was J.O. Irwin. In 1941 he made a ("not quite correct") reference to it in a discussion on a paper by Chambers and Yule (Irwin, 1941). Later he returned to it and described it in detail in his Presidential Address to the Royal Statistical Society (Irwin, 1963) and at a Symposium on Stochastic Models in Medicine and Biology in Madison (Irwin, 1964). Several years before these two presentations, he had distributed copies of the paper to a number of interested people including Feller. In the second edition of his book, Feller (1957) added the following remark: "It is very unfortunate that this remarkable paper passed practically unnoticed. In particular, it was unknown to the present author when he introduced various stochastic models for population growth in Die Grundlagen der Volterraschen Theorie des Kampfes ums Dasein in wahrscheinlichkeitstheoretischer Behandlung, *Acta Biotheoretica*, vol. 5 (1939). pp. 11–40." Oliviera-Pinto and Connolly (1982) reprinted it in a collection of selected papers in applicable mathematics together with Kermack and

McKendrick's (1927) "Contribution to the mathematical theory of epidemics" as the only examples for the relevance of mathematics to medicine and consider these two papers to "... have a strong claim to be the most remarkable of the whole book." Gani (1982) describes some results of McKendrick's pioneering papers emphasizing the use of stochastic methods. The *Dictionary and Classified Bibliography of Statistical Distributions in Scientific Work* by Patil et al. (1984) lists eight quotations in journals but not those in books such as Bartlett (1962) and Bailey (1975). Most of the 83 entries in the *Science Citation Index* for this paper refer to McKendrick's first-order partial differential equation describing the size of a cohort which is subject to time and age-specific mortality. The other papers quote McKendrick's contributions to various discrete distributions and their fit to data. In interpreting citation counts, we must remember that not all authors who are directly or indirectly using a result are quoting the original source and that not all those who quote a paper have actually read it.

Kostitzin (1934) is usually credited with devising an infinite system of ordinary differential equations describing the number of individuals with different (discrete) numbers of parasites (see Anderson and May (1991)). McKendrick's paper, however, not only contains similar systems of equations for a one-dimensional variable but also infinite systems for two-dimensional variables (see Equation (41)).

McKendrick's paper is divided into eight sections which together contain eleven examples. In this introduction, we comment only on a selection of these.

It is remarkable that his (last) Section 8 entitled "Vorticity in Statistical Problems" has apparently been completely ignored until now. Here McKendrick proposes a (linearized) dynamic system for the interaction between microbes and antibodies within a host.

EXAMPLE 11. "During the course of a microbic disease a battle is raging between microbes on the one hand and protective 'antibodies' on the other. These antibodies are themselves called into being by the action of the microbes. The process is reversible for both variables...."

He gives conditions in terms of inequalities for the model parameters under which we get periodic or damped solutions or augmentation of the amplitude. It was nearly 50 years later that such predator–prey-type equations were used again to describe an immune response (Bell, 1973; see also Hellriegel, 1992).

Between 1927 and 1939 McKendrick wrote a series of five *Contributions to the Mathematical Theory of Epidemics* (Kermack and McKendrick, 1927, 1932, 1933, 1937, 1939) which had a tremendous influence on later developments of deterministic epidemic theory. The *Science Citation Index* lists nearly 200 citations of the first of these papers. Because of the great impact of these papers the name of McKendrick is mostly associated with his deterministic epidemic modeling; his fundamental stochastic and statistical contributions are still largely ignored today.

The McKendrick Equation

An appreciation of McKendrick's equation (Example 9) and its influence on age-specific demographic models is given by Hoppensteadt (1975, 1984). Let $v_{t,\theta}$ denote the number of individuals aged θ at time t and let $f_{t,\theta}$ denote the rate of dying at age θ and time t. Then the equation

$$\frac{\partial v}{\partial t}+\frac{\partial v}{\partial \theta}=-f_{t,\theta}v_{t,\theta}$$

allow us to determine $v_{t,\theta}$ for suitable initial and boundary conditions. McKendrick considers the time-invariant case where f depends only on the chronological age. He calculates the stable age distribution either in exponentially growing, stationary, or exponentially decreasing populations. This work seems to have been derived independently by Lotka (see, e.g., Dublin and Lotka (1925)). Whereas Lotka uses only integral equations, McKendrick shows how his partial differential equation can be transformed into an integral equation. In Example 10 (with Kermack) McKendrick replaces chronological age by class age meaning the time since infection of a newly infected individual. For exactly this problem, such equations had already been introduced by Ross and Hudson (1917, p. 234, Eqn. (87)).

von Foerster (1959) used the same equation, apparently independently, in the context of cell dynamics. The name "von Foerster equation" was popularized by Trucco (1965). Some authors use the name "McKendrick equation" (see, e.g., Cushing (1983)) or "McKendrick–von Foerster equation" (see Wood (1994)).

The original equations either neglected the birth-rate or assumed a linear function of the population size. Gurtin and MacCamy (1974) have studied nonlinear generalizations. Hoppensteadt (1974) presented age-dependent epidemic model equations, where he distinguished chronological and class age, yet without obtaining any results. By using an endemic model which takes into account the age-specific prevalence of susceptibility, Dietz (1975) derived an estimate for the threshold parameter of an infection on the basis of the ratio of life expectancy and the average age at infection, for diseases with homogeneous mixing and life-long immunity. Metz and Diekmann (1986) provide a general framework for age- and time-dependent models, where other physiological variables like body mass are also considered. Hadeler and Dietz (1983) investigate epidemiological models which combine McKendrick's approach to age-dependence with his stochastic approach to the accumulation of parasites. Grenfell et al. (1995) also integrate into these models a discrete immune response of the host. The infinite system of partial differential equations is reduced to an approximating finite system of ordinary differential equations for the first and second moments of the underlying discrete time- and age-dependent random variables. (See also Isham (1995).)

Discrete Distributions

Estimating the Zero Class from a Truncated Poisson Sample

In Section 1 of his paper McKendrick not only derives the Negative Binomial Distribution as a solution for the linear birth process and its limiting Poisson distribution if the birth rate tends to zero, but also fits the resulting models to data on the household distribution of a cholera epidemic in India. He notices that there are more houses without a case than would be expected if one fitted a Poisson distribution to the zero-truncated data. From this he draws the practical conclusion that the cholera infection was only spread from one well to which 93 out of 223 houses were exposed. Irwin (1959) refers to this method and provides an explicit formula for the estimate in terms of a Lagrange expansion. Johnson and Kotz (1969) point out that this estimation method is based on the first two factorial moments. Dahiya and Gross (1973) provide a method to calculate a confidence interval for the missing zero class. Griffiths (1973) suggests that McKendrick's method can be applied to any truncated discrete distribution and that it can be modified by combining iteration with the maximum-likelihood method.

Correlated Poisson Variables

In Section 3 ("Two-Dimensional Cases and Correlation") McKendrick presents models for correlated Poisson variables. Kemp and Kemp (1965) have shown the relation of the sum of two correlated Poisson variables with the Hermite distribution. Aitchison and Ho (1989) point out the connection of the bivariate Poisson distribution with the multivariate Poisson lognormal distribution.

Burnt Fingers Model

In Section 5 ("Restricted Cases") McKendrick introduces a discrete distribution without reference to earlier work by Greenwood and Yule (1920) in the context of accident statistics. The underlying assumption is that an individual who has experienced at least one accident is more prudent, so that the subsequent rate of accidents is reduced. This model has been called the "burnt fingers model" (Arbous and Kerrich, 1951). McKendrick applies this distribution to the situation where the rate of occurrence of events is increased after at least one event, and interprets this additional rate as the relapse rate in the case of malaria, in addition to the superinfection rate. He describes the distributions of spleen sizes of young children assuming that the spleen size "bears a direct relation to the number of attacks which the patient has experienced." An account of these results in the context of the data from McKendrick is given in Appendix A of Dietz (1988).

The Size of an Epidemic in an Infinite Population

In Example 7, McKendrick derives the distribution of the total size of an epidemic in an infinite population starting with one initial case and a constant probability of recovery and infection. Kemp and Kemp (1968) have discussed the derivation of this distribution and its relationship to other stochastic models. The size of the epidemic has the same distribution as the number of lost games in the classical problem of gambler's ruin, the number of customers in the busy period of the M/M/1 queue, or the total number of progeny until extinction in the linear birth-and-death process. (See also Patil and Boswell (1975).) McKendrick uses k for the contact rate and l for the recovery rate. If we denotes

$$q = 1 - p = \frac{k}{k+l},$$

then the distribution is given by

$$p_i = \binom{2i-1}{i} \frac{p^{i-1}q^i}{(2i-1)}.$$

It would be justifiable to associate the name of McKendrick with this distribution.

The General Stochastic Epidemic in a Finite Population

Equation (39) in Section 5, Example 6 describes the so-called general stochastic epidemic in a population of finite size. McKendrick is concerned with the distribution of the final size of the epidemic and gives some numerical results for a household with six individuals. He notices the U-shaped distribution indicating that the introduction of one case into a completely susceptible homogeneously mixing population soon leads to either:

(a) extinction of the outbreak with few secondary cases; or
(b) a large epidemic which affects nearly the entire population.

Daniels (1967) later looked at the same problem and derived exact and approximate results. Already in this paper he had noticed the potential role of the Gontcharoff polynomials for the derivation of explicit formulas, a property which has been successfully exploited by Lefèvre and Picard (1990a, b, 1995).

Acknowledgment

I thank Joe Gani for letting me see his forthcoming paper on McKendrick prior to publication and for his helpful remarks on the first draft of this intoduction.

References

Aitchison, J. and Ho, C.H. (1989). The multivariate Poisson-log normal distribution. *Biometrika*, **76** 643–653.

Aitchison, J. and Watson, G.S. (1990). A not-so-plain tale from the raj: A.G. McKendrick, IMS. In *The Influence of Scottish Medicine* (D.A. Dow, ed.). Parthenon, Carnforth, pp. 115–128.

Anderson, R.M. and May, R.M. (1991). *Infectious Diseases of Humans: Dynamics and Control*. Oxford University Press, Oxford.

Arbous, A.G. and Kerrich, J.E. (1951). Accident statistics and the concept of accident-proneness. *Biometrics*, **7**, 340–432.

Bailey, N.T.J. (1975). *The Mathematical Theory of Infectious Diseases and its Applications*, 2nd ed. Griffin, London.

Bartlett, M.S. (1962). *An Introduction to Stochastic Processes with Special Reference to Methods and Applications*. Cambridge University Press, Cambridge.

Bell, G.I. (1973). Predator–prey equations simulating an immune response. *Math. Biosci.*, **16**, 291–314.

Cushing, J.M. (1983). Bifurcation of time periodic solutions of the McKendrick equations with applications to population dynamics. *Comput. Math. Appl.*, **9**, 459–478.

Dahiya, R.C. and Gross, A.J. (1973). Estimating the zero class from a truncated Poisson sample. *J. Amer. Statist. Assoc.*, **68**, 731–733.

Daniels, H.E. (1967). The distribution of the total size of an epidemic. In *Proceedings of the 5th Berkeley Symposium on Mathematical Statistics and Probability*, Vol IV: *Biology and Problems of Health* (L. LeCam and J. Neyman, eds.). University of California Press. Berkeley, CA, pp. 281–293.

Dietz, K. (1975). Transmission and control of arbovirus diseases. In *Epidemiology. Proceedings of a SIMS Conference on Epidemiology, Alta, Utah, July 8–12, 1974* (D. Ludwing and K.L. Cooke, eds.). SIAM, Philadelphia, PA, pp. 104–121.

Dietz, K. (1988). Mathematical models for transmission and control of malaria. In *Malaria: Principles and Practice of Malariology* (W.H. Wernsdorfer and I. McGregor, eds.), Vol. **2**. Churchill Livingstone, Edinburgh, pp. 1091–1133.

Dublin, L.I. and Lotka, A.J. (1925). On the true rate of natural increase as exemplified by the population of the United States, 1920. *J. Amer. Statist. Assoc.*, **20**, 305–339.

Feller, W. (1957). *An Introduction to Probability Theory and its Applications*, Vol. I, 2nd ed. Wiley, New York, p. 404.

Foerster, von H. (1959). Some remarks on changing populations. In *The Kinetics of Cellular Proliferation* (F. Stohlman, Jr., ed.). Grune and Stratton, New York, pp. 382–407.

Gani, J. (1982). The early use of stochastic methods: An historical note on McKendrick's pioneering papers. In *Statistics and Probability: Essays in Honor of C.R. Rao* (G. Kallianpur, P.R. Krishnaiah, and J.K. Ghosh, eds.). North-Holland, Amsterdam, pp. 263–268.

Gani, J. (1997). McKendrick, founder of stochastic epidemic modelling. In *Statisticians of the Centuries* (C.C. Heyde, ed.) (to appear).

Greenwood, M. and Yule, G.U. (1920). An inquiry into the nature of frequency distributions representative of multiple happenings with particular reference to the occurrence of multiple attacks of disease or of repeated accidents. *J. Roy. Statist. Soc.*, **83**, 255–279 [also in *Statistical Papers of George Udny Yule* (A.S. Stuart and M.G. Kendall, eds.), 1971. Griffin, London, pp. 269–293].

Grenfell, B.T., Dietz, K., and Roberts, M.G. (1995). Modelling the immuno-epidemiology of macroparasites in naturally-fluctuating host populations. In *Ecology of Infectious Diseases in Natural Populations*, (B.T. Grenfell and A.P. Dobson, eds.). Cambridge University Press, Cambridge.

Griffiths, D.A. (1973). Maximum likelihood esimation for the beta-binomial distribution and an application to the household distribution of the total number of cases of a disease. *Biometrics*, **29**, 637–648.

Gurtin, M.E. and MacCamy, R.C. (1974). Non-linear age-dependent population dynamics. *Arch. Rational Mech. Anal.*, **54**, 281–300.

Hadeler, K.P. and Dietz, K. (1983). Nonlinear hyperbolic partial differential equations for the dynamics of parasite populations. *Comput. Math. Appl.*, **9**, 415–430.

Harvey, W.F. (1943a). Anderson Gray McKendrick, 1876–1943. *Roy. Soc. Edinburgh Year Book*, pp. 23–24.

Harvey, W.F. (1943b). Anderson Gray McKendrick, 1876–1943. *Edinburgh Med. J.*, **50**, 500–506.

Hellriegel, B. (1992). Modelling the immune response to malaria with ecological concepts: Short-term behaviour against long-term equilibrium. *Proc. Roy. Soc. London Ser. B*, **250**, 249–256.

Hoppensteadt, F. (1974). An age dependent epidemic model. *J. Franklin Inst.*, **297**, 213–225.

Hoppensteadt, F. (1975). *Mathematical Theories of Populations: Demographics, Genetics and Epidemics*. SIAM, Philadelphia, PA.

Hoppensteadt, F. (1984). Some influences of population biology on mathematics. In *Essays in the History of Mathematics* (A. Schlissel, ed.). *Mem. Amer. Math. Soc.*, **48**, 25–29.

Irwin, J.O. (1941). Discussion on the paper by Chambers and Yule: Theory and observation in the investigation of accident causation. *Suppl. J. Roy. Statist. Soc.*, **7**, 101–107.

Irwin, J.O. (1959). On the estimation of the mean of a Poisson distribution from a sample with the zero class missing. *Biometrics*, **15**, 324–326.

Irwin, J.O. (1963). The place of mathematics in medical and biological statistics. *J. Roy. Statist. Soc. Ser. A*, **126**, 1–44.

Irwin, J.O. (1964). The contributions of G.U. Yule and A.G. McKendrick to stochastic process methods in biology and medicine. In *Stochastic Models in Medicine and Biology* (J. Gurland, ed.). University of Wisconsin Press, Madison, pp. 147–165.

Isham, V. (1995). Stochastic models of host–macroparasite interaction. *Ann. Appl. Probab.*, **5**, 720–740.

Johnson, N.L. and Kotz, S. (1969). *Discrete Distributions*. Houghton Mifflin, Boston.

Kemp, A.W. and Kemp, C.D. (1968). On a distribution associated with certain stochastic processes. *J. Roy. Statist. Soc. Ser. B*, **30**, 160–163.

Kemp, C.D. and Kemp, A.W. (1965). Some properties of the "Hermite" distribution. *Biometrika*, **52**, 381–394.

Kermack, W.O. and McKendrick, A.G. (1927). A contribution to the mathematical theory of epidemics. *Proc. Roy. Soc. London Ser. A*, **115**, 700–721 [also in Oliviera-Pinto and Conolly, 1982, pp. 222–247, and in *Bull. Math. Biol.*, **53**, 1990, 33–55].

Kermack, W.O. and McKendrick, A.G. (1932). Contributions to the mathematical theory of epidemics: II. The problem of endemicity. *Proc. Roy. Soc. London Ser. A*, **138**, 55–83 [also in *Bull. Math. Biol.*, **53**, 1990, 57–87].

Kermack, W.O. and McKendrick, A.G. (1933). Contributions to the mathematical theory of epidemics: III. Further studies of the problem of endemicity. *Proc. Roy. Soc. London Ser. A*, **141**, 94–122 [also in *Bull. Math. Biol.*, **53**, 1990, 89–118].

Kermack, W.O. and McKendrick, A.G. (1937). Contributions to the mathematical theory of epidemics: IV. Analysis of experimental epidemics of the virus disease mouse ectromelia. *J. Hygiene*, **37**, 172–187.

Kermack, W.O. and McKendrick, A.G. (1939). Contributions to the mathematical theory of epidemics: V. Analysis of experimental epidemics of the mouse-typhoid; a bacterial disease conferring incomplete immunity. *J. Hygiene*, **39**, 271–288.

Kostitzin, V.A. (1934). Symbiose, parasitisme et évolution (Étude Mathématique). In *Exposés de Biométrie et de Statistique Biologique* (G. Teissier, ed.). Actualités Scientifiques et Industrielles, vol. **96**. Herman, Paris, pp. 1–44 [also translated in *The Golden Age of Theoretical Ecology: 1923–1940* (F.M. Scudo and J.R. Ziegler), 1978. Springer-Verlag, Berlin, pp. 369–412].

Lefèvre, C. and Picard, P. (1990a). The final size distribution of epidemics spread by infectives behaving independently. In *Stochastic Processes in Epidemic Theory*, (J.-P. Gabriel, C. Lefèvre, and P. Picard, eds.). Lecutre Notes in Biomathematics, vol. **86**. Springer-Verlag, New York, pp. 155–169.

Lefèvre, C. and Picard P. (1990b). A non-standard family of polynomials and the final size distribution of Reed–Frost epidemic processes. *Adv. in Appl. Probab.*, **22**, 25–48.

Lefèvre, C. and Picard, P. (1995). Collective epidemic processes: A general modelling approach to the final outcome of SIR infectious diseases. In *Epidemic Models: Their Structure and Relation to Data* (D. Mollison, ed.). Cambridge University Press, Cambridge, pp. 53–70.

Metz, J.A.J. and Diekmann, O. (eds.) (1986). *The Dynamics of Physiologically Structured Populations*. Lecture Notes in Biomathematics, vol. **68**. Springer-Verlag, Berlin.

Oliviera-Pinto, G. and Connolly, B.W. (1982). *Applicable Mathematics of Non-physical Phenomena*. Ellis Horwood, Chichester.

Patil, G.P. and Boswell, M.T. (1975). Chance mechanisms for discrete distributions in scientific modelling. In *A Modern Course on Statistical Distributions in Scientific Work*, Vol. **2**: *Model Building and Model Selection* (G.P. Patil, S. Kotz, and J.K. Ord, eds.). Reidel, Dordrecht, pp. 11–24.

Patil, G.P., Boswell, M.T., Joshi, S.W., and Ratnaparkhi, M.V. (1984). *Dictionary and Classified Bibliography of Statistical Distributions in Scientific Work*, Vol. **1**: *Discrete Models*. International Co-operative Publishing House, Burtonsville, MD.

Ross, R. and Hudson, H. (1917). An application of the theory of probabilities to the study of *a priori* pathometry—Part III. *Proc. Roy. Soc. London Ser. A*, **93**, 225–240.

Trucco, E. (1965). Mathematical models for cellular systems: The von Foerster equation, Parts I, II. *Bull. Math. Biophys.*, **27**, 285–304, 449–471.

Wood, S.N. (1994). Obtaining birth and mortality patterns from structured population trajectories. *Ecological Monographs*, **64**, 23–44.

Added in Proof. In the latest issue of *Statistical Methods in Medical Research*, Vol. 6, pages 3–23 (1997) there is an article by Xiao-Li Meng on "The EM algorithm and medical studies: A historical link," which provides a detailed analysis of McKendrick's contribution to the EM algorithm, based on the 1926 paper under discussion.

Applications of Mathematics to Medical Problems

A.G. McKendrick
Laboratory of the Royal College of Physicians, Edinburgh

In the majority of the processes with which one is concerned in the study of the medical sciences, one has to deal with assemblages of individuals, be they living or be they dead, which become affected according to some characteristic. They may meet and exchange ideas, the meeting may result in the transference of some infectious disease, and so forth. The life of each individual consists of a train of such incidents, one following the other. From another point of view each member of the human community consists of an assemblage of cells. These cells react and interact amongst each other, and each individual lives a life which may be again considered as a succession of events, one following the other. If one thinks of these individuals, be they human beings or be they cells, as moving in all sorts of dimensions, reversibly or irreversibly, continuously or discontinuously, by unit stages or *per saltum*, then the method of their movement becomes a study in kinetics, and can be approached by the methods ordinarily adopted in the study of such systems.

It is the object of this communication to approach this field in a systematic manner, to find solutions for some of the variations which may arise, and to illustrate certain of these by examples.

1. One Dimension, Irreversible

I have been in the habit of employing vector diagrams for the representation of such problems. They have the advantage that the hypotheses which are adopted are clearly visualized as well by the non-mathematical reader as by the mathematical, and they also aid in helping one to realise the various modifications which may occur, and so to treat the study of the general problem systematically. To fix ideas let us consider a simple case; the relation of an assemblage of individuals to common colds. In the following series of compartments are classified at any instant the numbers of individuals who have experienced, 0, 1, 2, 3, ... attacks of this complaint. The history of each individual consists of a series of unit steps, originating in the compartment which describes his initial condition. The arrows in the diagram indicate the chance of passage from one compartment to the next—that is to say the chance of experiencing a further attack during the infinitesimal period of time dt.

In Figure 1 these arrows are of equal size, and by this we understand that the successive chances were of constant value; in Figure 2 the arrows increase in size, denoting an increase of susceptibility with each attack; in Figure 3 they decrease, which denotes that the individual is becoming decreasingly liable, or in medical parlance he is developing an immunity.

Guided by the diagram, and using the nomenclature $v_x =$ the number of individuals who have experienced x attacks (or shortly "of grade x"); $f_{t,x}\,dt =$ the probability that an individual of grade x will pass to grade $x+1$ in the time dt, and noting that the variation of the number in any grade is the difference between the number of incomers into that grade, and the number who go out from that grade, we have

$$dv_x = (f_{t,x-1}v_{x-1} - f_{t,x}v_x)\,dt. \tag{1}$$

In this case and in what follows, for the sake of conciseness, the solutions will be given for instantaneous point sources; other initial conditions may be obtained by summation.

In the first place let us assume that $f_{t,x}$ is of the form $\phi_t f_x$, that is to say,

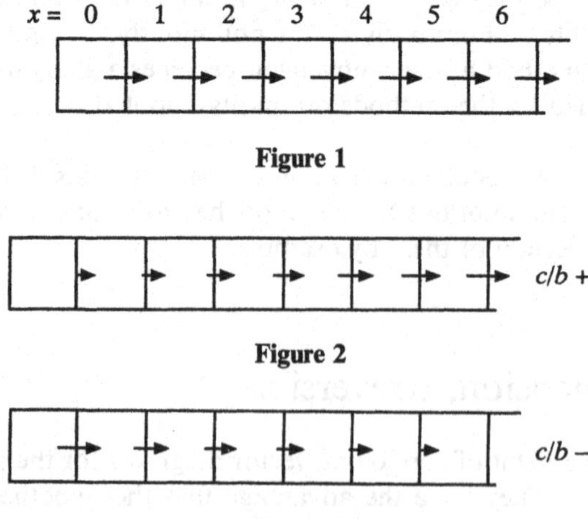

Figure 1

Figure 2

Figure 3

that the time function applies generally to the probability of exit from all compartments. (The general case will be considered later in dealing with two-dimensional problems.)

Let us adopt the nomenclature $\mu_r \equiv \sum_{x=0}^{\infty} \frac{(x-\mu)^r v_x}{N}$ where μ is the mean $\left(\sum_0^{\infty} \frac{x v_x}{N}\right)$, and N is the total number of individuals.

When $f_x = b + cx$ (a first approximation), we find

$$v_x = N\frac{b}{c}\left(\frac{b}{c}+1\right)\cdots\left(\frac{b}{c}+x-1\right)\frac{\left(1-\frac{\mu}{\mu_2}\right)^x}{x!}\left(\frac{\mu}{\mu_2}\right)^{b/c}. \tag{2}$$

(The values for the moments are obtained by differentiating the particular moment and making use of equation (1).)

Thus v_x is the $(x+1)^{\text{th}}$ term of the expansion of the binomial

$$N\left\{\frac{\mu_2}{\mu} - \left(\frac{\mu_2}{\mu} - 1\right)\right\}^{-b/c}$$

also

$$\frac{c}{b} = \frac{\mu_2 - \mu}{\mu^2} \qquad \text{and} \qquad \mu_3\mu + \mu_2\mu = 2\mu_2^2. \tag{3}$$

If we write

$$\lambda_r \equiv \sum_0^{\infty} x^r v_x, \qquad N = \frac{\lambda_1^2(\lambda_2 - \lambda_1)}{2\lambda_2^2 - \lambda_1(\lambda_3 + \lambda_2)}. \tag{4}$$

In the case where c/b tends to zero, the solution reduces to

$$v_x = Ne^{-\mu}\frac{\mu^x}{x!}, \tag{5}$$

i.e., Poisson's limit of the binomial. It is interesting to note that the time function ϕ_t has been eliminated, and does not affect the relative distributions given in (2) and (5). This is of importance in dealing with many problems, for example: (a) the effects of seasonal variations which apply generally to individuals of all grades, and which probably operate in all epidemics; (b) variations in the virulence of the organism during the course of the epidemic to which it gives rise; and (c) any variations depending upon the values μ_r or λ_r, which are themselves functions of the time and consequently may be expressed as ϕ_t, are eliminated and do not affect the distribution given by the solution.

Example 1. The following figures denote the number of houses in two suburbs of the town of Luckau in Hanover, in which x cases of cancer had occurred during the period 1875–1898.

	Observed	Calculated
Houses with 0 case	64	65
Houses with 1 case	43	40
Houses with 2 cases	10	12
Houses with 3 cases	2	2.5
Houses with 4 cases	1	0.4

The value of c/b was negligible (−0.009). The calculated figures were obtained by equation (5). Thus the figures afford no evidence that the occurrence of a late case was influenced by previous cases. Behla, however, from whose communication the table was obtained, writes with regard to those houses which experienced 3 and 4 cases during the period "das kann kein zufall sine."

EXAMPLE 2. The following figures refer to an epidemic of cholera in a village in India.

	Observed	Calculated
Houses with 0 case	168	37
Houses with 1 case	32	34
Houses with 2 cases	16	16
Houses with 3 cases	6	5
Houses with 4 cases	1	1
	—	—
	223	93

If one assumes that the value of c/b was zero, then $N = \lambda_1^2/(\lambda_2 - \lambda_1)$, and the calculated values of v_x are as tabulated. This suggests that the disease was probably water borne, that there were a number of wells, and that the inhabitants of 93 out of 223 houses drank from one well which was infected. On further local investigation it was found that there was one particular infected well from which a certain section of the community drank.

EXAMPLE 3. A modification of this method depends upon the fact that if conditions be constant over a number of years, the problem of one community over a series of years may be replaced by that of a number of communities over a single year. Bowley in his *Elements of Statistics* gives the following figures for deaths from human anthrax in Great Britain.

	Observed	Calculated
Years with 0–3 deaths	1	0.9
Years with 4–7 deaths	6	5.6
Years with 8–11 deaths	7	7.3
Years with 12–15 deaths	4	4.2
Years with 16–23 deaths	2	1.9

The closeness of the fit suggests that conditions had been constant, and the value for $c/b(+0.089)$ suggests that the disease is slightly epidemic in the sense that the chance of occurrence of fresh cases in any year increases with the number which have already occurred in that year.

The method has been extensively employed in various directions epidemiological and other. For further examples the reader is referred to References [2]–[8].

2. One Dimension, Reversible

The scheme for reversible cases is given in Figure 4. The variation in this case is the sum of the variations of the forward and backward movements, consequently,

$$dv_x = (f_{x-1}v_{x-1} - f_x v_x)\,dt + (f'_{x+1}v_{x+1} - f'_x v_x)\,dt. \tag{6}$$

Where $f = f' =$ constant, $\mu = 0$, and

$$v_x = N_e^{-\mu_2} I_x(\mu_2). \tag{7}$$

Where $I_x(\mu_2)$ is the Bessel function of the xth order for an imaginary argument.

When $f \neq f'$, but both are constant, $\mu = \mu_3$, and

$$v_x = N_e^{-\mu_2}\left(\frac{\mu_2 + \mu_3}{\mu_2 - \mu_3}\right)^{x/2} I_x(\sqrt{\mu_2^2 - \mu_3^2}). \tag{8}$$

(The value μ_3 is written here in place of μ, as the position for $x = 0$ may not be known.)

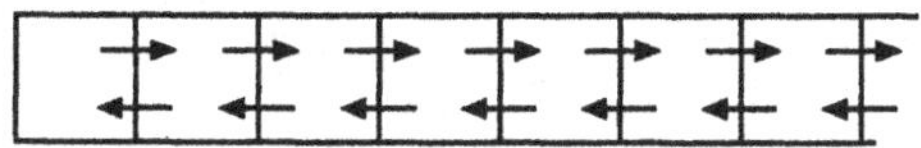

Figure 4

Where $f_x = f'_x = b + cx$,

$$v_x = Nb(b+c)\cdots(b+\overline{cx-1})(1+ct)^{-(2(b/c)+x)}\frac{t^x}{x!}$$

$$F\left(-\frac{b}{c}, -\frac{b}{c}+1, x+1, c^2t^2\right). \quad (9)$$

3. Two-Dimensional Cases and Correlation

The schema for these, if the variables x and y are independent, is obviously as shown in Figure 5.

There are now two movements, one in the x dimension, and one in the y dimension. Consequently,

$$dv_{x,y} = (f_{x-1,y}v_{x-1,y} - f_{xy}v_{xy})\,dt + (g_{x,y-1}v_{x,y-1} - g_{x,y}v_{x,y})\,dt. \quad (10)$$

When $f_{x,y} = b + cx$ and $g_{x,y} = d + ey$,

$$v_{x,y} = v_{0,0}\frac{b}{c}\left(\frac{b}{c}+1\right)\cdots\left(\frac{b}{c}+x-1\right)\cdot$$

$$\cdot\frac{\left(1-\frac{\mu_{x1}}{\mu_{x2}}\right)^x}{x!}\cdot\frac{d}{e}\left(\frac{d}{e}+1\right)\cdots\left(\frac{d}{e}+y-1\right)\cdot\frac{\left(1-\frac{\mu_{y1}}{\mu_{y2}}\right)^y}{y!}, \quad (11)$$

where

$$\mu_{x1} \equiv \sum_0^\infty \sum_0^\infty x\, v_{xy} \div N,$$

$$\mu_{xr} \equiv \sum_0^\infty \sum_0^\infty (x-\mu_{x1})^r v_{x,r} \div N,$$

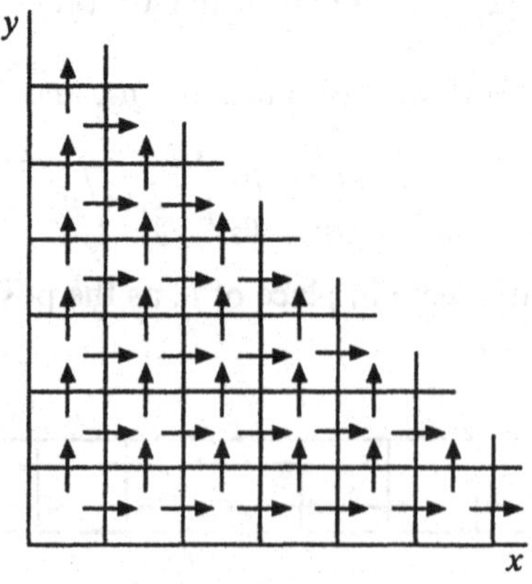

Figure 5

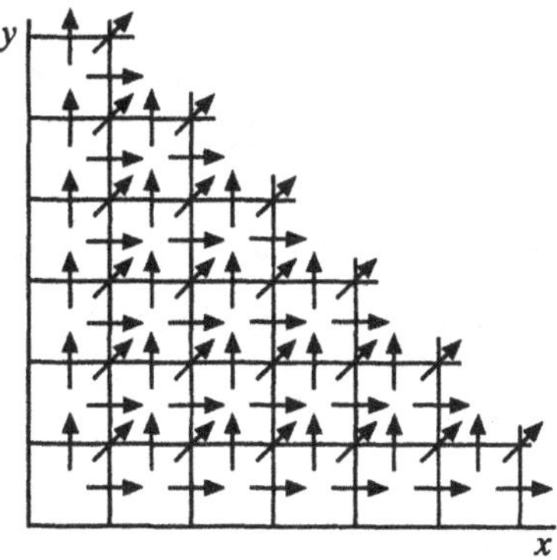

Figure 6

also

$$\frac{c}{b} = \frac{\mu_{x2} - \mu_{x1}}{\mu_{x1}^2} \qquad \text{and} \qquad \frac{e}{d} = \frac{\mu_{y2} - \mu_{y1}}{\mu_{y1}^2}. \tag{12}$$

In the case where c/b and e/d tend to zero,

$$v_{x,y} = N_e^{-(\mu_{x1}+\mu_{y1})} \frac{(\mu_{x1})^x}{x!} \frac{(\mu_{y1})^y}{y!}. \tag{13}$$

There is, however, a third direction of movement and this is illustrated in Figure 6: individuals may move form the compartment x, y into the compartment $x+1$, $y+1$. In Figure 5 where there is no oblique movement of this sort, we assumed that the probabilities of movement $f_{xy}\,dt$ and $g_{xy}\,dt$ were so small that in comparison with them their product is negligible. This is true of events which are independent. But if when an event of the one sort is likely to happen, an event of the other sort is also likely to happen, then oblique movement is no longer negligible. This relation between the two types of events is the logician's definition of correlation. I have been in the habit of calling it "oblique" correlation for the following reason. It is obvious that when no correlation exists between x and y the numbers v_{xy} will, when a sufficient period of time has elapsed, be so arranged that equal values of $v_{x,y}$ will lie on contours which have their major and minor axes parallel, the one to the x coordinate and the other to the y coordinate. When oblique correlation exists (as denoted by the oblique arrows) there will be a tendency for these *ovoid* forms to be dilated in the $x = y$ direction and so oblique *ovoid* forms will result. But this type of oblique contour may be also brought about by mathematical considerations of another nature. Thus if $f_{x,y}$ is an increasing function of y, a "shear" movement will tend to push the upper portion of a non-correlated distribution in the x direction, and similarly if $g_{x,y}$ is an increasing function of x there will be a shear movement pushing the portion on the right in an upward direction. This type of move-

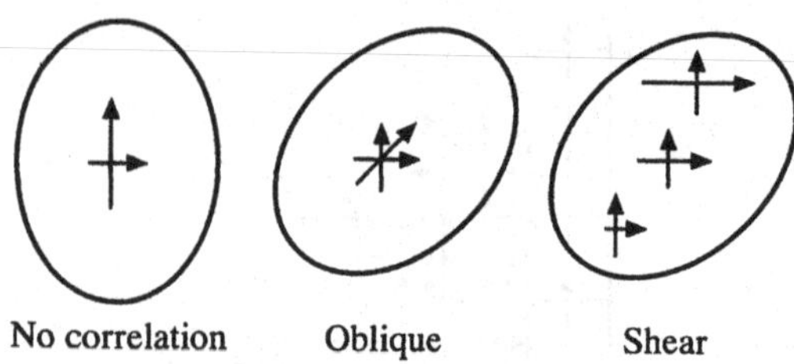

Figure 7

ment is shown in the third diagram of Figure 7. I have been in the habit of denoting this variety by the name "shear" correlation. Statistically it means that the chance of an occurrence of one sort depends upon the number of previous occurrences of the other sort.

"Shear" correlation may be illustrated in the following example from vital statistics. If x be the number of males, y the number of females, and $v_{x,y}$ denote the number of communities containing x males and y females, then the probability of a child (male or female) being born will depend (subject to the type of marriage which is in operation) upon the number of parents of both sexes, that is to say it will be a function of xy. In the resulting distribution the largest values of $v_{x,y}$ will lie along the diagonal $x = y$, or in other words there will tend to be an equality of sexes in the communities.

The general equation, which includes both types of correlation, is

$$\begin{aligned}\frac{dv_{x,y}}{dt} &= f_{x-1,y}v_{x-1,y} - f_{x,y}v_{xy} \\ &\quad + g_{x,y-1}v_{x,y-1} - g_{x,y}v_{xy} \\ &\quad + h_{x-1,y-1}v_{x-1,y-1} - h_{x,y}v_{x,y}. \end{aligned} \tag{14}$$

In the case where f, g and h are all constants, that is to say a simple case of oblique correlation, with no shear, the solution is

$$v_{x,y} = Ne^{-(\mu_{x2}+\mu_{y2}-\mu_{yx})} \sum_{s=0} \frac{(\mu_{x2}-\mu_{xy})^{x-s}}{(x-s)!} \frac{(\mu_{y2}-\mu_{xy})^{y-s}}{(y-s)!} \frac{\mu_{xy}^s}{s!} \tag{15}$$

where μ_{xy} is written for

$$\sum_0^\infty \sum_0^\infty (xy - \mu_{x1}\mu_{y1})x_{xy} \div N,$$

and the upper limit of the summation in the expression for v_{xy} is $s = x$ or y whichever is least.

The moments may be obtained as follows. We have for v_{00} the relation

$$\frac{dv_{00}}{dt} = -(f_{00} + g_{00} + h_{00})v_{00}$$

or if m be written for $\log N/v_{00}$

$$\frac{dm}{dt} = f_{00} + g_{00} + h_{00}.$$

In the case where

$$f_{x,y} = \alpha_1 + \alpha_2 x + \alpha_3 y,$$
$$g_{x,y} = \beta_1 + \beta_2 x + \beta_3 y,$$
$$h_{x,y} = \gamma_1 + \gamma_2 x + \gamma_3 y,$$

if we also write

$$P_r \equiv \frac{\alpha_r}{\alpha_1 + \beta_1 + \gamma_1},$$
$$Q_r \equiv \frac{\beta_r}{\alpha_1 + \beta_1 + \gamma_1},$$
$$R_r \equiv \frac{\gamma_r}{\alpha_1 + \beta_1 + \gamma_1},$$

equation (14) takes the form

$$\begin{aligned}\frac{dv_{x,y}}{dm} &= (P_1 + P_2(x-1) + P_3 y)v_{x-1,y} - (P_1 + P_2 x + P_3 y)v_{x,y} \\ &\quad + (Q_1 + Q_2 x + Q_3(y-1))v_{x,y-1} - (Q_1 + Q_2 x + Q_3 y)v_{x,y} \\ &\quad + (R_1 + R_2(x-1) + R_3(y-1))v_{x-1,y-1} - (R_1 + R_2 x + R_3 y)v_{x,y}.\end{aligned}$$

Hence by differentiating the expressions for the moments we have

$$\left.\begin{aligned}\frac{d\mu_{x1}}{dm} &= (P_1 + R_1) + (P_2 + R_2)\mu_{x1} + (P_3 + R_3)\mu_{y1}, \\ \frac{d\mu_{y1}}{dm} &= (Q_1 + R_1) + (Q_2 + R_2)\mu_{x1} + (Q_3 + R_3)\mu_{y1}, \\ \frac{d\mu_{x2}}{dm} &= (P_1 + R_1) + (P_2 + R_2)\mu_{x1} + (P_3 + Q_3)\mu_{y1} \\ &\quad + 2(P_2 + R_2)\mu_{x2} + 2(P_3 + R_3)\mu_{xy}, \\ \frac{d\mu_{y2}}{dm} &= (Q_1 + R_1) + (Q_2 + R_2)\mu_{x1} + (Q_3 + R_3)\mu_{y1} \\ &\quad + 2(Q_2 + R_2)\mu_{xy} + 2(Q_3 + R_3)\mu_{y2}, \\ \frac{d\mu_{xy}}{dm} &= R_1 + R_2\mu_{x1} + R_3\mu_{y1} + (Q_2 + R_2)\mu_{x2} + (P_3 + R_3)\mu_{y2} \\ &\quad + (P_2 + R_2 + Q_3 + R_3)\mu_{xy}.\end{aligned}\right\} \quad (16)$$

(We may remind the reader at this point that oblique correlation depends upon the values R_1, R_2 and R_3, whereas shear correlation depends upon P_3 and Q_2.) These equations may be solved by the usual methods.

In the particular case where f, g and h are all constants, we have from the above,

$$\left.\begin{aligned} \mu_{x1} &= \mu_{x2} = (P_1 + R_1)m, \\ \mu_{y1} &= \mu_{y2} = (Q_1 + R_1)m, \\ \mu_{xy} &= R_1 m, \end{aligned}\right\} \tag{17}$$

whence

$$\frac{\mu_{xy}}{\sqrt{\mu_{x2}\mu_{y2}}} = \frac{R_1}{\sqrt{(P_1 + R_1)(Q_1 + R_1)}} \tag{18}$$

an expression from which m and consequently t has been eliminated. This is of course the ordinary Pearsonian correlation coefficient r.

It is easy to show, by summations of the values of $v_{x,y}$, that if the statistics are arranged in the four fold table

	not A	A
not B	a	b
B	c	d

$$r = \frac{\log \dfrac{a(a+b+c+d)}{(a+b)(a+c)}}{\sqrt{\log \dfrac{a+b+c+d}{a+b} \cdot \log \dfrac{a+b+c+d}{a+c}}} \tag{19}$$

and it is interesting to note that when $\dfrac{c+d}{a+b}$, $\dfrac{b+d}{a+c}$, and $\dfrac{ad-bc}{(a+b)(a+c)}$ are so small that their squares may be neglected, this value of r reduces to $\dfrac{ad-bc}{\sqrt{(a+b)(a+c)(b+d)(c+d)}}$ which is the Pearsonian coefficient ϕ.

4. Summations

Before leaving this part of the subject attention may be drawn to the following schema Figure 8 which shows the effect of collecting the values of v_{xy} in either diagonal direction, i.e., according to the relation $n = x + y$ or $n = x - y$. Such summations may be written in the forms

$$S_n = \sum_{s=0}^{n} v_{n-s,s},$$

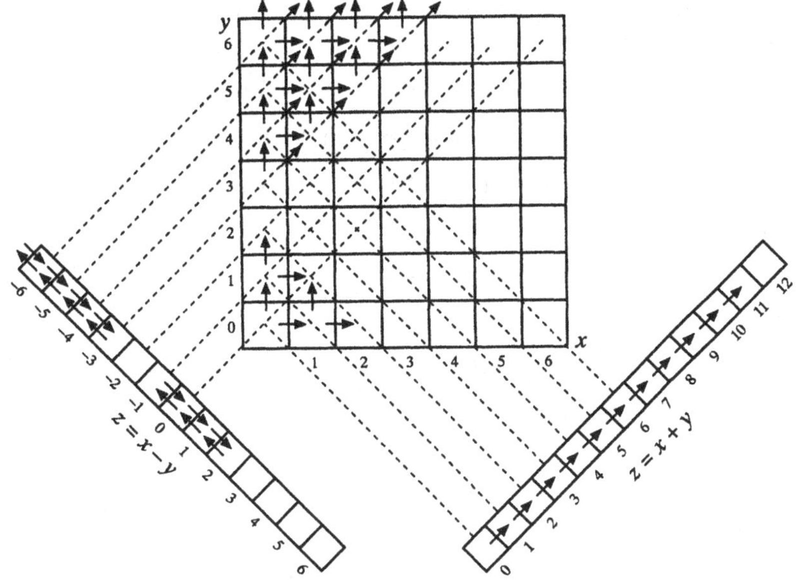

Figure 8

and

$$T_n = \sum_{s=0}^{\infty} v_{n+s,s},$$

$$T_{-n} = \sum_{s=0}^{\infty} v_{s,n+s}.$$

These are important in medical statistical problems. For example if x denote the number of fresh infections of a disease, and y denote the number of relapses, then $x + y = n$ denotes the number of attacks of the disease. The functions $f_{xy}\,dt$, and $g_{xy}\,dt$, for fresh infections and relapses respectively, are certainly not identical, and as it is impossible to differentiate clinically between fresh infection and relapse, statistics are only available in the form S_n.

It is at once apparent from the figure, that where f, g and h are constants the result of the summation $n = x + y$, is to convert an irreversible two-dimensional schema into a reversible schema in one dimension (the effect of oblique correlation does not appear), and the solution is the Bessel form which has already been dealt with.

Summation according to $n = x + y$ gives the equation

$$\frac{dS_n}{dm} + (P + Q)(S_{n-1} - S_n) + R(S_{n-2} - S_n), \tag{20}$$

where for convenience the suffixes of P_1, Q_1 and R_1 are dropped, whence for the moments according to n we have

$$\left.\begin{aligned} \mu &= (P+Q+2R)m = (1+R)m, \\ \mu_2 &= (P+Q+4R)m = (1+3R)m, \\ \mu_3 &= (P+Q+8R)m = (1+7R)m. \end{aligned}\right\} \tag{21}$$

Consequently

$$R = \frac{\mu - \mu_2}{\mu_2 - 3\mu} \tag{22}$$

and

$$\mu_3 = 3\mu_2 - 2\mu. \tag{23}$$

From the latter expression we also obtain the relation

$$N = \frac{3\lambda_1(\lambda_2 - \lambda) \pm \lambda_1\sqrt{(9\lambda_2^2 - 7\lambda_1^2 - 8\lambda_1\lambda_3 + 6\lambda_1\lambda_2)}}{2(\lambda_3 - 3\lambda_2 + 2\lambda_1)}. \tag{24}$$

Also after some reduction we find

$$r^2 = \frac{R^2}{(P+R)(Q+R)} = \frac{\left(2 + \dfrac{P}{Q} + \dfrac{Q}{P}\right)R^2}{1 + \left(\dfrac{P}{Q} + \dfrac{Q}{P}\right)R + R^2} \tag{25}$$

(and thus when

$$P = Q, \qquad r = \frac{2R}{1+R} = \frac{\mu_2}{\mu} - 1). \tag{26}$$

Finally we have

$$S_n = N_e^{-(3\mu-\mu_2)/2} \sum_{s=0}^{n/2} \frac{(2\mu - \mu_2)^{n-2s}}{(n-2s)!} \frac{\left(\dfrac{\mu_2 - \mu_1}{2}\right)^s}{s!}. \tag{27}$$

Example 4. In a phagocytic experiment it seemed likely that bacteria which were being ingested were not all discrete, some of them were united into pairs. If one considers for the moment that they were of two types, and that a pair consisted of one of each type, then the above analysis is applicable.

The figures were as follows:

	Observed	Calculated
Leucocytes containing 0 bacteria	269	268
Leucocytes containing 1 bacteria	4	7
Leucocytes containing 2 bacteria	26	23
Leucocytes containing 3 bacteria	0	0.6
Leucocytes containing 4 bacteria	1	1.1

and, since in this case $P_1 = Q, r = 0.86$.

This example rests upon incomplete assumptions, and upon insufficient data. It is introduced only to show how, when $P = Q$, the correlation coefficient may be calculated from one-dimensional data.

5. Restricted Cases

So far we have been dealing with the complete case in which entry into all compartments in the schema is possible. Let us now turn to particular cases in which certain limitations are introduced.

EXAMPLE 5. The problem of house infection may be approached in a manner rather different from that suggested in 1. In the accompanying Figure 9 the coordinate x denotes infections arising from without the house, and the coordinate y denotes infections originating within the house itself. There can be no entrance into the upper compartments of the zero column, for an internal infection is only possible after an infection from the exterior has taken place. The chance of an external infection occurring is proportional to the number of cases in all houses except the house with which one is dealing, and the probability of occurrence of an internal infection depends upon the number of cases which have already occurred within the house itself.

Thus, in its simplest form, we have

$$\frac{dv_{x,y}}{dt} = k\{(\sum\sum(x+y)v_{x,y} - n + 1)v_{x-1,y} - (\sum\sum(x+y)v_{x,y} - n)v_{x,y}\} + l\{(n-1)v_{x,y-1} - nv_{x,y}\}, \tag{28}$$

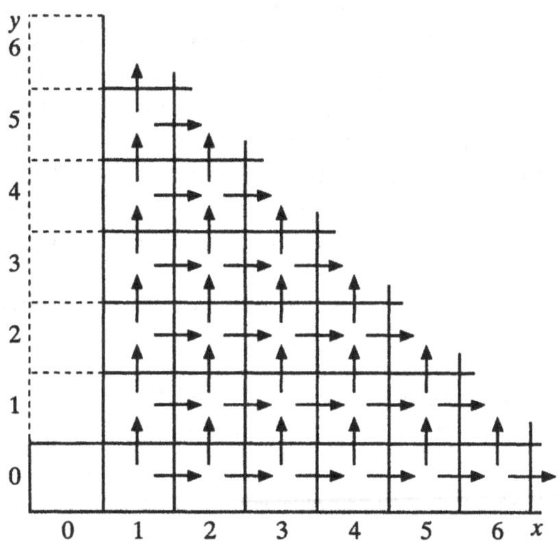

Figure 9

where $k\,dt$ and $l\,dt$ denote the chances of external and internal infections respectively.

Hence for the value S_n we have

$$\frac{dS_n}{dt} = kN\mu(S_{n-1} - S_n) + (l - k)\{(n-1)S_{n-1} - nS_n\}, \qquad (29)$$

where the moment μ is taken with reference to n.

Consequently

$$\frac{dN\mu}{dt} = \{k(N-1) + l\}N\mu$$

and

$$\frac{l}{k} = \frac{N(\mu - \mu_0)}{\log \dfrac{N}{S_0}} - N + 1, \qquad (30)$$

where μ_0 is the initial value of μ.

That is to say from the total number of cases ($N\mu$), the total number of houses (N), and the number of infected houses S_0, we can find the ratio of the probability of external infection to that of internal infection, as in most cases μ_0 is small as compared with μ.*

In the case of bubonic plague, from four epidemics in consecutive years in a certain village, $l/k = 199$. For a similar number of epidemics in a neighbouring village the value was 231. Allowing for a large error, the probability of internal infection would appear to be about 200 times as great as that of external infection. In the case of cancer, from one group of statistics a value of 10 was obtained, and from another a value of −9. Each of these numbers is the difference between two large numbers and has a large error. The high value of the ratio l/k in the case of plague may be easily understood when one remembers that the disease is transmitted by fleas, a species of animal which does not as a rule travel far from its own neighbourhood. In the case of cancer the figures afford no evidence either of insect infection or of infection by contagion. Present theories regarding the transmission of *kala azar* point to transmission through either the bedbug or the sand fly. The former is a very local insect, the latter is supposed to be distinctly localised, though not as strictly so as either the flea or the bed-bug. From figures which have been placed at my disposal the value of $1/k$ was of the order 70, suggesting that the sand fly is the more likely carrier.

Example 6. The problem of infection and relapse in malaria is similar to that dealt with in the last example. The figures which follow relate to sizes of

* If the whole distribution is given we can use the expressions

$$\frac{l}{k} = 1 + N\frac{\mu_2 - \mu}{\mu_2 + \mu} \quad (31) \qquad \text{and} \qquad \frac{3\mu}{\mu_2} - \frac{2\mu}{\mu_3} = 1 \quad (32).$$

In the case of the cancer statistics one thus obtains $l/k = 0.966$ and for the left side of the second expression (32) the value 1.23.

spleen of young children, and those who have studied the disease consider that the size of the spleen bears a direct relation to the number of attacks which the patient has experienced. As in this case the disease is endemic (that is to say it is in a more or less steady state) the number of sources may be taken as constant (a).

The equations are consequently

$$\frac{dv_{00}}{dt} = -kar_{00}, \qquad \frac{dv_{10}}{dt} = kav_{00} - (ka + l)v_{10},$$

and for values of y greater than zero,

$$\frac{dv_{x,y}}{dt} = ka(v_{x-1,y} - v_{xy}) + l(v_{x,y-1} - v_{x,y}), \tag{33}$$

where $k\,dt$ is the probability of external infection, and $l\,dt$ is the probability of relapse.

From this we deduce the series of equations.

$$\frac{dS_0}{dt} = -kaS_0,$$

$$\frac{dS_1}{dt} = kaS_0 - (ka + l)S_1, \tag{34}$$

and for values of n greater than unity,

$$\frac{dS_n}{dt} = (ka + l)(S_{n-1} - S_n).$$

Thus we find

$$\frac{l}{ka} = \frac{m - \mu}{1 - m - e^{-m}} \tag{35}$$

and

$$S_n = S_0 \frac{ka}{l}\left(\frac{ka}{l} + 1\right)^{n-1}\left[1 - \sum_0^{n-1} \frac{\left(\frac{lm}{ka}\right)^r}{r!} e^{-lm/ka}\right]. \tag{36}$$

The figures are as follows:

Values of n =	0	1	2	3	4	5	6	7	l/ka
A. Aden Malarious Villages:									
obs.	89	38	36	26	10	9	1	1	
calc.	(89)	39	35	24	13	6	2	1	1.78
B. Aden Sheikh Othman:									
obs.	1006	38	23	13	7	0	4		
calc.	(1006)	34	25	16	7	3	1		25.6

C. Punjab (1909) Spleen rates 30–40%:

obs.	264	67	44	15	3	2		
calc.	(264)	70	39	16	5	1		2.23

D. Punjab (1909) Spleen rates 50–60%:

obs.	81	38	26	16	6	3	2	
calc.	(81)	35	28	17	8	3	1	1.56

EXAMPLE 7. The problem of the behaviour of epidemics may be considered from various points of view, one of which is as follows. Let us trace what is likely to be the distribution of a number of epidemics in similar communities each starting from a single case, and let us adopt as variables the relation

$$\text{case}(n) = \text{infections}(x) - \text{recoveries}(y).$$

The course of affairs is illustrated in Figure 10.

We notice that each epidemic starts initially in the compartment (1, 0), i.e., one case which has not yet recovered, and also that when a compartment $n = x - y = 0$ is arrived at, the epidemic has come to an end for there are no more cases. Thus when the probability of recovery exceeds the probability of infection all epidemics must come to rest in one or other of the compartments $n = 0$. If we assume that $k\,dt$ is the chance that an individual may convey infection, then the probability that an epidemic of n cases will receive an additional case is $kn\,dt$. Similarly the chance that it will lose a case is $ln\,dt$. We assume also, in the first instance, that the population concerned in each epidemic is unlimited. Consequently from the schema

$$\frac{dv_{x,y}}{dt} = k\{(\overline{x-1}-y)v_{x-1,y} - (x-y)v_{x,y}\} + l\{(x-\overline{y-1})v_{x,y-1} - (x-y)v_{x,y}\}. \tag{37}$$

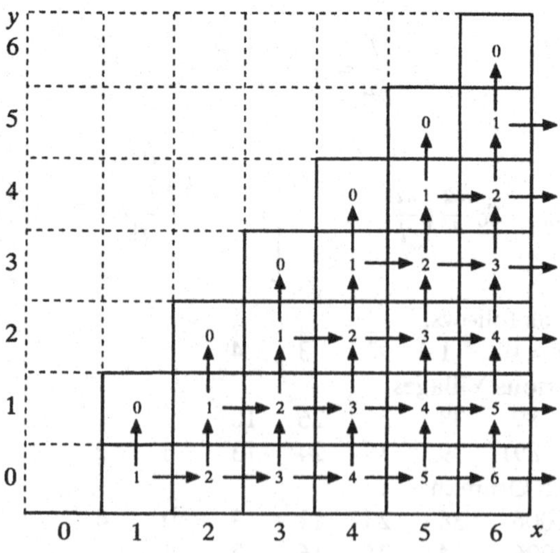

Figure 10

I have obtained solutions of this equation for serial values of x and y up to $v_{4,4}$, but the expressions are not concise, and do not suggest a general form. If, however, one is content with finding the values $v_{1,1}$, $v_{2,2}$, $v_{3,3}$, ... after an infinite time—that is to say if the problem is to find the distribution of a group of epidemics, which have reached finality, classified according to the number of cases which they have experienced, then the problem is less refractory, and the solution is

$$v_{n,n} = a_n\left(\frac{k}{k+l}\right)^{n-1}\left(\frac{l}{k+l}\right)^{n} N, \tag{38}$$

where

$$a_1 = 1,$$
$$a_2 = a_1a_1 = 1,$$
$$a_3 = a_1a_2 + a_2a_1 = 2,$$
$$\vdots$$
$$a_n = a_1a_{n-1} + a_2a_{n-2} + a_3a_{n-3} + \cdots a_{n-1}a_1.$$

These values may be obtained more concisely by means of the formula $a_n = \frac{2n-3}{n}2a_{n-1}$, for which I am indebted to Dr W.O. Kermack. It follows also from equation (37) as well as from equation (38) that $\frac{l}{k} = \frac{\mu}{\mu - 1}$. It will be seen that the coefficients a_n are the number of different paths by which an individual moving either east or north may pass from the compartment (1, 0) to the compartment (n, n). The resulting curves for $v_{n,n}$ are of the J type, that is to say they are like exponential curves with a negative index, in which the tail is unduly prolonged.

During the great epidemic of influenza in 1918, very accurate statistics were collected by the Australian Government,[9] of epidemics which had occurred in incoming ships during their voyage to Australia. The figures are as follows:

Number of epidemics with 1 cases	–	34
Number of epidemics with 2 cases	–	15
Number of epidemics with 3 cases	–	5
Number of epidemics with 4 cases	–	3
Number of epidemics with 5 cases	–	8
Number of epidemics with 6 cases	–	1
Number of epidemics with 7 cases	–	3
Number of epidemics with 8 cases	–	4
Number of epidemics with 9 cases	–	2
more than 9 cases	–	17

I have been unable to fit these figures by means of equation (38); on the one hand the approximations are probably insufficient, and on the other the crews of the ships varied very considerably. But the *J* character of the distribution is evident. It is also interesting to note that in 82 percent of the ship epidemics the number of cases was less than ten. Now influenza is a peculiarly difficult disease to diagnose in the individual case; consequently epidemics of less than ten cases are in ordinary circumstances seldom recognised and reported. The conclusion is suggested that as our limited experience of epidemic influenza is based upon statistics which may relate only to a small selected minority of the total number of epidemics, it may be in no sense representative, and may even be misleading.

If in the second place we wish to investigate the distribution of epidemics according to their duration, we have to find the value of $S_0 = \sum_0^\infty v_{n,n}$ as a function of the time. The problem then reduces to one in a single dimension n, in which there is an instantaneous point source at $n = 1$, and in which (for $l > k$) all epidemics finally enter the compartment S_0.

Let us now turn to the case where the population affected by each epidemic is limited, and of constant value p.

The equation is

$$\frac{dv_{x,y}}{dt} = k\{(p - \overline{x-1})(\overline{x-1} - y)r_{x-1,y} - (p-x)(x-y)v_{x,y}\} \\ + l\{(x - \overline{y-1})v_{x,y-1} - (x-y)v_{x,y}\}. \tag{39}$$

For magnitude of epidemics we have as before to find the value of $v_{n,n}$, when t tends to infinity.

These are found to be

$$\left.\begin{aligned} v_{11} &= \alpha_{p-1}N, \\ v_{22} &= \beta_{p-1}\alpha_{p-2}^2 N, \\ v_{33} &= \beta_{p-1}\beta_{p-2}\alpha_{p-2}^2 NS_3, \\ v_{44} &= \beta_{p-1}\beta_{p-2}\beta_{p-3}\alpha_{p-4}^2 NS_4, \\ &\text{etc.} \end{aligned}\right\} \tag{40}$$

where α_{p-r} is written for $\dfrac{l}{k(p-r)+l}$ and β_{p-r} for $\dfrac{k(p-r)}{k(p-r)+l}$.

The values of $S_3, S_4 \ldots$ are related as follows:

$$S_3 = \alpha_{p-3} + \alpha_{p-2},$$

$$S_4 = \alpha_{p-4}(\alpha_{p-4} + p_{-3} + \alpha_{p-2}) + \alpha_{p-3}(\alpha_{p-3} + \alpha_{p-2}), \text{ etc.}$$

These may be written symbolically in the form

$$S_3 = (3+2),$$

$$S_4 = 4(4+3+2) + 3(3+2), \text{ etc.}$$

Let us further, taking the suffix 5 as an example, use the notation

$$A_5 = 5 + 4 + 3 + 2,$$

$$B_5 = 5(5 + 4 + 3 + 2) + 4(4 + 3 + 2) + 3(3 + 2),$$

$$C_5 = 5\{5(5 + 4 + 3 + 2) + 4(4 + 3 + 2) + 3(3 + 2)\}$$
$$+ 4\{4(4 + 3 + 2) + 3(3 + 2)\},$$

whence

$$B_5 = 5A_5 + 4A_4 + 3A_3,$$

$$C_5 = 5B_5 + 4B_4.$$

Similarly

$$B_6 = 6A_6 + 5A_5 + 4A_4 + 3A_3,$$

$$C_6 = 6B_6 + 5B_5 + 4B_4,$$

$$D_6 = 6C_6 + 5C_5.$$

Then

$$S_3 = A_3,$$

$$S_4 = 4A_4 + 3A_3 = B_4,$$

$$S_5 = 5B_5 + 4B_4 = C_5,$$

$$S_6 = 6C_6 + 5C_5 = D_6,$$

$$S_7 = 7D_7 + 6D_6 = E_7, \text{ etc.}$$

In the first instance the values $A_3, A_4, \ldots$, etc., are calculated, and from these are obtained successively the B's, then the C's and so on. The character of the curves for the values of $v_{n,n}$ may be seen from the following numerical examples in the first of which $N = 1, l/k = 2$, and $p = 5$: and in the second $N = 1, l/k = 2, p = 6$.

$v_{1,1}$	0.33	0.29
$v_{2,2}$	0.11	0.08
$v_{3,3}$	0.09	0.06
$v_{4,4}$	0.13	0.08
$v_{5,5}$	0.34	0.13
$v_{6,6}$		0.37
	1.00	1.01

The distributions are of the U type.

6. Generalisation

The equations of all the examples with which we have been engaged may be written generally in the form

$$\begin{aligned}\frac{dv_{x,y}}{dt} &= f_{x-1,y}v_{x-1,y} - f_{x,y}v_{x,y} + f'_{x+1,y}v_{x+1,y} - f'_{x,y}v_{x,y} \\ &\quad + g_{x,y-1}v_{v,y-1} - g_{x,y}v_{x,y} + g'_{x,y+1}v_{x,y+1} - g'_{x,y}v_{x,y} \\ &\quad + h_{x-1,y-1}v_{x-1,y-1} - h_{x,y}v_{x,y} + h'_{x+1,y+1}v_{x+1,y+1} - h'_{x,y}v_{x,y} \\ &\quad + i_{x-1,y+1}v_{x-1,y+1} - i_{x,y}v_{x,y} + i'_{x+1,y-1}v_{x+1,y-1} - i'_{x,y}v_{x,y} \qquad (41)\end{aligned}$$

in which f, g, h and i refer to forward translations, and f', g', h' and i' refer to backward translations, and this generalisation may obviously be extended to any number of dimensions. If in place of a continuous variation through time we are dealing with a succession of events, then for $-dv/dt$ we write $v_{t-1} - v_t$.

Thus generally for any two variables x and y, a general equation may be built up by equating the right-hand side of the preceding equation to zero. This may be represented schematically in the form

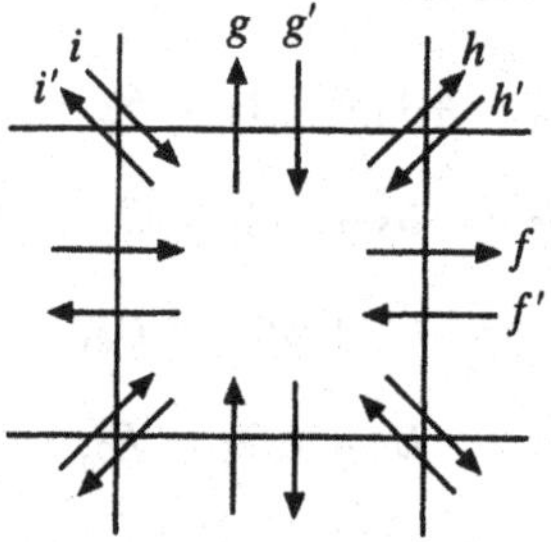

Figure 11

The arrows no longer represent translations, but are now differences between the numbers in the compartments which they connect. The schema, and its corresponding equation, now indicate the relationship which exists between the numbers in a particular compartment and those in the compartments with which it is contiguous. In two dimensions the number of nearest neighbours is 8, in three it is 26, and in n dimensions it is $3^n - 1$. So long as the variables are discontinuous, and proper units are employed, the above schema is generally descriptive, and the problem becomes one of a matrix of pure numbers. In certain cases which I have examined, more distant compartments also have an influence, but these cannot be dealt with here.

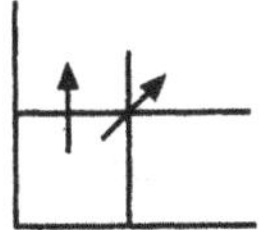

Figure 12

EXAMPLE 8. For a purely discontinuous example we may turn to ordinary probabilities. Let p be the probability of r successes in n events. The variables in this case are $y = n$ and $x = r$. Also $h = p$, and $g = 1 - p$. In an event an individual may either have experienced a failure in which case he moves up one compartment in the n direction, or he may have been successful in which case he moves obliquely in a north-easterly direction (Figure 12). There can obviously be no success without an event, so no horizontal arrows are drawn. This then is a problem of correlation, and its equation is

$$q(v_{n-1,r} - v_{n,r}) + p(v_{n-1,r-1} - v_{n,r}) = 0. \tag{42}$$

The solution of this equation, in the case in which initially no events, and no successes have occurred, is the well-known expression,

$$v_{n,r} = N\frac{n!}{(n-r)!r!}(1-p)^{n-r}p^r. \tag{43}$$

For the coefficient of correlation r we find

$$r^2 = \frac{p}{1-(1-p)\dfrac{n-4}{n+2}}, \tag{44}$$

which shows that r is not in this case a satisfactory coefficient. Again if a success entails a gain, and a failure a loss, then if y denotes the number of events, and x the number of gains, $h = p$ and $i' = i - p$. The schema is as in Figure 13, and the well-known result follows.

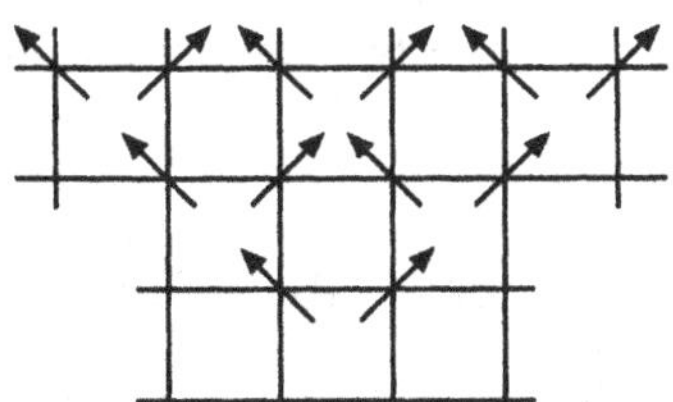

Figure 13

7. Continuous Variables

A consideration of continuous variables leads us from the foregoing general equation, on the one hand into the domain of mathematical physics, and on the other into wider statistical fields. Each uncorrelated term of equation may in this case be written as

$$\frac{f_{x\pm k}v_{x\pm k} - f_{x,y}v_{x,y}}{k}$$

which after expansion becomes

$$\pm\frac{\partial fv}{\partial x} + \frac{k}{2!}\frac{\partial^2 fv}{\partial x^2} \pm \frac{k^2}{3!}\frac{\partial^3 fv}{\partial x^3} + \cdots.$$

Similarly for the correlated terms we have

$$\frac{h_{x\pm k,y\pm k}v_{x\pm k,y\pm k} - h_{x,y}v_{x,y}}{k}$$

$$= \left(\pm\frac{\partial hv}{\partial x} \pm \frac{\partial hv}{\partial y}\right) + \frac{k}{2!}\left(\frac{\partial^2 hv}{\partial x^2} + 2\frac{\partial^2 hv}{\partial x\,\partial y} + \frac{\partial^2 hv}{\partial y^2}\right) \pm \cdots.$$

It is to be noted that in considering correlation a movement in the oblique direction consists of a movement of one unit in each dimension; hence if a unit is denoted by k it is the same for both x and y.

By making use of the above forms of expansion, the general equation may be built up. Thus for example in a reversible case in the three dimensions of space, when k is so small that its square may be neglected, and the functions of f, g etc., are equal constants

$$\frac{\partial v}{\partial t} = f\frac{k}{2!}\left(\frac{\partial^2 v}{\partial x^2} + \frac{\partial^2 v}{\partial y^2} + \frac{\partial^2 v}{\partial z^2}\right), \tag{45}$$

Which is the ordinary equation for diffusion (of matter or of heat). In this case k is the length of the mean freepath of molecule or electron as the case may be. If k tended to zero there would be no diffusion, and if k were large it would be necessary to include higher differentials. Thus Fourier's equation contains in its essence the idea of heterogeneity or discreteness.

Similarly in a reversible case in two dimensions in which the variables x and y are correlated we have

$$\frac{\partial v}{\partial t} = \frac{k}{2!}\left(\frac{\partial^2 fv}{\partial x^2} + \frac{\partial^2 gv}{\partial y^2} + \frac{\partial^2 hv}{\partial x^2} + 2\frac{\partial^2 hv}{\partial x\,\partial y} + \frac{\partial^2 hv}{\partial y}\right) \tag{46}$$

which when f, g and h are constant is the equation of the correlated frequency surface.

Again in the general case in three dimensions without oblique correlation, if h tends to zero we have

$$\frac{\partial v}{\partial t}+\frac{\partial(f-f')v}{\partial x}+\frac{\partial(g-g')v}{\partial y}+\frac{\partial(e-e')v}{\partial z}=0, \quad \text{or} \quad =\frac{dv}{dt}. \tag{47}$$

The left-hand side of the equation is equal to zero if there is no variation of material (i.e. of v), and if there be a variation it is equal to dv/dt, where this differential expresses the rate of variation of substance in following the element in its movement. If the functions f, g, e, f', g', e', are constants this is the ordinary equation of hydrodynamics. In actual fact k does not tend to zero. Diffusion must always take place as an element of the fluid flows onwards, but the place left empty is filled up from neighbouring elements, and the diffusion is so to speak neutralised.

Example 9. Let us now turn to a problem in vital statistics. Let the variables be the time (t) and the age (θ)—thus $v_{t,\theta}$ denotes the number of individuals aged θ at the time t. Time and age are continuous ($k \to 0$), are absolutely correlated ($r = 1$) and the rate of movement through both time and age are the same and are constant. The translation is irreversible (Figure 14). Thus if there is no death rate

$$\frac{v_{t-k,\theta-k}-v_{t,\theta}}{k}=0 \qquad \text{or} \qquad \frac{\partial v}{\partial t}+\frac{\partial v}{\partial \theta}=0.$$

If there be a death rate then

$$\frac{\partial v}{\partial t}+\frac{\partial v}{\partial \theta}=\frac{dv}{dt}, \tag{48}$$

where dv/dt has the same significance as in the hydrodynamical equation, viz., it is the variation in the element as one follows it in its movement.

Writing $dv/dt = -f_{t,\theta}v_{t,\theta}$, where $f_{t,\theta}\,dt$ is the probability of dying at age θ and at time t, we have

$$\frac{\partial v}{\partial t}+\frac{\partial v}{\partial \theta}=-f_{t,\theta}v_{t,\theta}.$$

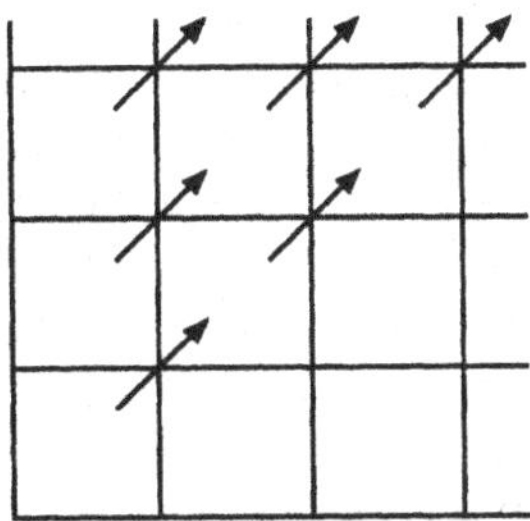

Figure 14

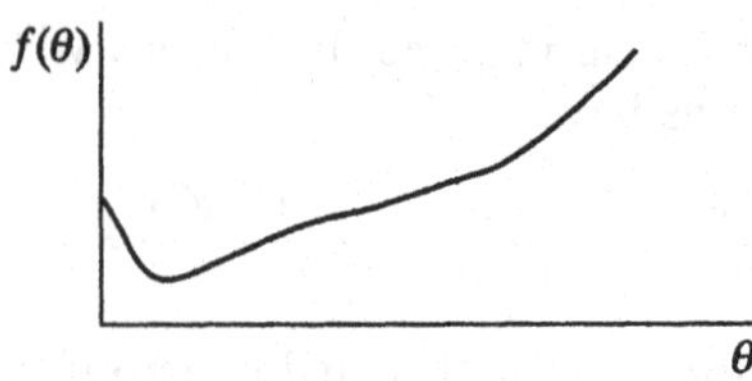

Figure 15

Whence

$$v_{t,\theta} = v_{t-\theta,0}e^{-\int_0^\theta f(t-\theta+\xi,\xi)\,d\xi} \tag{49}$$

which may be translated as follows: the number of persons aged θ at the time t, is equal to the number born at the time $t-\theta$ reduced exponentially by the successive death rates (taken for convenience discontinuously)

for age 0 at the time $t-\theta$,
for age 1 at the time $t-\theta+1$,
for age 2 at the time $t-\theta+2$ and so on up to $\xi=\theta$.

If we neglect variations in the death rate with the time, we know that $f(\theta)$ has roughly the form drawn in Figure 15.

Thus for a number of pairs of values θ_1 and θ_2, $f(\theta_1) = f(\theta_2)$. Now if a population is decreasing according to the relation $\partial v/\partial t = -cv$, where c is constant, the equation takes the form

$$\frac{\partial v_\theta}{\partial \theta} = -(f_\theta - c)v_\theta.$$

If c is less than the minimal value of $f(\theta)$, then v_θ will always decrease with increasing θ.

If c is equal to the minimal value of $f(\theta)$ there will be one level point, and if c is greater than the minimal value but less than $f(0)$ there will be two level points.

Thus age distribution curves will have forms roughly as in Figure 16.

In this diagram the dotted curve a is intended to represent the form when the population is steady, and the dotted line b is the locus of the level points. The curves above a are those of increasing populations, and those below it represent the age distributions of populations which are on the decrease. The hump on the curve occurs between the ages of 20 and 40, the child-bearing age, consequently the more a population is on the decrease, the greater is the tendency for the earlier part to be pushed up by the birth of children. This is an interesting example of a natural "governor" mechanism.

Example 10 (with Dr. W.O. Kermack). The problem of the course of an epidemic of a contagious disease is similar to the last. The variables are as

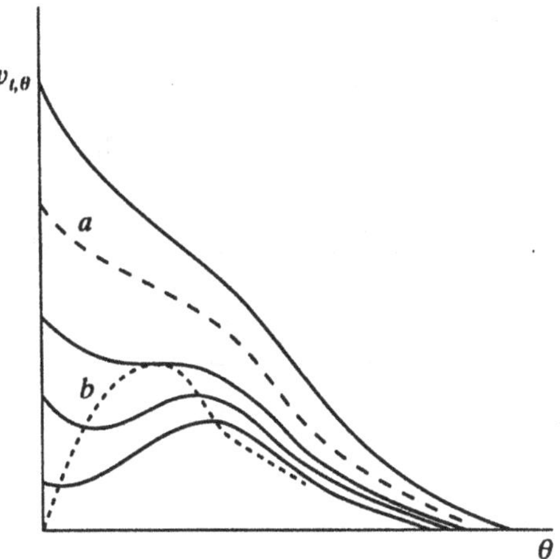

Figure 16

before, the time (t) and the age (θ), and these are absolutely correlated (Figure 14). We have to deal with an infective rate $\phi(t, \theta)$, and a rate of removal (including death and recovery) $\psi(t, \theta)$. We shall suppose that these are independent of the time. Let us treat the problem in the first instance as if t and θ were discontinuous variables.

The equation is then

$$v_{t-1,\theta-1} - v_{t,\theta} = \psi(\phi)v_{t,\theta}, \tag{50}$$

whence

$$r_{t,\theta} = \frac{v_{t-1,\theta-1}}{1+\psi(\theta)} = \frac{v_{t-2,\theta-2}}{(1+\psi(\theta))(1+\psi(\theta-1))} = v_{t-\theta,0}B_\theta, \tag{51}$$

where B_θ is written for the reciprocal of

$$(1+\psi(\theta))(1+\psi(\theta-1))\cdots(1+\psi(1)).$$

Now the number $v_{t,0}$ denotes the number of persons who became infected at the time t, and this by hypothesis is equal to $\sum_1^t \phi_\theta v_{t,\theta}$, whence

$$v_{t,0} = \sum_1^t \phi_\theta \beta_\theta v_{t-\theta,0} = \sum_1^t A_\theta v_{t-\theta,0}, \tag{52}$$

where for conciseness A_θ is written for $\phi_\theta B_\theta$.

Hence

$$\sum_0^\infty x^t v_{t,0} = \sum_0^\infty x^t \sum_1^t A_\theta v_{t-\theta,0} + N_0 \qquad \text{where} \qquad N_0 \equiv v_{00}$$

$$= \sum_0^\infty x^t v_{t,0} \sum_1^\infty x^\theta A_\theta + N_0,$$

x being chosen so that the series are convergent,

$$\therefore \quad \sum_0^\infty x^t v_{t,0} = \frac{N_0}{1 - \sum_1^\infty x^\theta A_\theta}. \tag{53}$$

Let N_t denote the number of persons infected at the time t, then

$$N_t = \sum_0^t v_{t,\theta} = \sum_0^t B_\theta v_{t-\theta,0}$$

and

$$\begin{aligned} \sum_0^\infty x^t N_t &= \sum_0^\infty x^t \sum_0^t B_\theta v_{t-\theta,0} \\ &= \sum_0^\infty x^t v_{t,0} \sum_0^\infty x^\theta B_\theta, \\ &= \frac{N_0 \sum_0^\infty x^\theta B_\theta}{1 - \sum_1^\infty x^\theta A_\theta}, \end{aligned} \tag{54}$$

$v_{t,0}$ and N_t are the coefficients of x^t in the expansions of the above expressions (53) and (54).

By cross multiplication, expansion and equating like powers of x it is easy to show also that

$$N_t = B_t N_0 + \sum_1^t A_\theta N_{t-\theta} \tag{55}$$

Passing to the continuous form; from equation (53) we have

$$\int_0^\infty x^t v_{t,0}\, dt = \frac{N_0}{1 - \int_0^\infty x^\theta A_\theta\, d\theta}$$

(where $A_\theta = \phi_\theta B_\theta = \phi_\theta e^{-\int_0^\theta \psi(\xi)\, d\xi}$)
or

$$\int_0^\infty e^{-zt} v_{t,0}\, dt = \frac{N_0}{1 - \int_0^\infty e^{-z\theta} A_\theta\, d\theta}, \tag{56}$$

whence using Laplace's transformation

$$v_{t,0} = \frac{1}{2\pi i} \int_{a-i\infty}^{a+i\infty} e^{zt} \int_0^\infty e^{-zt} v_{t,0}\, dt\, dz = \frac{1}{2\pi i} \int_{a-i\infty}^{a+i\infty} \frac{N_0 e^{tz}}{1 - \int_0^\infty e^{-\theta z} A_\theta\, d\theta}\, dz,$$

or shortly

$$v_{t,0} = L\left(\frac{N_0}{1 - \int_0^\infty e^{-\theta z} A_\theta \, d\theta}\right). \tag{57}$$

This then must under the conditions of the case be also a solution by equation (52) in its integral form,

$$v_{t,0} = \int_0^t A_\theta \, v_{t-\theta,0} \, d\theta. \tag{58}$$

This solution appears to hold only when singularities exist in $v_{t,0}$. In the case which we are discussing there is a singularity at $t = 0$, since one assumes that there is an instantaneous infinite infection rate at the origin. If a singularity of similar type exists at a point $t = t_0$ then it can be shown that the solution is

$$v_{t,0} = L\left(\frac{N_0 e^{-zt_0}}{1 - \int_0^\infty e^{-\theta z} A_\theta \, d\theta}\right). \tag{59}$$

Similarly

$$N_t = L\left(\frac{N_0 \int_0^\infty e^{-\theta z} B_\theta \, d\theta}{1 - \int_0^\infty e^{-\theta z} A_\theta \, d\theta}\right) \tag{60}$$

and this is by equation (55) a solution of the integral equation

$$N_t = B_t N_0 + \int_0^t A_\theta N_{t-\theta} \, d\theta, \tag{61}$$

a result which has been obtained by Fock.[10]

Thus if the infection and removal rates are known and can be expressed as continuous functions, both $v_{t,0}$ and N_t can be obtained. In actual experience $\phi(\theta)$ and $\psi(\theta)$ are not available, but statistics regarding $v_{t,0}$ are readily accessible, and values of N_t could also be obtained. The above operations must therefore be reversed.

Equation (56) may be written in the form

$$\int_0^\infty e^{-z\theta} A_\theta \, d\theta = 1 - \frac{N_0}{\int_0^\infty e^{-zt} v_{t,0} \, dt},$$

whence

$$\phi_\theta B_\theta = A_\theta = L\left(1 - \frac{N_0}{\int_0^\infty e^{-zt} v_{t,0} \, dt}\right), \tag{62}$$

where the exponential within the operator L is in this case $e^{z\theta}$.

Also from equation (54) in its continuous form, and from (56)

$$\int_0^\infty e^{-z\theta} B_\theta \, d\theta = \frac{\int_0^\infty e^{-zt} N_t \, dt}{\int_0^\infty e^{-zt} v_{t,0} \, dt}, \tag{63}$$

whence

$$B_\theta = L\left(\frac{\int_0^\infty e^{-zt} N_t \, dt}{\int_0^\infty e^{-zt} v_{t,0} \, dt}\right). \tag{64}$$

Thus if N_t and $v_{t,0}$ can be expressed as known functions, the values of $\phi(\theta)$ and $\psi(\theta)$ may be deduced. For the numerical solution of the problem the reader is referred to Whittaker (*Proc. Roy. Soc.*, 94, p. 367, 1918). From which it will be seen that if a series of numerical values of $v_{t,0}$ are known the corresponding values of A_θ may be determined from equation (58) in the form

$$v_{t,0} = \int_0^t A_{t-\theta} v_{\theta,0} \, d\theta.$$

The values of B_θ may then be determined directly by means of equation (61).

Thus the values of the functions $\phi(\theta)$ and $\psi(\phi)$ may be calculated either formally or numerically from statistics giving the values of $v_{t,0}$ and N_t.

In the above we have considered the case in which there is no seasonal effect on the ϕs and the ψs. If that be not the case then ϕ and ψ are both functions of t and θ. If $\phi(t,\theta) = \alpha_t\beta_\theta$, and $\psi(t,\theta) = \alpha'_t\beta'_\theta$, then the above method of treatment can be applied, but not otherwise. In many instances the product form may reasonably be used. For example, if α_t depends upon the number of mosquitos, and the disease is malaria, then it would be reasonable to suppose that the effect of a multiplication of the number of mosquitos would be to multiply the chances of infection at the various case ages θ.

These results may obviously be applied to the previous example if the birth rate can be expressed in the form $\phi(\theta)$.

8. Vorticity in Statistical Problems

Returning to equations (16) which also hold when x and y are continuous, and also if the constants be suitably modified when the progression is reversible, and omitting the R coefficients we have for the equations of the coordinates of the centre of gravity of the moving system

$$\frac{d\mu_x}{dm} = P_1 + P_2\mu_x + P_3\mu_y,$$

$$\frac{d\mu_y}{dm} = Q_1 + Q_2\mu_x + Q_3\mu_y,$$

whence

$$\frac{d^2\mu_x}{dm^2} - (P_2 + Q_2)\frac{d\mu_x}{dm} + (P_2Q_3 - P_3Q_2)\mu_x = P_3Q_1 - P_1Q_3,$$

and

$$\frac{d^2\mu_y}{dm^2} - (P_2 + Q_2)\frac{d\mu_y}{dm} + (P_2Q_3 - P_3Q_2)\mu_y = Q_2P_1 - Q_1P_2.$$

The motion is periodic if

$$-4P_3Q_2 > (P_2 - Q_3)^2,$$

therefore periodicity depends primarily upon the condition that one, but not both, of P_3 and Q_2 must be negative.

Again if $(P_2 + Q_3)$ is negative the motion will be damped, whilst if $(P_2 + Q_3)$ is positive there will be augmentation of amplitude.

Example 11. During the course of a microbic disease a battle is raging between microbes on the one hand and protective "antibodies" on the other. These antibodies are themselves called into being by the action of the microbes. The process is reversible for both variables (x = microbes and y = antibodies).

Now microbes multiply by compound interest, hence P_2 is positive ($= a_2$), and are destroyed at a rate proportional to the number of antibodies present, hence P_3 is negative ($= -a_3$). The state of affairs is as shown in Figure 17.

Antibodies are produced at a rate proportional to the number of microbes hence Q_2 is positive ($= b_2$), and are destroyed in the general

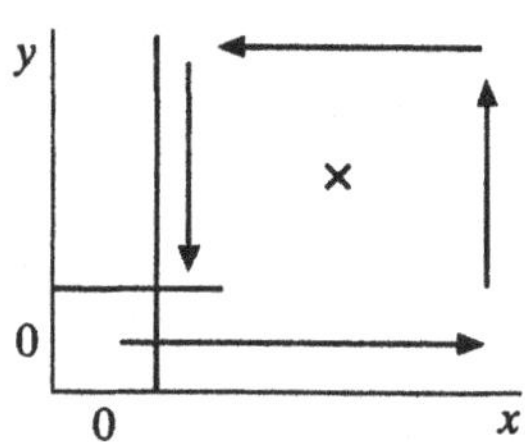

Figure 17

metabolism, in such a manner that the more the antibody the more the destruction, hence Q_3 is negative ($= -b_3$).

Hence if $4a_3b_2 > (a_2 + b_3)^2$ a vortex is formed, and if $a_2 - b_3 > 0$ there will be augmentation of amplitudes, whilst if $b_3 - a_2 > 0$ there will be damping.

Also when $t = \infty$

$$\mu_x = \frac{P_3Q_1 - P_1Q_3}{P_2Q_3 - P_3Q_2} = \frac{P_1b_3 - Q_1a_3}{a_3b_2 - a_2b_3}$$

and

$$\mu_y = \frac{P_1Q_2 - P_2Q_1}{P_2Q_3 - P_3Q_2} = \frac{P_1b_2 - Q_1a_2}{a_3b_2 - a_2b_3}.$$

From these considerations arrived at from the approximate equations, we can deduce at once:

(1) Augmentation of amplitude ($a_2 > b_3$) will lead to total and relatively sudden extinction of organisms at the end of an attack (or series of attacks) for the motion will enter the column $x = 0$.
(2) Damping ($b_3 > a_2$) will lead to a final state at which both organism and antibodies will continue in a steady state, their relative proportions depending upon the above relations.
(3) Relapses will occur when $4a_3b_2 > (a_2 + b_3)^2$. They will seldom, if ever, occur where amplitudes augment ($a_2 > b_3$) they will be a prominent feature as a_2 and b_3 tend to equality, and they will be absent when b_3 is very much greater than a_2, as in this case the motion becomes a-periodic.

This is only a rough approximation but it places in evidence the main features of infections diseases, viz., (a) termination by "crisis" and complete extinction of the disease; or (b) gradual decline by "lysis," with continued "carriage" of the disease. It draws attention also to (c) the occurrence of relapses in the intermediate types. This latter consideration is exemplified in Malta Fever, in which there may occur a series of relapses, and in which both types ($a_2 > b_3$) and ($b_3 > a_2$) are found to occur.

A better approximation is obtained by introducing a third variable, the temperature. The differential equations for the coordinates of the centre of gravity of the system are then of the third order, and more in accordance with experience (see also reference [11]).

In conclusion the author desires to thank Dr. W.O. Kermack for continued help and criticism, and Mr. E.T. Copson for encouragement and advice in many directions.

References

[1] Behla. Cent. f. Bact. (orig.) 24, Abt. 1.1898.
[2] McKendrick, *Science Progress.* Jan. 1914.

[3] McKendrick, *Proc. Lond. Math. Soc.*, Ser. 2, Vol. 13, p. 401. 1914.
[4] McKendrick, *Indian Journal of Medical Research*, Vol. 2, p. 882. 1915.
[5] McKendrick, *Indian Journal of Medical Research*, Vol. 3, p. 271. 1915.
[6] McKendrick, *Indian Journal of Medical Research*, Vol. 3, p. 266. 1915.
[7] McKendrick, *Indian Journal of Medical Research*, Vol. 3, p. 667. 1916.
[8] McKendrick, *Nature*, Vol. 104, p. 660. 1920.
[9] *Commonwealth of Australia Quarantine Service, Publication* 18. 1919.
[10] Fock, *Math. Zeit*, 21, 161. 1924.
[11] McKendrick, *Indian Journal of Medical Research*, Vol. 6, p. 614. 1919.

Introduction to Yates and Cochran (1938) The Analysis of Groups of Experiments

D.R. Cox

F. Yates (1901–1994) and W.G. Cochran (1910–1980) were two of the most influential and conceptually original statisticians of the middle portion of the twentieth century. Although Cochran worked for some years in Yates's department at Rothamsted Experimental Station this is their only joint publication.

The paper is a remarkable one for a number of reasons. It has at least four major ideas expressed lucidly in nonmathematical language. It is illustrated by cogent examples. It reveals strikingly the subtlety of understanding of analysis of variance that had been reached toward the end of one phase of development of statistical design of experiments.

Tracing the direct influence of the paper is not easy mainly because, although the *Journal of Agricultural Science* was and is a major journal, its impact on the general statistical world is indirect. The influence of the paper is, I think, primarily via the subsequent work of the authors and their associates, and in particular the much later book by Cochran and G.M. Cox (1950, 1957), and indeed by the impact of the whole Rothamsted tradition on statistics, many aspects of which are encapsulated in the paper.

The paper begins by distinguishing two aims of agronomic experiments, technical and scientific, although stressing that these aims are not exclusive. It is then argued that both, but especially the former, demand small trials replicated in time and space. The possibility that different treatment recommendations may be required under different circumstances and the difficulties of treating, for example, places as a random sample of the relevant population of places are discussed. There are clear parallels with similar problems in other fields, for example, with the distinction between pragmatic and explanatory clinical trials, with the contrast between single center and multicenter clinical trials, and with the issues of analysis involved in overviews (meta-analysis). The discussion of error estimation in the paper is, however, more careful than that in much of the current literature.

Analysis is then discussed in detail, covering the possibility that the error variance is not the same in all places. When there is an interaction, varieties (or treaments) × places, it is recognized that there are "endless complications" in drawing a conclusion about variety differences across a population of places.

The possibility that the error variance varies substantially between places is covered and, very interestingly, the possibility of heterogeneity in the interaction varieties × places is stressed. This is illustrated via an example in which a substantial interaction is accounted for by the anomalous behavior of one variety. The anomaly is then elucidated via a regression of each variety on the mean yield of all varieties at that center and clarified in a striking graph. The argument is rather parallel to that used by Tukey (1949) in developing his one degree of freedom for nonadditivity. The discussion of this example ends with a careful discussion of the assignment of error to varietal comparisons. Yates's missing value technique (Yates, 1933), a precursor of the EM algorithm, is used to check the effect of removing a potential outlier.

The paper is a model of concise clarity with a slight lapse in Section 5. This analyzes a set of thirteen 3 × 3 Latin squares when it is suspected that the error variances at each place, estimated with two degrees of freedom, are not constant and that this has biased the examination of treatment × place interaction. This is technically the least satisfactory part of the paper in that the variance of the observed Student t statistics is likely to be an insensitive statistic unless the degrees of freedom are large. The section ends with a rather mysterious formula for a chi-squared statistic containing two undefined symbols.

Finally, there is a discussion of the estimation of overall means, raising issues that are of much current interest in connection with overviews. Of necessity, the account mostly concentrates on normal theory problems with the same design at each place, but the key issues are addressed. The advantages and limitations of unweighted means, of means weighted by empirical estimates of local variance, of semiweighted means, in which the weights are determined by the sum of a local variance and a constant interaction variance, of maximum likelihood, and of empirically weighted means with a bound on the maximum weight are carefully if qualitatively discussed and then illustrated by a cogent example. Cochran returned to these issues nearly 15 years later in two major papers (Cochran and Carroll, 1953; Cochran, 1954). It is noted that if the error variance is associated with the effect under estimation, then crude weighting will produce a biased estimate.

The paper is also available in *Experimental Design, Selected Papers of Frank Yates* (Griffin, 1970); and *Contributions to Statistics*, W.G. Cochran (Wiley, 1982).

References

Cochran, W.G. (1954). The combination of estimates from different experiments. *Biometrics*, **10**, 101–129.

Cochran, W.G. and Carroll, S.P. (1953). A sampling investigation of the efficiency of weighting inversely as the estimated variance. *Biometrics*, **9**, 447–459.

Cochran, W.G. and Cox, G.M. (1950, 2nd ed., 1957). *Experimental Designs*. Wiley, New York.

Tukey, J.W. (1949). One degree of freedom for nonadditivity. *Biometrics*, **5**, 232–242.

Yates, F. (1933). The analysis of replicated experiments when the field results are incomplete. *Empire J. Exper. Agric.* **1**, 129–142.

The Analysis of Groups of Experiments

F. Yates and W.G. Cochran
Statistical Department, Rothamsted Experimental Station, Harpenden

1. Introduction

Agricultural experiments on the same factor or group of factors are usually carried out at a number of places and repeated over a number of years. There are two reasons for this. First, the effect of most factors (fertilizers, varieties, etc.) varies considerably from place to place and from year to year, owing to differences of soil, agronomic practices, climatic conditions and other variations in environment. Consequently the results obtained at a single place and in a single year, however accurate in themselves, are of limited utility either for the immediate practical end of determining the most profitable variety, level of manuring, etc., or for the more fundamental task of elucidating the underlying scientific laws. Secondly, the execution of any large-scale agricultural research demands an area of land for experiment which is not usually available at a single experimental station, and consequently much experimental work is conducted co-operatively by farmers and agricultural institutions which are not themselves primarily experimental.

The agricultural experimenter is thus frequently confronted with the results of a set of experiments on the same problem, and has the task of analysing and summarizing these, and assigning standard errors to any estimates he may derive. Though at first sight the statistical problem (at least in the simpler cases) appears to be very similar to that of the analysis of a single replicated trial, the situation will usually on investigation be found to be more complex, and the uncritical application of methods appropriate to single experiments may lead to erroneous conclusions. The object of this paper is to give illustrations of the statistical procedure suitable for dealing with material of this type.

2. General Considerations

Agronomic experiments are undertaken with two different aims in view, which may roughly be termed the technical and the scientific. Their aim may be regarded as scientific insofar as the elucidation of the underlying laws is attempted, and as technical insofar as empirical rules for the conduct of practical agriculture are sought. The two aims are, of course, not in any sense mutually exclusive, and the results of most well-conducted experiments on technique serve to add to the structure of general scientific law, or at least to indicate places where the existing structure is inadequate, while experiments on questions of a more fundamental type will themselves provide the foundation of further technical advances.

Insofar as the object of a set of experiments is technical, the estimation of the average response to a treatment, or the average difference between varieties, is of considerable importance even when this response varies from place to place or from year to year. For unless we both know the causes of this variation and can predict the future incidence of these causes we shall be unable to make allowance for it, and can only base future practice on the average effects. Thus, for example, if the response to a fertilizer on a certain soil type and within a certain district is governed by meteorological events subsequent to its application the question of whether or not it is profitable to apply this fertilizer, and in what amount, must (in the absence of any prediction of future meteorological events) be governed by the average response curve over a sufficiently representative sample of years. In years in which the weather turns out to be unfavourable to the fertilizer a loss will be incurred, but this will be compensated for by years which are especially favourable to the fertilizer.

Any experimental programme which is instituted to assess the value of any particular treatment or practice or to determine the optimal amount of such treatment should therefore be so designed that it is capable of furnishing an accurate and unbiased estimate of the average response to this treatment in the various combinations of circumstances in which the treatment will subsequently be applied. The simplest and indeed the only certain way of ensuring that this condition shall be fulfilled is to choose fields on which the experiments are to be conducted by random selection from all fields which are to be covered by the subsequent recommendations.

The fact that the experimental sites are a random sample of this nature does not preclude different recommendations being made for different categories included in this random sample. We may, for instance, find that the response varies according to the nature of the previous crop, in which case the recommendations may be correspondingly varied. Moreover, in a programme extending over several years, the recommendations may become more specific as more information is accumulated, and the experiments themselves may be used to determine rules for the more effective application of the treatments tested, as in fertilizer trials in which the chemical examination of soil samples may lead to the evolution of practical chemical tests for fertilizer requirements.

At present it is usually impossible to secure a set of sites selected entirely at random. An attempt can be made to see that the sites actually used are a "representative" selection, but averages of the responses from such a collection of sites cannot be accepted with the same certainty as would the averages from a random collection.

On the other hand, comparisons between the responses on different sites are not influenced by lack of randomness in the selection of sites (except insofar as an estimate of the variance of the response is required) and indeed for the purpose of determining the exact or empirical natural laws governing the responses, the deliberate inclusion of sites representing extreme conditions may be of value. Lack of randomness is then only harmful insofar as it results in the omission of sites of certain types and in the consequent arbitrary restriction of the range of conditions. In this respect scientific research is easier than technical research.

3. The Analogy Between a Set of Experiments and a Single Replicated Trial

If a number of experiments containing the same varieties (or other treatments) are carried out at different places, we may set out the mean yields of each variety at each place in the form of a two-way table. The marginal means of this table will give the average differences between varieties and between places. The table bears a formal analogy to the two-way table of individual plot yields, arranged by blocks and varieties, of a randomized block experiment, and we can therefore perform an analysis of variance in the ordinary manner, obtaining a partition of the degrees of freedom in the case of six places and eight varieties, for example) as follows:

	Degrees of freedom
Places	5
Varieties	7
Remainder	35
Total	47

The remainder sum of squares represents that part of the sum of squares which is due to variation (real or apparent) of the varietal differences at the different places. This variation may reasonably be called the *interaction* between varieties and places. It will include a component of variation arising from the experimental errors at the different places.

If the experiments are carried out in randomized blocks (or in any other type of experiment allowing a valid estimate of error) the above analysis may be extended to include a comprehensive analysis of the yields of the individual plots. If there are five replicates at each place, for example, there will be 240 plot yields, and the partition of the degrees of freedom will then be as follows:

	Degrees of freedom
Places	5
Varieties	7
Varieties × places	35
Blocks	24
Experimental error	168
Total	239

It should be noted that in this analysis the sums of squares for varieties and for varieties × places are together equal to the total of the sums of squares for varieties in the analyses of the separate experiments. Similarly the sums of squares for blocks and for experimental error are equal to the totals of these times in the separate analyses. If, as is usual, the comprehensive analysis is given in units of a single plot yield, the sums of squares derived from the two-way table of places and varieties must be multiplied or divided by 5 according as means or totals are there tabulated.

The first point to notice about this comprehensive analysis of variance is that the estimates of error from all six places are pooled. If the errors of all experiments are substantially the same, such pooling gives a more accurate estimate than the estimates derived from each separate experiment, since a larger number of degrees of freedom is available. If the errors are different, the pooled estimate of the error variance is in fact the estimate of the mean of the error variances of the separate experiments. It will therefore still be the correct estimate of the error affecting the mean difference (over all places) of two varieties, but it will no longer be applicable to comparisons involving some of the places only. Moreover, as will be explained in more detail below, the ordinary tests of significance, even of means over all places, will be incorrect.

If the errors of all the experiments are the same, the other mean squares in the analysis of variance table may be compared with the mean square for experimental error by means of the z test. The two comparisons of chief interest are those for varieties and for varieties × places. The meaning of these will be clear if we remember that there is a separate set of varietal means at each place, and that the differences between these means are not necessarily the same at all places. If the mean square for varieties is significant, this indicates the significance of the average differences of the varieties *over the particular set of places chosen.* If varieties × places is also significant, a significant variation from place to place in the varietal differences is indicated. In this latter case it is clear that the choice of places must affect the magnitude of the average difference between varieties: with a different set of places we might obtain a substantially different set of average differences. Even if varieties × places is not significant, this fact cannot be taken as indicating *no* variation in the varietal differences, but only that such variation is likely to be smaller than an amount which can be determined by the arguments of fiducial probability.

We may, therefore, desire to determine the variation that is likely to occur in the average differences between the varieties when different sets of places are chosen, and in particular whether the average differences actually obtained differ significantly from zero when variation from place to place is allowed for. Endless complications affect this question, and with the material ordinarily available a definite answer is usually impossible. The various points that arise will be made clear by an actual example, but first we may consider the situation in the ideal case where the chosen places are a strictly random selection from all possible places.

At first sight it would appear to be legitimate in this case to compare the mean square for varieties with that for varieties × places by means of the z test. There is, however, no reason to suppose that the variation of the varietal differences from place to place is the same for each pair of varieties. Thus the eight varieties of our example might consist of two sets of four, the varieties of each set being closely similar among themselves but differing widely from those of the other set, not only in their average yield, but also in their variations in yield from place to place.

The sums of squares for varieties and for varieties × places would then have large components derived from one degree and five degrees of freedom respectively, while the remaining components might be of the order of experimental error. In the limiting case, therefore, when the experimental error is negligible in comparison with the differences of the two sets and the average difference over all possible places is zero, the z derived from the two mean squares will be distributed as z for 1 and 5 degrees of freedom instead of as z for 7 and 35 degrees of freedom. Verdicts of significance and of subnormal variation will therefore be reached far more often than they should be.

The correct procedure in this case is to divide the sums of squares for varieties and for varieties × places into separate components, and compare each component separately. Thus we shall have:

	Degrees of freedom
Varieties: Sets	1
Within sets	6
Varieties × places: Sets	5
Within sets	30

The one degree of freedom between sets can now legitimately be compared with the five degrees of freedom for sets × places, but the degrees of freedom within sets may require further subdivision before comparison.

It is worth noting that the test of a single degree of freedom can be made by the t test, by tabulating the differences between the means of the two sets for each place separately. This test is in practice often more convenient than the z test, of which it is the equivalent.

4. An Actual Example of the Analysis of a Set of Variety Trials

Table I, which has been reproduced by Fisher (1935) as an example for analysis by the reader, gives the results of twelve variety trials on barley conducted in the State of Minnesota and discussed by Immer et al. (1934). The trials were carried out at six experiment stations in each of two years, and actually included ten varieties of which only five (those selected by Fisher) are considered here.

The experiments were all arranged in randomized blocks with three replicates of each variety. When all ten varieties are included there are therefore 18 degrees of freedom for experimental error at each station.

The error mean squares for the twelve experiments were computed from the yields of the separate plots, which have been given in full by Immer. They are shown inn Table II.

If the errors at all stations are in fact the same, these error mean squares, when divided by the true error variance and multiplied by the number of

Table I. Yields of Barley Varieties in Twelve independent Trials. Totals of Three Plots, in Bushels per Acre.

Place and year	Manchuria	Svansota	Velvet	Trebi	Peatland	Total
University Farm 1931	81.0	105.4	119.7	109.7	98.3	514.1
1932	80.7	82.3	80.4	87.2	84.2	414.8
Waseca 1931	146.6	142.0	150.7	191.5	145.7	776.5
1932	100.4	115.5	112.2	147.7	108.1	583.9
Morris 1931	82.3	77.3	78.4	131.3	89.6	458.9
1932	103.1	105.1	116.5	139.9	129.6	594.2
Crookston 1931	119.8	121.4	124.0	140.8	124.8	630.8
1932	98.9	61.9	96.2	125.5	75.7	458.2
Grand Rapids 1931	98.9	89.0	69.1	89.3	104.1	450.4
1932	66.4	49.9	96.7	61.9	80.3	355.2
Duluth 1931	86.9	77.1	78.9	101.8	96.0	440.7
1932	67.7	66.7	67.4	91.8	94.1	387.7
Total	1132.7	1093.6	1190.2	1418.4	1230.5	6065.4

Table II. Error Mean Squares of Barley Experiments.

	Mean square		Approximate χ^2	
	1931	1932	1931	1932
University Farm	21.25	15.98	16.43	12.36
Waseca	26.11	25.21	20.19	19.49
Morris	18.62	20.03	14.40	15.49
Crookston	30.27	21.95	23.40	16.97
Grand Rapids	26.28	26.40	20.32	20.41
Duluth	27.00	20.28	20.88	15.68

degrees of freedom, 18, will be distributed as χ^2 with 18 degrees of freedom. If we take the mean of the error mean squares, 23.28, as an estimate of the true error variance, the distribution obtained will not be far removed from the χ^2 distribution. The actual values so obtained are shown in Table II, and their distribution is compared with the χ^2 distribution in Table III. Variation in the experimental error from station to station would be indicated by this distribution having a wider dispersion than the χ^2 distribution.

In this case it is clear that the agreement is good, and consequently we shall be doing no violence to the data if we assume that the experimental errors are the same for all the experiments. This gives 23.28 as the general estimate (216 degrees of freedom) for the error variance of a single plot, and the standard error of the values of Table I is therefore $\sqrt{(3 \times 23.28)}$ or ± 8.36.

The analysis of variance of the values of Table I is given in Table IV in units of a single plot. The components due to places and years have been separated in the analysis in the ordinary manner.

Every mean square, except that for varieties $\times$ years, will be found, on testing by the z test, to be significantly above the error mean square. Examination of Table I indicates that variety Trebi is accounting for a good deal of the additional variation due to varieties and varieties $\times$ places, for the mean yield of this variety over all the experiments is much above that of the other four varieties, but at University Farm and Grand Rapids, two of the lowest yielding stations, it has done no better than the other varieties.

In order to separate the effect of Trebi it is necessary to calculate the difference between the yield of Trebi and the mean of the yields of the other four varieties for each of the twelve experiments, and to analyse the variance of these quantities. For purposes of computation the quantities:

$$5 \times \text{yield of Trebi} - \text{total yield of station}$$
$$= 4(\text{yield of Trebi} - \text{mean of other varieties})$$

are more convenient.

Table III. Comparison with the Theoretical χ^2 Distribution.

P	1.0	0.99	0.98	0.95	0.90	0.80	0.70	0.50	0.30	0.20	0.10	0.05	0.02	0.01	0
χ^2	0	7.02	7.91	9.39	10.86	12.86	14.44	17.34	20.60	22.76	25.99	28.87	32.35	34.80	∞
No. observed	0	0	0	0	1	1	4	4	1	1	0	0	0	0	
No. expected	0.12	0.12	0.36	0.6	1.2	1.2	2.4	2.4	1.2	1.2	0.36	0.36	0.12	0.12	

Table IV. General Analysis of Variance (Units of a Single Plot).

	Degrees of freedom	Sum of squares	Mean square
Places	5	7073.64	1414.73
Years	1	1266.17	1266.17
Places × years	5	2297.96	459.59
Varieties	4	1769.99	442.50
Varieties × places	20	1477.67	73.88
Varieties × years	4	97.27	24.32 } 42.72
Varieties × places × years	20	928.09	46.40 }
Total	59	14910.79	
Experimental error	216		23.28

The analysis of variance is similar to that which gives places, years and places × years in the main analysis. The divisor of the square of a single quantity is $3 \times (4^2 + 1 + 1 + 1 + 1) = 60$ and the square of the total of all twelve quantities, divided by 720, gives the sum of squares representing the average difference between Trebi and the other varieties over all stations.

The four items involving varieties in the analysis of variance are thus each split up into two parts, representing the difference of Trebi and the other varieties, and the variation of these other varieties among themselves. This partition is given in Table V. The second part of each sum of squares can be derived by subtraction, or by calculating the sum of squares of the deviations of the four remaining varieties from their own mean.

Study of this table immediately shows that the majority of the variation in varietal differences between places is accounted for by the difference of Trebi from the other varieties. The mean square for varieties × places has been reduced from 73.88 to 35.97 by the elimination of Trebi, and this latter is in itself not significantly above the experimental error. The mean square for varieties × places × years has not been similarly reduced, however, in

Table V. Analysis of Variance, Trebi Versus Remainder.

	Degrees of freedom	Sum of squares	Mean square
Varieties: Trebi	1	1463.76	1463.76
Remainder	3	306.23	102.08
Varieties × places: Trebi	5	938.09	187.62
Remainder	15	539.58	35.97
Varieties × years: Trebi	1	7.73	7.73
Remainder	3	89.54	29.85
Varieties × places × years: Trebi	5	162.10	32.42
Remainder	15	765.97	51.06

fact it is actually increased (though not significantly), and the last three remainder items taken together are still significantly above the experimental error. There is thus still some slight additional variation in response from year to year and place to place. The place to place variation appears to arise about equally from all the remaining three varieties, but the differences between the different years are almost wholly attributable to the anomalous behaviour at Grand Rapids of Velvet, which yielded low in 1931 and high in 1932. If the one degree of freedom from varieties × years and varieties × places × years arising from this difference is eliminated by the "missing plot" technique (Yates, 1933) we have:

	Degrees of freedom	Sum of squares	Mean squares
Velvet at Grand Rapids	1	453.19	453.19
Remainder	23	572.16	24.88

Thus the remainder is all accounted for by experimental error.

It has already been noted that Trebi yielded relatively highly at the high-yielding centres. The degree of association between varietal differences and general fertility (as indicated by the mean of all five varieties) can be further investigated by calculating the regressions of the yields of the separate varieties on the mean yields of all varieties. The deviations of the mean yields of the six stations from the general mean in the order given are nearly proportional to

$$-2, +10, +1, +2, -6, -5.$$

The sum of the squares of these numbers is 170. Multiplying the varietal totals at each place by these numbers we obtain the sums:

Manchuria	1004.6
Svansota	1196.2
Velvet	1137.8
Trebi	1926.8
Peatland	736.3
Total	6001.7

The sum of the squares of the deviations of these sums, divided by 170×6, gives the part of the sum of squares accounted for by the differences of the regressions. This can be further subdivided as before, giving:

	Degrees of freedom	Sum of squares	Mean square
Varieties × Places:			
Differences of regressions: Trebi	1	646.75	646.75
Remainder	3	123.17	41.06
Total	4	769.92	
Deviations from regressions: Trebi	4	291.34	72.84
Remainder	12	416.39	34.70
Total	16	707.73	

Thus the greater part of the differences between Trebi and the remaining varieties is accounted for by a linear regression on mean yield. There is still a somewhat higher residual variation (M.S. 72.84) of Trebi from its own regression than of the other varieties from their regressions, though the difference is not significant. Of the four remaining varieties Peatland appears to have a lower regression than the others, giving significantly higher yields at the lower yielding stations only, the difference in regressions being significant when Peatland is tested against the other three varieties.

The whole situation is set out in graphical form in Figure 1, where the stations are arranged according to their mean yields and the calculated regressions are shown. It may be mentioned that of the remaining five varieties not included in the discussion, three show similar regressions on mean yield.

This artifice of taking the regression on the mean yield of the difference of one variety from the mean of the others is frequently of use in revealing relations between general fertility and varietal differences. A similar procedure can be followed with response to fertilizers or other treatments. The object of taking the regression on the mean yield rather than on the yield of the remaining varieties is to eliminate a spurious component of regression which will otherwise be introduced by experimental errors. If the variability of each variety at each centre is the same, apart from the components of variability accounted for by the regressions, the regression so obtained will give a correct impression of the results. This is always the situation as far as experimental error is concerned (except in those rare cases in which one variety is more variable than the others). It may or may not be the case for the other components of variability. In our example, as we have seen, the deviations of Trebi from its regression are somewhat greater than those of the other varieties. In such case we should theoretically take repressions on a weighted mean yield, but there will be little change in the results unless the additional components of variance are very large.

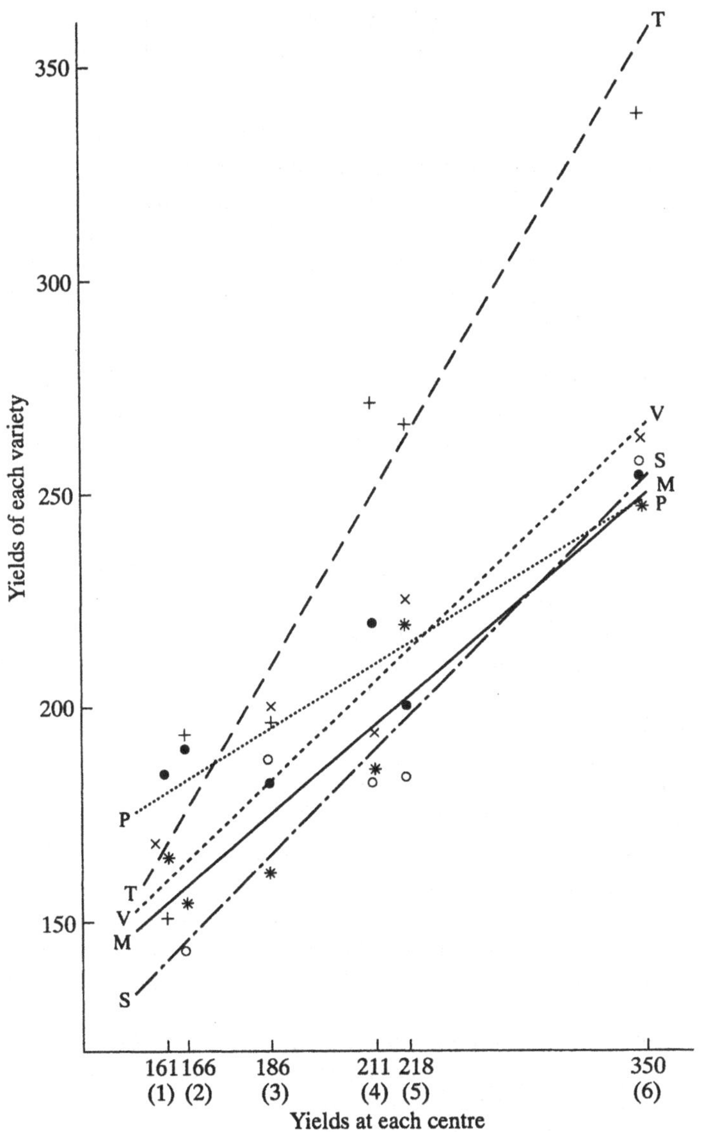

(1) Grand Rapids, (2) Duluth, (3) University Farm, (4) Morris, (5) Crookston, (6) Waseca

Trebi + ———— Velvet × - - - - Svansota ○ —·—·—

Manchuria ∗ ———— Peatland ● ....

Figure 1. Regressions on mean yield. The yields shown are totals of the two years. The mean yields per plot (bushels per acre) are one-sixth of these totals.

We can also examine how far it is possible to make practical recommendations from the results of these experiments.* The following points are of importance:

(1) How far is the superiority of Trebi over the remaining four varieties likely to recur in future years at the same set of stations?
(2) How far is Trebi likely to be superior to the other varieties at other stations in years similar to 1931 and 1932, and in particular with what degree of confidence may Trebi be recommended for general adoption in preference to the other four varieties in some or all of the districts of the state of Minnesota?

The answer to question (1) is clearly limited by the fact that only 2 years' results are available. Therefore, we cannot make any general statement as regards years which are radically different in weather conditions to 1931 and 1932. If, however, 1931 and 1932 themselves differed considerably as regards weather conditions, and if, moreover, the weather conditions varied considerably from station to station in the same year, the weather conditions over the twelve experiments might be regarded as an approximation to a random sample from all possible conditions, in which case the pooled estimate of varieties $\times$ years and varieties $\times$ places $\times$ years (these not being significantly different) might be regarded as an appropriate estimate of the variance due to weather conditions, to differences between fields at the same station, and to experimental error. In this respect Trebi is no more variable than the remaining varieties, and if the anomalous variation of Velvet at Grand Rapids be excluded the pooled estimate (23 degrees of freedom) is only slightly above experimental error. In default of any special explanation of this anomalous variation, however, it will be more reasonable not to exclude this degree of freedom, in which case we should assess the variance due to the above causes as 42.72 (24 degrees of freedom) and this agrees closely with the similar estimate (which cannot be attributed to any one outstanding difference) from the varieties which have been omitted from the present analysis.

Since three plots go to make up the total at any one place in 1 year, the additional variance of a single varietal mean (on a single plot basis) due to weather conditions and differences between fields at the same station, over that arising from experimental error, is

$$\tfrac{1}{3}(42.72 - 23.28) = 6.48,$$

and the variance of the difference of two varieties due to these causes will be double this, i.e., 12.96.

* It should be noted that in practice there are many other factors besides yield which must be taken into consideration. In this instance Dr. Immer informs me that the malting quality of Trebi is poor.

In addition to this real variation in subsequent years of the varietal differences at a place, the errors of the estimates obtained from the 2 years' experimental results must also be taken into account. These are calculated in the ordinary manner from the mean square, 42.72, for varieties × years, and varieties × places × years. Thus the error variance of the estimate of the difference of two varieties at any one place is twice $\frac{1}{6}(42.72) = 14.24$. (If varieties × years and varieties × places × years were different further subdivision of the components, similar to that illustrated below when considering place to place variation, would be necessary.)

In one sense, therefore, the variance of the expected difference of any two varieties, say of Trebi and Peatland at Waseca, in any subsequent year (with similar weather conditions) is $12.96 + 14.24 = 27.20$, but it must be remembered that if a whole series of subsequent years is taken the actual differences will not be distributed about the estimated difference 14.2 with variance 27.20, but about some unknown "true" difference with variance 12.96, the unknown true difference having itself a fiducial distribution about the estimated difference given by the variance 14.24.

The answer to the second question depends on how far the actual stations may be regarded as a random sample of all stations. If this is the case, the estimate of varieties × places for Trebi will be the appropriate estimate of the variance from place to place, including one-half the variance due to differences between fields at the same station, to experimental error and to weather conditions except in so far as they are constant over all stations in each year. This is based on only five degrees of freedom and is therefore ill determined, but accepting the value 187.62, the variance of the difference of Trebi and the remaining varieties due to places only is

$$\tfrac{1}{6}(1 + \tfrac{1}{4})(187.62 - 42.72) = 30.19,$$

and, therefore, that due to place, field and weather conditions (but excluding experimental error) is

$$30.19 + (1 + \tfrac{1}{4})6.48 = 38.29.$$

In addition to this variation the error of the estimated mean difference must be taken into account. The mean difference of Trebi from the mean of the other varieties is 7.1 per plot, and this has an estimated variance due to places, fields, differences in weather conditions from place to place, and experimental error, of

$$\tfrac{1}{36}(1 + \tfrac{1}{4})187.62 = 6.51,$$

and also an additional undetermined component of variance due to differences between years.

Hence, the difference to be expected in a single field is subject to a variation about the true mean having an estimated variance of 38.29, and the estimate of the true mean 7.1 has an error variance of 6.51. Consequently it will frequently happen in default of other information that Trebi will yield less than some other variety that might have been grown. At Grand Rapids, Trebi yielded 15–20% less than Peatland in the 2 years of the experiment. It is poor consolation to the farmer of a farm similar to this to be told that Trebi is giving substantially higher yields on other farms.

It would be rash, however, to recommend Peatland for the whole of the Grand Rapids and Duluth districts and Trebi for the whole of the other districts, in particular the Waseca district, until we known how far variation in the varietal differences depends on factors common to the whole of a district or soil type and how far on factors exclusive to individual farms, such as variations in manuring and cultivation practices and differing crop rotations. Only parallel experiments in the same district on farms which may themselves be reasonably regarded as a random selection from all farms in the district will separate these two sources of variation, and it is therefore impossible from the general analysis of variance to say with any confidence whether Trebi is particularly suited to the district of Waseca or whether its high yield here is due to special conditions at the experimental station. As we have seen, however, the superiority of Trebi is associated with high general level of fertility. If, therefore, we know that the Waseca district is as a whole high yielding we may confidently recommend Trebi for general adoption in the district (with a reservation as to weather conditions). On the other hand, if the general yield of the district is only average, the experimental station being outstanding, then we should only be justified in recommending Trebi for farms of high fertility in this district but might also include farms of high fertility in other districts.

Immer does not report the soil types of the various stations, but it is noteworthy that Peatland, which proved the best variety at the low-yielding stations, has (as its name implies) been specially selected for peat soils, which are likely to be low yielding.

5. Example of Variation in Experimental Error

If the experimental errors of the different experiments are substantially different the use of the z test in conjunction with the pooled estimate of error may be misleading, in just the same way as the pooling of all the degrees of freedom from varieties × places was misleading in the set of varietal trials already considered.

The following is an example in which the z test indicates an almost significant interaction between a treatment effect and places, whereas proper tests show that there is no indication of any such variation in the treatment effect. The example is particularly interesting in that on a first examination of the data the results of the z test led the experimenter to draw false conclusions.

The experiments consisted of a series of thirteen 3×3 Latin squares, described by Lewis and Trevains (1934) and carried out in order to test the effectiveness of, and difference between, an ammonium phosphate mixture and an ordinary fertilizer mixture on sugar beet. Large responses to the fertilizers were shown in all the experiments. The question arose as to whether there was any significant difference between the two forms of fertilizer.

The yields are shown in Table VI. The mean yields of the two forms of fertilizer over all experiments are practically identical, indicating an absence of any consistent difference between the two forms. The analysis of variance, using a pooled estimate of error from all squares, is given in Table VII.

The z between treatment $\times$ centres and error is 0.361, which is almost significant, the 5% point being 0.382. Inspection of the differences between the two forms shows that eight of these are small (≤ 0.30), while the remaining five range numerically from 0.83 to 1.94. These five are all associated with large error mean squares. The values of t for the separate experiments have therefore been calculated and are given in Table VI. These thirteen observed values are compared with the theoretical t distribution for 2 degrees of freedom in Table VIII. The two distributions agree excellently, not one of the values of t being below the 0.1 level of probability. We must conclude, therefore, that there is no evidence from the experiments that the two mixtures behaved differently at any centre.

The above method requires modification if the true difference μ between the two forms of fertilizer is appreciably different from zero, for the quantities $t' = (x - \mu)/S$ will then conform to the t distribution, instead of the t's calculated as above The quantity μ is not exactly known, but if the centres are at all numerous the use of the mean difference $\bar{x}$, or some form of weighted mean difference, such as one of those discussed in the next section, will give quantities which closely approximate to the t distribution.

Although inspection of Table VIII shows quite conclusively in the present example that the observed t's are in no way abnormal, border-line cases will arise in which a proper test of significance is desirable. An obvious form of test would be that based on the variance of the observed t's, or of some analogous function. One such function is the "weighted sum of squares of deviations,"

$$Q = Sw(x - \bar{x}_w)^2 = Swx^2 - \bar{x}_w \cdot Swx,$$

where the weights w are the reciprocals of the estimates of the error variances of the differences x, and $\bar{x}_w$ is the weighted mean of the x's, i.e.,

$$\bar{x}_w = \frac{Swx}{Sw}.$$

If we then calculate

$$\chi'^2 = (k-1) + \sqrt{\frac{n-4}{n-1}}\left\{\frac{n-2}{n}Q - (k-1)\right\},$$

Table VI. Experiments on Sugar Beet.

Centre	Yields of roots, tons per acre: No fertilizer	Yields of roots, tons per acre: Amm. phos. mixture	Yields of roots, tons per acre: Ordinary mixture	Mean response	Amm. phos. − ordinary (x)	Error mean square per plot (s^2)	χ^2	t
1	7.44	15.69	13.75	+7.28	+1.94	0.8599	3.74	+2.56
2	7.19	12.28	11.32	+4.61	+0.96	0.3543	1.54	+1.98
3	10.07	13.93	13.10	+3.44	+0.83	0.5329	2.32	+1.39
4	7.74	10.97	11.89	+3.69	−0.92	1.1528	5.02	−1.05
5	11.88	13.96	15.06	+2.63	−1.10	0.2638	1.15	−2.68
6	11.94	14.35	14.36	+2.42	−0.01	1.7249	7.51	−0.01
7	6.20	10.27	10.02	+3.94	+0.25	0.4803	2.09	+0.44
8	8.99	11.17	11.47	+2.33	−0.30	0.1107	0.482	−1.10
9	9.46	12.54	12.46	+3.04	−0.08	0.0184	0.0801	+0.72
10	7.42	10.93	10.79	+3.44	+0.14	0.0046	0.0200	+2.53
11	3.70	5.46	5.38	+1.72	+0.08	0.0073	0.0318	+1.15
12	9.62	12.72	13.01	+3.24	−0.29	0.1920	0.836	−0.81
13	9.47	13.53	13.72	+4.16	−0.19	0.2706	1.18	−0.45
Mean	8.55	12.14	12.02	+3.53	+0.1115	0.4594	2.00	

Table VII. Analysis of Variance of Responses to Fertilizer and of Difference Between Mixtures (Single-Plot Basis).

	Degrees of freedom	Sum of squares	Mean square	z
Response to fertilizer:				
Mean response	1	324.7605	324.7605	
Response × centres	12	45.8462	3.8205	
Differences between mixtures:				
Mean difference	1	0.2493	0.2493	
Difference × centres	12	11.3542	0.9462	0.361
Error	26	11.9450	0.4594	

χ'^2 will be distributed approximately as χ^2 with $k-1$ degrees of freedom. The relation of Q to the ordinary expression for the variance of t' is shown by the alternative form

$$Q = S(t'^2) - \{S(t'\sqrt{w})\}^2/S(w).$$

This test should not be used if n is less than 6. Actual comparison of the distribution of the t''s with the t distribution should then be resorted to, the unweighted mean $\bar{x}$ being used as the estimate of μ.

An example of the calculation of Q and χ' is given in §7.

6. Methods of Estimating the Average Response

As has been pointed out in the second section, the average response to a treatment over a set of experiments is frequently of considerable importance, even when the response varies from experiment to experiment. The problem of how it may best be estimated from the results of the separate experiments must therefore be considered.

If the experiments are all of equal precision the efficient estimate is clearly the ordinary mean of the apparent responses in each experiment, whether the true responses are the same or vary from experiment to experiment. If, on the other hand, some of the experiments are more precise than others, the ordinary mean, by giving equal weight to both the less and the more accurate results, may appear at first sight to furnish a considerably less precise estimate than might be obtained by more refined statistical processes. As will appear from what follows, however, there are several factors which increase the advantages of the ordinary mean in relation to other possible estimates, so that unless the experiments differ widely in accuracy, or the conditions are somewhat different from those ordinarily met with in agri-

Table VIII. Comparison with Theoretical t and χ^2 Distributions.

P	1.0	0.9	0.8	0.7	0.6	0.5	0.4	0.3	0.2	0.1	0.05	0.02	0.01	0
t	0	0.14	0.29	0.44	0.62	0.82	1.06	1.39	1.89	2.92	4.30	6.96	9.92	∞
Expected	1.3	1.3	1.3	1.3	1.3	1.3	1.3	1.3	1.3	0.65	0.39	0.13	0.13	
Observed	1	0	$\frac{1}{2}$	$1\frac{1}{2}$	2	1	$2\frac{1}{2}$	$\frac{1}{2}$	4	0	0	0	0	

P	1.0	0.99	0.98	0.95	0.90	0.80	0.70	0.50	0.30	0.20	0.10	0.05	0.02	0.01	0
χ^2	0	0.0201	0.0404	0.103	0.211	0.446	0.713	1.39	2.41	3.22	4.60	5.59	7.82	9.21	∞
Expected	0.13	0.13	0.39	0.65	1.3	1.3	2.6	2.6	1.3	1.3	0.65	0.39	0.13	0.13	
Observed	1	1	1	0	0	1	3	3	0	1	1	1	0	0	

culture, the ordinary mean is in practice the most satisfactory as well as the most straightforward estimate to adopt.

The simplest alternative to the ordinary mean is the weighted mean mentioned at the end of the last section, in which the weights are inversely proportional to the error variances of the estimates derived from the various experiments. This weighted mean would be the efficient estimate if there were no variation in the true response from experiment to experiment, and if, moreover, the error variances of the experiments were accurately known. If the error variances are only estimated from a small number of degrees of freedom, however, the weighted mean loses greatly in efficiency and is frequently less efficient than the unweighted mean.

If the true response varies from experiment to experiment, having a variance of σ_0^2, and the error variances are accurately known, the efficient estimate of the mean response is provided by a weighted mean with weights inversely proportional to $\sigma_0^2 + \sigma_1^2, \sigma_0^2 + \sigma_2^2 \ldots,$ where $\sigma_1^2, \sigma_2^2, \ldots,$ are the error variances of the estimates from the various experiments. This has been called the *semi-weighted mean*, since the weights are intermediate between those of the weighted mean and the equal weights of the ordinary mean.

If the response does not vary from experiment to experiment, but the error variances are not accurately known, being estimated from $n_1, n_2, \ldots,$ degrees of freedom, the efficient estimate is obtained by the solution of the maximum likelihood equation:

$$\mu = S\frac{(n_i+1)x_i}{n_i s_i^2 + (x_i - \mu)^2} \Big/ S\frac{(n_i+1)}{n_i s_i^2 + (x_i - \mu)^2}.$$

This solution has the effect of giving lower weights to the more discrepant values than would be given by the ordinary weighted mean. It is not difficult to solve the equation by successive approximation, starting with a value of μ equal to the unweighted mean, but since in agricultural experiments cases in which the response can confidently be asserted not to vary are rare, the additional numerical work is not ordinarily justifiable, except when exact tests of a significance are required and when the n_i are small.

Thus the available rigorous methods of weighting are not of much use in the reduction of the results of the type ordinarily met with. On the other hand, when a set of experiments of widely varying accuracy is encountered, some method of discounting the results of the less accurate experiments is required. The simplest method would be to reject the results of the less accurate experiments entirely, but this involves the drawing of an arbitrary line of division. Anyone who has attempted this will know how easy it is in certain cases to produce substantial changes in the mean response by the inclusion or exclusion of certain border-line experiments.

An alternative procedure is that of fixing an upper limit to the weight assignable to any one experiment. All experiments having error variances which give apparent weights greater than this upper limit are treated as of equal weight. Experiments having a lesser accuracy are weighted inversely as their error variances. The efficiency of this procedure is discussed in Cochran (1937), where it is shown to be substantially more efficient than the use of the ordinary weighted mean if the numbers of degrees of freedom for error are small. Quite large changes in the choice of the upper limit do not seriously affect the efficiency, and equally will not produce any great changes in the resultant estimate. In most agricultural field experiments the upper limit given by an error variance corresponding to a standard error of 5–7% per plot would seem appropriate in cases in which there is no evidence of variation in the response from experiment to experiment.

A further alternative procedure which produces much the same effect is provided by the use of the semi-weighted mean, assigning some arbitrary value to σ_0^2. This procedure has the advantage of being easily adaptable to cases in which there is evidence of variation in response from experiment to experiment. If, for instance, there are eight replicates of each treatment, the value of σ_0^2 corresponding to 4% per plot will be $(0.04)^2(\frac{1}{8}+\frac{1}{8})$, i.e., 0.0004 times the square of the mean yield. This will produce about the same effect as taking a lower limit corresponding to a standard error of 5% per plot. If in addition the estimated variance of the response from centre to centre is 0.0006 times the square of the mean yield, then we might reasonably take a value of σ_0^2 corresponding to 0.0010 times the square of the mean yield.

If the error variances of the various experiments are accurately known the error variance of any form of weighted mean is given by

$$\frac{w_1'^2\sigma_1^2 + w_2'^2\sigma_2^2 + \cdots}{(w_1' + w_2' + \cdots)^2},$$

where $w_1', w_2', \ldots$ represent the weights actually adopted. If $w_1', w_2', \ldots$ are equal to $1/\sigma_1^2$, $1/\sigma_2^2, \ldots$, this expression reduces to the expression for the error variance of the full weighted mean, namely $1/(w_1' + w_2' + \cdots)$, and if all the weights are equal the error variance of the unweighted mean of k estimates

$$\frac{1}{k^2}(\sigma_1^2 + \sigma_2^2 + \cdots)$$

is obtained.

If, however, the error variances are estimated, and the weights depend on these estimates, the above expression will not be correct. In particular the estimated error variance of the fully weighted mean in a group of experiments each with n degrees of freedom for error will be $n/(n-4)$ times the expression given above (the variances being replaced by their estimates throughout). The error variance of any semi-weighted mean, or weighted mean with upper limit, will have to be similarly increased. No exact expressions are available, but in general the additional factor must lie between $n/(n-4)$ and unity. In the case of the weighted mean with upper limit to the weights the inclusion of the factor $n/(n-4)$ in the terms which have weights below the upper limit is likely to give a reasonable approximation.

The mean (weighted or unweighted) may be tested for significance by means of the t test, using the estimated standard error. The test is not exact, since the number of degrees of freedom is not properly defined, but if a number somewhat less than the total number of degrees of freedom for error in the whole set of experiments is chosen the test will be quite satisfactory.

There is one further point that must be examined before using any form of mean in which the weights depend on the relative precision of the various experiments. If the precision is associated in any way with the magnitude of the response, such a weighted mean will produce biased estimates and must not be used. Thus, for example, the response to a fertilizer might be greater on poor land, and this land might be more irregular than good land, so that experiments on poor land would give results of lower precision. In such a case any of the above weighted means would lead to an estimate of the average response which would be smaller than it should be.

To see whether association of this type exists the experiments may be divided into two or more classes according to accuracy, and the differences between the mean response in each class examined. Alternatively the regression of the responses on the standard errors of the experiments may be calculated.

7. Example of the Analysis of a Set of Experiments of Unequal Precision

Table IX gives the responses (in yield of roots) to the three standard fertilizers in a set of $3 \times 3 \times 3$ experiments on sugar beet. These experiments were conducted in various beet growing districts in England. The results shown are those of the year 1934 and are reported in full in the Rothamsted Report (1934). It cannot be claimed that the sites were selected at random (practical considerations precluded this course) and consequently any values obtained for the average responses must be accepted with caution, but the results will serve to illustrate the statistical points involved.

At Bardney, Brigg and Poppleton there were two complete replications, i.e., fifty-four plots, while at each of the remaining centres there was a single replication, twenty-seven plots, only, the error being estimated from the interactions of the quadratic components of the responses and from the unconfounded second order interactions. At all centres the experiments were arranged in blocks of nine plots.

The size of the plot varied. It is immediately apparent, from inspection, or by application of the process described in §4, that the experiments are of very varying precision. In general the larger plots, as might be expected,

Table IX. Responses to Fertilizers in a Series of Experiments on Sugar Beet.

	Washed roots (tons per acre)						
		Linear response to					
Station	Mean yield	N	P	K	Standard error	Wt.	Degrees of freedom
Allscott	10.97	−0.24	+0.63	+0.57	±0.519	3.7	15
Bardney	11.44	+1.23†	+0.35	+0.01	±0.285	12.3	22
Brigg	13.42	+0.11	−0.38	−0.21	±0.603	2.8	22
Bury	13.83	+2.08†	−0.05	−0.22	±0.351		15
Cantley	12.90	+0.20	+0.32	+0.14	±0.453	4.9	15
Colwick	10.12	+1.05†	+0.87†	−0.07	±0.287	12.2	15
Ely	12.46	−1.14	+0.80	−0.08	±0.886	1.3	15
Felstead	11.28	+3.34†	+0.11	−0.23	±0.356	7.9	15
Ipswich	12.45	+1.64†	+0.57	−0.34	±0.344	8.5	15
King's Lynn	19.54	+0.52	+0.12	−0.57	±0.481	4.3	15
Newark	14.10	+1.37†	−0.54*	−0.33	±0.198	25.5	15
Oaklands	12.84	0.00	−0.14	+0.40	±0.622	2.6	15
Peterborough	17.99	−0.14	+1.02	−1.34*	±0.618	2.6	15
Poppleton	14.21	+2.72†	−0.21	−0.18	±0.357	7.8	22
Wissington	14.55	+3.32†	+0.19	+0.38	±0.443	5.1	15
Mean	13.47	+1.07	+0.32	−0.06	±0.125	109.6	246

* 5% significance. † 1% significance.

gave the more accurate results, though the gain in precision was not proportional to the increase in area.

(a) The Response to Nitrogen

The response to nitrogen clearly varies significantly from centre to centre, this variation being large in comparison with experimental error. The ordinary analysis of variance of these responses is given in Table X.

Table X. Analysis of Variance of Response to Nitrogen.

	Degrees of freedom	Sum of squares	Mean square
Average response	1	17.1949	17.1949
Response × centres	14	25.5891	1.8278
Pooled estimate of error			0.2345

The pooled estimate of error is equal to the mean of the squares of the standard errors given in Table IX. The estimate of the standard error of the average response is therefore

$$\sqrt{(0.2345/15)} = \pm 0.125.$$

Since the errors are unequal the t distribution will not be exactly followed, the actual 5% point being subject to slight uncertainty, but in any case intermediate between those given by t for 15 and for 246 degrees of freedom.

The variation of the response from centre to centre has therefore an estimated variance of

$$1.8278 - 0.2345 = 1.5933,$$

excluding variance due to error. This method of estimation is not fully efficient, but may be used in cases such as the present in which the variation is large in comparison with the experimental errors.

It is clear that even were the precision of the experiments known with exactitude, the standard errors being those of Table IX, the semi-weighted mean of §6, which could then be used, would differ little from the unweighted mean, since the weights would only range from

$$\frac{1}{1.5933 + 0.0392} \quad \text{to} \quad \frac{1}{1.5933 + 0.7850},$$

i.e., from 0.61 to 0.42. The unweighted mean is therefore the only estimate of the average response to nitrogen that need be considered. It may be noted that in this set of experiments there appears to be some association between degree of accuracy and magnitude of response to nitrogen. This is an additional reason for not using any form of weighted mean.

(b) The Response to Superphosphate

The responses to superphosphate are of much smaller magnitude than those to nitrogen. Only two, those of Colwick and Newark, are significant, but eleven out of the fifteen are positive, and consequently there is some evidence for a general response.

The unweighted mean of the responses is $+0.32$ and the standard error of the quantity is, as before, ± 0.125. The unweighted mean is therefore significant.

The weights corresponding to the estimated standard errors are given in Table IX. The sum of the products of these weights and the responses to phosphate, divided by the sum of the weights, gives the weighted mean, $+0.365$. This differs somewhat from the unweighted mean, and inspection shows that the difference is largely due to the fact that the two stations which gave significant results received high weights. The weight assigned to Newark is nearly one-quarter of the total weight. The estimated error of this experiment is 3% per plot, which would appear to be lower than is likely to be attained in practice. Fixing a lower limit of error at 5% per plot, which is equivalent to a weight of 10.0 for the experiments of twenty-seven plots and of 20.0 to experiments with fifty-four plots, we obtain the weighted mean with upper limit, 0.323.

Following the rule given in §6, the estimated standard error of the weighted mean will be given by the square root of

$$\frac{3.7 \times \frac{15}{11} + 12.3 \times \frac{22}{18} + \cdots}{(3.7 + 12.3 + \cdots)^2}.$$

This gives the value ± 0.111. Similarly the estimated standard error of the weighted mean with upper limit to the weights is given by the square root of

$$\frac{3.7 \times \frac{15}{11} + 12.3 \times \frac{22}{18} + \cdots + \frac{10^2}{12.2} + \cdots}{(3.7 + 12.3 + \cdots + 10 + \cdots)^2}.$$

This gives a value of ± 0.112. This is presumably somewhat of an overestimate, as this mean is likely to be somewhat more accurate than the weighted mean.

In order to test whether there is any evidence of variation in the phosphate response from centre to centre the weighted sum of squares of deviations Q may be calculated by the formula given at the end of §5. For convenience in this calculation it is best to tabulate the products wx of the weights and the responses separately for each centre. The values obtained are

$$Swx^2 = 27.6068,$$

$$\bar{x}_w Swx = 14.6044,$$

$$Q = 13.0024.$$

The value of χ'^2 may now be calculated. In the present case the number of degrees of freedom for error varies from experiment to experiment. We will take n to be equal to the mean 16.4 of these numbers. This procedure will be satisfactory if the numbers do not differ too widely and are reasonably large in all experiments. Using this value we have

$$\chi'^2 = 14 + \sqrt{\frac{12.4}{15.4}}\left(\frac{14.4}{16.4} 13.0024 - 14\right) = 11.7.$$

Clearly there is no evidence of any variation in response from centre to centre.

It may, however, be considered that although there is no evidence of variation in response, such variation should not be precluded, and that consequently the upper limit of the weights should be lower than the values of 10 and 20 taken above. Fixing the limit at 7.8, so as to give seven of the fifteen experiments equal weight, we obtain a mean of +0.301, with an estimated standard error of ± 0.109. In fertilizer experiments such as the present, where from the nature of the treatments, constancy of response would seem unlikely, this last estimate of the mean response appears to be the most satisfactory, since it gives equal weight to all the more accurate experiments and at the same time prevents the less accurate experiments from unduly influencing the results.

(c) The Response to Potash

The effect of potash shows no significance, either in mean response or in variation in response from centre to centre. The significant depression at Peterborough can consequently be reasonably attributed to chance.

The analysis follows the lines already given and need not be set out in detail here. The weighted mean with upper limit 7.8 has the value -0.03 ± 0.109. The value of χ'^2 for testing the significance of the variation in response is 11.6.

The results discussed here are, of course, only a part of the full results of the experiments. No consideration has been given to the curvature of the response curves, or to the interactions of the different fertilizers. The whole set of experiments provides an excellent illustration of the power of factorial design to provide accurate and comprehensive information. It will be noted, among other things, that the mean responses to the three fertilizers are determined with a standard error of less than 1% of the mean yield.

Summary

When a set of experiments involving the same or similar treatments is carried out at a number of places, or in a number of years, the results usually require comprehensive examination and summary. In general, each set of results must be considered on its merits, and it is not possible to lay down rules of procedure that will be applicable in all cases, but there are certain preliminary steps in the analysis which can be dealt with in general terms. These are discussed in the present paper and illustrated by actual examples. It is pointed out that the ordinary analysis of variance procedure suitable for dealing with the results of a single experiment may require modification, owing to lack of equality in the errors of the different experiments, and owing to non-homogeneity of the components of the interaction of treatments with places and times.

References

Cochran, W.G. (1937). *J. Roy. Statist. Soc.*, Suppl. **4**, 102–18.
Fisher, R.A. (1935). *The Design of Experiments.* Edinburgh.
Immer, F.R., Hayes, H.K., and Powers, Le Roy (1934). *J. Amer. Soc. Agron.*, **26**, 403–19.
Lewis, A.H. and Trevains, D. (1934). *Empire J. Exper. Agric.*, **2**, 244.
Rep. Rothamst. Exp. Sta. (1934), p. 222.
Yates, F. (1933). *Empire J. Exper. Agric.*, **1**, 129–42.

Introduction to
Rao (1948) Large Sample Tests of Statistical Hypotheses Concerning Several Parameters with Applications to Problems of Estimation

Pranab K. Sen
University of North Carolina, Chapel Hill

1. Preamble

Professor Calyampudi Radhakrishna Rao has emerged as one of the most outstanding mathematical statisticians of our time, and his fundamental research contributions covering a wider spectrum of mathematical statistics, design of experiment, combinatorial mathematics, multivariate analysis, information theory, sample surveys, biometry, econometrics, and a variety of other fields, span over a period of little more than 50 years. While C.R. Rao, currently in his mid-seventies, has been embarking on new research projects, the classical work he accomplished in the 1940s is gaining new momentum through vigorous extensions, fruitful applications in diverse fields, and novel interpretations. In *Breakthroughs in Statistics*, Volume 1, Dr. P.K. Pathak has written an Introduction to another outstanding (1945) article of C.R. Rao, relating to the classical *Cramér–Rao Inequality*, wherein he has also made detailed comments on Rao's significant contributions in related fields. In view of this, we will avoid the duplications, and mainly confine ourselves to the scientific aspects of the current article and its profound impact on up-to-date developments in statistical inference and its applications.

2. Emergence of Score Statistics

Statistical inference (covering both estimation and hypothesis testing problems) rests on some underlying probability distributional models which may or may not be of known functional forms, and generally involve multiple parameters. In a parametric setup, one may be interested in a subset of such parameters, treating the rest as *nuisance parameters*. In a semiparametric or nonparametric setup the situation may even be more complex in the sense that part of the underlying model may itself be regarded as a nuisance parametric function. In testing a simple null versus a simple alternative hypothesis, the classical *Neyman–Pearson fundamental lemma* yields an optimal test, while the *maximum likelihood estimators* (MLEs) are known to have some optimal properties under fairly general regularity assumptions. In both cases, the *likelihood function* of the sample observations plays the key role. This simple scenario may change in multiparameter models where the likelihood function may become involved with nuisance parameters or other parametric restraints. Some works in this area, prior to Rao (1948), are due to Wilks (1938, 1939), and Wald (1941, 1943), among others. Whereas Abraham Wald concentrated on asymptotically optimal tests of statistical hypotheses, Samuel Wilks attacked both the problems of hypothesis testing and confidence set estimation; the binding force behind all these developments has been the asymptotic behavior of the classical likelihood ratio statistics and the derived MLE. In Wald's treatment of asymptotic tests, there is an asymptotic equivalence result on the likelihood ratio statistic and a suitable quadratic form in the MLE, so that the asymptotic multinormality of the MLE and the Cochran theorem on quadratic forms in multinormal variables provide the access to the desired asymptotics. Wilks's papers also exploited this idea to a certain extent. The ingenuity of Rao (1948) lies in the incorporation of the so-called *score statistics* which are comparatively easier to handle, and may often require less stringent regularity assumptions than the MLE.

Let us denote the likelihood function based on n independent sample observations $\mathbf{X} = (X_1, \dots, X_n)'$ by $L_n(\boldsymbol{\theta}|\mathbf{X}) = L_n(\boldsymbol{\theta})$, say, where $\boldsymbol{\theta} = (\theta_1, \dots, \theta_\kappa)'$ is the vector of unknown parameters appearing in the joint density of the given observations. Suppose that we desire to test a null hypothesis

$$H_0 : \boldsymbol{\theta} \in \Omega_0 \ (\in \boldsymbol{\Theta}), \tag{1}$$

against an alternative

$$H_1 : \boldsymbol{\theta} \in \Omega_1 \ (\in \boldsymbol{\Theta}), \tag{2}$$

where Ω_0 and Ω_1 are disjoint subsets of the parameter space $\boldsymbol{\Theta}$. Usually, Ω_0 can be expressed in terms of a lower-dimensional subspace of $\boldsymbol{\Theta}$, and sometimes through appropriate parametric transformations, it may even be expressed in terms of a component of θ of a specified value, treating the complementary part as a nuisance parameter. The classical likelihood ratio test statistic for this hypothesis testing problem is given by

$$\mathscr{L}_n = \frac{\sup\{L_n(\boldsymbol{\theta}) : \boldsymbol{\theta} \in \Omega_1\}}{\sup\{L_n(\boldsymbol{\theta}) : \theta \in \Omega_0\}}, \tag{3}$$

so that it entails the computation of two sets of MLEs (i.e., under H_0 and H_1). Keeping the general composite hypothesis testing problem in mind, we consider the following hypotheses:

$$H_0: h(\boldsymbol{\theta}) = \mathbf{0} \quad \text{versus} \quad H_1: h(\boldsymbol{\theta}) \neq \mathbf{0}, \tag{4}$$

where $\mathbf{h}: \mathbf{R}^k \to \mathbf{R}^r$ is a possibly vector-valued function such that the $(k \times r)$-matrix $\mathbf{H}(\boldsymbol{\theta}) = (\partial/\partial\theta)\mathbf{h}(\boldsymbol{\theta})$ exists, is continuous in $\boldsymbol{\theta}$, and has rank equal to $r\ (\leq k)$. Then let $\hat{\boldsymbol{\theta}}_n$ be the unrestricted MLE of $\boldsymbol{\theta}$, and let $\mathbf{I}(\boldsymbol{\theta})$ be the Fisher *information matrix* per unit observation, the observations being assumed to be independent and identically distributed (i.i.d.). The Wald test statistic for the above hypothesis testing problem is given by

$$W_n = n[\mathbf{h}(\hat{\boldsymbol{\theta}}_n)]'\mathbf{Q}^{-1}[\mathbf{h}(\hat{\boldsymbol{\theta}}_n)], \tag{5}$$

where

$$\mathbf{Q} = [\mathbf{H}(\hat{\boldsymbol{\theta}}_n)]'[\mathbf{I}(\hat{\boldsymbol{\theta}}_n)]^{-1}[\mathbf{H}(\hat{\boldsymbol{\theta}}_n)]. \tag{6}$$

Therefore, computationally, W_n entails the unrestricted MLE and the derivative and information matrices. In this setup, the Rao score statistics are defined as

$$\mathbf{U}_n(\boldsymbol{\theta}) = (\partial/\partial\boldsymbol{\theta})L_n(\boldsymbol{\theta}), \tag{7}$$

and this vector can often be evaluated at the (restricted RMLE $\tilde{\boldsymbol{\theta}}_n$ wherein $\boldsymbol{\theta}$ is confined to Ω_0. Rao showed how to make good use of this score statistics vector in testing suitable hypotheses as well as in estimating the parameters under suitable restraints, without sacrificing any statistical information in an asymptotic setup. When Ω_0 can be characterized as a specified subvector of the parameters, with possibly others treated as nuisance parameters, the computation of Rao's score statistics becomes simpler. Moreover, the MLE are generally not expressible algebraically in terms of the sample observations (the O, A, B, AB bloodgroup model cited in Rao (1948) is a classical example), and hence, verification of their asymptotic properties (such as the asymptotic normality) may often entail additional regularity assumptions over those required for $\mathbf{U}_n(\boldsymbol{\theta})$. In this respect too, Rao's score statistics has an advantage over the likelihood ratio test as well as the Wald test statistics. Technically speaking, the asymptotics for the Wilks likelihood ratio test statistic were formulated more extensively by Wald (1943), and he established an asymptotic equivalence relation of his test statistic and the log-likelihood ratio test statistic. In that way, in the Wilks likelihood ratio test statistic, we do not need the computation of the information matrix or the derivative matrix, but it entails two sets of MLE. Thus, in either case, from a computational point of view, Rao's score statistics may be simpler. In a general setup of constructing asymptotically optimal tests for composite hypotheses, Neyman (1959) proposed some tests which are known as the $C(\alpha)$ tests. Hall and Mathiason (1990) extended the Neyman $C(\alpha)$ test to the general multiparameter case, and termed it the Neyman–Rao test.

Again, it would not be improper to put due emphasis on the Rao (1948) paper where most of the ideas in subsequent publications germinated. Since Rao (1948) placed a good deal of emphasis on multinomial distributions, it may be pertinent to bring in here the relevance of Neyman (1949), who advocated the use of *minimum chi-square* and *modified minimum chi-square* test statistics and estimators. In principle, the form of such test statistics is similar to the Wald statistic, but instead of the unrestricted MLE, here we use the minimum chi-square or modified minimum chi-squared estimators, which are all known to be BAN (*best asymptotically normal*) under fairly general regularity conditions. Bhapkar (1966) showed that for such multinomial models, the Wald and Neyman test statistics are algebraically the same under linearized constraints. This suggests that the Rao score statistics may easily be adapted to BAN estimators instead of the MLE. We will make more comments on it later on.

3. Relative Standing of the Score Statistics

Recall that the Rao–Wald–Wilks developments took place before the asymptotic theory of statistical inference emerged on its strong footing. Nor were the Scheffé–Tukey–Roy–Bose–Dunnett-type simultaneous confidence sets available at that time. In the light of these later developments, more can be said about the relative merits and demerits of the Rao–Wald–Wilks procedures. Most notably, for local alternatives, the likelihood ratio, Wald's, and Rao's score test statistics all have asymptotically a common noncentral chi-squared distribution, and hence they are asymptotically power equivalent for local alternatives. However, the picture may not be so much isomorphic for nonlocal alternatives. Hoeffding (1965) initiated the *large deviation* probability approach for the study of asymptotically optimal tests for fixed alternatives. For multinomial distributions, he was able to show that for a fixed alternative, the likelihood ratio test is asymptotically more powerful (in the sense of the rate of convergence of the asymptotic power to 1 with the increase in the sample size), than any other test when the latter statistic is not functionally related to the former. In that sense, in general, the Rao (1948) procedure may not be asymptotically optimal. Similar conclusions can be drawn for other underlying probability distributions. To illustrate this point, let us consider a particular model, like the classical *logistic regression model*, where the information matrix $\mathbf{I}(\boldsymbol{\theta})$ may vary considerably with the unknown parameter point over the parameter space. Since the Rao score statistics use the MLE under the null hypothesis, the adjustments they entail for the asymptotic chi-square law under the null hypothesis may not generally work out under alternatives. In the most simple case, if the null hypothesis H_0 relates to $\boldsymbol{\theta} = \mathbf{0}$, say, against $\boldsymbol{\theta} \neq \mathbf{0}$, then the information matrix evaluated at the null point, needed to construct the test statistic, may differ considerably for another point not very close to it, and this may slow down the rate of convergence of the power function (to 1) or may even pull down it at points away from the null value. This is typically the case with nonlinear models. In some cases, the Rao procedure may not have a monotone power function when the parameter point is away from the null value. However, this feature should not be overemphasized; for the exponential family of densities, all three procedures have close affinity in a broader domain of the parameter space. Let me iterate another aspect of multiparameter hypothesis testing problems where such score statistics may have a different perspective than the Wald–Wilks-type tests. These are characterized by a restricted parameter space under the alternative hypothesis. For example, in a multisample model we may be interested in the homogeneity of the associated parameters (vectors) against an ordered alternative. In this context, often, the construction of likelihood-ratio tests may encounter considerable computational difficulties, whereas the *union–intersection* (UI-) principle of Roy (1953) in conjunction with Rao's score statistics may lead to alternative solutions which are generally easier to handle operationally, and which are asymptotically equivalent to the Wald–Wilks-type tests for local alternatives. Basically, scores tests are more adaptable in a UI-testing setup, and thereby remain more applicable in various nonstandard problems. We will emphasize these aspects more in the next two sections.

4. Scores Tests in Sequential Analysis

Let us discuss the role of Rao's scores tests in *sequential analysis* and some related problems in *clinical trials* and medical studies. For testing a simple null versus a simple alternative hypothesis, the Wald (1947) classical *sequential probability ratio test* (SPRT), having a direct link with the classical likelihood ratio test in the fixed sample size case, is known to possess some optimality properties, particularly in terms of the *average sample number* (ASN). The situation is somewhat different when there are some nuisance parameters. Even for the normal mean testing problem when the variance is unknown, generalizations of the SPRT considered by a host of workers are not completely free from some arbitrariness. In an asymptotic setup wherein the distance between the parameter points under the null and alternative hypotheses is made to converge to 0, Rao's score statistic adjusted for nuisance parameters (as in this classical paper) provides a simpler procedure having an asymptotically locally most powerful property. Rao (1950) actually elaborated this sequential score statistic, and Bradley (1953) considered some nice applications in biometric contexts. Asymptotically locally most powerful properties of such sequential score tests were studied by Berk (1975). The host of nonparametric sequential tests (and estimates) for location and regression parameters, presented in Chapters 9 and 10 of Sen (1981), also bears distinctly the profound impact of score statistics in nonparametrics. More along this line, with greater emphasis on *robustness*, has also been discussed in Jurečková and Sen (1996, Chaps. 9–10).

In medical investigations, particularly, in clinical trials, a follow-up scheme is generally encountered whereby the dataset is accumulated over the period of study. The classical assumptions of independence and homogeneity of increments may not hold here, and as a result, the classical sequential analysis may not be appropriate in this context. Moreover, from medical ethics or an operational point of view, it is quite natural to adopt some *interim analysis* schemes, whereby we look (statistically) into the accumulating dataset at regular time-intervals or even monitor them continuously. *Group sequential procedures* (GSP), *time-sequential* procedures, and *progressively censoring schemes* (PCS) have been incorporated to provide statistical methodology to carry out such interim analyses without increasing the risk of Type I errors. This results in a *repeated significance test* (RST) on the accumulating dataset, and modifications of the usual likelihood-ratio-type of tests are therefore of prime interest. However, in a majority of such applications, we do have composite null and alternative hypotheses, and often, semiparametric or nonparametric models are judged more appropriate from the validity-robustness point of view. For these reasons it is often quite difficult to prescribe an RST version of the classical likelihood ratio tests; in fact, even in the most simple cases, the related asymptotics may become quite involved, but the prospects for score statistics are far better. We may again refer to Chapters 9–11 of Sen (1981) where nonparametric tests and estimates have been considered along with their parametric contenders, and the foundation has been laid by the score statistics.

5. Nonstandard Models

In the above discussion we have confined ourselves to the case of i.i.d. random variables. The question naturally arises: What happens when the i.i.d. clause may not hold? Rao (1948), himself, addressed the case of independent but not necessarily i.i.d. random variables. Also, significant developments have taken place in relaxing the independence assumption by structered dependence patterns, and further, the identity of distributions may be replaced by weaker conditions, such as the stationarity of mean and variance. *Econometric* or *time-series* models and *semiparametric* models are classical examples of this type. In econometric/time-series models, serial dependence of observations is a common phenomenon. In complete generality, this may introduce a large number of nuisance parameters, even when stationarity may be true, at least in a weak sense. For this reason, *m-dependence, autoregressive patterns* and, more generally, *ARMA* models have been extensively used in the literature. Under such structured dependence, the classical Wald–Wilks-type of test statistics may be constructed, but their exact distribution theory may become quite complex. On top of that, nonstationarity may arise due to trend-effects in mean as well as dispersion of the variates over time. The effect of heteroscedasticity may be quite comparable to the *Neyman–Scott* problem where there are too many nuisance parameters. Because of that, the MLE may not be consistent, and hence, the Wald–Wilks-type of tests may lose their natural appeal too. For this reason, some ARCH (*autoregressive conditional heteroscedastic*) models have been proposed in the literature where score tests work out well. Bera and Ullah (1991), along with an extensive bibliography, provided a thorough survey of Rao's score tests in econometrics. In a related scenario, it may be plausible to assume a parametric model for a part of the dataset while allowing a more general (i.e., nonparametric) model for the complementary part: This provides the genesis of the so-called semiparametric models. In such a case it may often be difficult to write down the complete likelihood function so as to construct an appropriate Wald–Wilks-type test. Nevertheless, it may be possible to make good use of Rao's score statistics and construct some test statistics having some asymptotic optimality properties. To illustrate this point, let us consider a hypothesis-testing problem in a typical *life-testing* model where the observable random variables are the successive order statistics along with other concomitant variates. The construction of the Wald–Wilks-type test encounters computational, as well as conceptual, dificulties, but the score test works out better. We may refer to Sen (1976, 1979) for some details. The basic trick is that there is an inherent (though inhomogeneous) *Markov property* of these scores, and hence, suitable martingale characterizations work out well. These in turn provide the access to general asymptotics for such score tests. A more notable example is the *proportional hazard* model of Cox (1972), which has initiated a vigorous growth of research literature on semiparametric models in statistical inference. If we look into the *partial likelihood* model of Cox (1975), we can immediately appreciate the advantages of the score statistics over their contenders. The martingale structure and its role in the general asymptotics in this context have been illustrated in the last chapter of Sen (1981), and in a more abstract *counting processes* approach in Andersen et al. (1993), among others. The winner is clearly the test based on score statistics. Such counting processes have made their way in various fields of applications: survival analysis, reliability networks, neural networks, and image analysis are the most prominent ones. In such applications, it may not be very wise to make some precise parametric model assumptions, and hence, martingale representations have been adopted to a greater extent. Again, in this setup we may often adapt the Rao score statistics in a generalized fashion, and draw statistical inference in a reasonable way. It will not be out of the way to mention that in the classical nonparametrics too, the very way *locally most powerful rank* (LMPR) tests have been constructed by a host of research workers clearly signals the victory of the score statistics over others. We refer to the classical text by Hajék and Sidák (1967) for a detailed account of these LMPR tests. These score test statistics have made their way boldly in nonparametric and robust procedures in statistical inference. Whereas Wald–Wilks-type test statistics can only be constructed under additional regularity assumptions, score tests turn out to be more manageable and asymptotically equally informative. In nonparametrics, we may refer to normal scores and locally optimal score test statistics which retain their validity over a large class of underlying distributions, and yet share their asymptotic optimality under specific parametric alternatives. This *global robustness* perspective of *pseudo-score* tests in nonparametrics has mainly been motivated through the asymptotic properties of Rao's score statistics, with amendments from rank theory and inputs from martingale methodology; we may refer to Sen (1981) for details. In the context of *local robustness* arising due to *error contamination* or gross errors and outliers, similarly the classical robust estimators, considered by Huber (1981) and others, have a distinct score statistics flavor. The *estimating equations* (EE) for such robust estimators are analogues of the classical score statistics in the MLE case, where the scores are derived from some local robustness considerations. In the same vein, we may refer to Sen (1982) for an M-test for a simple regression model, and to Jurečková and Sen (1996, Chap. 10) for a more general treatment of such M-tests based on general score statistics. Even in the case of linear combinations of functions of order statistics, or L-*statistics*, optimality properties can mostly be related through such score statistics. In general, for linear models, there are certain general asymptotic equivalence results on L-, M-, and R-estimators and derived test statistics. Since such equivalence results have mostly been derived through a parametric consideration (albeit in a semiparametric mold), Rao's score statistics provide the general motivation for these conclusions; we refer to Jurečková and Sen (1996), Chaps. 3–7).

6. Concluding Remarks

We may note that in a general multiparameter setting, even in the simplest exponential family of densities, an unique best test may not exist. The notion of A-, D-, E-*optimality* and Kiefer's *universal optimality* provide bench rules for competing tests. Nevertheless, in many applications, it may be difficult to obtain a single test statistic or (point or confidence set) estimator which is uniquely best even in the above sense. For example, in the context of confidence sets in (multinormal) multivariate linear models, Wijsman (1980) established an optimal property of Roy's (1953) procedure based on the UI-principle. Thus, in this sense, what Rao (1948) suggested for interval estimation of a subset of parameters goes over in the much more general framework of Roy (1953) and Wijsman (1980); the asymptotic normality of the MLE and related score statistics provides the access to such results. The second noteworthy point in favor of Rao's score statistics is their flexibility for partial (or pseudo) likelihood functions where a Wald–Wilks-type of test statistic may be more difficult to comprehend and to work out in applications. For this reason, in modern semiparametrics, nonparametrics, and in inference problems relating to various stochastic processes arising in diverse applications, generally such score statistics work out better. Of course, if in a simpler situation, the Wald–Wilks-type test can be incorporated, we would advocate that not only on the grounds of local optimality in an asymptotic setup, but also on the grounds of large deviational optimality, already discussed.

References

Andersen, P.K., Borgan, O., Gill, R.D., and Keiding, N. (1993). *Statistical Methods Based on Counting Processes*. Springer-Verlag, New York.

Bera, A.K. and Ullah, A. (1991). Rao's score test in econometrics. *J. Quantit. Econ.*, **7**, 189–220.

Berk, R. (1975). Locally most powerful sequential tests. *Ann. Statist.*, **3**, 373–381.

Bhapkar, V.P. (1966). A note on the equivalence of two test criteria for hypotheses in categorical data. *J. Amer. Statist. Assoc.*, **61**, 228–235.

Bradley, R.A. (1953). Some statistical methods in taste testing and quality evaluation. *Biometrics*, **9**, 22–38.

Cox, D.R. (1972). Regression models and life tables (with discussion). *J. Roy. Statist. Soc. Ser. B*, **74**, 187–220.

Cox, D.R. (1975). Partial likelihood. *Biometrika*, **62**, 269–276.

Hájek, J. and Sidák, Z. (1967). *Theory of Rank Tests*. Academic Press, New York.

Hall, W.J. and Mathiason, D.J. (1990). On large-sample estimation and testing in parametric models. *Internat. Statist. Rev.*, **58**, 77–97.

Hoeffding, W. (1965). Asymptotically optimal tests for multinomial distributions (with discussion). *Ann. Math. Statist.*, **36**, 369–408.

Huber, P.J. (1981). *Robust Statistics*. Wiley, New York.

Jurečková, J. and Sen, P.K. (1996). *Robust Statistical Procedures: Asymptotics and Interrelations*. Wiley, New York.

Neyman, J. (1949). Contributions to the theory of χ^2 test. In *Proc. Berkeley Symp. Math. Statist. Probab.* University of California, Berkeley, pp. 239–273.

Neyman, J. (1959). Optimal asymptotic tests of composite statistical hypothesis. In *Probability and Statistics (Herald Carmér Volume)* (U. Grenander, ed.). Wiley, New York, pp. 212–234.

Rao, C.R. (1945). Information and the accuracy attainable in the estimation of statistical parameters. *Bull. Calcutta Math. Soc.*, **37**, 81–91.

Rao, C.R. (1948). Large sample tests of statistical hypotheses concerning several parameters with applications to problems of estimation. *Proc. Cambridge Philos. Sco.*, **44**, 50–57.

Rao, C.R. (1950). Sequential tests of null hypotheses. *Sankhyá*, **10**, 361–370.

Roy, S.N. (1953). On a heuristic method of test construction and its use in multivariate analysis. *Ann. Math. Statist.*, **24**, 220–238.

Sen, P.K. (1976). Weak convergence of progressively censored likelihood ratio statistics and its role in asymptotic theory of life testing. *Ann. Statist.*, **4**, 1247–1257.

Sen, P.K. (1979). Weak convergence of some quantile processes arising in progessively censored tests. *Ann. Statist.*, **7**, 414–431.

Sen. P.K. (1981). *Sequential Nonparametrics: Invariance Principles and Statistical Inference*. Wiley, New York.

Sen, P.K. (1982). On M-tests in linear models. *Biometrika*, **69**, 245–248.

Wald, A. (1941). Asymptotically most powerful tests of statistical hypotheses. *Ann. Math. Statist.*, **12**, 1–19.

Wald, A. (1943). Tests of statistical hypothese concerning several parameters when the number of observations is large. *Trans. Amer. Math. Soc.*, **54**, 426–482.

Wald, A. (1947). *Sequential Analysis*. Wiley, New York.

Wijsman, R.A. (1980). Smallest simultaneous confidence sets with applications in multivariate analysis. In *Multivariate Analysis*, V (P.R. Krishnaiah, ed.). North-Holland, Amsterdam, pp. 483–498.

Wilks, S.S. (1938). The large sample distribution of the log-likelihood ratio for testing composite hypotheses. *Ann. Math. Statist.*, **9**, 60–62.

Wilks, S.S. (1939). Optimum fiducial regions for simultaneous estimation of several population parameters from large samples. *Ann. Math. Statist.*, **10**, 85.

Large Sample Tests of Statistical Hypotheses Concerning Several Parameters with Applications to Problems of Estimation

C. Radhakrishna Rao

1. Introduction

If the probability differential of a set of stochastic variates contains k unknown parameters, the statistical hypotheses concerning them may be simple or composite. The hypothesis leading to a complete specification of the values of the k parameters is called a simple hypothesis, and the one leading to a collection of admissible sets a composite hypothesis. In this paper we shall be concerned with the testing of these two types of hypotheses on the basis of a large number of observations from any probability distribution satisfying some mild restrictions and their use in problems of estimation.

If we have a number of samples whose probability densities involve a set of parameters, we may have to test whether a single set is relevant to all the samples before combining them to arrive at the best estimates. This test, which may be called the test of homogeneity of parallel samples, involves a composite hypothesis. A general test of homogeneity different from the χ^2 test of independence of samples each arranged in some categories has been proposed and applied to test for agreement in gene frequencies between two samples giving the distribution in *O, A, B* and *AB* blood-group classes.

Another important group of problems is the estimation of parameters subject to restrictions which are sometimes derived from empirical considerations. The validity of these restricitons may be formally tested before giving final estimates. The use of such empirical relations among the parameters to be estimated, when known, enhances the precision of the estimates, although they may not be strictly accurate. A slightly inaccurate relationship may introduce bias in the estimates, but such estimates are more useful than the less efficient estimates so long as the bias, in any case, is small in comparison with its standard error. This, in some way, is secured when the test for a hypothesis specifying some restrictions indicates close agreement with the observations. It is also of importance to satisfy oneself that the increase in efficiency is of such a magnitude as to justify the use of some restrictions, although they may introduce errors which are smaller in comparison with standard errors of estimates. An instance to this point is the empirical formula $y_{12} = (y_1 + y_2)/(1 + 4y_1y_2)$ suggested by Kosambi (1944) giving the relation connecting the recombination fractions y_1 and y_2 for two successive segments of a chromosome with y_{12} that for the combined segment. The use of this has been found to enhance considerably the precision of the estimates of the recombination fractions.

Methods for determining the confidence regions in the case of several parameters have also been discussed in the light of the new tests proposed above.

2. The Problem of Distribution

There are two problems of distribution which are useful in deriving tests of significance for simple and composite hypotheses. Let

$$x_1, \dots, x_p, \quad y_1, \dots, y_q, \quad \dots,$$

be independent sets of observations from probability laws with densities represented by $f_1(x|\theta), f_2(y|\theta), \dots$, such that each function contains at least one of the unknown parameters $\theta_1, \theta_2, \dots, \theta_k$. The likelihood of the parameters which is the same as the probability density at the observed sets of data is given by

$$L = f_1(x|\theta) \times f_2(y|\theta) \cdots.$$

We define, following Fisher (1935), the quantities

$$\phi_i = \frac{\partial \log L}{\partial \theta_i} \qquad (i = 1, 2, \dots, k),$$

as efficient scores. The mean values of these scores are zero. Their covariance matrix is represented by (α_{ij}) and its reciprocal by (α^{ij}). We shall assume that there exist positive quantities η such that

$$E\left(\frac{1}{f_i}\frac{\partial f_i}{\partial \theta_j}\right)^{2+\eta} \tag{2.1}$$

are finite. Under these conditions, if the non-vanishing terms in the sequence $\partial \log f_i/\partial \theta_j (i = 1, 2, \dots)$ for any j form a sufficiently large set, it follows from general limit theorems that the limiting distribution of $\phi_1, \dots, \phi_k$ at the true values $\theta_1, \dots, \theta_k$ tends to the multivariate normal form with zero mean and covariance matrix (α_{ij}). From this it follows that the statistic

$$\chi^2 = \sum\sum \alpha^{ij}\phi_i\phi_j$$

is distributed, in large samples, as χ^2 with k degrees of freedom when the true values of the parameters are $\theta_1, \theta_2, \ldots, \theta_k$.

In the case where the probability densities $f_1, f_2, \ldots$ are the same, it is enough for the limiting properties to hold that

$$E\left(\frac{1}{f}\frac{\partial f}{\partial \theta_j}\right)^2$$

is finite for every j, which is less restrictive than the condition (2.1). I am grateful to Mr. Bartlett for drawing my attention to this.

Suppose that the θ's are subject to s restrictions defined by s independent relations

$$\psi_i(\theta_1, \ldots, \theta_k) = 0 \qquad (i = 1, 2, \ldots, s). \tag{2.2}$$

The maximum likelihood estimates are given by

$$\left.\begin{aligned} \phi_i + \sum_j \lambda_j \frac{\partial \psi_j}{\partial \theta_i} = 0 \qquad (i = 1, 2, \ldots, k), \\ \psi_i = 0 \qquad (i = 1, 2, \ldots, s), \end{aligned}\right\} \tag{2.3}$$

where λ's are Lagrangian multipliers. Let $\dot\theta_1, \ldots, \dot\theta_k$ be the maximum likelihood estimates. Since the set of equations (2.3) involves $(k - s)$ linear restrictions on $\phi_i(\dot\theta)$, it is expected that the statistic

$$\chi^2 = \sum\sum \alpha^{ij}(\dot\theta)\phi_i(\dot\theta)\phi_j(\dot\theta)$$

is distributed as χ^2 with s degrees of freedom which is $k - s$ less than the degrees of freedom for true values $\theta_1, \ldots, \theta_k$.

This can be demonstrated if we assume that the restrictions (2.2) specify s of the parameters which may be taken as $\theta_{k-s+1}, \ldots, \theta_k$ as functions of the $k - s$ free parameters $\theta_1, \ldots, \theta_{k-s}$, so that the likelihood is an explicit function of these parameters only, and further that the joint distribution of $\dot\theta_1, \ldots, \dot\theta_{k-s}$ tends to the multivariate normal form in large samples with variances and covariances of $O(n^{-1})$. It is known that the latter assumption is true provided the probability laws satisfy the condition (2.1), and further that the maximum likelihood estimates are *uniformly consistent* (Wald, 1943; Doob, 1934). I have omitted this latter condition by mistake in an earlier paper (Rao, 1947) in establishing some optimum properties of the maximum likelihood estimates. This does not seem to be a necessary condition, and the approach to normality is probably true under less stringent conditions.

Let us take the case of two parameters and one restriction which may be taken as $\theta_2 = w(\theta_1)$. The differential coefficient $d\theta_2/d\theta_1$ is denoted by $\lambda(\theta_1)$.

The maximum likelihood estimates satisfy

$$\phi_1(\dot{\theta}) + \lambda(\dot{\theta}_1)\phi_2(\dot{\theta}) = 0, \qquad \dot{\theta}_2 - w(\dot{\theta}_1) = 0. \tag{2.4}$$

If the given relation is true, then the statistic

$$\chi_0^2 = \sum\sum \alpha^{ij}(\theta)\phi_i(\theta)\phi_j(\theta) \tag{2.5}$$

depends only on θ_1, and is distributed as χ^2 with two degrees of freedom at the true value of θ_1. The expression (2.5) treated as a function of θ_1 may be expanded in the neighbourhood of $\dot{\theta}_1$. The first term is

$$\chi_1^2 = \sum\sum \alpha^{ij}(\dot{\theta})\phi_i(\dot{\theta})\phi_j(\dot{\theta}). \tag{2.6}$$

The second term is

$$\begin{aligned} &2(\theta_1 - \dot{\theta}_1)[\phi_1(\theta)\{\alpha^{11}(\alpha_{11} + \lambda\alpha_{12}) + \alpha^{12}(\alpha_{12} + \lambda\alpha_{22})\} \\ &\quad + \phi_2(\theta)\{\alpha^{22}(\alpha_{12} + \lambda\alpha_{22}) + \alpha^{12}(\alpha_{11} + \lambda\alpha_{12})\}] \\ &\qquad = 2(\theta_1 - \dot{\theta}_1)[\phi_1(\dot{\theta}) + \lambda\phi_2(\dot{\theta})] = 0, \end{aligned} \tag{2.7}$$

in virtue of (2.4). In the expression (2.7) terms of the order (n^0) only have been retained, $\partial\phi_i/\partial\theta_j$ being replaced by α_{ij} and terms of the type

$$\frac{\partial\alpha_{ij}}{\partial\theta_i}(\theta_1 - \dot{\theta}_1)\phi_i\phi_j$$

being omitted as they are of $O(n^{1/2})$.

The third term can be easily shown to be

$$\chi_2^2 = (\theta_1 - \dot{\theta}_1)^2[\alpha_{11}(\dot{\theta}) + 2\lambda\alpha_{12}(\dot{\theta}) + \lambda^2\alpha_{22}(\dot{\theta})].$$

Neglecting terms of higher order of smallness we get

$$\chi_0^2 = \chi_1^2 + \chi_2^2.$$

Since

$$1/V(\theta_1 - \dot{\theta}_1) \sim \alpha_{11}(\dot{\theta}) + 2\lambda\alpha_{12}(\dot{\theta}) + \lambda^2\alpha_{22}(\dot{\theta}),$$

it follows that χ_2^2 is distributed in large samples as χ^2 with one degree of freedom. It can be demonstrated by expanding $\phi_i(\dot{\theta})$ in powers of $(\theta_1 - \dot{\theta}_1)$ that $(\theta_1 - \dot{\theta}_1)$ and $\phi_i(\dot{\theta})$ tend to be uncorrelated in large samples, so that χ_1^2 and χ_2^2 are independently distributed in the limiting case.

Since χ_0^2 is distributed as χ^2 with two degrees of freedom and χ_2^2 with one degree of freedom, it follows that the residual part χ_1^2 is distributed as χ^2 with one degree of freedom.

In the case of s relations and $k(\geq s)$ parameters χ_0^2 can be expressed as a function of $(k - s)$ parameters and split into two portions, one of which is a χ_2^2 with $(k - s)$ degrees of freedom measuring the discrepancy of the $(k - s)$ estimated parameters from their true values, and another a χ_1^2 with s degrees of freedom measuring the departures from the assigned relationships.

3. Derivation of Statistics for Simple and Composite Hypotheses

In the case of a single parameter, to test the simple hypothesis $\theta = \theta^0$, the statistic $\phi_1^2(\theta^0)/\alpha_{11}(\theta^0)$ is used as χ^2 with one degree of freedom. The quantity $\phi_1(\theta^0)$ has been called by Fisher (1935, 1946) the efficient score at the assigned value, and its use leads to elegant analysis in statistical tests. The optimum properties of this test have been discussed by Wald (1941) and Rao and Poti (1946).

In the multiparameter case let us consider the set of values $\theta_1^0 + h_1, \ldots, \theta_k^0 + h_k$, where $h_1, \ldots, h_k$ are small as alternatives to $\theta_1^0, \ldots, \theta_k^0$. The proportionate increase in the likelihood is given by

$$h_1\phi_1 + \cdots + h_k\phi_k. \tag{3.1}$$

The best test of the hypothesis in the sense that it affords the maximum discrimination when the alternatives differ from the assigned values by small quantities is provided by the statistic

$$w = h_1\phi_1 + \cdots + h_k\phi_k, \tag{3.2}$$

which leads to the use of the statistic

$$\chi^2 = w^2 \Big/ \sum\sum h_i h_j \alpha_{ij}(\theta^0) \tag{3.3}$$

as χ^2 with one degree of freedom.

If the ratios of $h_1, \ldots, h_k$ can be assigned from *a priori* considerations, which is sometimes possible, the test can be carried out with exactitude. On the other hand, we may have to determine $h_1, \ldots, h_k$ from the departures of the assigned values $\theta_1^0, \ldots, \theta_k^0$ from those values indicated by the data and introduce suitable changes in judging the significance of the derived statistic. This may be done by finding the ratios of $h_1, \ldots, h_k$ such that χ^2 of (3.3) is maximum. The maximum value comes out as

$$\chi^2 = \sum\sum \alpha^{ij}(\theta^0)\phi_i(\theta^0)\phi_j(\theta^0). \tag{3.4}$$

large samples this can be used, as shown in the previous section, as χ^2 with k degrees of freedom to test the hypothesis that the values of $\theta_1, \ldots, \theta_k$ are $\theta_1^0, \ldots, \theta_k^0$ respectively. This differs from the statistic proposed by Wald (1943), wherein he uses

$$\chi^2 = \sum\sum \alpha_{ij}(\dot{\theta})(\dot{\theta}_i - \theta_i^0)(\dot{\theta}_j - \theta_j^0)$$

to test the above hypothesis, where the $\dot{\theta}$'s are the maximum likelihood estimates. The test associated with (3.4) besides being simpler than Wald's has some theoretical advantages as shown in §5.

A composite hypothesis specifies that the admissible sets of values lie on the intersections of surfaces

$$\psi_j(\theta_1, \ldots, \theta_k) = 0 \qquad (j = 1, 2, \ldots). \tag{3.5}$$

If $s \leq k$ of these functions are independent the composite hypothesis is said to have $(k - s)$ degrees of freedom. Since a single set is responsible for the observed sample we may find their best estimates subject to the restrictions (3.5) and change the problem to that of testing a simple hypothesis whether these estimates agree with the data.

If the best estimates under the above restrictions are $\dot{\theta}_1, \ldots, \dot{\theta}_k$, the statistic

$$\chi^2 = \sum\sum \alpha^{ij}(\dot{\theta})\phi_i(\dot{\theta})\phi_j(\dot{\theta}) \tag{3.6}$$

can be used as χ^2 as shown in (2.4) with s degrees of freedom to test the composite hypothesis that the parameters satisfy s conditions. The $(k - s)$ degrees of freedom have been lost in constructing a suitable simple hypothesis from the composite hypothesis. As a general rule we may say that the degrees of freedom of χ^2 for testing a composite hypothesis is k, the number of parameters $-f$, the degrees of freedom of the hypothesis, which is the same as $(k - f)$ the number of restrictions they obey.

4. A General Test of Homogeneity of Parallel Samples

The test of agreement of parallel samples, where each sample consists of observations arranged in mutually exclusive classes, can be treated as a test of independence in a contingency table if nothing is specified about the nature of the distribution in the various classes. Thus if we have r samples each arranged in p classes, the χ^2 test of independence has $(r - 1)(p - 1)$ degrees of freedom. If the distribution in the p classes can be specified by a probability law involving $k \leq (p - 1)$ parameters, then the test of agreement in parallel samples is equivalent to a test of a composite hypothesis which specifies $k(r - 1)$ (relations among the rk parameters, and these are exactly the degrees of freedom of the χ^2 test of composite hypothesis. The disagreement in parallel samples is specified by $k(r - 1)$ degrees of freedom, and a test for their significance need only be carried out. The exact expression for the χ^2 statistic is

$$\sum_{s=1}^{r}\sum\sum_{i,j=1}^{k} \alpha_s^{ij}(\dot{\theta})\phi_i^s(\dot{\theta})\phi_j^s(\dot{\theta}), \tag{4.1}$$

where $\dot{\theta}$'s are obtained from the equations $\sum_s \phi_i^s(\theta) = 0 (i = 1, 2, \ldots, k)$ and (α_s^{ij}) is the matrix inverse to the information matrix for the sth sample and

Table 1. Blood-Group Frequencies in Two Samples of Christians (Indian).

	O	A	B	AB	Total
Army cadets	56	60	18	6	140
Other Christians	120	122	42	11	295
Total	176	182	60	17	435

ϕ_i^s is the ith efficient score for the sth sample. This test is applicable in all cases whether the variables are continuous or discontinuous provided the sample size is large. The test is illustrated with an example given below, and the calculations are similar in any analogous situation.

The distributions in the four O, A, B and AB blood-group classes of 140 Christians who are army cadets and 295 other Christians are given in Table 1. The problem is to test whether the two samples agree in the gene frequencies.

If p, q, r are the A, B, O gene frequencies, then the probabilities and their derivatives are:

	Probabilities and derivatives		
	π	$\partial\pi/\partial p$	$\partial\pi/\partial q$
O	r^2	$-2r$	$-2r$
A	$p(p+2r)$	$2r$	$-2p$
B	$q(q+2r)$	$-2q$	$2r$
AB	$2pq$	$2q$	$2p$

On the given hypothesis the maximum likelihood values are to be obtained from the combined sample. Fairly approximate solutions as obtained from Bernstein's (1925) method are

$$p = 0.26449, \qquad q = 0.09317, \qquad r = 0.64234.$$

The probabilities and coefficients for the calculation of efficient scores are:

		Coefficients for scores	
	Probability π	$(1/\pi)(\partial\pi/\partial p)$	$(1/\pi)(\partial\pi/\partial q)$
O	0.41260	−3.11362	−3.11362
A	0.40974	3.13543	−1.27104
B	0.12838	−1.45217	10.00685
AB	0.04928	3.75086	10.73307

The information matrix for a single observation is

$$I_{pp} = 9.00315, \qquad I_{pq} = 2.47676,$$

$$I_{pq} = 2.47676, \qquad I_{qq} = 23.21612.$$

The elements of the inverse matrix are

$$I^{pp} = 0.114430, \qquad I^{pq} = -0.012208,$$

$$I^{pq} = -0.012208, \qquad I^{qq} = 0.044376.$$

The efficient scores for each sample are obtained by multiplying the observed frequencies with the coefficients for scores given above and adding up over all the classes:

	ϕ_p	ϕ_q
Sample 1	10.30918	−7.30340
Sample 2	−10.51362	7.21019
Total = Φ	−0.20444	−0.09321

The small additive corrections to the approximate values p and q are given by

$$dp = \frac{(I^{pp}\Phi_p + I^{pq}\Phi_q)}{N} = -0.0000,5116,$$

$$dq = \frac{(I^{pq}\Phi_p + I^{qq}\Phi_q)}{N} = -0.0000,0377.$$

The efficient scores and informations matrix at these values are needed for the test. They can be obtained by slight adjustments if the approximations are good to start with. The changes in the elements of the information matrix are negligible. The χ^2 is 0.17258, which is small for two degrees of freedom, thus indicating close agreement.

Table 2. Adjusted Efficient Scores and χ^2.

Sample	n	$\phi_1' = \phi_1 - \frac{n}{N}\Phi_1$	$\phi_2' = \phi_2 - \frac{n}{N}\Phi_2$	$\chi^2 = \frac{1}{n}\sum\sum I^{ij}\phi_i'\phi_j'$
1	140	10.37497	−7.27341	0.11704
2	295	−10.37497	7.27341	0.05554
Total	435	0	0	0.17258
				2 D.F.

We can in such cases give the best estimates of gene frequencies as derived from the combined sample:

$$\dot{p} = 0.26444, \qquad V(\dot{p}) = \frac{I^{pp}}{N} = 0.00026305,$$

$$\dot{q} = 0.09317, \qquad V(\dot{q}) = \frac{I^{qq}}{N} = 0.00010202,$$

$$(\dot{r}) = 0.64239, \qquad V(\dot{r}) = \frac{I^{pp} + 2I^{pq} + I^{qq}}{N} = 0.00030893.$$

The general formula for χ^2 in the case of two samples can be written as

$$\chi^2 = \left(\frac{1}{n_1} + \frac{1}{n_2}\right) \sum \sum I^{ij} \phi_i \phi_j,$$

where I^{ij} are elements of the matrix inverse to the information matrix for a single observation and n_1 and n_2 are sample sizes, and ϕ's are efficient scores at the estimated values for one of the samples. The degrees of freedom in this case are equal to the number of parameters under consideration.

5. Confidence Regions and Intervals

It has been shown in §2 that the statistic

$$\chi^2 = \sum \sum \alpha^{ij}(\theta)\phi_i(\theta)\phi_j(\theta)$$

considered as a funciton of the observations and the unknown parameters $\theta_1, \ldots, \theta_k$ is distributed, in the limit, independently of the parameters. When such pivotal quantities as defined by Fisher (1945) exist it is possible to divide the set of parameters into two groups S_1 and S_2 such that any hypothesis assigning a set of parameters belonging to only one of the groups S_1 (say) is rejected by the observed data on a desired probability level $\alpha\%$. The groups S_1 and S_2 are defined by the inequalities

$$\sum \sum \alpha^{ij}(\theta)\phi_i(\theta)\phi_j(\theta) \geq \alpha\% \text{ value of } \chi^2 \text{ with } k \text{ degrees of freedom,}$$
$$< \alpha\% \text{ value of } \chi^2 \text{ with } k \text{ degrees of freedom,}$$

respectively. The region defined by the group S_2 in a space of k dimensions in which the sets of parameters may be represented, is called the confidence region. The regions so constructed from the observations satisfy the property that in repeated samples they exclude the true set of parameters only $\alpha\%$ of times. Some optimum properties of these regions are mentioned in an abstract of a paper by Wilks (1939).

The confidence region constructed above is useful only when all the parameters are considered simultaneously. If the confidence interval for a single parameter (say) θ_1 irrespective of the others is required then the following procedure is necessary. If θ_1 is considered known the maximum likelihood estimates $\dot{\theta}_2, \ldots, \dot{\theta}_k$ can be determined as functions of θ_1 and the observations. This amounts to estimating the parameters with the restriction that the value of θ_1 is given. Under such circumstances it has been shown in §2 that the statistic

$$\chi^2 = \sum\sum \alpha^{ij}(\theta_1, \dot{\theta})\phi_i(\theta_1, \dot{\theta})\phi_j(\theta_1, \dot{\theta})$$

is distributed as χ^2 with $k - (k - 1)$ degrees of freedom. The α% confidence interval for θ_1 is defined by the inequality

$$\sum\sum \alpha^{ij}(\theta_1, \dot{\theta})\phi_i(\theta_1, \dot{\theta})\phi_j(\theta_1, \dot{\theta})$$

$$< \alpha\% \text{ value of } \chi^2 \text{ with one degree of freedom.}$$

Similarly confidence regions for any subset of s parameters can be determined. The χ^2 to be used in this case has $k - (k - s)$ degrees of freedom.

References

Bernstein, F. Zusammenfassende Betrachtungen über die erblichen Blutstrukturen des Menschen. *Z. Indukt. Abstamm. -u. VererbLehre*, **37** (1925), 237–270.

Doob, J.L. Probability and statistics. *Trans. Amer. Math. Soc.*, **36** (1934), 759–775.

Fisher, R.A. The detection of linkage with dominant abnormalities. *Ann. Eugen. London*, **6** (1935), 187–201.

Fisher, R.A. The logical inversion of the notion of the random variable. *Sankhyā*, **7** (1945), 130–133.

Fisher, R.A. A system of scoring linkage data with special reference to pied factors in mice. *Amer. Natur.*, **80** (1946), 568–578.

Kosambi, D.D. The estimation of map distance from recombination values. *Ann. Eugen. London*, **12** (1944), 172–176.

Rao, C.R. Minimum variance and the estimation of several parameters. *Proc. Cambridge Philos. Soc.*, **43** (1947), 280–283.

Rao, C.R. and Poti, S.J. On locally most powerful tests when alternatives are one sided. *Sankhyā*, **7** (1946), 439.

Wald, A. Some examples of asymptotically most powerful tests. *Ann. Math. Statist.*, **12** (1941), 396–408.

Wald, A. Tests of statistical hypotheses concerning several parameters when the number of observations is large. *Trans. Amer. Math. Soc.*, **54** (1943), 426–482.

Wilks, S.S. Optimum fiducial regions for simultaneous estimation of several population parameters from large samples. *Ann. Math. Statist.*, **10** (1939), 85.

Introduction to
Cornfield (1951) A Method of Estimating Comparative Rates from Clinical Data. Applications to Cancer of the Lung, Breast, and Cervix

Mitchell H. Gail

A major goal of epidemiology is to quantify how disease risk varies with exposure to possible disease-causing agents. The most direct way to obtain such information is to follow a cohort of exposed (E) individuals and an otherwise comparable cohort of unexposed individuals ($\bar{E}$) and to determine for each cohort the probability of developing disease (D). Letting $P(D|E)$ and $P(D|\bar{E})$ denote, respectively, the probabilities of disease in the exposed and unexposed cohorts, we commonly measure the effects of exposure either as the risk difference, $P(D|E) - P(D|\bar{E})$, or as the relative risk $r = P(D|E)/P(D|\bar{E})$. For rare diseases, the relative risk may be large while the risk difference is small. For example, if $P(D|E) = 0.001$ and $P(D|\bar{E}) = 0.0001$, the relative risk is 10, whereas the risk difference is only 0.0009. The relative risk is often used in etiologic research aimed at identifying potential causal risk factors, while risk differences are mainly used to estimate the impact of an exposure on population disease burden.

Despite the simplicity of the cohort design and the advantage that exposure can be measured at the beginning of follow-up, cohort studies are not the most widely-used method for studying chronic diseases, especially diseases with low incidence rates. Indeed, of the 26 papers on cancer etiology, published in Volume 86 in 1994 in the *Journal of the National Cancer Institute*, only 9 (35%) reported on cohort studies. This is because cohort studies of such diseases require that large numbers of subjects be followed for long periods of time in order to acquire the needed information on risk differences or relative risk. Such studies are therefore expensive and time-consuming, unless one has access to previously collected data on cohort exposure and disease outcome, as occurs in a "retrospective cohort study."

The majority of etiologic studies on chronic diseases rely instead on the "case-control" design. We assemble subjects with a given disease (the "cases") and compares their exposure distribution with that of a sample of noncases, the "controls" ($\bar{D}$). If cases tend to be more exposed than controls, we have established an association between exposure and disease status. For many years, investigators have found etiologic leads by studying cases. In 1775, Percivall Pott discovered that scrotal cancer incidence was extraordinarily high among chimney sweepers by studying scrotal cancer cases. In the 1850s, John Snow demonstrated that cholera was a waterborne disease by his painstaking study of the location and opportunities for exposure to contaminated water of the cholera cases. But, according to Breslow and Day (1980), formal case-control studies are a comparatively recent development. They cite a study of breast cancer by Lane-Claypon (1926) as the first case-control study.

Case-control studies are less time-consuming, less costly, and require fewer subjects than cohort studies in most instances. Data from case-control studies can be used to determine whether there is an association between exposure and disease risk, by comparing the exposure distributions of cases and controls. It is doubtful, however, that such studies would have achieved their current predominance as the main tool of etiologic epidemiology without a firm theoretical link to the risk of disease in the underlying source population. Cornfield's (1951) paper provides such a link.

Cornfield (1951) stressed the importance of selecting representative cases and controls from the underlying source population. In his section on "Pitfalls," Cornfield questions whether the controls used in Lane-Claypon's study of breast cancer were over-represented in women with large numbers of children, and, commenting on another hospital-based case-control study of breast cancer, he noted that the cases were older than expected from the age-specific incidence of breast cancer in the general population. Thus, Cornfield emphasized that in order for results from a case-control study to estimate risk in a general population, we had to begin with representative sampling of cases and controls from that population.

The main problem that Cornfield solved was how to relate quantities like $P(E|D)$ and $P(E|\bar{D})$, which can be estimated from case-control data, to the desired quantities $P(D|E)$, $P(D|\bar{E})$, and r, which would be directly estimable from cohort data. Using Bayes' theorem he showed that

$$P(D|E) = P(D)P(E|D)/(P(D)P(E|D) + P(\bar{D})P(E|\bar{D})) \tag{1}$$

and

$$P(D|\bar{E}) = P(D)P(\bar{E}|D)/(P(D)P(\bar{E}|D) + P(\bar{D})P(\bar{E}|\bar{D})). \tag{2}$$

Thus, if we know the incidence of disease in the general population, $P(D)$, we can combine this information with information on the exposure distribution in cases and controls to estimate the absolute exposure-specific risks, $P(D|E)$ and $P(D|\bar{E})$. Cornfield gave a method for constructing confidence intervals for such estimates of absolute risk under the assumption that $P(D)$ is a known constant.

Often $P(D)$ is not known but can be estimated from "population-based" case-control data. In a population-based study, the number of subjects in the source population is known, and there is a census of the cases that arise in a given time period. Thus, $P(D)$ can be estimated, and sampling frames can be constructed from which to sample representative cases and controls. Confidence intervals for absolute exposure-specific risk that take uncertainty of estimates of $P(D)$ into account were given by Greenland (1987) and by Benichou and Wacholder (1994), and Benichou and Gail (1995) gave methods of inference on absolute risk for multivariate risk models.

Many case-control studies are not population-based, however. Instead, cases with the disease of interest are often obtained from among patients in a hospital, and other patients with other diseases thought not to be related to the exposure are selected as controls from that hospital. The overall disease risk, $P(D)$, is not estimable from such data. Cornfield (1951) nonetheless showed that the relative risk, r, is well approximated for diseases with small risk by the exposure odds ratio $(P(E|D)/P(\bar{E}|D))/(P(E|\bar{D})/P(\bar{E}|\bar{D}))$, which is estimable from case-control data. The validity of this estimate depends on the assumption that the cases and controls are representative (with respect to the distributions of exposure) of cases and controls in the general population. (Schlesselman (1982), however, gives conditions under which cases and controls are each nonrepresentative but the biases compensate in such a way as to yield unbiased estimates of r.) It is this remarkable ability of the case-control study to yield estimates of relative risk, a parameter of evident relevance to etiologic studies, that accounts for the wide popularity of the method.

Although Cornfield was the first to show formally that the relative risk could be estimated from a case-control study and that absolute exposure-specific risks could be estimated if $P(D)$ were known in addition, Doll and Hill (1950) gave an arithmetic argument indicating that relative risks of lung cancer for various levels of smoking could be estimated from case-control data, and, in 1952, Doll and Hill, using overall lung cancer disease rates in the source population of "Greater London," computed estimates of the absolute risk of lung cancer according to level of daily cigarette consumption. Doll and Hill also stressed the importance of representativeness of the case-control samples and of interpreting the data in relation to risk in the underlying source population. In their efforts to understand the role of smoking as a risk factor for lung cancer, Cornfield (1951) and Doll and Hill (1950, 1952) clarified the case-control method and put in on a sound theoretical basis (Gail, 1996).

Cornfield made other important contributions to the methodology and understanding of the case-control study. He developed an accurate asymptotic approximation to exact conditional confidence intervals on the odds ratio (Cornfield, 1956) for inference on the relative risk. Recent advances in computing algorithms and technology yield the exact conditional confidence intervals (Mehta et al. 1985).

Cornfield et al. (1959) argued that the relative risk, r, enjoys several advantages for etiologic studies. One advantage is that elevations in relative risk are unlikely to represent artifacts from the use of misclassified cases. To be precise, Cornfield proved: "If a causal agent A increases the risk for disease I and has no effect on the risk for disease II, then the relative risk of developing disease I, alone, is greater than the relative risk of developing diseases I and II combined, while the absolute risk is unaffected." Another advantage is that the relative risk is not increased by exposure to an additional uncorrelated causal agent. That is, "If two uncorrelated agents, A and B, each increase the risk of a disease, and if the risk of the disease in the absence of either agent is small (in a sense to be defined), then the apparent relative risk for A, r, is less than the risk for A in the absence of B."

A profoundly important problem facing epidemiologists is the possibility that an observed association between exposure and outcome results from a hidden factor ("confounder") that is associated with both disease status and exposure. For example, if older people tend to have more money and also tend to have higher cancer risk, then an analysis of the association of wealth with cancer risk that fails to take age into account may lead to the false conclusion that wealth is a causal risk factor for cancer. If the potential confounders have been measured in cases and controls, analytical methods based on stratification by levels of the confounders (Mantel and Haenszel, 1959) or by the use of logistic regression models (see, e.g., Breslow and Day (1980)) can be used to determine whether the association between exposure and disease status persists after adjusting for the potential confounders.

There remains the possibility of unidentified or unmeasured confounders. For example, in arguing against the evidence that smoking caused lung cancer, Fisher (1957, 1958) raised the possibility that an unmeasured genetic "constitutional factor" both increased lung cancer risk and increased the propensity to smoke. Cornfield et al. (1959) adduced several arguments against the "constitutional hypothesis," but one of the most powerful and generally useful arguments was based on an inequality due to Cornfield. The inequality stated: "If an agent A, with no causal effect upon the risk of disease, nevertheless, because of a positive correlation with some other causal agent, B, shows an apparent risk, r, for those exposed to A, relative to those not so exposed, then the prevalence of B among those exposed to A, relative to the prevalence among those not so exposed, must be greater than r." In other words, if the relative risk for lung cancer is ten among smokers compared to nonsmokers, then a putative "constitutional factor" must be at least ten times more prevalent among smokers than among nonsmokers in order to explain the association between smoking and lung cancer. No specific factors satisfying this condition had been identified or proposed by Fisher. Today many epidemiologists are reluctant to draw causal inferences from observational data with relative risks below two because of the possibility of unidentified confounders, but, if the relative risk is three or more, many investigators regard the evidence as stronger on the basis of the Cornfield inequality.

Cornfield et al. (1959) noted: "Nothing short of a series of independently conducted, controlled, experiments on human subjects, continued for 30 to 60 years, could provide a clear-cut and unequivocal choice between" the constitutional hypothesis and the hypothesis that smoking caused lung cancer. Such an experiment is unthinkable, and one must rely instead on the best available epidemiologic case-control and cohort data and on a careful examination of other possible explanations for the strong association between smoking and lung cancer risk. In a masterful review of the available evidence, Cornfield et al. (1959) considered various possible artifacts that might have produced such an association. One of these is "recall bias" that results when a subject with disease tends to exaggerate his previous exposure, compared to a control without disease. In some case-control studies, smoking history had been systematically documented in hospital records before the advent of lung cancer, eliminating the possibility of recall bias. Moreover, smoking history was established before the onset of disease in several cohort studies, all of which confirmed the strong association between smoking and lung cancer risk.

Cornfield et al. (1959) adduced a wide range of data and facts to appraise and in some cases eliminate alternative explanations for the association between smoking and lung cancer risk. For example, in criticizing the constitutional hypothesis, Cornfield et al. noted that lung cancer had increased continuously in the previous 50 years. The constitutional hypothesis would require that a new carcinogen is acting on a genetically susceptible subpopulation or that a new mutation is spreading rapidly in the population to explain the rising lung cancer rates. Tobacco smoke contains substances that cause cancer when applied to the skin of mice and rats. The constitutional hypothesis would require that such substances not be carcinogenic for human lungs. Cigarettes cause mainly lung cancer, whereas pipes and cigars cause mainly mouth and throat cancers. There would need to be two constitutional make-ups, one for cigarette smokers and another for pipe or cigar smokers. Finally, lung cancer mortality is reduced in those who quit smoking, compared to those who continue smoking. To explain this fact, the constitutional factor must decrease with age, allowing one to stop smoking, and, at the same time to experience decreasing risk from lung cancer unrelated to smoking. Cornfield et al. concluded: "No one of these considerations is perhaps sufficient by itself to counter the constitutional hypothesis, *ad hoc* modification of which can accommodate each additional piece of evidence. A point is reached, however, when a continuously modified hypothesis becomes difficult to entertain seriously."

Thus, Cornfield (1951) showed how to estimate measures of disease risk in the underlying population from case-control data, and he warned against sources of bias. In subsequent work (Cornfield, 1956), he developed asymptotic methods of inference for the relative risk. Motivated by scientific and public health concerns, he and his colleagues (Cornfield et al., 1959) considered many alternative explanations for the well-established association between smoking and lung cancer risk and concluded that the consistency of the epidemiologic and experimental evidence "supports the conclusion of a causal relationship with cigarette smoking, while there are serious inconsistencies in reconciling the evidence with other hypotheses which have been advanced." The inequality he developed is an important tool for evaluating the possibility that an association results from unmeasured confounders, and his other results on properties of the relative risk argue for its use in etiologic studies.

References

Benichou, J. and Gail, M.H. (1995). Methods of inference for estimates of absolute risk derived from population-based case-control studies. *Biometrics*, **51**, 182–194.

Benichou, J. and Wacholder, S. (1994). A comparison of three approaches to estimate exposure-specific incidence rates from population-based case-control data. *Statist. Medicine*, **13**, 651–661.

Breslow, N.E. and Day, N.E. (1980). *Statistical Methods in Cancer Research*, Volume 1. *The Analysis of Case-Control Studies*. International Agency for Research on Cancer, Lyon, France.

Cornfield, J. (1951). A method of estimating comparative rates from clinical data. Applications to cancer of the lung, breast and cervix. *J. National Cancer Inst.*, **11**, 1269–1275.

Cornfield, J. (1956). A statistical problem arising from retrospective studies. In J. Neyman (ed.), *Third Berkeley Symposium*, Volume 4, pp. 135–148. University of California Press, Berkeley, CA.

Doll, R. and Hill, A.B. (1950). Smoking and carcinoma of the lung: Preliminary report. *British Medical J.*, **2**, 739–748.

Doll, R. and Hill, A.B. (1952). A study of the aetiology of carcinoma of the lung. *British Medical J.*, **2**, 1271–1286.

Fisher, R.A. (1957). Dangers of cigarette smoking. *British Medical J.*, **2**, 297–298.

Fisher, R.A. (1958). Lung cancer and cigarettes? *Nature*, **182**, 108.

Gail, M.H. (1996). Statistics in action. *J. Amer. Statist. Assoc.* (in press).

Greenland, S. (1987). Estimation of exposure-specific rates from case-control data. *J. Chronic Diseases*, **40**, 1087–1094.

Lane-Claypon, J.E. (1926). A further report on cancer of the breast. *Reports on Public Health and Medical Subjects*, **32**, Ministry of Health, London.

Mantel, N. and Haenszel, W. (1959). Statistical aspects of the analysis of data from retrospective studies of disease. *J. National Cancer Inst.*, **22**, 719–748.

Metha, C.R., Patel, N.R., and Gray, R. (1985). On computing an exact confidence interval for the common odds ratio in several 2×2 contingency tables. *J. Amer. Statist. Assoc.*, **80**, 969–973.

Schlesselman, J.J. (1982). *Case-Control Studies: Design, Conduct, Analysis*. Oxford University Press, New York.

A Method of Estimating Comparative Rates from Clinical Data. Applications to Cancer of the Lung, Breast, and Cervix

Jerome Cornfield
National Cancer Institute, National Institutes of Health,
U.S. Public Health Service, Bethesda, MD

A frequent problem in epidemiological research is the attempt to determine whether the probability of having or incurring a stated disease, such as cancer of the lung, during a specified interval of time is related to the possession of a certain characteristic, such as smoking. In principle, such a question offers no difficulty. One selects representative groups of persons having and not having the characteristic and determines the percentage in each group who have or develop the disease during this time period. This yields a true rate. The difference in the magnitudes of the rates for those possessing and lacking the characteristic indicates the strength of the association. If it were true, for example, that a very large percentage of cigarette smokers eventually contracted lung cancer, this would suggest the possibility that tobacco is a strong carcinogen.

An investigation that involves selecting representative groups of those having and not having a characteristic is expensive and time consuming, however, and is rarely if ever used. Actual practice in the field is to take two groups presumed to be representative of persons who do and do not have the disease and determine the percentage in each group who have the characteristic. Thus rather than determine the percentage of smokers and nonsmokers who have cancer of the lung, one determines the percentage of persons with and without cancer of the lung who are smokers. This yields, not a true rate, but rather what is usually referred to as a relative frequency. Relative frequencies can be computed with comparative ease from hospital or other clinical records, and in consequence most investigations based on clinical records yield nothing but relative frequencies. The difference in the magnitudes of the relative frequencies does not indicate the strength of the association, however. Even if it were true that there were many more smokers among those with lung cancer than among those without it, this would not by itself suggest whether tobacco was a weak or a strong carcinogen. We are consequently interested in whether it is possible to deduce the rates from knowledge of the relative frequencies.

A General Method

To fix our ideas we may illustrate how the general problem can be attacked with some data recently published by Schrek, Baker, Ballard, and Dolgoff [1]. They report that 77 percent of the white males studied, aged 40–49, with cancer of the lung, smoked 10 or more cigarettes per day, while only 58 percent of a group of white males, aged 40–49, presumed to be representative of the non-lung-cancer population, smoked that much. Can we estimate from these data the frequency with which cancer of the lung occurs among smokers and nonsmokers?

Denote by p_1 $(= 0.77)$ the proportion of smokers among those with cancer of the lung, by p_2 $(= 0.58)$ the proportion of smokers among those without cancer of the lung, and by X the proportion of the general population that has cancer of the lung during a specified period of time. We may then summarize the relevant information for the general population in a two-by-two table showing the proportion of the population falling in each of the four possible categories.

Characteristic	Having cancer of the lung	Not having cancer of the lung
Smokers	p_1X	$p_2(1-X)$
Nonsmokers	$(1-p_1)X$	$(1-p_2)(1-X)$
Total	X	$1-X$

One can now compute that the percentage of the general population that smokes is $p_2 + X(p_1 - p_2)$, that the proportion of smokers having cancer of the lung is

$$p_1X/[(p_2 + X(p_1 - p_2)]. \tag{1}$$

Similarly, the proportion of nonsmokers having cancer of the lung is

$$(1 - p_1)X/[(1 - p_2) - X(p_1 - p_2)]. \tag{2}$$

Formulas (1) and (2) yield the true rates we seek.

Given the appropriate data, formulas (1) and (2) are easy to complete. They are somewhat cumbersome algebraically, however. The following approximation to the true rates, therefore, seems useful. If the proportion of the general population having cancer of the lung, X, is small relative to both the proportion of the control group smoking and not smoking, p_2 and $1 - p_2$, the contribution of the term $X(p_1 - p_2)$ to the denominator of formulas (1) and (2) is trivial and may be neglected. In that case the approximate rate of cancer of the lung among smokers becomes $\frac{p_1 X}{p_2}$ and the corresponding rate for nonsmokers $\frac{(1 - p_1)X}{1 - p_2}$. Whenever $p_1 - p_2$ is greater than zero, p_1/p_2 is greater than unity. We may conclude from the approximation, therefore, that whenever a greater proportion of the diseased than of the control group possess a characteristic, the incidence of the disease is always higher among those possessing the characteristic. This is the intuition on which the procedures used in such clinic studies are based. Although it has frequently been questioned, it can now be easily seen to be correct.

It also follows from this analysis, however, that if one knows X, the prevalence of cancer of the lung in the general population, one can compute its prevalence among the smoking and nonsmoking population. Hospital or clinical records usually cannot furnish an estimate of X, however, since one seldom knows the size of the population exposed to risk from which the actual cases are drawn. Its value is frequently known, at least approximately, from other sources. Thus, we have estimated from Dorn's data [2] that the annual prevalence of cancer of the lung among all white males aged 40–49 is 15.5 per 100,000.* X consequently is equal to 0.155×10^{-3}. We may now construct a table showing the proportion of the population in each of the four categories from the data of Schrek et al.

	Having cancer of the lung	Not having cancer of the lung	Total
Smokers	0.119×10^{-3}	0.579910	0.580029
Nonsmokers	0.036×10^{-3}	0.419935	0.419971
Total	0.155×10^{-2}	0.999845	1.000000

The proportion of smokers who have cancer of the lung using formulas (1) and (2) is thus 0.205×10^{-3} as contrasted with 0.086×10^{-3} for nonsmokers. The corresponding rates are 20.5 and 8.6 per 100,000 per year. These rates clearly provide a sounder basis for appraising the effect of cigarette smoking

* Dorn's published data show an annual prevalence rate in the period 1937–1939 of 29.7 per 100,000 for cancer of all respiratory organs among white and colored males, aged 40–49. In the North 52.1 percent of the respiratory cases in all age groups for both males and females was accounted for by lung cancer. The estimate of 15.5 ($= 29.7 \times 0.521$), is consequently somewhat rough.

than does the knowledge that 77 percent of those with cancer of the lung and 58 percent without it smoke.

If one is interested only in knowing the relative amount by which the prevalence of the disease is augmented by the possession of the attribute, one may calculate this without knowledge of X, since the ratio of the two rates is $\frac{p_1}{p_2}\frac{(1-p_2)}{(1-p_1)}$ when X is small. One can thus conclude from the Schrek data alone that the prevalence of cancer of the lung among white males aged 40–49 is 2.4 times as high among those who smoke 10 or more cigarettes a day as among those who do not.

The more extensive, but age-standardized, data of Levin, Goldstein, and Gerhardt [3] on the same subject may be used to illustrate the same calculation. They show that 66.1 percent of all (presumably white) males at all age groups who had cancer of the lung smoked some cigarettes as compared with 44.1 percent smoking among the control group. Setting $0.661 = p_1$ and $0.441 = p_2$, we have $\frac{p_1(1-p_2)}{p_2(1-p_1)} = 2.5$. The prevalence of lung cancer, according to these data is 2.5 times as high among cigarette smokers as among nonsmokers. (The agreement with the Schrek data is closer than would be expected in view of differences in the population covered, definitions used, and number of cases studied. The application of the present method to other studies of lung cancer and tobacco yields much more divergent results.)

The calculations may also be applied to multiple classifications such as the data on cancer of the cervix in Cardiff, Wales, recently published by Maliphant [4]. In Table 1, the first column gives the percent distribution of women who develop cancer of the cervix by marital status and number of children borne, while the second column shows the same distribution for all

Table 1. Distribution of Women with and without Cervical Cancer by Marital Status and Number of Children.

	Women contracting cancer of the cervix, 100 p_1	All women, 100 p_2	Incidence rate per 100,000, p_1X/p_2
Unmarried	1.3	10.5	9.9
Married:			
No children	5.0	13.0	30.7
1 or more children, total	93.7	76.5	97.6
1 child	13.3	15.3	69.3
2 children	18.3	17.0	85.8
3 children	15.0	13.0	92.0
4 children	11.0	9.6	91.3
5 children	9.2	6.4	114.6
6 or more	26.9	15.2	141.6
Total	100.0	100.0	

women. Women under 40 have been excluded. From other data given by Maliphant we have estimated that the incidence rate of cervical cancer for women over 40 in Cardiff was 79.7 per 100,000 (somewhat below the corresponding rate in this country.) This yields X and we accordingly have been able to calculate the incidence rates by marital status and number of children shown in the third column. The relation between cervical cancer and number of children born is obviously shown more clearly and usefully by the rates in the third column than by the relative frequencies in the first two.

Tests of Significance on the Computed Rate

Since most clinical studies are based on limited numbers of cases, it is of some importance to be able to estimate the limits of error of rates calculated according to this procedure. The approximate formula for the variance of a ratio sometimes used is inappropriate for this purpose, since it will sometimes show p_2X/p_1 differing significantly from X when a test on the difference $p_2 - p_1$ shows that it does not differ significantly from zero. To avoid this we employ a test of Fieller's [5]. Thus, writing the computed prevalence rate as $p_1X/p_2 = r$ and denoting by:

n_1 = the number of disease cases,
n_2 = the number of control cases,
t_α = the value of t in the normal curve corresponding to the 100α-percent probability level,
pq = the unbiased estimate of the unknown population value PQ

$$\left(= \frac{n_1p_1 + n_2p_2}{n_1 + n_2 - 1}\left[1 - \frac{n_1p_1 + n_2p_2}{n_1 + n_2}\right]\right)$$

the upper and lower confidence limits for the 100α-percent probability level of the estimate r are given by

$$\frac{r \pm \dfrac{t_\alpha X}{p_2}\sqrt{\dfrac{pq}{n_1}}\left[1 + \dfrac{1}{n_2p_2^2}(n_1p_1^2 - t_\alpha^2 pq)\right]^{1/2}}{1 - \dfrac{t_\alpha^2 pq}{n_2p_2^2}},$$

when X is considered free from sampling error. We may use the Schrek data to illustrate the use of this formula. Thus, letting $n_1 = 35$, $n_2 = 171$, setting $t_\alpha = 2$, and using p, X, and r as previously calculated, we compute the upper limit to the rate as 25.6 per 100,000 and the lower limit as 16.1 per 100,000. Since the value of X used, 15.5 per 100,000, falls outside these limits, we conclude that the rates for smokers and nonsmokers differ significantly at the 5 percent probability level. Whenever p_1 and p_2 differ significantly at the 100α-percent level, the limits computed in this fashion will not include X, and *vice versa*. Thus, if one simply wishes to test significance, it is sufficient to test the difference between p_1 and p_2. If one wishes to express error limits in the same units that the prevalence rate is expressed, however, one must use the formula given.*

Pitfalls

Our major purpose in preparing this note has been to show that any set of data that furnishes estimates of relative frequencies can be used to obtain estimates of rates. The procedure suggested, however, has assumed that the diseased and control groups used are representative of these same groups in the general population. If this assumption is not satisfied, then neither the rates, the relative frequencies, nor any other statistics calculated from the data will have applicability beyond the particular group studied.

We may illustrate the difficulties that can arise on this score with two examples. The first relates to Lane-Claypon's study of cancer of the breast [6]. In this study a detailed questionnaire was filled in for 508 women with breast cancer and 509 control women, who were being treated by the co-operating hospitals for "some trouble, other than cancer." We reproduce in Table 2 the percent distribution by number of children ever borne for each group. Only women having passed the menopause are included. We do not know X, the prevalence rate of breast cancer in the United Kingdom at the time the data were collected, and have therefore confined ourselves to computing relative prevalence.

If the data are to be taken at their face value, one must conclude that lowered prevalence of breast cancer is associated with increasing numbers of children. Greenwood in an analysis of Lane-Claypon's data [6] in fact concludes, "we think then that an etiological factor of importance has now been fully demonstrated." At the very beginning of his analysis, however, he points out, without attaching any significance to it, that the control group had borne an average of about 25 percent more children than had all women in England and Wales with the same duration of marriage. This would appear to provide definite evidence for the unrepresentative character of the control group and to cast doubt on the adequacy of the evidence.

* The procedure discussed in the text yields a two-sided test of significance; i.e., it tests the hypothesis that the rate for smokers is significantly different from that for nonsmokers. It would be more realistic to use a one-sided test; i.e., test the hypothesis that the rate for smokers is significantly *higher* than that for nonsmokers. To do this one uses the same formula but calculates only a lower limit, using a value of t_α appropriate to the one-sided test. Thus, for $\alpha = 0.05$, $t_\alpha = 1.645$.

In testing whether p_1 and p_2 are drawn from the same population it is appropriate to compute a pooled variance as has been done. When the results of such a test of significance suggest that p_1 and p_2 could not have been drawn from the same population, however, the use of a pooled variance to compute error limits is no longer correct. In fact, exact confidence limits can no longer be calculated for this case. The results yielded by the formula will nevertheless be sufficiently accurate for most practical purposes.

Table 2. Distribution of Women with and without Breast Cancer by Marital Status and Number of Children.

Characteristic	Cancer group, p_1	Control group, p_2	$\frac{p_1}{p_2}$	Relative prevalence
Unmarried	20.91	16.42	1.273	100
Married:				
No children	14.55	10.45	1.392	109
1 to 3 children	29.09	24.78	1.174	92
4 to 6 children	21.21	22.39	0.947	74
7 or more	14.24	25.97	0.548	43
Total	100.00	100.00		

The basic difficulty in this example is the unrepresentative nature of the control group. Since there is always some doubt whether or not a control group selected from among hospital patients can provide an accurate estimate of the frequency of a characteristic in the population at large, the difficulty may be quite general. The possibility that the diseased group is not representative either, cannot be entirely disregarded, however. We reproduce in Table 3 the distribution by age of 413 patients with adenocarcinoma

Table 3. Actual and Expected Distribution of Breast Cancer Cases by Age—Ellis Fischel State Cancer Hospital.

Age	Number of breast cancer cases[1]: Reported (1)	Number of breast cancer cases[1]: Expected (2) $(3)\frac{\text{Tot.}(1)}{\text{Tot.}(3)}$	(3) (4) × (5)	Incidence of breast cancer per 100,000[2] (4)	Percent distribution Missouri population 1940[3] (5)
Less than 30	3	7	1.1	2.2	0.4825
30–34	10	13	1.9	25.1	0.0795
35–39	24	25	3.7	50.7	0.0774
40–44	26	41	6.1	91.1	0.0734
45–49	40	54	8.0	122.9	0.0674
50–54	51	51	7.5	129.0	0.0648
55–59	54	57	8.4	169.6	0.0580
60–64	54	53	7.8	190.4	0.0495
65–69	59	45	6.6	193.3	0.0410
70–74	48	34	5.1	205.5	0.0344
75 and over	44	33	4.9	184.5	0.0516
Total	413	413	61.1		1.0000

[1] Chi square for difference = 24.5, $P < 0.01$.
[2] As estimated by Dorn [2].
[3] U.S. Bureau of the Census, Population, vol. II, pt. 4, Table 7.

of the breast admitted to the Ellis Fischel State Cancer Hospital in the years 1940–46 as given by Ackerman and Regato [7]. For comparison we give the expected distribution on the basis of known incidence rates by age.

It is obvious from inspection that an excess number in the older age groups were encountered, and that to some extent the hospital was functioning as a home for the aged. An epidemiological investigation the results of which would be sensitive to the age distribution of the persons studied might consequently be adversely affected.

Any set of hospital or clinical data that is worth analyzing at all is worth analyzing properly. It is from this point of view that the technique proposed seems useful. The preceding two examples suggest, however, that the results of even the most carefully analyzed set of such data may be open to question, and that these doubts can be resolved only by methods of data collection that provide representative samples of diseased and nondiseased persons.

References

[1] Schrek, R., Baker, L.A., Ballard, G.P., and Dolgoff, S. Tobacco smoking as an etiologic factor in disease. I. Cancer. *Cancer Res.*, **10**, 49–58, 1950.

[2] Dorn, H.F. Illness from cancer in the United States. Public Health Report 59, Nos. 2, 3, and 4, 1944.

[3] Levin, M.L., Goldstein, H., and Gerhardt, P.R. Cancer and tobacco smoking. *J.A.M.A.*, **143**, 336–338, 1950.

[4] Maliphant, R.G. The incidence of cancer of the uterine cervix. *British Medical J.*, **I**, 978–982, 1949.

[5] Fieller, E.C. The biological standardization of insulin. *Supp. J. Roy. Statist. Soc.*, **7**, 1–64, 1940.

[6] Lane-Claypon, J.E. A further report on cancer of the breast. Reports on Public Health and Medical Subjects, No 32, Ministry of Health, London, 1926.

[7] Ackerman, L.V. and del Regato, J.A. *Cancer, Diagnosis, Treatment and Prognosis*. C.V. Mosby, St. Louis, 1947, p. 927.

[8] Cornfield, J., Haenszel, W., Hammond, E.C., Lilienfeld, A.M., Shimkin, M.B., and Wynder, E.L. (1959). Smoking and lung cancer: Recent evidence and a discussion of some questions. *J. National Cancer Inst.*, **22**, 173–203.

Introduction to

Metropolis, Rosenbluth, Rosenbluth, Teller, and Teller (1953) Equations of State Calculations by Fast Computing Machines. *J. Chem. Phys.*, **21**, 1087–1092. and Geman and Geman (1984) Stochastic Relaxation, Gibbs Distributions, and the Bayesian Restoration of Images. *IEEE Trans. Pattern Anal. Machine Intelligence*, **6**, 721–741.*

Peter J. Huber
Universität Bayreuth, Germany

The breakthrough to be discussed here occurred in two steps, separated in time by over three decades. The first, conceptually decisive step, but lacking direct relevance to mainstream statistics, was the invention of Markov Chain Monte Carlo methods by Metropolis et al. (1953). The second step, decisive for applications in statistics, occurred when Geman and Geman (1984) wrestled the seminal idea of Metropolis et al. from its statistical mechanics surroundings, modified it, and applied it to Bayesian modeling and the computation of posterior distributions in otherwise intractable situations.

The 1953 paper by Nicholas Metropolis and two husband-and-wife teams, Arianna and Marshall Rosenbluth, and Augusta and Edward Teller, was ground-breaking for equation of state calculations in statistical mechanics. Assume you are given a mechanical system with a large but finite set S of possible states, $s \in S$ having energy E_s. If this system forms a so-called Gibbs ensemble, that is, if it is in thermal equilibrium with its surroundings, then it obeys a Boltzmann distribution (also called Gibbs distribution): state s has probability

$$p_s = \exp\left(-\frac{E_s}{kT}\right) \Big/ Z, \qquad (1)$$

* Regrettably, permission to reprint this paper could not be obtained.

where T is absolute temperature, k is the Boltzmann constant, and

$$Z = \sum_s \exp\left(-\frac{E_s}{kT}\right)$$

normalizes the sum to 1. Typically there are too many states in S for brute force calculations of expectations and other functionals, and naive Monte Carlo sampling (i.e., choosing states randomly, with equal probability) does not work either, since for the temperatures of interest the probability measure (1) is concentrated on a very small corner of the state space. Metropolis et al. however noticed that (1) can be interpreted as the limiting distribution of certain Markov chains, and that this permitted more efficient sampling schemes.

Assume that π_{rs}, is the transition matrix of *any* symmetric and irreducible random walk on S, that is, $\pi_{rs} = \pi_{sr}$, and every state can ultimately be reached from any other state. This random walk now is modified as follows. Assume the system is in state r and the random walk model tells you to perform a random move to state s. If $E_s \leq E_r$, the move is performed unconditionally. If $E_s > E_r$, the system either stays in r or moves to s, the move being performed with conditional probability

$$\exp\left(-\frac{E_s - E_r}{kT}\right). \tag{2}$$

The resulting Markov process is aperiodic and has (1) as its unique stationary distribution. Thus, such a Markov chain can be used to calculate ensemble averages

$$\langle f \rangle = \sum f(s) p_s \tag{3}$$

of state functions by forming time averages over sufficiently long Monte Carlo simulations.

Of course, this is interesting only if the following two conditions hold:

(a) S is so large that a direct evaluation of (1) and (3) is out of the question.
(b) The energy function E_s and the random walk are such that for pairs of states with $\pi_{rs} > 0$ the energy difference $E_s - E_r$ is easy to calculate.

Now, what is new in the paper by Geman and Geman (a team of brothers in this case)? First, there is an infusion of ideas from the theory of Markov Random Fields (MRF), which had been developed by Dobrushin, Spitzer and others in the late 1960s and early 1970s.

A finite MRF can be defined as follows. We are given a finite set Λ of sites (say a square grid of pixels) with a given neighborhood structure (the neighborhood of a pixel might be formed by the eight surrounding pixels with coordinates differing at most by 1). A state s then is a collection of values $\{s(i), i \in \Lambda\}$ (say gray values, integers in the range $0 \leq s(i) < 64$). Note that for a fixed temperature any arbitrary probability distribution on the set S of possible states can be written in the form (1). Such a distribution is an MRF if the conditional distribution of $s(i)$, given the values $s(j)$ *at all* sites $j \neq i$, is the same as when we condition only on the *neighboring* sites. While Metropolis et al. in principle permitted arbitrary transitions $r \rightarrow s$ in their random walk model, it becomes clear from the MRF property that in order to satisfy condition (b) the transitions ought to be local in the sense of the neighborhood system on Λ. This holds in particular if the state transitions are done by changing the value $s(i)$ at a single site. If the distribution (1) is to remain stationary under such changes, then we must make sure that the new random value at i still has the same conditional distribution, given the values on the sites $j \neq i$. That is, we should sample from the conditional distribution. But the latter can be calculated easily thanks to the MRF property. This idea almost automatically leads to what has been called the Gibbs sampler by Geman and Geman; its formal description is given by the conditional probability formulas (4.8), (4.9), and Sections XI and XII of the paper.

Second, with the increased availability of inexpensive computing power, another obvious application of the original Metropolis idea had become feasible: Optimization by simulated annealing. See Kirkpatrick et al. (1983) for the published version of the still relevant survey cited by Geman and Geman. If we drive the temperature in (1) to zero, the Boltzmann distribution will concentrate on the states with the lowest energy. However, temperature must be lowered very slowly, because otherwise the system is likely to be trapped in a local minimum. Geman and Geman apparently were the first to prove rigorously that a logarithmic scheme, i.e., driving the temperature to 0 such that $T(n) \geq C/\log(n)$ in the nth step of the iteration, where C is a sufficiently large constant, will do the job. A logarithmic scheme is the best possible: if there is a genuine local minimum distinct from the global minimum, then it is straightforward to show by applying a Borel–Cantelli-type argument to (2) that an annealing scheme with $T(n) = c/\log(n)$, where $c > 0$ now is a suitably small constant, will cause the Markov chain to be trapped in this local minimum with nonzero probability.

In their paper, Geman and Geman join these strands together and give them an original new twist. They expand the lattice of pixel sites by adding edge sites. Then they reinterpret the statistical mechanics setup in a Bayesian framework. The prior distribution on this lattice is an MRF, lacking specific image information. Instead of energy minimization, the task is to find states (i.e., images) with maximal a posteriori probability, given the observed, degraded version of the picture.

In retrospect it is clear that image reconstruction is a most suggestive entry point into MRF and Markov chain simulation methods in statistical work, because, like the initial MRF work, it has to do with random processes on rectangular grids. Though, in my opinion, the image reconstruction problem should be considered as a mere paradigm—admittedly a good one, but very likely not the best possible application of the idea. The predicament with image reconstruction is that images usually come in droves, with ever-increasing resolution, and simulated annealing remains overly computer-intensive for routine use with high-volume data.

The importance and real breakthrough of the paper by Geman and Geman is that it exemplifies an extremely powerful method for solving hard problems in the general area of Bayesian modeling, thereby opening the conceptual door for an amazingly wide range of applications. In analogy to the original Metropolis method, the approach can and will be used for computing posterior distributions in cases where direct calculations are out of the question, and where naive simulations do not work because probabilities are inhomogeneous and highly concentrated. While in the Geman and Geman paper the statistical mechanics (Ising model) parentage of the idea still is clearly recognizable, it now has begun to fade away. I need not dwell on those applications in detail since an excellent and easily accessible survey has recently been given by Besag et al. (1995).

References

Besag, J., Green, P., Higdon, D., and Mengersen, K. (1995). Bayesian computations and stochastic systems. *Statist. Sci.*, **10**, 1–66.

Kirkpatrick, S., Gelatt, C.D., Jr., and Vecchi, M.P. (1983). Optimization by simulated annealing. *Science*, **220**, 671–680.

Equation of State Calculations by Fast Computing Machines

Nicholas Metropolis, Arianna W. Rosenbluth,
Marshall N. Rosenbluth, and Augusta H. Teller
Los Alamos Scientific Laboratory, Los Alamos, New Mexico
and
Edward Teller*
Department of Physics, University of Chicago,
Chicago, Illinois

Abstract

A general method, suitable for fast computing machines, for investigating such properties as equations of state for substances consisting of interacting individual molecules is described. The method consists of a modified Monte Carlo integration over configuration space. Results for the two-dimensional rigid-sphere system have been obtained on the Los Alamos MANIAC and are presented here. These results are compared to the free volume equation of state and to a four-term virial coefficient expansion.

I. Introduction

The purpose of this paper is to describe a general method, suitable for fast electronic computing machines, of calculating the properties of any substance which may be considered as composed of interacting individual molecules. Classical statistics is assumed, only two-body forces are considered, and the potential yield of a molecule is assumed spherically symmetric. These are the usual assumptions made in theories of liquids. Subject to the above assumptions, the method is not restricted to any range of temperature or density. This paper will also present results of a preliminary two-dimensional calculation for the rigid-sphere system. Work on the two-dimensional case with a Lennard–Jones potential is in progress and will be reported in a later paper. Also, the problem in three dimensions is being investigated.

* Now at the Radiation Laboratory of the University of California, Livermore, California.

II. The General Method for an Arbitrary Potential Between the Particles

In order to reduce the problem to a feasible size for numerical work, we can, of course, consider only a finite number of particles. This number N may be as high as several hundred. Our system consists of a square† containing N particles. In order to minimize the surface effects we suppose the complete substance to be periodic, consisting of many such squares, each square containing N particles in the same configuration. Thus we define $\bar{d}_{AB}$, *the minimum distance between particles A* and *B*, as the shortest distance between A and any of the particles B, of which there is one in each of the squares which comprise the complete substance. If we have a potential which falls off rapidly with distance, there will be at most one of the distances AB which can make a substantial contribution; hence we need consider only the minimum distance $\bar{d}_{AB}$.

Our method in this respect is similar to the cell method except that our cells contain several hundred particles instead of one. One would think that such a sample would be quite adequate for describing any one-phase system. We do find, however, that in two-phase systems the surface between the phases makes quite a perturbation. Also, statistical fluctuations may be sizable.

If we know the positions of the N particles in the square, we can easily calculate, for example, the potential energy of the system,

$$E = \frac{1}{2} \sum_{\substack{i=1 \\ i \neq j}}^{N} \sum_{j=1}^{N} V(\bar{d}_{ij}). \tag{1}$$

(Here V is the potential between molecules, and $\bar{d}_{ij}$ *is the minimum distance between particles i* and *j* as defined above.)

In order to calculate the properties of our system we use the canonical ensemble. So, to calculate the equilibrium value of any quantity of interest F,

$$\bar{F} = \left[\int F \exp\left(-\frac{E}{kT}\right) d^{2N}p d^{2N}q\right] \Big/ \left[\int \exp\left(-\frac{E}{kT}\right) d^{2N}p d^{2N}q\right], \tag{2}$$

where $(d^{2n}Pd^{2n}q)$ is a volume element in the $4N$-dimensional phase space. Moreover, since forces between particles are velocity-independent, the momentum integrals may be separated off, and we need perform only the integration over the $2N$-dimensional configuration space. It is evidently impractical to carry out a several hundred-dimensional integral by the usual

† We will use the two-dimensional nomenclature here since it is easier to visualize. The extension to three dimensions is obvious.

numerical methods, so we resort to the Monte Carlo method.‡ The Monte Carlo method for many-dimensional integrals consists simply of integrating over a random sampling of points instead of over a regular array of points.

Thus the most naive method of carrying out the integration would be to put each of the N particles at a random position in the square (this defines a random point in the $2N$-dimensional configuration space), then calculate the energy of the system according to (1), and give this configuration a weight $\exp(-E/kT)$. This method, however, is not practical for close-packed configurations, since with high probability we choose a configuration where $\exp(-E/kT)$ is very small; hence a configuration of very low weight. So the method we employ is actually a modified Monte Carlo scheme, where, instead of choosing configurations randomly, then weighting them with $\exp(-E/kT)$, we choose configurations with a probability $\exp(-E/kT)$ and weight them evenly.

This we do as follows: We place the N particles in any configuration, for example, in a regular lattice. Then we move each of the particles in succession according to the following prescription:

$$\begin{aligned} X &\to X + \alpha\xi_1, \\ Y &\to Y + \alpha\xi_2, \end{aligned} \tag{3}$$

where α is the maximum allowed displacement, which for the sake of this argument is arbitrary, and ξ_1 and ξ_2 are random numbers§ between (-1) and 1. Then, after we move a particle, it is equally likely to be anywere within a square of side 2α centered about its original position. (In accord with the periodicity assumption, if the indicated move would put the particle outside the square, this only means that it re-enters the square from the opposite side.)

We then calculate the change in energy of the system ΔE, which is caused by the move. If $\Delta E < 0$, i.e., if the move would bring the system to a state of lower energy, we allow the move and put the particle in its new position. If $\Delta E > 0$, we allow the move with probability $\exp(-\Delta E/kT)$; i.e., we take a random number ξ_3 between 0 and 1, and if $\xi_3 < \exp(-\Delta E/kT)$, we move the particle to its new position. If $\xi_3 > \exp(-E/kT)$, we return it to its old position. Then, whether the move has been allowed or not, i.e., whether we are in a different configuration or in the original configuration, we consider

‡ This method has been proposed independently by J.E. Mayer and by S. Ulam. Mayer suggested the method as a tool to deal with the problem of the liquid state, while Ulam proposed it as a procedure of general usefulness. B. Alder, J. Kirkwood, S. Frankel, and V. Lewinson discussed an application very similar to ours.

§ It might be mentioned that the random numbers that we used were generated by the middle square process. That is, if ξ^n is an m digit random number, then a new random number ξ_{n+1} is given as the middle m digits of the complete $2m$ digit square of ξ_n.

that we are in a new configuration for the purpose of taking our averages. So

$$\bar{F} = \frac{1}{M} \sum_{j=1}^{M} F_j, \tag{4}$$

where F_j is the value of the property F of the system after the jth move is carried out according to the complete prescription above. Having attempted to move a particle we proceed similarly with the next one.

We now prove that the method outlined above does choose configurations with a probability $\exp(-E/kT)$. Since a particle is allowed to move to any point within a square of side 2α with a finite probability, it is clear that a large enough number of moves will enable it to reach any point in the complete square.|| Since this is true of all particles, we may reach any point in configuration space. Hence, the method is ergodic.

Next consider a very large ensemble of systems. Suppose for simplicity that there are only a finite number of states¶ of the system, and that ν_r is the number of systems of the ensemble in state r. What we must prove is that after many moves the ensemble tends to a distribution

$$\nu_r \propto \exp(-E_r/kT).$$

Now let us make a move in all the systems of our ensemble. Let the *a priori* probability that the move will carry a system in state r to state s be P_{rs}. [By the *a priori* probability we mean the probability before discriminating on $\exp(-\Delta E/kT)$.] First, it is clear that $P_{rs} = P_{sr}$, since according to the way our game is played a particle is equally likely to be moved anywhere within a square of side 2α centered about its original position. Thus, if states r and s differ from each other only by the position of the particle moved and if these positions are within each other's squares, the transition probabilities are equal; otherwise they are zero. Assume $E_r > E_s$. Then the number of systems moving from state r to state s will be simply $\nu_r P_{rs}$, since all moves to a state of lower energy are allowed. The number moving from s to r will be $\nu_s P_{sr} \exp(-(E_r - E_s)/kT)$, since here we must weigh by the exponential factor. Thus the net number of systems moving from s to r is

$$P_{rs}(\nu_s \exp(-(E_r - E_s)/kT) - \nu_r). \tag{5}$$

So we see that between any two states r and s, if

$$(\nu_r/\nu_s) > [\exp(-E_r/kT)/\exp(-E_s/kT)], \tag{6}$$

on the average more systems move from state r to state s. We have seen already that the method is ergodic; i.e., that any state can be reached from

|| In practice it is, of course, not necessary to make enough moves to allow a particle to diffuse evenly throughout the system since configuration space is symmetric with respect to interchange of particles.

¶ A state here means a given point in configuration space.

any other, albeit in several moves. These two facts mean that our ensemble must approach the canonical distribution. It is, incidentally, clear from the above derivation that after a forbidden move we must count again the initial configuration. Not to do this would correspond in the above case to removing from the ensemble those sysems which tried to move from s to r and were forbidden. This would unjustifiably reduce the number in state s relative to r.

The above argument does not, of course, specify how rapidly the canonical distribution is approached. It may be mentioned in this connection that the maximum displacement α must be chosen with some care; if too large, most moves will be forbidden, and if too small, the configuration will not change enough. In either case it will then take longer to come to equilibrium.

For the rigid-sphere case, the game of chance on $\exp(-\Delta E/kT)$ is, of course, not necessary since ΔE is either zero or infinity. The particles are moved, one at a time, according to (3). If a sphere, after such a move, happens to overlap another sphere, we return it to its original position.

III. Specialization to Rigid Spheres in Two Dimensions

A. The Equation of State

The virial theorem of Clausius can be used to give an equation of state in terms of $\bar{n}$, the average density of other particles at the surface of a particle. Let $\mathbf{X}_i^{(\text{tot})}$ and $\mathbf{X}_i^{(\text{tot})}$ represent the total and the internal force, respectively, acting on particle i, at a position $\mathbf{r}_i$. Then the virial theorem can be written

$$\left\langle \sum_i \mathbf{X}_i^{(\text{tot})} \cdot \mathbf{r}_i \right\rangle_{\text{Av}} = 2PA + \left\langle \sum_i \mathbf{X}_i^{(\text{int})} \cdot \mathbf{r}_i \right\rangle_{\text{Av}} = 2E_{\text{kin}}. \qquad (7)$$

Here P is the pressure, A the area, and E_{kin} the total kinetic energy,

$$E^*_{\text{kin}} = Nm\bar{v}^2/2$$

of the system of N particles.

Consider the collisions of the spheres for convenience as represented by those of a particle of radius d_0, twice the radius of the actual spheres, surrounded by $\bar{n}$ point particles per unit area. Those surrounding particles in an area of $2\pi d_0 v \cos\phi\, \Delta t$, traveling with velocity v at an angle ϕ with the radius vector, collide with the central particle provided $|\phi| < \pi/2$. (See Figure 1.) Assuming elastic recoil, they each exert an average force during the time Δt on the central particle of

$$2mv \cos\phi/\Delta t.$$

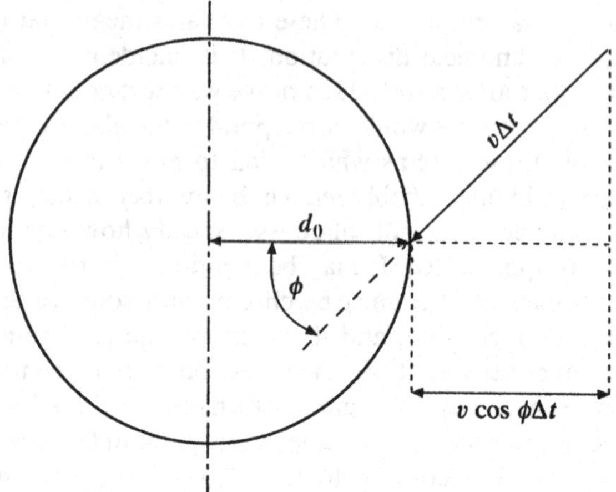

Figure 1. Collisions of rigid spheres.

One can see that all ϕ's are equally probable, since for any velocity-independent potential between particles the velocity distribution will just be Maxwellian, hence isotropic. The total force acting on the central particle, averaged over ϕ, over time, and over velocity, is

$$\bar{F}_i = m\bar{v}^2 \pi d_0 \bar{n}. \tag{8}$$

The sum

$$\left\langle \sum_i \mathbf{X}_i^{(\text{int})} \cdot r_i \right\rangle_{\text{Av}}$$

is

$$-\tfrac{1}{2} \sum_i \left\{ \sum_{\substack{j \\ i \neq j}} r_{ij} F_{ij} \right\},$$

with F_{ij} the magnitude of the force between two particles and r_{ij} the distance between them. We see that $r_{ij} = d_0$ and $\sum_j F_{ij}$ is given by (8), so we have

$$\left\langle \sum_i \mathbf{X}_i^{(\text{int})} \cdot \mathbf{r}_i \right\rangle_{\text{Av}} = -(Nm\bar{v}^2/2)\pi d_0^2 \bar{n}. \tag{9}$$

Substitution of (9) into (7) and replacement of $(N/2)m^2$ by E_{kin} gives finally

$$PA = E_{\text{kin}}(1 + \pi\, d_0^2 \bar{n}/2) \equiv NkT(1 + \pi\, d_0^2 \bar{n}/2). \tag{10}$$

This equation shows that a determination of the one quantity $\bar{n}$, according to (4) as a function of A, the area, is sufficient to determine the equation of state for the rigid spheres.

B. The Actual Calculation of $\bar{n}$

We set up the calculation on a system composed of $N = 224$ particles ($i = 0, 1 \ldots 223$) placed inside a square of unit side and unit area. The particles were arranged initially in a trigonal lattice of fourteen particles per row by 16 particles per column, alternate rows being displaced relative to each other as shown in Figure 2. This arrangement gives each particle six nearest neighbors at approximately equal distances of $d = 1/14$ from it.

Instead of performing the calculation for various areas A and for a fixed distance d_0, we shall solve the equivalent problem of leaving $A = 1$ fixed and changing d_0. We denote by A_0 the area the particles occupy in close-packed arrangement (see Figure 3). For numerical convenience we defined an auxiliary parameter ν, which we varied from zero to seven, and in terms of which the ratio (A/A_0) and the forbidden distance d_0 are defined as follows:

$$d_0 = d(1 - 2^{\nu-8}), \qquad d = (1/14), \tag{11a}$$

$$(A/A_0) = 1/(3^{1/2}d_0^2N/2) = 1/0.98974329(1 - 2^{\nu-8})^2. \tag{11b}$$

The unit cell is a parallelogram with interior angle 60°, side d_0, and altitude $3^{1/2}d_0/2$ in the close-packed system.

Every configuration reached by proceeding according to the method of the preceding section was analyzed in terms of a radial distribution function

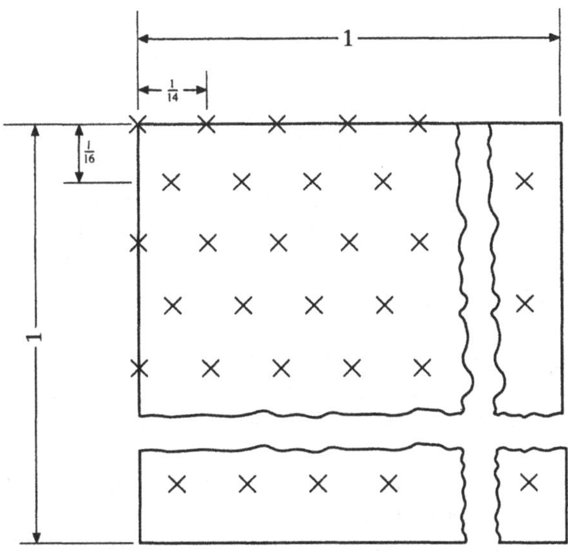

Figure 2. Initial trigonal lattice.

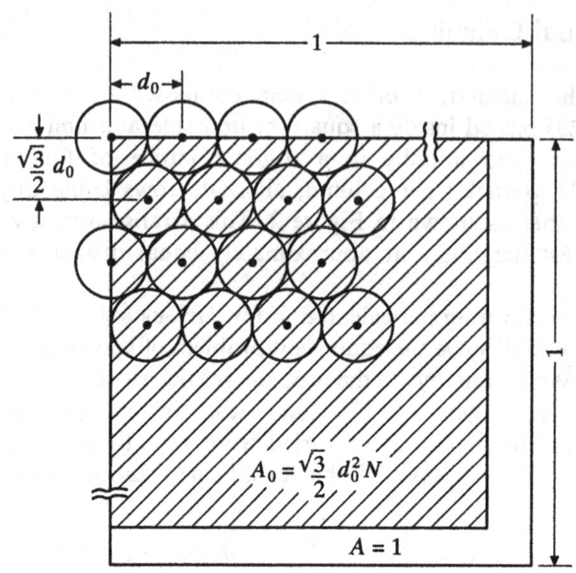

Figure 3. The close-packed arrangement for determining A_0.

$N(r^2)$. We chose a $K > 1$ for each ν and divided the area between πd_0^2 and $K^2 \pi d_0^2$ into 64 zones of equal area ΔA^2,

$$\Delta A^2 = (K^2 - 1)\pi d_0^2/64.$$

We then had the machine calculate for each configuration the number of pairs of particles N_m $(m = 1, 2, \ldots, 64)$ separated by distances r which satisfy

$$(m-1)\Delta A^2 + \pi d_0^2 < \pi r^2 \leq m\Delta A^2 + \pi d_0^2. \tag{12}$$

The N_m were averaged over successive configurations according to (4), and after every 16 cycles (a cycle consists of moving every particle once) were extrapolated back to $r^2 = d_0^2$ to obtain $N_{1/2}$. This $N_{1/2}$ differs from $\bar{n}$ in (10) by a constant factor depending on N and K.

The quantity K was chosen for each ν to give reasonable statistics for the N_m. It would, of course, have been possible by choosing fairly large K's, with perhaps a larger number of zones, to obtain $N(r^2)$ at large distances. The oscillatory behavior of $N(r^2)$ at large distances is of some interest. However, the time per cycle goes up fairly rapidly with K and with the number of zones in the distance analysis. For this reason only the behavior of $N(r^2)$ in the neighborhood of d_0^2 was investigated.

The maximum displacement α of (3) was set to $(d - d_0)$. About half the moves in a cycle were forbidden by this choice, and the initial approach to equilibrium from the regular lattice was fairly rapid.

IV. Numerical Results for Rigid Spheres in Two Dimensions

We first ran for something less than 16 cycles in order to get rid of the effects of the initial regular configuration on the averages. Then about 48 to 64 cycles were run at

$$\nu = 2,\ 4,\ 5,\ 5.5,\ 6,\ 6.25,\ 6.5, \text{ and } 7.$$

Also, a smaller amount of data was obtained at $\nu = 0, 1$, and 3. The time per cycle on the Los Alamos MANIAC is approximately 3 minutes, and a given point on the pressure curve was obtained in 4 to 5 hours of running. Figure 4 shows $(PA/NkT) - 1$ versus $(A/A_0) - 1$ on a log–log scale from our results (curve A), compared to the free volume equation of Wood* (curve B) and to the curve given by the first four virial coefficients (curve C).

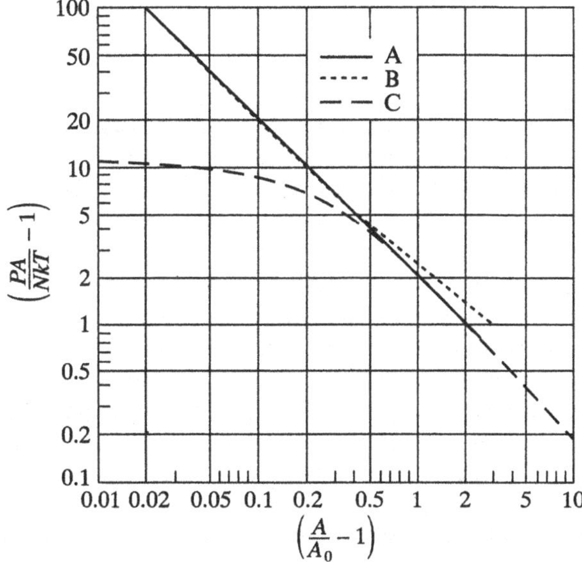

Figure 4. A plot of $(PA/NkT) - 1$ versus $(A/A_0) - 1$. Curve A (Solid line) gives the results of this paper. Curves B and C (dashed and dot–dashed lines) give the results of the free volume theory and of the first four virial coefficients, respectively.

* William W. Wood. *J. Chem. Phys.*, 20, 1334 (1952).

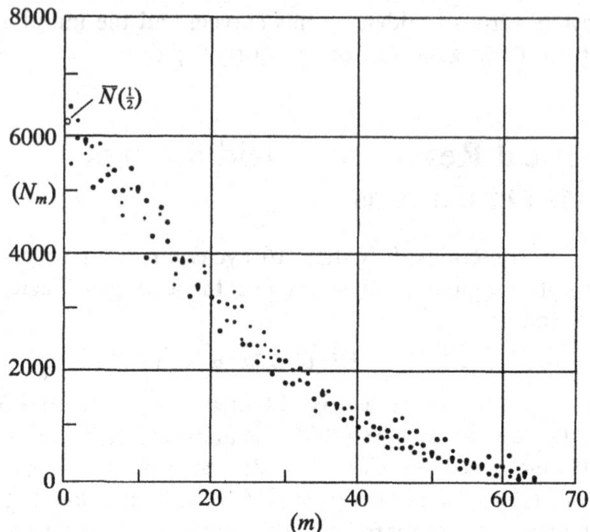

Figure 5. The radial distribution function N_m for $\nu = 5$, $(A/A_0) = 1.31966$, $K = 1.5$. The average of the extrapolated values of $N_{1/2}$ in $\bar{N}_{1/2} = 6301$. The resultant value of $(PA/NkT) - 1$ is $64\bar{N}_{1/2}/N^2(K^2 - 1)$ or 6.43. Values after 16 cycles, ●; after 32, ×; and after 48, ○.

The last two virial coefficients were obtained by straightforward Monte Carlo integration on the MANIAC (see Section V). It is seen that the agreement between curves A and B at small areas and between curves A and C at large areas is good. Deviation from the free volume theory begins with a fairly sudden break at $\nu = 6(A/A_0 \simeq 1.8)$.

A sample plot of the radial distribution function for $\nu = 5$ is given in Figure 5. The various types of points represent values after 16, 32, and 48 cycles. For $\nu = 5$, least-square fits with a straight line to the first 16 N_m values were made, giving extrapolated values of $N_{1/2}^{(1)} = 6367$, $N_{1/2}^{(2)} = 6160$, and $N_{1/2}^{(3)} = 6377$. The average of these three was used in constucting PA/NkT. In general, least-square fits of the first sixteen to twenty N_m's by means of a parabola, or, where it seemed suitable, a straight line, were made. The errors indicated in Figure 4 are the root-mean-square deviations for the three or four $N_{1/2}$ values. Our average error seemed to be about 3 percent.

Table I gives the results of our calculations in numerical form. The columns are ν, A/A_0, $(PA/NkT) - 1$, and, for comparison purposes, $(PA/NkT - 1)$ for the free volume theory and for the first four coefficients in the virial coefficient expansion, in that order, and finally PA_0/NkT from our results.

Table I. Results of this Calculation for $(PA/NkT) - 1 = X_1$ Compared to the Free Volume Theory (X_2) and the Four-Term Virial Expansion (X_3). Also (PA_0/NkT) from Our Calculations.

v	(A/A_0)	X_1	X_2	X_3	(PA_0/NkT)
2	1.04269	49.17	47.35	9.77	48.11
4	1.14957	13.95	13.85	7.55	13.01
5	1.31966	6.43	6.72	5.35	5.63
5.5	1.4909	4.41	4.53	4.02	3.63
6	1.7962	2.929	2.939	2.680	2.187
6.25	2.04616	2.186	2.323	2.065	1.557
6.5	2.41751	1.486	1.802	1.514	1.028
7	4.04145	0.6766	0.990	0.667	0.4149

V. The Virial Coefficient Expansion

One can show* that

$$\begin{aligned}(PA/NkT) - 1 &= C_1(A_0/A) + C_2(A_0/A)^2 \\ &\quad + C_3(A_0/A)^3 + C_4(A_0/A)^4 + 0(A_0/A)^5, \\ C_1 &= \pi/3^{1/2}, \qquad C_2 = 4\pi^2 A_{3,3}/9, \\ C_3 &= \pi^3(6A_{4,5} - 3A_{4,4} - A_{4,6})/3^{1/2}, \\ C_4 &= (8\pi^3/135)\cdot[12A_{5,5} - 60A'_{5,6} - 10A''_{5,6} \\ &\quad + 30A'_{5,7} + 60A''_{5,7} + 10A'''_{5,7} - 30A'_{5,8} \\ &\quad - 15A''_{5,8} + 10A_{5,9} - A_{5,10}].\end{aligned} \tag{13}$$

The coefficients $A_{i,k}$ are cluster integrals over configuration space of i particles, with k bonds between them. In our problem a bond is established if the two particles overlap. The cluster integral is the volume of configuration space for which the appropriate bonds are established. If k bonds can be distributed over the i particles in two or more different ways without destroying the irreducibility of the integrals, the separate cases are distinguished by primes. For example, A_{33} is given schematically by the diagram

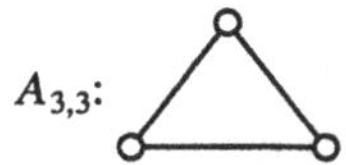

* J.E. Mayer and M.G. Mayer, *Statistical Mechanics* (Wiley, New York, 1940), pp. 277–291.

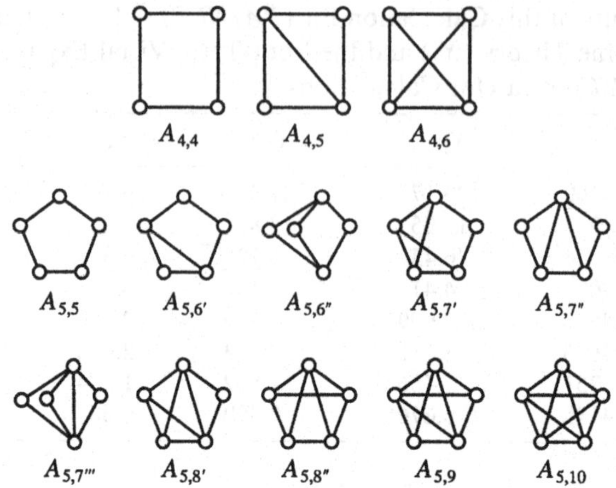

Figure 6. Schematic diagrams for the various area integrals.

and mathematically as follows: if we define $f(r_{ij})$ by

$$f(r_{ij}) = 1 \quad \text{if} \quad r_{ij} < d,$$
$$f(r_{ij}) = 0 \quad \text{if} \quad r_{ij} > d,$$

then

$$A_{3,3} = \frac{1}{\pi^2 d^4} \int \cdots \int dx_1\, dx_2\, dx_3\, dy_1\, dy_2\, dy_3 (f_{12} f_{23} f_{31}).$$

The schematics for the remaining integrals are indicated in Figure 6.

The coefficients $A_{3,3}$, $A_{4,4}$, and $A_{4,5}$ were calculated algebraically, the remainder numerically by Monte Carlo integration. That is, for $A_{5,5}$ for example, particle 1 was placed at the origin, and particles 2, 3, 4, and 5 were put down at random, subject to $f_{12} = f_{23} = f_{13} = f_{15} = 1$. The number of trials for which $f_{45} = 1$, divided by the total number of trials, is just $A_{5,5}$.

The data on $A_{4,5}$ is quite reliable. We obtained

$$A_{4,6}/A_{4,4} = 0.752(\pm 0.002).$$

However, because of the relatively large positive and negative terms in C_4 of (13), the coefficient C_4, being a small difference, is less accurate. We obtained

$$C_4 = 8\pi^3(0.585)/135 \quad (\pm \sim 5 \text{ percent}).$$

Our final formula is

$$\begin{aligned}(PA/NkT) - 1 = 1.813799(A_0/A) \\ + 2.57269(A_0/A)^2 + 3.179(A_0/A)^3 \\ + 3.38(A_0/A)^4 + 0(A_0/A)^5.\end{aligned} \tag{14}$$

This formula is plotted in curve C of Figure 4 and tabulated for some values of (A/A_0) in column 5 of Table I. It is seen in Figure 4 that the curves agrees very well with our calculated equation of state for $(A/A_0) > 2.5$. In this region both the possible error in our last virial coefficients and the contribution of succeeding terms in the expansion are quite small (less than our probable statistical error) so that the virial expansion should be accurate.

VI. Conclusion

The method of Monte Carlo integrations over configuration space seems to be a feasible approach to statistical mechanical problems which are as yet not analytically soluble. At least for a single-phase system a sample of several hundred particles seems sufficient. In the case of two-dimensional rigid spheres, runs made with 56 particles and with 224 particles agreed within statistical error. For a computing time of a few hours with presently available electronic computers, it seems possible to obtain the pressure for a given volume and temperature to an accuracy of a few percent.

In the case of two-dimensional rigid spheres our results are in agreement with the free volume approximstion for $A/A_0 < 1.8$ and with a five-term virial expansion for $A/A_0 > 2.5$. There is no indication of a phase transition.

Work is now in progress for a system of particles with Lennard–Jones-type interactions and for three-dimensional rigid spheres.

Introduction to Whittle (1953) The Analysis of Multiple Stationary Time Series

Matthew Calder
Colorado State University

Richard A. Davis
Colorado State University

1. Introduction

During the 1940s and 1950s the subject of statistical time series analysis matured from a scattered collection of methods into formal branch of statistics. Much of the mathematical groundwork was laid down and many of the practical methods were developed during this period. The foundations of the field were developed by individuals whose names are familiar to most statisticians: Yule, Wold, Wiener, and Kolmogorov. Interestingly, many of the methodological techniques that originated 40 to 50 years ago still play a principal role in modern time-series analysis. A key participant in this golden-age of time-series analysis was Peter Whittle.

Statisticians have looked at temporally sequential data, or time series, since the recording of numbers began. However, it was not until Yule developed the concept of the autoregressive sequence that the modeling of time series was considered to be of a substantially different character than other statistical modeling problems. Prior to that, time series were modeled as the sum of a deterministic function and a random noise component, in much the same way as any other regression problem. Yule's autoregressive model, on the other hand, relates the present value of a series directly to it's past. This seemingly obvious step raises a host of fundamental questions concerning the existence, uniqueness, and characterization of such processes.

Many of these questions were answered by the spectral theory of Wiener and Kolmogorov. The spectral theory connects the autocorrelation function of a series to a spectral distribution function, via the Fourier transform. This leads to the spectral representation of a stationary process which underpins much of modern time-series analysis. Complimenting the spectral theory was a time-domain theory that culminates in the decomposition theorem of

Wold. This alternative representation decomposes a stationary time series into a superposition of an infinite-order moving average and a stationary deterministic process. Both of these theories put Yule's autoregressive models on a firm mathematical footing and set the stage for the flurry of activity that was to come.

2. Historical Background

One of the main results in "The Analysis of Multivariate Stationary Time Series" is an approximation to the likelihood of a stationary multivariate Gaussian process. This approximation is a generalization of the univariate case previously derived in Whittle (1951). The development of the result in the univariate case parallels that of the multivariate case, and the form of the results are similar. A quick synopsis of the univariate approximation will shed some light on the more general results.

Consider a set of observations $X_1, \ldots, X_N$ to be taken from a univariate stationary Gaussian process $\{X_t\}$ with mean zero. Assume that the correlation function of the time series can be parametrized by the vector $\boldsymbol{\beta} = (\beta_1, \ldots, \beta_m)'$ and that the spectral density $(\sigma^2/2\pi)g(\cdot\,;\boldsymbol{\beta})$ exists, i.e.,

$$\mathrm{Cov}(X_s, X_t) = \frac{\sigma^2}{2\pi}\int_0^{2\pi} e^{i(s-t)\omega} g(\omega;\boldsymbol{\beta})\,d\omega.$$

Then, ignoring constants, -2 [log of the likelihood of the parameter vector $\boldsymbol{\theta} = (\boldsymbol{\beta}, \sigma^2)$ based on the observation vector $\mathbf{X} = (X_1, \ldots, X_N)'$] is given by

$$L(\boldsymbol{\theta}) = \sigma^{-2}\mathbf{X}'G^{-1}\mathbf{X} + \log|\sigma^2 G|, \tag{1}$$

where $\sigma^2 G = \sigma^2 G(\boldsymbol{\beta})$ is the covariance matrix of $\mathbf{X}$. The maximum likelihood estimate, $\hat{\boldsymbol{\theta}}$, is found by maximizing $L(\boldsymbol{\theta})$ over the allowable parameter space. Explicit solutions to this maximization problem rarely exist, and hence maximization of the likelihood is performed using numerical optimization techniques, requiring multiple evaluations of $L(\boldsymbol{\theta})$. These evaluations can be daunting due to the complicated matrix operations involved, and must have been especially so in 1953 before the advent of high-speed digital computers and sophisticated optimization techniques.

Whittle's idea was to approximate the quadratic form in such a way as to avoid computing the explicit inverse and determinant of G. The approximation is based on the observation that G can be approximated by a circulant matrix, $\tilde{G}$, which in turn can be easily diagonalized by the matrix U consisting of rows,

$$\mathbf{u}_j = N^{-1/2}(1, e^{i\omega_j}, e^{2i\omega_j}, \ldots, e^{(N-1)i\omega_j}),$$

where $\omega_j = 2\pi j/N$ represent the Fourier frequencies (see Brockwell and Davis (1991), pp. 134–136). In particular,

$$UGU' \approx U\tilde{G}U' = D = \operatorname{diag}(g(\omega_0;\boldsymbol{\beta}),\ldots,g(\omega_{N-1};\boldsymbol{\beta}))$$

and

$$U\mathbf{X} = \mathbf{Z} = (Z_0,\ldots,Z_{N-1})',$$

where $Z_j = N^{-1/2}\sum_{t=0}^{N-1} e^{it\omega_j}X_{t+1}$ is the jth ordinate of the discrete Fourier transform of $\mathbf{X}$. Using these properties, it follows that

$$\mathbf{X}'G^{-1}\mathbf{X} \approx \mathbf{X}'\tilde{G}^{-1}\mathbf{X} = \mathbf{X}'U'D^{-1}U\mathbf{X} = \sum_{j=0}^{N-1}\frac{|Z_j|^2}{g(\omega_j;\boldsymbol{\beta})}$$

and

$$\log|\sigma^2 G| \approx \log|\sigma^2 UD^{-1}U'| = \sum_{j=0}^{N-1}\log(\sigma^2 g(\omega_j;\boldsymbol{\beta})).$$

These approximations suggest replacing the log-likelihood in (1) by

$$L_W(\boldsymbol{\theta}) = \sum_{j=0}^{N-1}\left(\log(\sigma^2 g(\omega_j;\boldsymbol{\beta})) + \frac{|Z_j|^2}{\sigma^2 g(\omega_j;\boldsymbol{\beta})}\right). \tag{2}$$

In the univariate case the likelihood can be further simplified by first maximizing over the parameter σ^2. This can be done independently of the other parameters, giving the estimate,

$$\hat{\sigma}^2 = \frac{1}{N}\sum_{j=0}^{N-1}\frac{|Z_j|^2}{g(\omega_j;\boldsymbol{\beta})}.$$

Substituting this estimate back into (2) gives the concentrated likelihood,

$$L_W(\boldsymbol{\beta},\hat{\sigma}^2) = N\log\left(\frac{1}{N}\sum_{j=0}^{N-1}\frac{|Z_j|^2}{g(\omega_j;\boldsymbol{\beta})}\right) + \sum_{j=0}^{N-1}\log(g(\omega_j;\boldsymbol{\beta})), \tag{3}$$

which is a function of $\boldsymbol{\beta}$ only. Equation (2) is known as the Whittle approximation to the likelihood. Calculation of $L_W(\boldsymbol{\theta})$, and hence numerical optimization of $L_W(\boldsymbol{\theta})$, is substantially simpler than that based on $L_W(\boldsymbol{\theta})$ since the quadratic form, $\mathbf{X}'G^{-1}\mathbf{X}$, and the determinant, $|G|$, are not explicitly calculated. Also the parameters come in only through the spectral density, which for many types of models has an explicit formula. The periodogram must be calculated, but it too can be calculated quite easily via the fast Fourier transform.

The estimation procedure for multivariate processes derived in "The Analysis of Multiple Stationary Processes" minimizes a quantity analogous to (2) above. Although there are some fundamental differences between the univariate and multivariate cases, it is comforting to see that Whittle's approximation to the likelihood takes essentially the same form for both.

3. Summary of the Paper

In Section 1 of the paper, Whittle stresses the need to develop methods for the analysis of multivariate time series. Much of the theory of time series is concerned with a single sequence of measurements, however, data obtained in practice are often multidimensional. Economic time series might consist of price indices, interest rates, and cost indicators all measured simultaneously. Radar data are often generated by antennae arrays, with each antennae producing its own time series. And as any audiophile will tell, stereo sound is much more interesting than mono.

For a standard linear model, the least squares estimate is the same as that obtained by maximizing the likelihood assuming a Gaussian model. Analogously, Whittle describes his estimator as "least squares" because it is derived by maximizing the Gaussian likelihood even though the underlying model is not necessarily Gaussian. In the vernacular of modern time series, such estimates are called "maximum (Gaussian) likelihood" while "least squares" refers to estimates that are obtained by minimizing the quadratic form occurring in the Gaussian likelihood.

A first step in Whittle's approximation to the Gaussian likelihood is to express the process $\{\mathbf{X}_t\}$ as an infinite order moving average (MA),

$$\mathbf{X}_t = \sum_{k=0}^{\infty} C_k \boldsymbol{\eta}_{t-k}, \tag{4}$$

where $\{C_k\}$ is a sequence of matrices with C_0 equal to the identity, and $\{\boldsymbol{\eta}_t\}$ is a sequence of uncorrelated random vectors with covariance matrix $\boldsymbol{\Sigma}$. Assuming (4) can be inverted, this yields the autoregressive (AR) representation

$$\mathbf{X}_t - \sum_{k=0}^{\infty} D_k \mathbf{X}_{t-k} = \boldsymbol{\eta}_t.$$

Apart from end effects, $(\mathbf{X}_1, \ldots, \mathbf{X}_N)$ can be expressed in terms of $(\boldsymbol{\eta}_1, \ldots, \boldsymbol{\eta}_N)$ and vice versa so that $-2\log$ of the Gaussian likelihood of $(\mathbf{X}_1, \ldots, \mathbf{X}_N)$ is approximately

$$\sum_{t=1}^{N} \boldsymbol{\eta}_t' \boldsymbol{\Sigma}^{-1} \boldsymbol{\eta}_t + N \log |\boldsymbol{\Sigma}|.$$

In the remainder of the Introduction, Whittle points out some of the difficulties that arise in the multivariate case that are not present in the univariate case. One major hurdle is the lack of separation of the parameters describing the "essential features" of the process, from those describing the "variances of the residuals." For example, in the MA representation in (4), the estimation of $\boldsymbol{\Sigma}$ cannot be done independently of the estimation of the $\{C_k\}$ as it was in the univariate case (see (3) above). In addition, if the matrices C_k are functions of a parameter vector $\boldsymbol{\beta}$, then the asymptotic variance of the maximum likelihood estimate of $\boldsymbol{\beta}$ will typically depend on $\boldsymbol{\Sigma}$.

In Section 2, Whittle discusses the Wold decomposition, similar to that given in (4). It is noted that, in general, the MA coefficient matrices $\{C_k\}$ are not uniquely determined by the spectral matrix function $F(\omega)$. That is, it may not be possible to discriminate between different parametrizations of the model from a given F. Whittle also gives conditions necessary to invert the MA representation to obtain an AR representation. He further connects the spectral matrix function to the transfer functions corresponding to the respective AR and MA filters.

In Section 3, Whittle derives an expression for the prediction variance of a multivariate stationary process. In the multivariate case there is no scalar prediction error, but rather a vector of prediction errors (the $\boldsymbol{\eta}_t$ in (4)) with an associated covariance matrix. The determinant of this matrix is taken as the measure of prediction error, and is found to be

$$V = \exp\left\{\frac{1}{2\pi}\int_0^{2\pi} \log|F(\omega)|\, d\omega\right\}. \tag{5}$$

This is analogous to Kolmogorov's formula for the prediction variance of a univariate stationary process. Also, $V^{N/2}$ is approximately the Jacobian of the transformation from the likelihood of the observations $\{\mathbf{X}_t\}$ to the likelihood of the standardized moving average noise terms $\{\boldsymbol{\Sigma}^{-1/2}\boldsymbol{\eta}_t\}$. This fact is used to derive the estimation equations in Section 5.

In Section 4, Whittle derives asymptotic formulas for the cumulants of arbitrary linear functions of the sample cross-covariances of a multivariate stationary Gaussian process. Corresponding to any such linear function ξ Whittle shows there exists a Hermitian matrix function $Q(z)$ such that

$$\xi = N\int_0^{2\pi} \operatorname{tr}(Q(e^{i\omega})f(\omega))\, d\omega,$$

where $f(\omega)$ is the periodogram matrix function

$$f(\omega) = \frac{1}{N}\left(\sum_{t=1}^{N}\mathbf{X}_t e^{it\omega}\right)\left(\sum_{s=1}^{N}\mathbf{X}_s e^{is\omega}\right)'.$$

The cumulants $k^{(r)}(\xi)$ are then shown to satisfy

$$k^{(r)}(\xi) \sim \frac{2^{r-1}(r-1)!N}{2\pi} N\int_0^{2\pi} \operatorname{tr}(Q(e^{i\omega})F(\omega))^r\, d\omega,$$

which are subsequently used to derive the limiting behavior of the likelihood function and its derivatives.

In Section 5, the approximation to the likelihood is derived in a surprisingly direct manner, using the infinite moving average representation of the process. The sum of squares of the noise terms in expressed as

$$\sum_{t=1}^{N} \boldsymbol{\eta}_t' \boldsymbol{\Sigma}^{-1} \boldsymbol{\eta}_t \approx \frac{N}{2\pi} \int_0^{2\pi} \operatorname{tr}(f(\omega) F^{-1}(\omega)) \, d\omega,$$

which, when combined with (4), implies that -2 log of the likelihood can be approximated by

$$L_W(\boldsymbol{\theta}) = \frac{N}{2\pi} \int_0^{2\pi} \log(|F(\omega; \boldsymbol{\theta})|) + \operatorname{tr}(f(\omega) F^{-1}(\omega; \boldsymbol{\theta})) \, d\omega. \tag{6}$$

Compare this expression (6) for $L_W(\theta)$ to that obtained in the univariate case (3).

In Section 6, Whittle argues that when the process with true parameter $\boldsymbol{\theta} = \boldsymbol{\theta}_0$ is Gaussian, $\partial L_W(\boldsymbol{\theta}_0)/\partial \boldsymbol{\theta}$ is asymptotically normal with mean 0, and covariance matrix

$$2M = \left[\frac{N}{4\pi} \int_0^{2\pi} \operatorname{tr} \left[\frac{\partial F}{\partial \theta_j} F^{-1} \frac{\partial F}{\partial \theta_k} F^{-1} \right] d\omega \right]_{j,k},$$

and

$$\frac{\partial^2 L_W(\boldsymbol{\theta})}{\partial \boldsymbol{\theta}^2} = M + o_p(N).$$

In now follows from standard Taylor series expansions of $L_W(\boldsymbol{\theta})$ and assuming consistency of the estimators that $\hat{\boldsymbol{\theta}}$ is asymptotically normal with mean $\boldsymbol{\theta}_0$ and covariance matrix $2M^{-1}$.

Section 7 uses expansions similar to those in Section 6 to develop testing procedures. A generalized likelihood ratio test based upon the prediction variance V of Section 3 is derived and an example is given demonstrating its use.

In Section 8, a more complete example involving a bivariate series of sunspot measurements is given. The model used in the example is of a simplified symmetrical type, but both fitting and testing procedures are demonstrated.

4. Subsequent Developments

The approximation given in (6) is often referred to as the Whittle likelihood in the literature. It has taken a progression of researchers to nail down all of the details and to formalize the *theorems* of Whittle in both the univariate case (Walker (1964) and Hannan (1973)) and in the multivariate case (Dunsmuir and Hannan (1976); Deistler, Dunsmuir, and Hannan (1978); Pötscher (1987); and Dahlhaus and Pötscher (1989)).

Much of this work centered on questions of identifiability, that are intimately connected with establishing consistency of the estimators. Whittle pays little attention to this important problem which for the case of multivariate autoregressive moving average (ARMA) models is not a simple matter. Consider, for example, the following bivariate AR(1) model:

$$X_{1t} = \phi X_{2,t-1} + \eta_{1t},$$

$$X_{2t} = \eta_{2t},$$

where $\boldsymbol{\eta}_t = (\eta_{1t}, \eta_{2t})'$ are i.i.d. random vectors. This model also has the MA(1) representation,

$$X_{1t} = \eta_{1t} + \phi\eta_{2,t-1},$$

$$X_{2t} = \eta_{2t},$$

and hence the likelihood cannot discriminate between either of these two models.

A more subtle but important point not addressed by Whittle is the quality of the approximation to the exact Gaussian likelihood. For example, it is not readily apparent that the estimate based on Whittle's approximation, and that based on the exact Gaussian likelihood are asymptotically equivalent. Subsequent results have shown this asymptotic equivalence for a wide class of time-series models and extended the results of Sections 6 and 7 to cases where the underlying distribution is non-Gaussian.

Although the Whittle likelihood provides a simple approximation to the Gaussian likelihood, it does not appear to have been used much for ARMA modeling. This may be in part due to the work being "done too early ... to reap its due recognition" (Whittle (1986)) and hence simply overlooked, or perhaps because of the development of alternative methods, such as the Box–Jenkins backcasting algorithm. Today there is no reason to use either of these methods for this purpose. Recursive techniques based upon the Kalman filter and state-space representations can be used to compute exact Gaussian likelihoods efficiently.

There has been a resurgence of interest in the Whittle approximation to the likelihood. For fractionally integrated ARMA models and certain classes of spatial models, finite state-space representations do not exist which preclude the use of recursive methods to calculate the likelihood efficiently. Nevertheless, these models do have explicit expressions for the spectral density and therefore Whittle's approximation becomes an indispensable tool.

5. Biography

Peter Whittle was born in Wellington, New Zealand in 1927. He became interested in stochastic processes and time series while working for the New Zealand Department of Scientific and Industrial Research. After completing

B.Sc. degrees in mathematics and physics, and an M.Sc. degree in mathematics, in 1948, he began post-graduate work with Dr. Herman Wold at the University of Uppsala, Sweden. During this time he, in his own words, "constructed the asymptotic likelihood theory for the stationary Gaussian process, essentially if unrigorously."

After earning a Doctorate of Philosophy in 1951, he returned to New Zealand and the Department of Scientific Research. He continued work in time series and also worked on problems involving nonlinear models, spatial models, random partial differential equations, and reversibility. After spending a year in Canberra, he began working on problems of prediction and optimization. It was at this time, 1959, that he moved to Britain to join the Statistical Laboratory at Cambridge. In 1961, he became the head of the Statistical Laboratory at Manchester University, replacing Maurice Bartlett, for whom he has a deep reverence. In 1966 he moved on to the Churchill Chair in the Mathematics of Operational Research at Cambridge.

References

Brockwell, P.J. and Davis, R.A. (1991). *Time Series: Theory and Methods*, 2nd ed. Springer-Verlag, New York.

Dahlhaus, R. and Pötscher, B.M. (1989). Convergence results for maximum likelihood type estimators in multivariable ARMA models II. *J. Multivariate Anal.*, **30**, 241–244.

Deistler, M., Dunsmuir, W., and Hannan, E.J. (1978). Vector linear time senes models: Corrections and extensions. *Adv. in Appl. Probab.*, **10**, 360–372.

Dunsmuir, W.J.M. and Hannan, E.J. (1976). Vector linear time series models. *Adv. in Appl. Probab.*, **8**, 339–364.

Hannan, E.J. (1973). The asymptotic theory of linear time series models. *J. Appl. Probab.*, **10**, 130–145.

Pötscher, B.M. (1987). Convergence results for maximum likelihood type estimators in multivariable ARMA models. *J. Multivariate Anal.*, **21**, 29–52.

Walker, A.M. (1964). Asymptotic properties of least squares estimates of parameters of the spectrum of a stationary non-deterministic time series. *J. Austral. Math. Soc.*, **4**, 363–384.

Whittle, P. (1951). *Hypothesis Testing in Time Series Analysis*. Thesis, Uppsala University, Almqvist and Wiksell, Uppsala.

Whittle, P. (1986). In the late afternoon. In *The Craft of Probabilistic Modelling* (J. Gani, ed.). Springer-Verlag, New York.

The Analysis of Multiple Stationary Time Series

P. Whittle
University Institute of Statistics, Uppsala

Summary

After some preparatory work, the least square estimation equations are derived for a purely nondeterministic stationary multiple process (Th. 6). The asymptotic covariances of the parameter estimates are calculated for a normal process (Th. 9) and a test of fit derived (Th. 10). The testing of a sunspot model provides an illustration of the methods developed.

1. Introductory

In a series of recent publications (Whittle, 1951–1952c) the author has considered the application of the classical least square principle to the analysis of a single stationary time series, in an attempt to obtain a uniform treatment of the estimation and test problems which arise. (By "least square estimation equations" is meant here those equations which would have been obtained on the maximum likelihood principle if the variates had been assumed normally distributed.) The method depends essentially upon the natural complement which least square theory forms to the spectral and autocovariance theory of stationary processes, concerning as they both do only the second moments of the observations. Thus, quantities such as the residual sum of squares may be expressed in terms of the theoretical spectral function and the observed autocovariances or the periodogram.

From the theoretical point of view, this method should lead in most case to estimates and tests with the restricted optimum properties enjoyed by least square statistics in general, at least asymptotically; from the practical point of view, experience has shown it to be as simple and direct as the subject would seem to allow.

However, it is seldom that one has occasion to analyse a single series. The vast majority of practical problems require the analysis of a multiple series—meteorological and economic data movide perhaps the most obvious examples. In this paper an attempt is therefore made to extend the least square theory to this more general case.

Now, the difficulty of analysing a q-tuple series may be said to increase roughly as q^2 (the number of auto- and cross-correlograms which must be calculated, and the order of the number of parameters to be estimated), while the number of observations increases only as q. These facts are undoubtedly discouraging, and offset only in part by the fact that the stochastic relations of a multiple process should in general be fairly simple in nature. Indeed, intuition would lead us to believe that if account were taken of all the relevant variates of a self-contained system, then the process would be Markovian (see Moyal, 1949, p. 200.) Further, if the variates fall into a well-defined causal chain, there are good reasons for believing that the set of stochastic relations reduces to a so-called recursive system (see Wold, 1952, p. 64)

Another complication introduced by the multiplicity of the series is that of *nonidentifiability* (Frisch, 1938). That is, the number of unknown parameters exceeds the number of distinct estimation equations available, so that the solution for the parameter estimates is indeterminate. The indeterminacy generally takes the form that the separate equations of the fitted scheme may be arbitrarily added and subtracted, and still be consistent with the estimation equations. (Indeterminacies of another type can arise in the single series case, cf. Wold, 1938, p. 126, but these are not so serious.)

To attain identifiability it is necessary to reduce the number of parameters by requiring them to obey extra relations—relations which may often be quite arbitrary. However, since it is certain that statistics alone cannot provide a solution of the identifiability problem, we shall assume that identifiability has been achieved, so that the estimation equations always yield determinate solutions.

We note a last difficulty which will become more apparent in the course of the paper: that those parameters referring to essential features of the process and those referring simply to the variances of the residual variates are in general so intermixed as to be inseparable (separation is possible in the single series case). This not only complicates the estimation equations, but also affects the asymptotic properties of the parameter estimates.

The aim of the paper is to provide a practical solution of the problem rather than a theoretical treatment of the many points which arise. It has nevertheless been impossible to avoid rather much mathematics, so much that it has been considered best to give the main results in theorem form. This is done solely in the interests of clarity, however, and not to provide a pretence of rigour where no particular degree of rigour has been attempted.

Notation. The complex conjugate of z is denoted by $\bar{z}$. A matrix $\mathbf{A}$ with elements a_{jk} will sometimes be written (a_{jk}). Its determinant, transpose and trace will be denoted $|\mathbf{A}|$, $\mathbf{A}'$, and tr $\mathbf{A}$ respectively. The complex conjugate of the transpose, $\mathbf{A}'$, will be written $\mathbf{A}^*$.

2. Specification of the Process

Suppose that we consider a q-variate process $\{x_{1t}, x_{2t}, \ldots, x_{qt}\}$, $(t = \ldots -2, -1, 0, 1, 2, \ldots)$, the variates for time t forming $q \times 1$ vector $\mathbf{X}_t$. We shall restrict ourselves to a purely nondeterministic process, so that $E(\mathbf{X}_t) = 0$. (The effect of a non-zero mean will be considered later.) We shall now define the theoretical autocovariances and spectral functions together with their empirical equivalents:

$$\Gamma_{jk}(s) = E(x_{j,t+s}x_{kt}), \tag{2.1}$$

$$F_{jk}(\omega) = \sum_{s=-\infty}^{\infty} \Gamma_{jk}(s)e^{i\omega s}, \tag{2.2}$$

$$\mathbf{F}(\omega) = (F_{jk}(\omega)), \tag{2.3}$$

$$C_{jk}(s) = (1/N)\sum_{t=1}^{N-s} x_{j,t+s}x_{kt}, \tag{2.4}$$

$$f_{jk}(\omega) = \sum_{s=-N}^{N} C_{jk}(s)e^{i\omega s}, \tag{2.5}$$

$$\mathbf{f}(\omega) = (f_{jk}(\omega)). \tag{2.6}$$

$\mathbf{F}(\omega)$ is thus the matrix of theoretical spectral functions $F_{jk}(\omega)$. $C_{jk}(s)$ is the empirical cross-covariance of x_{jt} and x_{kt}, lagged s steps, and is based upon a sample series of N terms, $(\mathbf{X}_1, \mathbf{X}_2, \ldots, \mathbf{X}_N)$. Strictly, the sum in (2.4) should be divided by $N - s$ instead of N, but we shall consistently assume N so large that end effects may be neglected. (It will later be apparent that we need only consider the earlier covariances, for which s is relatively small.) The $f_{jk}(\omega)$ are the empirical spectral functions, or periodograms, and $\mathbf{f}(\omega)$ is their matrix.

The first facts to be noted are that the matrices $\mathbf{F}(\omega)$, $\mathbf{f}(\omega)$ Hermitian and Hermitian positive semi-definite (see Cramér, 1940). That $\mathbf{F}(\omega)$ is Hermitian follows from the fact that $F_{jk}(\omega) = F_{kj}(-\omega) = \overline{F_{kj}(\omega)}$. That it is Hermitian positive definite follows from the fact that $\xi'\mathbf{F}(\omega)\bar{\xi}$ is the spectral function of the variate $\xi'\mathbf{X}_t$, (ξ an arbitrary vector), and is thus never negative. Similarly for $\mathbf{f}(\omega)$.

Now, Wold (1938) has given a canonical representation of the univariate stationary process, which in the case of a purely nondeterministic variate reduces to a one-sided moving average

$$x_t = b_0\varepsilon_t + b_1\varepsilon_{t-1} + \cdots \tag{2.7}$$

over a series of uncorrelated variates ε_t. (The interpretation of (2.7) is well known: if a series of uncorrelated disturbances $\ldots \varepsilon_{t-1}, \varepsilon_t, \varepsilon_{t+1}, \ldots$ is injected into a system, and ε_t has a lingering effect $b_s\varepsilon_t$ at time $t+s$, and the effects of the different disturbances are additive, then the total effect is given by (2.7). The series $b_0, b_1, b_2, \ldots$ is known as the *transient response* of the system.)

Zasuhin (1941) has generalized Wold's representation to the multivariate case, so that corresponding to (2.7) we have

$$x_{jt} = \sum_{m=0}^{\infty} b_{jkm}\varepsilon_{k,t-m} \tag{2.8}$$

$(j = 1, 2, \ldots, q; t = \ldots -1, 0, 1, \ldots)$ where all the ε's are uncorrelated, have zero mean and unit variance. A feature of Zasuhin's representation is that the b_{jkm} coefficients are not uniquely determined by $\mathbf{F}(\omega)$. This is just the nonidentifiability phenomenon mentioned before, however, and will not trouble us, since the constraints to which we have assumed the parameters subject will also serve to uniquely determine the b_{jkm} coefficients.

Let us now define a shift operator U for which

$$Ux_t = x_{t-1} \tag{2.9}$$

and define a matrix $\mathbf{B}(U)$.

$$\mathbf{B}(U) = [B_{jk}(U)],$$

$$B_{jk}(U) = \sum_{m=0}^{\infty} b_{jkm}U^m, \tag{2.10}$$

so that the elements of $\mathbf{B}(U)$ may be expanded in positive powers of U. Then (2.8) may be written more neatly

$$\mathbf{X}_t = \mathbf{B}(U)\boldsymbol{\varepsilon}_t, \tag{2.11}$$

where $\boldsymbol{\varepsilon}_t = (\varepsilon_{1t}, \varepsilon_{2t}, \ldots, \varepsilon_{qt})$. We shall term (2.11) the *moving average representation* of the process. If the equation in z, $|\mathbf{B}(z)| = 0$, has no roots on or within $|z| = 1$, then (2.11) may be inverted into an *autoregressive representation* of the process

$$\boldsymbol{\varepsilon}_t = [\mathbf{B}(U)]^{-1}\mathbf{X}_t = \mathbf{A}(U)\mathbf{X}_t \tag{2.12}$$

say. The reciprocal of $\mathbf{B}(U), \mathbf{A}(U)$, is to be obtained by first inverting $\mathbf{B}(U)$ as a matrix, and then expanding each element of the inverted matrix in powers of U. Only positive powers of U will occur (due to the condition on the roots of $|\mathbf{B}(z)|$) so that future values $\mathbf{X}_{t+1}, \mathbf{X}_{t+2}, \ldots,$ do not occur in the representation. The elements of $\mathbf{A}(U)$ will be denoted

$$\mathbf{A}(U) = [A_{jk}(U)],$$

$$A_{jk}(U) = \sum_{m=0}^{\infty} a_{jkm} U^m. \tag{2.13}$$

Given the autoregressive representation (2.12), a necessary and sufficient condition that it be invertable to the moving average representation (2.11) is that $|\mathbf{A}(z)| = 0$ have no roots on or within the unit circle (cf. Mann and Wald, 1943; Moran, 1949). This result corresponds to the usual condition that a simple autoregression generate a stationary process (Wold, 1938, p. 99).

There is a theorem connecting the spectral matrix with the coefficients of the two representations (cf. Bartlett, 1946, p. 92).

Theorem 1

$$\mathbf{F}(\omega) = \mathbf{B}(e^{i\omega})\mathbf{B}'(e^{-i\omega}) = [\mathbf{A}'(e^{-i\omega})\mathbf{A}(e^{i\omega})]^{-1}.$$

Writing $z = e^{i\omega}$, we have

$$\begin{aligned} F_{jk}(\omega) &= E\left(\sum_s x_{j,t+s} x_{kt} z^s\right) \\ &= E\left(\sum_s \sum_m \sum_n \sum_u \sum_v b_{jmu} b_{knv} \varepsilon_{m,t+s-u} \varepsilon_{n,t-v} z^s\right) \\ &= \sum_s \sum_m \sum_u b_{jmu} b_{km,u-s} z^s = \sum_m \sum_u \sum_v b_{jmu} z^u b_{kmv} z^{-v} \\ &= \sum_m B_{jm}(z) B_{km}(z^{-1}), \end{aligned}$$

which is equivalent to the desired result.

Example. Consider the double process

$$x_{1t} + a x_{1,t-1} + b x_{1,t-2} = c\varepsilon_{1t},$$

$$x_{2t} + d x_{1,t-1} = e(\varepsilon_{2t} - f\varepsilon_{2,t-1}),$$

or

$$\begin{bmatrix} 1 + aU + bU^2 & \cdot \\ dU & 1 \end{bmatrix} \mathbf{X}_t = \begin{bmatrix} c & \cdot \\ \cdot & e(1 - fU) \end{bmatrix} \varepsilon_t.$$

If we write the last equation as $\mathbf{P}\mathbf{X}_t = \mathbf{Q}\varepsilon_t$, then

$$\mathbf{A}(U) = \mathbf{Q}^{-1}\mathbf{P} = \begin{bmatrix} \dfrac{1 + aU + bU^2}{c} & \cdot \\ \dfrac{dU}{e(1 - fU)} & \dfrac{1}{e(1 - fU)} \end{bmatrix}$$

and

$$\mathbf{B}(U) = \mathbf{P}^{-1}\mathbf{Q} = \frac{1}{1 + aU + bU^2} \begin{bmatrix} c^2 & \cdot \\ -cdU & e(1 - fU)(1 + aU + bU^2) \end{bmatrix}.$$

We have

$$\mathbf{F}(\omega) = \mathbf{B}\mathbf{B}^* = \frac{1}{|1 + az + bz^2|^2} \begin{bmatrix} c^2 & -c^2dz^{-1} \\ -c^2dz & c^2d^2 + e|1 - fz|^2|1 + az + bz^2|^2 \end{bmatrix},$$

where $z = e^{i\omega}$.

3. The Predication Variance

Consider the autoregressive representation of the process (2.12). Let us denote the absolute term in the expansion of $\mathbf{A}(U)$ in powers of U by $\mathbf{A}_0$, so that (2.12) may be written

$$\mathbf{A}_0\mathbf{X}_t + (\text{past } \mathbf{X}_t \text{ values}) = \boldsymbol{\varepsilon}_t,$$

or

$$\begin{aligned} \mathbf{X}_t &= \mathbf{A}_0^{-1}\boldsymbol{\varepsilon}_t + (\text{past } \mathbf{X}_t \text{ values}) \\ &= \mathbf{B}_0\boldsymbol{\varepsilon}_t + (\text{past } \mathbf{X}_t \text{ values}), \end{aligned} \tag{3.1}$$

where $\mathbf{B}_0$ is correspondingly the absolute term in the expansion of $\mathbf{B}(U)$. Now, the bracketed term in (3.1) provides the best linear prediction of $\mathbf{X}_t$ in terms of $\mathbf{X}_{t-1}, \mathbf{X}_{t-2}, \ldots$, so that $\mathbf{B}_0\boldsymbol{\varepsilon}_t$ is the deviation of predication from reality, that part of $\mathbf{X}_t$ which cannot be deduced from past history, consisting as it does of the random variation which has been injected into the process since the last instant of time.

Corresponding to the variance of a single variate is the *total variance* of a vector variate, equal to the determinant of the covariance matrix (Wilks, 1932). In the present case

$$E[(\mathbf{B}_0\boldsymbol{\varepsilon}_t)(\mathbf{B}_0\boldsymbol{\varepsilon}_t)'] = \mathbf{B}_0\mathbf{B}_0', \tag{3.2}$$

so that the total variance of $\mathbf{B}_0\boldsymbol{\varepsilon}_t$, a quantity which we may term the *total prediction variance*, is

$$V = |\mathbf{B}_0\mathbf{B}_0'| = |\mathbf{B}_0|^2 = |\mathbf{A}_0|^{-2}. \tag{3.3}$$

V gives a measure of the total random variance entering the process at every step. It may be derived in another way: if the likelihood of the sample $(\mathbf{X}_1, \mathbf{X}_2, \ldots, \mathbf{X}_N)$ is expressed in terms of the likelihood of $(\boldsymbol{\varepsilon}_1, \boldsymbol{\varepsilon}_2, \ldots, \boldsymbol{\varepsilon}_N)$, then the Jacobian of the transformation is

$$\left| \frac{\partial(\mathbf{X}_1, \mathbf{X}_2, \ldots, \mathbf{X}_N)}{\partial(\boldsymbol{\varepsilon}_1, \boldsymbol{\varepsilon}_2, \ldots, \boldsymbol{\varepsilon}_N)} \right| = |\mathbf{B}_0|^N, \tag{3.4}$$

and we should expect this quantity to be equal to just $V^{N/2}$.

The total prediction variance may be expressed in terms of the spectral matrix of the process:

Theorem 2

$$V = \exp\left[\frac{1}{2\pi}\int_0^{2\pi} \log|\mathbf{F}(\omega)|\, d\omega\right].$$

In the case of a univariate process this is Kolmogoroff's result (1941). To prove it, we make use of the fact that, since $|\mathbf{B}_0|$ is the absolute term in the expansion of $|\mathbf{B}(z)|$, $(|z| \le 1)$, then $\log|\mathbf{B}_0|$ is the absolute term in the expansion of $\log|\mathbf{B}(z)|$. We have thus

$$\frac{1}{2\pi}\int_0^{2\pi} \log|\mathbf{F}(\omega)|\, d\omega = \frac{1}{2\pi}\int_0^{2\pi} \log|\mathbf{B}(z)|\, d\omega + \frac{1}{2\pi}\int_0^{2\pi} \log|\mathbf{B}(z^{-1})|\, d\omega$$

$$= \log|\mathbf{B}_0| + \log|\mathbf{B}_0| = \log V,$$

which is the required result.

As is the case in so many problems of this type, the multivariate process may be reduced in a standard fashion to q mutually uncorrelated univariate processes (cf. Wiener, 1949, p. 44). For, suppose that the matrix $\mathbf{F}(\omega)$ has eigenvalues $\lambda_1(\omega), \lambda_2(\omega), \ldots, \lambda_q(\omega)$, with corresponding normalized eigenvectors $\mathbf{C}_1(z), \mathbf{C}_2(z), \ldots, \mathbf{C}_q(z)$, so that

$$\mathbf{F}(\omega) = \sum_j \lambda_j \mathbf{C}_j \mathbf{C}_j^*, \tag{3.5}$$

$$\mathbf{C}_j^* \mathbf{C}_k \equiv \delta_{jk} \tag{3.6}$$

Theorem 3 (The Unitary Transformation).

$$y_{jt} = \mathbf{C}_j'(U^{-1})\mathbf{X}_t$$

produces q mutually uncorrelated processes $\{y_{jt}\}$ with respective spectral functions $\lambda_j(\omega)(j = 1, 2, \ldots, q)$.

For, if $\mathbf{Y}_t' = (y_{1t}, y_{2t}, \ldots, y_{qt})$ and $\mathbf{C}^*(z) = (\mathbf{C}_1, \mathbf{C}_2, \ldots, \mathbf{C}_q)$, then

$$\mathbf{Y}_t = \mathbf{C}(U)\mathbf{X}_t, \tag{3.7}$$

and, by the same argument as that used to establish Th. 1 from (2.11), we see that the matrix of spectral functions for the $\mathbf{Y}_t$ process is

$$\mathbf{F}^{(y)}(\omega) = \mathbf{C}(z)\mathbf{F}(\omega)\mathbf{C}^*(z)$$

$$= [\mathbf{C}_j'(z^{-1})\mathbf{F}(\omega)\mathbf{C}_k(z)] = (\delta_{jk}\lambda_j(\omega)) \tag{3.8}$$

by (3.5), (3.6). This is equivalent to the result of the theorem.

The $q\{y_{jt}\}$ processes, which may be termed the *canonical components* of the multivariate process, will have prediction variances

$$v_j = \exp\left[\frac{1}{2\pi}\int_0^{2\pi} \log \lambda_j(\omega)\, d\omega\right] \tag{3.9}$$

and these may be termed the *canonical prediction variances* of the process. Obviously, $\prod_l^q v_j = V$.

While the canonical transformation of Th. 3 has a meaning, and can be useful in theoretical work, it can very seldom be realised in practice, and we shall not find great use for it. There are, however, certain exceptional cases where the eigenvalues and eigenvectors of $\mathbf{F}(\omega)$ may be readily calculated, and where the eigenvectors do not contain any of the process parameters. These cases may be treated very elegantly, for then the q y series may be considered completely separately one from the other.

Example. Consider the symmetric double autoregression

$$\alpha(U)x_{1t} + \beta(U)x_{2t} = \varepsilon_{1t},$$

$$\beta(U)x_{1t} + \alpha(U)x_{2t} = \varepsilon_{2t},$$

for which

$$\mathbf{F}(\omega)^{-1} = \begin{bmatrix} \alpha(z)\alpha(z^{-1}) + \beta(z)\beta(z^{-1}) & \alpha(z)\beta(z^{-1}) + \alpha(z^{-1})\beta(z) \\ \alpha(z)\beta(z^{-1}) + \alpha(z^{-1})\beta(z) & \alpha(z)\alpha(z^{-1}) + \beta(z)\beta(z^{-1}) \end{bmatrix}.$$

We find the eigenvalues of $\mathbf{F}(\omega)$ to be $\lambda(\omega) = |\alpha(z) \pm \beta(z)|^2$ with corresponding eigenvectors $\sqrt{\frac{1}{2}}(1, \pm 1)$. The canonical components of the process are thus $\{(x_{1t} \pm x_{2t})/\sqrt{2}\}$. If $\alpha(z) = \alpha_0 + \alpha_1 z + \cdots$ and $\beta(z) = \beta_0 + \beta_1 z + \cdots$, then the canonical predication variances are $(\alpha_0 \pm \beta_0)^{-2}$.

4. Moments of Sample Functions

In this section we shall consider the distribution of linear functions (real coefficients) of the auto- and cross-covariances, on the assumption that the residual variates ε are normally distributed. First, we shall show that any linear function ξ of these covariances may be written)

$$\xi = N\mathscr{A}\ \mathrm{tr}[\mathbf{Q}(z)\mathbf{f}(\omega)] = (N/2\pi)\int_0^{2\pi} \mathrm{tr}[\mathbf{Q}(z)\mathbf{f}(\omega)]\, d\omega, \tag{4.1}$$

where $\mathbf{Q}(z)$ is a $q \times q$ Hermitian matrix whose elements are functions of $z = e^{i\omega}$, and $\mathscr{A}$ denotes the operation "absolute term in z in." For

$$\xi = N\mathscr{A}\ \mathrm{tr}[\mathbf{Q}(z)\mathbf{f}(\omega)]$$

$$= N\mathscr{A} \sum_j \sum_k Q_{kj}(z) f_{jk}(\omega) = N \sum_j \sum_k \sum_s q_{kj,-s} C_{jk}(s). \tag{4.2}$$

Now, to require that $\mathbf{Q}$ be Hermitian is to require that $q_{kj,-s} = q_{jks}$ for all appropriate j, k and s, since the q's are assumed real. But $q_{kj,-s}$ is the coefficient of $C_{jk}(s)$, and q_{jks} is the coefficient of $C_{kj}(-s) = C_{jk}(s)$ also. The restriction is thus only an apparent one, and (4.2) represents the general linear function of the covariances. We shall now prove our main result of this section.

Theorem 4. *If $Q(z)$ possesses a Laurent expansion in z on $|z| = 1$, then the rth cumulant of ξ is given asymptotically by*

$$k_r(\xi) = 2^{r-1}(r-1)!N\mathscr{A}\,\mathrm{tr}[\mathbf{Q}(z)\mathbf{F}(\omega)]^r$$
$$= \frac{2^{r-1}(r-1)!N}{2\pi}\int_0^{2\pi} \mathrm{tr}[\mathbf{Q}(z)\mathbf{F}(\omega)]^r\, d\omega.$$

We shall first prove the result for the case $\mathbf{F} = \mathbf{I}$, when the x's are all independent standard variates.

We shall introduce the auxiliary $N \times N$ circulant matrix

$$W = \begin{bmatrix} \cdot & 1 & \cdot & \cdots & \cdot \\ \cdot & \cdot & 1 & \cdots & \cdot \\ \cdots & \cdots & \cdots & \cdots & \cdots \\ \cdot & \cdot & \cdot & \cdots & 1 \\ 1 & \cdot & \cdot & \cdots & \cdot \end{bmatrix}. \tag{4.3}$$

Denoting the vector $(x_{j1}, x_{j2}, \ldots, x_{jN})$ by $\mathbf{x}_j'$, then

$$NC_{jk}(s) \simeq \mathbf{x}_j'\mathbf{W}^s\mathbf{x}_k \tag{4.4}$$

if s is small compared with N, so that we can write

$$\xi = N\mathscr{A}\sum_j\sum_k Q_{kj}(z)f_{jk}(\omega)$$
$$\simeq \sum_j\sum_k x_j'Q_{kj}(W)x_k. \tag{4.5}$$

By ordinary normal theory, the characteristic function of ξ is then

$$\Phi_\xi(\theta) \simeq |\mathbf{I} - 2i\theta\mathbf{Q}(\mathbf{W})|^{-1/2} \tag{4.6}$$

and

$$k_r(\xi) \simeq 2^{r-1}(r-1)!\ \mathrm{tr}\,\mathbf{Q}(\mathbf{W})^n \tag{4.7}$$

where $\mathbf{Q}(\mathbf{W})$ is the matrix of matrices

$$\mathbf{Q}(\mathbf{W}) = [Q_{jk}(\mathbf{W})]. \tag{4.8}$$

Since $\mathbf{Q}(z)$ is Hermitian

$$Q_{jk}(\mathbf{W}) = Q_{kj}(\mathbf{W}^{-1}) = Q_{kj}(\mathbf{W})' \tag{4.9}$$

so that $\mathbf{Q}(\mathbf{W})$ is symmetric, as it must be for (4.6) and (4.7) to hold. (This is the reason why $\mathbf{Q}(z)$ was required to be Hermitian). Now, if $P_1(z), P_2(z), \ldots$ are functions with Laurent expansions in z on $|z| = 1$, then it may be shown (Whittle, 1951, pp. 35–49) that

$$\operatorname{tr}[P_1(\mathbf{W})P_2(\mathbf{W})\ldots] \simeq (N/2\pi)\int_0^{2\pi} [P_1(z)P_2(z)\ldots]\,d\omega. \tag{4.10}$$

Setting this result in (4.7) we have

$$\begin{aligned}
k_r(\xi) &= 2^{r-1}(r-1)!\operatorname{tr}\sum_{v_1}\sum_{v_2}\cdots\sum_{v_r} Q_{v_1,v_2}(\mathbf{W})Q_{v_2v_3}(\mathbf{W})\ldots Q_{v_rv_1}(\mathbf{W})\\
&= \frac{2^{r-1}(r-1)!N}{2\pi}\int_0^{2\pi}\sum_{v_1}\sum_{v_2}\cdots\sum_{v_r} Q_{v_1v_2}(z)Q_{v_2v_3}(z)\ldots Q_{v_rv_1}(z)\,d\omega\\
&= \frac{2^{r-1}(r-1)!N}{2\pi}\int_0^{2\pi}\operatorname{tr}\mathbf{Q}(z)^r\,d\omega,
\end{aligned} \tag{4.11}$$

which is the result of the theorem when $\mathbf{F} = \mathbf{I}$. This result may be readily extended to the general else. We have

$$\begin{aligned}
Nf_{jk}(\omega) &= N\sum_s C_{jk}(s)z^s = \sum_s\sum_t x_{j,t+s}x_{kt}z^s\\
&= \sum_s\sum_t\sum_m\sum_n\sum_u\sum_v b_{jmu}b_{knv}\varepsilon_{m,t+s-u}\varepsilon_{n,t-v}z^s.
\end{aligned}$$

Setting now $t-v=t'$, $t+s-u=t'+s'$ (and immediately dropping the primes!), the expression becomes

$$\begin{aligned}
Nf_{jk}(\omega) &\simeq \sum_s\sum_t\sum_m\sum_n\sum_u\sum_v b_{jmu}z^u b_{knv}z^{-v}\varepsilon_{m,t+s}\varepsilon_{nt}z^s\\
&= N\sum_m\sum_n\sum_u\sum_v b_{jmu}z^u b_{knv}z^{-v}f_{mn}^{(\varepsilon)}(\omega)\\
&= N\sum_m\sum_n B_{jm}(z)B_{kn}(z^{-1})f_{mn}^{(\varepsilon)}(\omega).
\end{aligned} \tag{4.12}$$

Here $f_{mn}^{(\varepsilon)}(\omega)$ is the cross-periodogram of the two series $(\varepsilon_{m1}, \varepsilon_{m2}, \ldots, \varepsilon_{mN})$, $(\varepsilon_{n1}, \varepsilon_{n2}, \ldots, \varepsilon_{nN})$. Equation (4.12) may be written

$$\mathbf{f}(\omega) \simeq \mathbf{B}(z)\mathbf{f}^{(\varepsilon)}(\omega)\mathbf{B}(z^{-1})' = \mathbf{B}\mathbf{f}^{(\varepsilon)}\mathbf{B}^*. \tag{4.13}$$

Thus

$$\xi = N\mathscr{A}\operatorname{tr}(\mathbf{Q}\mathbf{f}) \simeq N\mathscr{A}\operatorname{tr}(\mathbf{Q}\mathbf{B}\mathbf{f}^{(\varepsilon)}\mathbf{B}^*) = N\mathscr{A}\operatorname{tr}(\mathbf{B}^*\mathbf{Q}\mathbf{B}\mathbf{f}^{(\varepsilon)}), \tag{4.14}$$

so that by (4.11)

$$\begin{aligned}
k_r(\xi) &\simeq 2^{r-1}(r-1)!N\mathscr{A}\operatorname{tr}(\mathbf{B}^*\mathbf{Q}\mathbf{B})^r = 2^{r-1}(r-1)!N\mathscr{A}\operatorname{tr}(\mathbf{Q}\mathbf{B}\mathbf{B}^*)^r\\
&= 2^{r-1}(r-1)!N\mathscr{A}\operatorname{tr}(\mathbf{Q}\mathbf{F})^r
\end{aligned}$$

which is the required result.

5. The Estimation Equations

We are now prepared to turn our attention to the sampling side of the question. The least square method of estimation consists in that one expresses the sum of squares of the residual variates (which will in this case not be the ε's, but a constant times the ε's) in terms of the observations $\mathbf{X}_t$, and then minimizes this expression with respect to the unknown parameter values subject to the condition that the Jacobian $|\mathbf{J}|$ of the transformation from the residuals to the $\mathbf{X}$'s be constant. This last condition, which fixes the relative scales of the observed and residual variates, is necessary if the minimization is to be non-trivial. It is not mentioned in classical theory because for uncorrelated variates $\mathbf{J} = \mathbf{I}$, and the condition is automatically fulfilled, but as soon as correlation is introduced explicit regard must be taken to it.

Now, the estimation equations obtained upon this principle are equivalent to those obtained upon the maximum likelihood principle if the variates are assumed normally distributed. We shall use this fact to simplify our presentation; that is, we shall derive our estimation equation by maximizing a Gaussian likelihood. This is a purely formal device—we do not necessarily assume a normal distribution.

We shall now deduce the Gaussian likelihood of the $\mathbf{X}$'s from that of ε's. In Th. 2 we have already expressed the Jacobian $|\partial(X)/\partial(\varepsilon)|$ in terms of the process parameters. All that remains then is to express $\sum_{j=1}^{q}\sum_{t=1}^{N}\varepsilon_{jt}^2$ in terms of these parameters and the observations X (That we can consider the ε's for the same time range as the $\mathbf{X}$'s, i.e., for $t = 1, 2, \ldots, N$, is a result of the facts that the dependence of $\mathbf{X}_t$ upon $\boldsymbol{\varepsilon}_{t-s}$ tends to zero as $s \to \infty$, and that we are neglecting end effects.)

Theorem 5. *If $F(\omega)^{-1}$ is analytic in $e^{i\omega}$ on $(0, 2\pi)$ then*

$$\sum_{j=1}^{q}\sum_{t=1}^{N}\varepsilon_{jt}^2 \doteq N\mathscr{A}\,\mathrm{tr}(\mathbf{f}\mathbf{F}^{-1}) = (N/2\pi)\int_0^{2\pi}\mathrm{tr}[\mathbf{f}(\omega)\mathbf{F}(\omega)^{-1}]\,d\omega.$$

The proof is direct; we have

$$\begin{aligned}\sum_{j=1}^{q}\sum_{t=1}^{N}\varepsilon_{jt}^2 &= \sum_j\sum_t\sum_m\sum_n\sum_u\sum_v a_{jmu}a_{jnv}x_{m,t-u}x_{n,t-v}\\ &\doteq N\sum_j\sum_m\sum_n\sum_u\sum_v a_{jmu}a_{jnv}C_{mn}(v-u)\\ &= N\mathscr{A}\sum_s\sum_j\sum_m\sum_n\sum_u\sum_v a_{jmu}z^u a_{jnv}z^{-v}C_{mn}(s)z^s\\ &= N\mathscr{A}\sum_j\sum_m\sum_n A_{jm}(z)A_{jn}(z^{-1})f_{mn}(\omega)\\ &= N\mathscr{A}\,\mathrm{tr}(\mathbf{f}\mathbf{A}^*\mathbf{A}) = N\mathscr{A}\,\mathrm{tr}(\mathbf{f}\mathbf{F}^{-1}),\end{aligned}$$

which is the required result. If $\mathbf{F}^{-1}$ has singularities on $(0, 2\pi)$, then the autoregressive representation of the series does not exist (cf. §2), so the proof breaks down.

The Gaussian likelihood is thus given, apart from end-effects, by

$$p(x) = (2\pi)^{-N_{q/2}} \left| \frac{\partial(\boldsymbol{\varepsilon}_1, \boldsymbol{\varepsilon}_2, \ldots, \boldsymbol{\varepsilon}_N)}{\partial(\mathbf{X}_1, \mathbf{X}_2, \ldots, \mathbf{X}_N)} \right| \exp\left[-\frac{1}{2} \sum_{j=1}^{q} \sum_{t=1}^{N} \varepsilon_{jt}^2 \right]$$

$$= (2\pi)^{-Nq/2} V^{-N/2} \exp\left[-\frac{N}{4\pi} \int_0^{2\pi} \operatorname{tr}(\mathbf{f}\mathbf{F}^{-1})\, d\omega \right] \tag{5.1}$$

where V is given by Th. 2.

Theorem 6. *The least square parameter estimates are obtained by minimizing*

$$L = (N/2\pi) \int_0^{2\pi} [\log |\mathbf{F}(\omega)| + \operatorname{tr}[\mathbf{f}(\omega)\mathbf{F}(\omega)^{-1}]]\, d\omega.$$

This is immediately seen from the fact that $L = -2 \log p$.

Theorem 6 yields the estimation equations in the form of various weighted integrals of the periodograms, not a very applicable form. If, however

$$\mathbf{F}(\omega)^{-1} = \left(\sum_s c_{jks} z^s \right), \tag{5.2}$$

then

$$L = (N/2\pi) \int_0^{2\pi} \log |\mathbf{F}(\omega)|\, d\omega + N \sum_{j=1}^{q} \sum_{k=1}^{q} \sum_{s=-\infty}^{\infty} c_{jks} C_{jk}(s), \tag{5.3}$$

so that the same values enter the estimation equations as linear functions of the covariances (hence the importance of §4). The sum in (5.3) is strictly infinite with respect to s, but the c's converge to zero as $s \to \pm\infty$ (ultimately on account of the nondeterminacy of the process), so that the weight falls on the earlier covariances in (5.3), and the infinite sum can be approximated by a finite one. The practical estimation equations are obtained by minimizing L as given by equation (5.3).

Example. Consider the example of §2. We have

$$A_0 = \begin{bmatrix} 1/c & \cdot \\ \cdot & 1/e \end{bmatrix},$$

so that $V = (ce)^2$. Further

$$\mathbf{F}(\omega)^{-1} = \begin{bmatrix} \dfrac{|1 + az + bz^2|^2}{c^2} + \dfrac{d^2}{e^2|1 - fz|^2} & \dfrac{dz^{-1}}{e^2|1 - fz|^2} \\ \dfrac{dz}{e^2|1 - fz|^2} & \dfrac{1}{e^2|1 - fz|^2} \end{bmatrix}$$

so that by (5.3)

$$L/N = 2\log(ce) + c^{-2}[(1+a^2+b^2)C_{11}(0) + 2(a+ab)C_{11}(1) + 2bC_{11}(2)]$$

$$+\frac{1}{e^2(1-f^2)}\sum_s f^{|s|}[d^2C_{11}(s) + dC_{12}(s-1) + dC_{21}(s+1) + C_{22}(s)].$$

Upon minimizing with respect to a, b and c we obtain the usual estimation equations for a simple autoregressive scheme, as might have been expected, since x_{1t} has no dependence upon x_{2t}. The estimates of d, e and f are more complicated, but iterative methods give the solution quite rapidly.

The next logical step would now be to specify the parameters somewhat more exactly, but at this stage a new difficulty arises. In the single series case the author defined a *normalized spectral function* $G(\omega) = F(\omega)/V$ (Whittle, 1951, p. 80), the aim of this procedure being to separate V from the parameters of $G(\omega)$. For in this case it is the G parameters which are the essential parameters of the process, defining as they do the linear operation to which the process is equivalent, while V is really only a scale factor. Further, the least square estimates of the G parameters have second moments which are asymptotically independent of the form of the distribution function of the residual variate (at least, if the residuals are statistically independent) but this is not true of V. In fact, all that can be said of V is that its least square estimate, $\hat{V}$, is asymptotically uncorrelated with the others. However, V provides the least square criterion of the fit of a given hypothesis (the maximized Gaussian likelihood $\hat{p} = \text{const.}\ \hat{V}^{-N/2}$).

Now, it is by no means obvious in what manner this step should be generalized to the present case. One possibility would be to define $\mathbf{G}(\omega) = V^{-1/q}\mathbf{F}(\omega)$, so that the estimation equations of Th. 6 become

$$\frac{1}{2\pi q}\int_0^{2\pi} \operatorname{tr}[\mathbf{f}\hat{\mathbf{G}}^{-1}]\,d\omega = \text{minimum} = \hat{V}^{1/q}, \tag{5.4}$$

$$\int_0^{2\pi} \log|\hat{\mathbf{G}}|\,d\omega = 0, \tag{5.5}$$

where the circumflex denotes "least square estimate." Equation (5.5) is a consequence of the normalization of $\mathbf{G}$, and is an identity in the parameter values.

However, the choice of V as a parameter may not necessarily be a natural one, as examination of a few special cases will show. Another possible choice would be to take the canonical prediction variances as q of the parameters of the process, but this is generally even less satisfactory from the computational point of view. The conclusion seems to be, then, that the choice of parameters is best governed by the particular case, and that general rules cannot be laid down.

Inserting equation (5.4) in

$$\hat{p}(x) = (2\pi)^{-Nq/2}\hat{V}^{-N/2}\exp\left[-\frac{N}{4\pi\hat{V}^{1/q}}\int_0^{2\pi}\operatorname{tr}[\mathbf{f}\hat{\mathbf{G}}^{-1}]\,d\omega\right] \tag{5.6}$$

we obtain

Theorem 7. *The maximized value of the Gaussian likelihood may be expressed*

$$\hat{p}(x) = (2\pi e)^{Nq/2}\hat{V}^{-N/2}.$$

Thus the only sample function entering $\hat{p}$ is $\hat{V}$. This has its importance in test theory.

We have hitherto assumed that all variates have zero mean, but this is obviously unrealistic. Substituting $x_{jt} - \mu_j$ for x_{jt} in the expression L of Th. 6 $(j = 1, 2, \ldots, q)$, and minimizing w.r.t. μ_j, we find that the least square estimate of the mean of x_{jt} is asymptotically

$$\hat{\mu}_j = \sum_{t=1}^{N} x_{1t}/N, \tag{5.7}$$

and is thus independent of $\mathbf{F}(\omega)$.

Lastly, it may be of interest to see what value of $\mathbf{F}(\omega)$ maximizes $p(x)$ absolutely. That is, we no longer require that $\mathbf{F}$ be undetermined only to within a finite number of parameters, but let it vary freely. This is of course a problem in the calculus of variations. Differentiating L of Th. 6 with respect to the *function* $F_{jk}(\omega)$ we obtain

$$\{F^{(jk)}/|F|\} - \operatorname{tr}(\mathbf{f}\mathbf{F}^{-1}(\delta_j\delta_k)\mathbf{F}^{-1}) = 0, \tag{5.8}$$

where $F^{(jk)}$ is the co-factor of $F_{jk}(\omega)$ in $\mathbf{F}(\omega)$. Now, the first term in (5.8) is the kjth element of the matrix $\mathbf{F}^{-1}$, and the second may be directly verified as being the jkth element of the matrix $(\mathbf{f}\mathbf{F}^{-1})'(\mathbf{F}^{-1})'$. Combining the q^2 relations (5.8) we have then

$$\mathbf{F}^{-1} = \mathbf{F}^{-1}\mathbf{f}\mathbf{F}^{-1} \tag{5.9}$$

with solution $\mathbf{F}(\omega) = \mathbf{f}(\omega)$, as could well have been expected. Setting $\mathbf{F} = \mathbf{f}$ in the expression for V, we see that the least possible value of the estimated total prediction variance which can be obtained by fitting a stationary non-deterministic model is

$$V_{min} = \exp\left[\frac{1}{2\pi}\int_0^{2\pi}\log|\mathbf{f}(\omega)|\,d\omega\right] \tag{5.10}$$

6. Second Moments of the Parameter Estimates

A careful treatment of the properties of the least square estimates (i.e., the establishment of conditions for their consistency, limited optimality, etc.) would require a paper to itself. For the moment, therefore, we shall content ourselves with assuming the consistency of the estimates. It may be noted, however, that a very complete treatment already exists (Mann and Wald, 1943) for one particular case: that of the estimation of the coefficients of a finite autoregression, with results also for the non-normal case. This treatment does not lend itself to immediate generalization, however; not so much on account of the restriction of autoregressivity, as of that implied by the particular choice of parameters to be estimated. We must in general consider the case where estimation of individual coefficients is unnecessary, the whole body of coefficients being determined by a relatively small number of parameters.

If the parameters of the process are $(\theta_1, \theta_2, \ldots, \theta_p)$, let us denote the $p \times 1$ vector of derivatives of L with respect to these parameters by $\mathbf{L}_1$ (the derivative being evaluated at the true parameter values), and the $p \times p$ matrix of corresponding second derivatives by $\mathbf{L}_{11}$.

Theorem 8. *If the sample variates are normally distributed, then*

$$E(\mathbf{L}_1) = 0, \qquad E(\mathbf{L}_1\mathbf{L}_1') = 2\mathbf{M}, \qquad \textit{and} \qquad E(\mathbf{L}_{11}) = \mathbf{M},$$

where $\mathbf{M}$ *is the matrix*

$$\mathbf{M} = \left[\frac{N}{2\pi}\int_0^{2\pi} \operatorname{tr}\frac{\partial \mathbf{F}}{\partial \theta_j}\mathbf{F}^{-1}\frac{\partial \mathbf{F}}{\partial \theta_k}\mathbf{F}^{-1}\right] d\omega\bigg].$$

We have

$$\frac{\partial L}{\partial \theta_j} = \frac{N}{2\pi}\int_0^{2\pi} \operatorname{tr}\left[\frac{\partial \mathbf{F}}{\partial \theta_j}\mathbf{F}^{-1} - \mathbf{f}\mathbf{F}^{-1}\frac{\partial \mathbf{F}}{\partial \theta_j}\mathbf{F}^{-1}\right] d\omega, \tag{6.1}$$

$$\frac{\partial^2 L}{\partial \theta_j \partial \theta_k} = \frac{N}{2\pi}\int_0^{2\pi} \operatorname{tr}\left[\frac{\partial^2 \mathbf{F}}{\partial \theta_j \partial \theta_k}\mathbf{F}^{-1} - \frac{\partial \mathbf{F}}{\partial \theta_j}\mathbf{F}^{-1}\frac{\partial \mathbf{F}}{\partial \theta_k}\mathbf{F}^{-1} + \mathbf{f}\mathbf{F}^{-1}\frac{\partial \mathbf{F}}{\partial \theta_j}\mathbf{F}^{-1}\frac{\partial \mathbf{F}}{\partial \theta_k}\mathbf{F}^{-1}\right.$$
$$\left. + \mathbf{f}F^{-1}\frac{\partial \mathbf{F}}{\partial \theta_k}\mathbf{F}^{-1}\frac{\partial \mathbf{F}}{\partial \theta_j}\mathbf{F}^{-1}\right] d\omega. \tag{6.2}$$

Regarded as linear functions of the covariances, these quantities are already in the form (4.1), and their expectations and covariances may be obtained by a direct application of Th. 4. The reader will readily verify the results of the theorem. (Note that $\operatorname{cov}(\xi_1, \xi_2)$ is the coefficient of 2λ in $k_2(\xi_1 + \lambda\xi_2)$.)

In may be remarked that Th. 8 only needs to be supplemented with a few regularity conditions, and we have sufficient to prove that the root $(\hat{\theta}_1, \hat{\theta}_2, \ldots, \hat{\theta}_p)$ of $\hat{\mathbf{L}} = 0$ falls with asymptotic probability 1 in any arbitrarily small region enclosing $(\theta_1, \theta_2, \ldots, \theta_p)$, i.e., the least square estimates are consistent as least for a normal process.

Theorem 9. *For a normal process the least square estimates* $\hat{\theta}_1, \hat{\theta}_2, \ldots, \hat{\theta}_p$ *have asymptotic covariance matrix*

$$2\mathbf{M}^{-1} = \left[\frac{N}{4\pi}\int_0^{2\pi} \operatorname{tr}\left[\frac{\partial \mathbf{F}}{\partial \theta_j}\mathbf{F}^{-1}\frac{\partial \mathbf{F}}{\partial \theta_k}\mathbf{F}^{-1}\right] d\omega\right]^{-1}. \tag{6.3}$$

Let us denote the minimized value of L by $\hat{L}$, and the corresponding value of $\mathbf{L}_1$ (i.e., $\mathbf{L}_1$ with the θ's replaced by $\hat{\theta}$'s) by $\hat{\mathbf{L}}_1$. Further, let the vectors of parameters and parameter estimates be denoted by $\boldsymbol{\theta}$, $\hat{\boldsymbol{\theta}}$ respectively. Then

$$\hat{L} = L + \mathbf{L}_1'(\hat{\boldsymbol{\theta}} - \boldsymbol{\theta}) + \tfrac{1}{2}(\hat{\boldsymbol{\theta}} - \boldsymbol{\theta})'\mathbf{L}_{11}(\hat{\boldsymbol{\theta}} - \boldsymbol{\theta}) + \cdots, \tag{6.4}$$

$$\hat{\mathbf{L}}_1 = 0 = \mathbf{L}_1 + \mathbf{L}_{11}(\hat{\boldsymbol{\theta}} - \boldsymbol{\theta}) + \cdots. \tag{6.5}$$

Since the estimates are consistent, $(\hat{\boldsymbol{\theta}} - \boldsymbol{\theta})$ will with asymptotic certainty be of small order in N usually $0(N^{-1/2})$, and we may as an approximation truncate (6.4) and (6.5) at the points indicated. From (6.5) we have then that

$$\hat{\boldsymbol{\theta}} - \boldsymbol{\theta} \doteq -\mathbf{L}_{11}^{-1}\mathbf{L}_1 \doteq (E\mathbf{L}_{11})^{-1}\mathbf{L}_1, \tag{6.6}$$

since $\mathbf{L}_{11}$ has a coefficient of variation of small order in N. Thus the covariance matrix of $\hat{\boldsymbol{\theta}}_1, \hat{\boldsymbol{\theta}}_2, \ldots, \hat{\boldsymbol{\theta}}_p$, is

$$\begin{aligned} E[(\hat{\boldsymbol{\theta}} - \boldsymbol{\theta})(\hat{\boldsymbol{\theta}} - \boldsymbol{\theta})'] &\doteq (E\mathbf{L}_{11})^{-1}E(\mathbf{L}_1\mathbf{L}_1')(E\mathbf{L}_{11})^{-1} \\ &\doteq \mathbf{M}^{-1}(2\mathbf{M})\mathbf{M}^{-1} = 2\mathbf{M}^{-1}, \end{aligned} \tag{6.7}$$

which is the result of the theorem.

It is interesting to note that $\mathbf{M}$ is invariant if $\mathbf{F}$ is replaced by $\mathbf{F}^{-1}$. That is, if for a given process the roles of $\mathbf{X}_t$ and $\boldsymbol{\varepsilon}_t$ are interchanged, then the parameter estimates of the new process thus formed will have the same covariance matrix as before, despite the fact that the parameters have radically changed in interpretation.

The form of the covariance matrix may be further specified for special choices of the parameters. Thus, if we take V as one of the parameters, we find the $\hat{V}$ has asymptotic variance $2qV^2/N$, and that it is uncorrelated with the remaining parameters, which have a covariance matrix identical with that of Th. 8, except that $\mathbf{F}$ is replaced by $\mathbf{G}$. If we choose the canonical prediction variances v_j as q of the parameters, we find that the estimates $\hat{v}_j$ have variances $2v_j^2/N$ and that they are correlated neither with one another nor with remaining parameter estimates. The estimates of the means, $\hat{\mu}_j$, are uncorrelated with remaining estimates, and have covariance matrix $\mathbf{F}(0)/N$.

The question of whether the above expressions still hold if the residuals are other than normally distributed is as yet an unsolved one. We can give an example of the type of result. Suppose that apart from the parameters which enter into the expression for V we have only a single parameter, θ, so that

$$\hat{\theta} - \theta \doteq -\frac{\partial L}{\partial \theta} \bigg/ \frac{\partial^2 L}{\partial \theta^2} = \frac{\varepsilon'\mathbf{P}\varepsilon}{\varepsilon'\mathbf{Q}\varepsilon}, \tag{6.8}$$

say, where we have written $\hat{\theta} - \theta$ as a quotient of two quadratic forms in the ε_{jt}'s $(j = 1, 2, \ldots, q; t = 1, 2, \ldots, N)$. If the estimate is to be unbiased we must have $\operatorname{tr}\mathbf{P} \equiv \sum_i p_{ii} = 0$, which it is. We readily find that the condition that var $(\hat{\theta})$ be asymptotically independent of $k_4(\varepsilon)$ is that $\sum_i p_{ii}^2 = 0$ or that all p_{ii} be zero. This is found to be equivalent to the condition that all the diagonal elements of

$$\int_0^{2\pi} [\mathbf{B}(z)]^{-1} \frac{\partial \mathbf{F}}{\partial \theta} [\mathbf{B}(z^{-1})']^{-1}\, d\omega \tag{6.9}$$

be zero, which might be interpreted as requiring that all the diagonal elements of

$$\int_0^{2\pi} \log \mathbf{F}(\omega)\, d\omega \tag{6.10}$$

be independent of θ. Condition (6.9) reduces in the single series case to the condition for unbiasedness, and so is then always fulfilled.

7. Tests of Fit

Substituting $(\hat{\theta} - \theta)$ from (6.6) into (6.4) we obtain

$$L - \hat{L} \simeq \tfrac{1}{2}\mathbf{L}_1'\mathbf{L}_{11}^{-1}\mathbf{L}_1 \simeq \tfrac{1}{2}\mathbf{L}_1'(E\mathbf{L}_{11})^{-1}\mathbf{L}_1 \tag{7.1}$$

and since $2E(L_{11}) = E(L_1 L_1')$ we have

$$L - \hat{L} \simeq \mathbf{L}_1' E(\mathbf{L}_1\mathbf{L}_1')^{-1}\mathbf{L}_1, \tag{7.2}$$

so that $L - \hat{L}$ is approximately equal to the sum of squares of p orthogonalized standard statistical variates, where p is the number of parameters fitted. Now, since the cumulants of $\mathbf{L}_1$ are all $O(N)$, we may expect that $\mathbf{L}_1$ will be asymptotically normally distributed, so that the same may be said of the standardized variates, and $L - \hat{L}$ is asymptotically distributed as χ^2 with p degrees of freedom.

This simple fact may be made the basis of a theory of fit testing. For, suppose that two hypotheses H_1 and H_2 involve respectively p_1 and p_2 parameters $(p_2 > p_1)$, and that p_1 of H_2's parameters are identical with H_1's parameters, a fact which we may express by saying that H_2 includes H_1. If now the minimized values of L are respectively $\hat{L}_1$ and $\hat{L}_2$, then $L - \hat{L}_j$ is distributed as χ^2 with p_j degrees of freedom, and by the partition theorem

$$(L - \hat{L}_2) - (L - \hat{L}_1) = \hat{L}_1 - \hat{L}_2 \tag{7.3}$$

is asymptotically distributed as χ^2 with $p_2 - p_1$ degrees of freedom, since the variates contributing to $L - \hat{L}_1(\partial L/\partial\theta_j; j = 1, 2, \ldots, p_1)$ also contribute to $L - \hat{L}_2$.

Theorem 10. *If $\hat{V}_1, \hat{V}_2$ are the least square estimates of the total prediction variance on Gaussian hypotheses H_1, H_2, then*

$$\psi^2 = (N - p_2/q)\log_e(\hat{V}_1/\hat{V}_2)$$

is asymptotically distributed as χ^2 with $p_2 - p_1$ degrees of freedom.

To see this, we note that $\log \hat{p} = \text{const.} - \frac{1}{2}\hat{L} = \text{const.} - (N/2)\log \hat{V}$ (cf. Th. 7), so that $\hat{L} = \text{const.} + N \log \hat{V}$, and $\hat{L}_1 - \hat{L}_2 = N \log (\hat{V}_1/\hat{V}_2)$. The second order term, $-p_2/q$, while not directly indicated in the derivation, has been added to allow for lost "degrees of freedom" in the estimation of the residual variance V_2 (cf. Whittle, 1952a and c; Walker, 1952).

Now, the ratio $\hat{V}_1/\hat{V}_2$ measures the improvement in fit obtained by introducing the $p_2 - p_1$ extra parameters which distinguish H_2 from H_1. Alternatively, we may say that it provides a measure of the fit of H_1 relative to the wider alternatives permitted by H_2. From this point of view we have a test of fit, and Th. 9 gives the approximate distribution of the test statistic.

For an interesting alternative approach to the test problem, see Bartlett and Rajalakshman (1953).

Example. Consider the hypotheses: H_1, that the variates $\mathbf{X}_t$ are uncorrelated with past values, and H_2, that $\mathbf{X}_t$ is linearly dependent upon $\mathbf{X}_{t-1}$. That is

$$H_1 \qquad \mathbf{A}_0\mathbf{X}_t = \boldsymbol{\varepsilon}_t,$$

$$H_2 \qquad \mathbf{A}_0\mathbf{X}_t + \mathbf{A}_1\mathbf{X}_{t-1} = \boldsymbol{\varepsilon}_t.$$

Estimating the elements of $\mathbf{A}_0$, and $\mathbf{A}_0$ and $\mathbf{A}_1$ by least squares (note the A_0 has only $q(q+1)/2$ statistically distinguishable elements) we obtain

$$\hat{V}_1 = |(C_{jk}(0))|$$

and

$$\hat{V}_2 = \begin{vmatrix} (C_{jk}(0)) & (C_{jk}(1)) \\ (C_{kj}(1)) & (C_{jk}(0)) \end{vmatrix} \bigg/ |C_{jk}(0)|.$$

The test function is

$$\psi^2 = (N - (3q+1)/2)\log_e(\hat{V}_1/\hat{V}_2),$$

and is approximately distributed as χ^2 with q^2 degrees of freedom. If means have been fitted, the factor multiplying the logarithm is best modified to $(N - (5q+1)/2)$.

8. An Application to Sunspot Data

To illustrate the application of the methods developed we shall describe part of an analysis of sunspot observations. An account of the complete investigation is to appear elsewhere (Whittle, 1953) so we shall restrict ourselves to that which is relevant in the present connection.

The observations give the total sunspot area for a series of 120 six-monthly periods, and for two belts of solar latitude (16°–21° N., and 16°–21° S.), so that we have two time series, whose terms we shall denote N_t and

S_t ($t = 1, 2, \ldots, 120$). A reasonable first assumption is that the spot generating mechanism is symmetric about the equator, so that it is plausible that the two variates follow a symmetric double autoregression

$$\alpha(U)N_t + \beta(U)S_t = \varepsilon_t,$$
$$\beta(U)N_t + \alpha(U)S_t = \varepsilon_t', \tag{8.1}$$

where ε_t and ε_t' may without loss of generality be assumed uncorrelated. Adding and subtracting these two equations (cf. the example of §3) we find that the two series, $x_t = N_t + S_t$ and $y_t = N_t - S_s$ develop independently one of the other.

It is most plausible that x_t and y_t depend largely upon immediately preceding values and upon values in the previous sunspot cycle (i.e., x_{t-22} and y_{t-22}). Fitting then simple autoregression with lags 1, 2 and 22 to the two series, we find by the test of fit of Th. 9 that lags 2 are superfluous in the presence of lags 22, which improve the fit vastly. The fitted autoregressions become then

$$x_t = 0.503x_{t-1} + 0.388x_{t-22},$$
$$y_t = 0.114y_{t-1} - 0.107y_{t-22}. \tag{8.2}$$

Adding these two equations we obtain

$$N_t = 0.309N_{t-1} + 0.141N_{t-22} + 0.200S_{t-1} + 0.248S_{t-22}. \tag{8.3}$$

Equation (8.3) corresponds to neither of the equations of (8.1) directly, but is one of the equations of the so-called *reduced form*, expressing N_t and S_t solely in terms of *past* values. It is just this reduced form which we shall require, however.

Alfvén (1950) has proposed a theory of sunspot generation which leads to the conclusion that the sunspot intensity in the one hemisphere should directly depend upon that in the opposite hemisphere in the previous cycle. Consequently, the coefficient of S_{t-22} in (8.3) should be relatively large, and this is the point we intend to test.

To test the significance of the S_{t-22} coefficient, we set up a model in which it is assumed zero, i.e., in which the coefficients of x_{t-22} and y_{t-22} in (8.2) are required to be equal, so that

$$x_t = ax_{t-1} + bx_{t-22},$$
$$y_t = cy_{t-1} + by_{t-22}, \tag{8.4}$$

say. By (5.3) the estimation equation for such a scheme are derived from the condition that

$$\log(vv') + [(1 + a^2 + b^2)C_0 - 2aC_1 - 2bC_{22} + 2abC_{21}]/v$$
$$+ [(1 + c^2 + b^2)C_0' - 2cC_1' - 2bC_{22}' + 2cbC_{21}']/v' \tag{8.5}$$

be a minimum, where v, v' are the prediction variances, and C_s, C_s' the sth observed autocovariances of the x_t and y_t series respectively. The total estimated prediction variance is then $\hat{v}\hat{v}'$. We find that when we take the more general model corresponding to (8.2) then $\hat{v} = 42.969$ and $\hat{v}' = 234.587$, while for the more restricted model (8.4) $\hat{v} = 529.969$ and $\hat{v}' = 236.791$. Thus (see Th. 9)

$$\psi^2 = \left(120 - \frac{4}{2}\right) \log \frac{(529.969)(236.791)}{(421.969)(234.587)} = 28.76.$$

On hypothesis (8.4) ψ^2 should be distributed as χ^2 with one degree of freedom. The probability of obtaining so large a χ^2 as 28.76 is vanishingly small, however, and we conclude that the S_{t-22} coefficients is extremely significant.

The coefficient of N_{t-22} is also significant, so that N_t has a significant direct dependence upon N_{t-22}. This would not have been expected upon Alfvén's original theory, and a certain reflection region must be introduced to account for it, a region whose existence Alfvén has independently deduced by purely physical considerations.

While it may be expected that model (8.3) reproduces the more important features of the mechanism, it is certainly a gross oversimplification of reality. In an attempt to investigate the "fine structure" of the mechanism the author has therefore fitted a double autoregression involving all lags from 1 to 26. The simple model (8.3) is confirmed, but evidence is also found for several new phenomena. (For a full description of these results see Whittle, 1953.)

Acknowledgment

The author would wish to express his indebtedness to Professor H. Wold, Uppsala, for a number of stimulating conversations on the subject of multiple series.

References

Alfvén, H. (1950). *Cosmical Electrodynamics.* Oxford.

Bartlett, M.S. (1946). *Stochastic Processes.* Mimeographed N. Carolina Lecture Notes.

Bartlett, M.S. and Rajalakshman, D.V. (1953). *J. Roy. Statist. Soc. Ser. B*, **15**, 107–124.

Cramér, H. (1940). *Ann. Math.*, **41**, 215–230.

Frisch, R. (1938). *Statistical Versus Theoretical Relationships in Economic Macrodynamics.* Mimeographed memorandum.

Kolmogoroff, A.N. (1941). *Bull. Univ. Moscou*, **5**, 3–14.

Mann, H. and Wald, A. (1943). *Econometrica*, **11**, 173–220.

Moran, P. (1949). *Biometrika*, **36**, 63–70.
Moyal, J. (1949). *J. Roy. Statist. Soc. Ser. B*, **11**, 150–210.
Walker, A. (1952). *J. Roy. Statist. Soc. Ser. B*, **14**, 117–134.
Whittle, P. (1951). *Hypothesis Testing in Time Series Analysis*. Almquist and Wicksell, Uppsala.
Whittle, P. (1952a). *Skand. Aktuarie Tidskr.*, 48–60.
Whittle, P. (1952b). *Trabajos de Estadistica*, **3**, 43–57.
Whittle, P. (1952c). *Biometrika*, **39**, 309–318.
Whittle, P. (1953). *Astrophysical J.*
Wiener, N. (1949). *The Extrapolation, Interpolation and Smoothing of Stationary Time Series*. Wiley, New York.
Wilks, S. (1932). *Biometrika*, **24**, 471–494.
Wold, H. (1938). *A Study in the Analysis of Stationary Time-Series*. Almquist and Wicksell, Uppsala.
Wold, H. (1952). *Demand Analysis*. Wiley, New York.
Zasuhin, V. (1941). *Comptes Rendus (Doklady) de l'Académie des Sciences de l'URSS*, **23**, 435–437.

Introduction to Daniels (1954) Saddlepoint Approximations in Statistics

Elvezio Ronchetti
University of Geneva

1. Importance of the Paper

Daniels's paper provides at least three important contributions to statistics. From a methodological point of view, it introduces the use of saddlepoint techniques in statistics. These have proved to be a general and valuable tool for deriving very accurate approximations to the distribution of general estimators and test statistics both in parametric and nonparametric situations. Secondly, it shows the relationship between saddlepoint approximations and other approximations based on the seemingly unrelated idea of conjugate density. Finally, it derives a very accurate approximation of the density of the arithmetic mean of n independently identically distributed random variables. The main properties of this approximation are its simplicity and a *relative error* of order n^{-1} which leads to a very accurate approximation in small sample sizes and in the tails of the distribution. The fundamental ideas contained in Daniels's paper carry over to complex situations and the characteristics of his approximation remain valid in a large variety of problems.

To summarize: Daniels's paper was influential because it provided a useful and general tool to statistical methodology which allows us to derive very accurate approximations to the distribution of general statistics.

This commentary is organized as follows. In Section 2 we review the main points of the paper and we discuss in detail the three contributions we mentioned above. Some historical remarks which highlight Daniels's contributions are presented in Section 3. In Section 4 we review some developments which arose from the paper and discuss some future directions.

2. Main Points of the Paper

2.1. Saddlepoint Techniques

The first step in Daniels's paper is formula (2.2) where the density of the mean of n independently identically distributed random variables is expressed as the integral obtained by Fourier inversion. Daniels realized that this integral was suitable for an approximation by the method of steepest descent. The main contribution to the integral in an expansion with respect to n will come from a neighborhood of a point T which is a zero of the derivative of the function in the exponent, that is, the solution of (2.3). This point is a saddlepoint of the surface $(x, y) \mapsto u(x, y)$, where $T = x + iy$ and $u(x, y)$ is the real part of the complex function $K(T) - T\bar{x}$. (Note that in Daniels's notation $\bar{x}$ is *not* the observed value of the arithmetic mean but just some fixed value where we want to approximate the density of the mean.) Plots of such saddlepoints are given in Field and Ronchetti (1990, pp. 24–25). Therefore, by choosing the path of integration to pass through the saddlepoint and by going down on the path of steepest descent of the surface, we can capture the main contribution to the integral. In fact, the dominant term $g_n(\bar{x})$of the final expansion given by (2.6) is basically the integrand evaluated at the saddlepoint T_0.

Two important aspects appear in Daniels's derivation of the approximation. First of all, it contains already the basic elements needed to go beyond the case of the mean. Indeed, formula (2.2) holds for an arbitrary statistic provided $K(T)$ is replaced by $K_n(nT)/n$, where $K_n(\cdot)$ is the cumulant generating function of the statistic whose density is to be approximated. Although K_n is typically unknown, an approximation obtained, for instance, by Taylor expansion can be used in (2.2). The following steps of the derivation remain the same; cf. Easton and Ronchetti (1986). The second remark concerns the style. Daniels writes the first part in the style of a good applied mathematician and this allows him to focus on the ideas. In the following sections he then proves the result and the order of the approximation, in particular, by means of Watson's lemma (Section 3). Fortunately, the paper was written in 1954 for I suspect such a style would not be acceptable nowadays in the *Annals of Statistics*!

2.2. Relationship with the Method of Conjugate Densities

In Section 4 of the paper, Daniels shows that saddlepoint approximations can be viewed probabilistically by means of the method of conjugate densities. This technique had been used by Esscher (1932), Cramér (1938), Khinchin (1949), and Feller (1971) to prove large deviations results. Conjugate or associate densities also arise naturally in information theory; cf.

Kullback (1960) and Section 3 of this paper. The idea can be summarized as follows.

Suppose we are interested in approximating the density of the arithmetic mean at some point $\bar{x}$. By defining the conjugate density $h(\cdot)$ of the underlying density $f(\cdot)$ (cf. the formula after (4.2), p. 635), we recenter the problem in such a way that the expectation under h is $\bar{x}$. At this point we can use locally a normal approximation which turns out to be very accurate (with an error of order n^{-1}) because under the conjugate density we are approximating a density at the center at its expected value. This procedure is repeated for each point $\bar{x}$. The final step is to relate the approximate density computed under the conjugate to the desired density, $f_n(\bar{x})$, of the mean under the original density f. This is provided by the formula under (4.2) in the middle of p. 635. It can be seen that the "transfer factor" is again given by the integrand of (2.2) evaluated at the saddlepoint. The saddlepoint equation (2.3) in turn defines the recentering procedure at $\bar{x}$.

By proving that the method of steepest descent and the technique based on the conjugate density lead to the same approximation, Daniels gave new insight into the properties and in particular the accuracy of the approximation. Whereas Daniels (and this author) shows a clear preference for the the derivation by means of the method of steepest descent, we can argue that it is the combination of both approaches which really leads to a deep understanding of the properties of the approxiamtion.

2.3. Properties of the Approximation

Sections 5 and 7 of the paper are devoted to investigate the accuracy of the saddlepoint approximation. Formula (2.6) already shows two key elements of the approximation. If we use the dominant term $g_n(\bar{x})$ of the expansion to approximate the true density $f_n(\bar{x})$, we see that $g_n(\bar{x})$ is nonnegative and that the *relative error* is of order n^{-1}.

These are the most important properties of this approximation and a major advantage with respect to Edgeworth expansions. The latter can become negative in the tails and have at best an absolute error of order n^{-1}. Moreover, in Section 7, Daniels showed that for a wide class of underlying densities, the coefficient of the term of order n^{-1} does not depend on $\bar{x}$. Thus in such cases the relative error is of order n^{-1} uniformly. Actually, the order of the error can be improved to $O(n^{-3/2})$ for values in the range $\bar{x} - EX = O(n^{-1/2})$ by renormalization, that is, by dividing $g_n(\bar{x})$ by $\int g_n(y)\,dy$. This fact does not appear in Daniels's paper but comes out naturally by using an alternative derivation of the saddlepoint approximation proposed by Hampel (1973) which is based on the expansion of $f_n'(\bar{x})/f_n(\bar{x})$ rather than $f_n(\bar{x})$.

The accuracy of the saddlepoint approximation is confirmed numerically by the computations presented in Section 5, and by many numerical studies undertaken after Daniels's paper for a large variety of situations. In fact "[saddlepoint approximations] always seem to be more accurate than one should expect. In fact they seem to embody the truth of the matter far better than the exact formula, which is itself often difficult to comprehend. The saddlepoint approximation somehow seems to capture the essence.... One does feel that there is some underlying reason why the approximation is so much better than expected." Daniels in a conversation with P. Whittle (1993, p. 350).

A last remark concerning the computation of saddlepoint approximations. In order to recenter the density, we have to compute for each point $\bar{x}$ the solution of the implicit equation (2.3). This may appear computationally intensive and perhaps it was in 1954, especially for multivariate problems. However, the computing power available today and the fact that we are typically interested in approximating marginal distributions rather than full joint densities, eliminate the computational obstacle and put saddlepoint approximations in the class of modern statistical techniques which rely on the computer. Also from this point of view Daniels's paper was ahead of its time.

3. Historical Remarks

Historically, the method of steepest descent in mathematics can be traced back to Riemann (1892) who found an asymptotic approximation to the hypergeometric function. Debye (1909) generalized the work of Riemann. Daniels was the first to realize the potential of these methods for statistics. Actually, his first use of saddlepoint techniques goes back to his Ph.D. thesis in 1945 when he applied these ideas to a problem arising out of textiles.

Although the method of conjugate densities had been used by Esscher (1932), Cramér (1938), Khinchin (1949), and Feller (1971) to prove large deviations results, Daniels shed new light on the technique by proving that it produced the same approximation as that obtained by saddlepoint methods. Moreover, instead of looking at rates of convergence as is typically the case in the large deviations literature, Daniels used the technique to approximate directly the density of the mean and was able to show the excellent accuracy of the approximation; cf. Field and Hampel (1982).

Another approximation closely related to the saddlepoint approximation was introduced independently by Hampel (1973) who aptly coined the expression *small sample asymptotics* to indicate the spirit of these techniques. His approach is based on the idea of recentering the original distribution combined with the expansion of the logarithmic derivative f_n'/f_n rather than the density f_n itself. The pure local characteristic of this expansion leads to better approximations, for instance, in the case of bimodal distributions.

A final remark about the spelling of the word saddlepoint. At a conference in 1995 on "New Approximations Techniques For Statistical Inference" held in Ascona, Switzerland, Daniels explained why he spelled "saddlepoint" as one word: "The reason is that in the original manuscript of my 1954 paper 'saddlepoint' appeared as a typist's error, but the editor of the *Annals of Mathematical Statistics* either didn't notice or didn't mind. So I decided to leave it as it matched up rather nicely with the German form Sattelpunkt which also appears as one word."

4. Developments

After Daniels's paper, and up to the mid-1960s, saddlepoint techniques were applied to several types of problems by Daniels and a few other authors. The apparent loss of interest in this area after the mid-1960s had been followed by a strong revival at the beginning of the 1980s. References can be found in the review article by Reid (1988) and in the books by Barndorff-Nielsen and Cox (1989, Chap. 6), and Field and Ronchetti (1990).

There is a considerable literature on saddlepoint approximations for the exponential family and transformation models; cf. Barndorff-Nielsen (1983, 1986) and Reid (1996). The recent book by Jensen (1995) gives a good account of these results. In particular, an important application of Daniels's technique is the approximation of the density of the maximum likelihood estimator in an exponential family. This leads to a nice, elegant, and accurate formula often referred to as Barndorff-Nielsen's formula. It is interesting to see that Daniels had noted this result in a discussion of a paper by Cox (1958).

In Daniels's approach it is assumed that the underlying distribution F is known. A recent new development is the empirical saddlepoint approximation. It is obtained by replacing the true underlying distribution F by the empirical distribution $F^{(n)}$ of the observations. It provides an alternatives to empirical methods such as the bootstrap and empirical likelihood. A large literature is emerging in this area. Also Bayesian approximations of posterior densities are related to saddlepoint approximations.

It is impossible to enumerate all the applications of saddlepoint techniques but the number, the diversity, and accuracy of the resulting approximations clearly show the great influence of Daniels's paper in statistics.

References

Barndorff-Nielsen, O.E. (1983). On a formula for the distribution of the maximum likelihood estimator. *Biometrika*, **70**, 343–365.

Barndorff-Nielsen, O.E. (1986). Inference on full or partial parameters based on the standardized signed log likelihood ratio. *Biometrika*, **73**, 307–322.

Barndorff-Nielsen, O.E. and Cox, D.R. (1989). *Asymptotic Techniques for Use in Statistics.* Chapman & Hall, New York.

Cox, D.R. (1958). The regression analysis of binary sequences. *J. Roy. Statist. Soc. Ser. B*, **20**, 215–242.

Cramér, H. (1938). Sur un nouveau théorème-limite de la théorie des probabilités. *Actualités Scientifiques et Industrielles*, vol. 736. Hermann, Paris.

Daniels, H.E. (1954). Saddlepoint approximations in statistics. *Ann. Math. Statist.*, **25**, 631–650.

Debye, P. (1909). Näherungsformeln für die Zylinderfunktionen für große Werte des Arguments und unbeschränkt verämderliche Werte des Index. *Math. Ann.*, **67**, 535–558.

Easton, G.S. and Ronchetti, E. (1986). General saddlepoint approximations with applications to L-statistics. *J. Amer. Statist. Assoc.*, **81**, 420–430.

Esscher, F. (1932). On the probability function in collective risk theory. *Scandinavian Actuarial J.*, **15**, 175–195.

Feller, W. (1971). *An Introduction to Probability and Its Applications.* Wiley, New York.

Field, C.A. and Hampel, F.R. (1982). Small-sample asymptotic distributions of M-estimators of location. *Biometrika*, **69**, 29–46.

Field, C.A. and Ronchetti, E. (1990). *Small Sample Asymptotics.* IMS Monograph Series, vol. 13, Hayward, CA.

Hampel, F.R. (1973). Some small-sample asymptotics. In *Proceedings of the Prague Symposium on Asymptotic Statistics* (J. Hájek, ed.). Charles University, Prague, pp. 109–126.

Jensen, J.L. (1995). *Saddlepoint Approximations.* Clarendon Press, Oxford.

Khinchin, A.I. (1949). *Mathematical Foundations of Statistical Mechanics.* Dover, New York.

Kullback, S. (1960). *Information Theory and Statistics.* Wiley, New York.

Reid, N. (1988). Saddlepoint methods and statistical inference. *Statist. Sci.*, **3**, 213–238.

Reid, N. (1996). Higher order asymptotics and likelihood: A review. *Canad. J. Statist.*, to appear.

Riemann, B. (1892). *Riemann's Gesammelte Mathematische Werke.* Dover, New York, pp. 424–430.

Whittle, P. (1993). A conversation with Henry Daniels. *Statist. Sci.*, **8**, 342–353.

Saddlepoint Approximations in Statistics*

H.E. Daniels
University of Cambridge and University of Chicago

1. Introduction and Summary

It is often required to approximate to the distribution of some statistic whose exact distribution cannot be conveniently obtained. When the first few moments are known, a common procedure is to fit a law of the Pearson or Edgeworth type having the same moments as far as they are given. Both these methods are often satisfactory in practice, but have the drawback that errors in the "tail" regions of the distribution are sometimes comparable with the frequencies themselves. The Edgeworth approximation in particular notoriously can assume negative values in such regions.

The characteristic function of the statistic may be known, and the difficulty is then the analytical one of inverting a Fourier transform explicitly. In this paper we show that for a statistic such as the mean of a sample of size n, or the ratio of two such means, a satisfactory approximation to its probability density, when it exists, can be obtained nearly always by the method of steepest descents. This gives an asymptotic expansion in powers of n^{-1} whose dominant term, called the saddlepoint approximation, has a number of desirable features. The error incurred by its use is $O(n^{-1})$ as against the more usual $O(n^{-1/2})$ associated with the normal approximation. Moreover it is shown that in an important class of cases the *relative* error of the approximation is uniformly $O(n^{-1})$ over the whole admissible range of the variable.

The method of steepest descents was first used systematically by Debye for Bessel functions of large order (Watson [17]) and was introduced by Darwin and Fowler (Fowler [9]) into statistical mechanics, where it has

* Research carried out partly under sponsorship of the Office of Naval Research.

remained an indispensable tool. Apart from the work of Jeffreys [12] and occasional isolated applications by other writers (e.g., Cox [2]), the technique has been largely ignored by writers on statistical theory.

In the present paper, distributions having probability densities are discussed first, the saddlepoint approximation and its associated asymptotic expansion being obtained for the probability density of the mean $\bar{x}$ of a sample of n. It is shown how the steepest descents technique is related to an alternative method used by Khinchin [14] and, in a slightly different context, by Cramér [5]. General conditions are established under which the relative error of the saddlepoint approximation is $O(n^{-1})$ uniformly for all admissible $\bar{x}$, with a corresponding result for the asymptotic expansion. The case of discrete variables is briefly discussed, and finally the method is used for approximating to the distribution of ratios.

2. Mean of n Independent Identically Distributed Random Variables

Let x be a continuously distributed random variable with distribution function $F(x)$. Assume that a density function $f(x) = F'(x)$ exists and suppose the moment-generating function

$$M(T) = e^{K(T)} = \int_{-\infty}^{\infty} e^{Tx} f(x)\, dx$$

converges for real T in some nonvanishing interval containing the origin. Let $-c_1 < T < c_2$ be the largest such interval, where $0 \leqq c_1 \leqq \infty$ and $0 \leqq c_2 \leqq \infty$ but $c_1 + c_2 > 0$. Thus either c_1 or c_2 may be zero, though not both, and the moments need not all exist.

Consider the mean $\bar{x}$ of n independent x's. Its density function $f_n(\bar{x}) = F_n'(\bar{x})$ is given by the usual Fourier inversion formula

$$f_n(\bar{x}) = \frac{n}{2\pi} \int_{-\infty}^{\infty} M^n(it) e^{-nit\bar{x}}\, dt \tag{2.1}$$

(More generally $\int_{-\infty}^{\infty}$ may be replaced by $\lim_{t\to\infty} \int_{-t}^{t}$, but the argument is unaffected.) It is convenient here to employ the equivalent inversion formula

$$f_n(\bar{x}) = \frac{n}{2\pi i} \int_{\tau-i\infty}^{\tau'+i\infty} e^{n[K(T)-T\bar{x}]}\, dT \tag{2.2}$$

where $-c_1 < \mathscr{R}(T) < c_2$ on the path of integration, and $K(T)$ is the cumulant-generating function.

When n is large, an approximation to $f_n(\bar{x})$ is found by choosing the path of integration to pass through a saddlepoint of the integrand in such a way that the integrand is negligible outside its immediate neighbourhood. The saddle-points are situated where the exponent has zero derivative, that is where

$$K'(T) = \bar{x}. \tag{2.3}$$

We shall prove in Section 5 that under general conditions (2.3) has a single real root T_0 in $(-c_1, c_2)$ for every value of $\bar{x}$ such that $0 < F_n(\bar{x}) < 1$, and that $K''(T_0) > 0$. Let us choose the path of integration to be a straight line through T_0 parallel to the imaginary axis. Since $K(T) - T\bar{x}$ has a minimum at T_0 for real T, the modulus of the integrand must have a maximum at T_0 on the chosen path. Now we can show by a familiar argument (cf. Wintner [18, p. 14]) that on *any* admissible straight line parallel to the imaginary axis the integrand attains its maximum modulus only where the line crosses the real axis. For on the line $T = \tau + iy$,

$$\begin{aligned} |M(T)e^{-T\bar{x}}| &= e^{-\tau\bar{x}} \left| \int_{-\infty}^{\infty} e^{(\tau+iy)x}\, dF(x) \right| \\ &\leqq e^{-\tau\bar{x}} M(\tau). \end{aligned}$$

Equality cannot hold for some $y \neq 0$, otherwise $\int_{-\infty}^{\infty} e^{(\tau+iy)x}\, dF(x) = M(\tau)e^{i\alpha}$ so that $\int_{-\infty}^{\infty} e^{\tau x}[1 - \cos(yx - \alpha)]\, dF(x) = 0$, which contradicts the existence of a density function. Moreover, since $M(\tau + iy) = O(|y|^{-1})$ for large $|y|$ by the Riemann–Lebesgue lemma, the integrand cannot approach arbitrarily near its maximum modulus as $|y|$ becomes large. Consequently, for the particular path chosen, only the neighbourhood of T_0 need be considered when n is large.

The argument then proceeds formally as follows. On the contour near T_0,

$$\begin{aligned} K(T) - T\bar{x} = K(T_0) - T_0\bar{x} - \tfrac{1}{2}K''(T_0)y^2 - \tfrac{1}{6}K'''(T_0)iy^3 \\ + \tfrac{1}{24} \cdot K^{iv}(T_0)y^4 + \cdots. \end{aligned} \tag{2.4}$$

Setting $y = v/[nK''(T_0)]^{1/2}$ and expanding the integrand we get

$$\begin{aligned} f_n(\bar{x}) \sim \frac{1}{2\pi}\left[\frac{n}{K''(T_0)}\right]^{1/2} e^{n[K(T_0)-T_0\bar{x}]} \\ \cdot \int_{-\infty}^{\infty} e^{-v^2/2}\left\{1 - \tfrac{1}{6}\lambda_3(T_0)\frac{iv^3}{n^{1/2}} + \frac{1}{n}\left[\tfrac{1}{24}\lambda_4(T_0)v^4 - \tfrac{1}{72}\lambda_3^2(T_0)v^6\right] + \cdots\right\} dv, \end{aligned} \tag{2.5}$$

where $\lambda_j(T) = K^{(j)}(T)/[K''(T)]^{j/2}$ for $j \geqq 3$. The odd powers of v vanish on integration and we obtain an expansion in powers of n^{-1},

$$f_n(\bar{x}) \sim g_n(\bar{x})\left\{1 + \frac{1}{n}\left[\tfrac{1}{8}\lambda_4(T_0) - \tfrac{5}{24}\lambda_3^2(T_0)\right] + \cdots\right\}, \tag{2.6}$$

where $g_n(\bar{x}) = [n/2\pi K''(T_0)]^{1/2} e^{n[K(T_0)-T_0\bar{x}]}$. We call $g_n(\bar{x})$ the *saddlepoint approximation* to $f_n(\bar{x})$.

3. The Method of Steepest Descents

It is not apparent from the above formal development that (2.6) is a proper asymptotic expansion in which the remainder is of the same order as the last term neglected. The asymptotic nature of an expansion of this type is usually established by the method of steepest descents with the aid of a lemma due to Watson [17], the path of integration being the curve of steepest descent through T_0, upon which the modulus of the integrand decreases most rapidly. An account of the method is given by Jeffreys and Jeffreys [13]. The analysis is simplified by using a "truncated" version of Watson's lemma introduced by Jeffreys and Jeffreys for this purpose.* The special form appropriate to the present discussion is as follows.

Lemma. *If $\psi(z)$ is analytic in a neighbourhood of $z = 0$ and bounded for real $z = w$ in an interval $-A \leqq w \leqq B$ with $A > 0$ and $B > 0$, then*

$$\left(\frac{n}{2\pi}\right)^{1/2}\int_{-A}^{B} e^{-nw^2/2}\psi(w)\,dw \sim \psi(0) + \frac{1}{2n}\psi''(0) + \cdots + \frac{1}{(2n)^r}\frac{\psi^{(2r)}(0)}{r!} + \cdots \tag{3.1}$$

is an asymptotic expansion in powers of n^{-1}.

To apply the lemma, deform the contour so that for $|T - T_0| \leqq \delta$ the line $T = T_0 + iy$ is replaced by the curve of steepest descent which is that branch of $\mathscr{I}\{K(T) - T\bar{x}\} = 0$ touching $T = T_0 + iy$ at T_0, when δ is chosen small enough to exclude possible saddlepoints other than T_0. The contour is thereafter continued along the orthogonal curves of constant $\mathscr{R}\{K(T) - T\bar{x}\}$. These can easily be shown to meet the original path in points $T_0 - i\alpha$ and $T_0 + i\beta$ where $\alpha > 0$ and $\beta > 0$, if δ is small enough, since T_0 is a simple root of (2.3). The rest of the contour remains as before.

On the steepest descent curve, $K(T) - T\bar{x}$ is real and decreases steadily on each side of T_0. Make the substitution

$$\begin{aligned} -\tfrac{1}{2}w^2 &= K(T) - T\bar{x} - K(T_0) + T_0\bar{x} \\ &= \tfrac{1}{2}K''(T_0)(T - T_0)^2 + \tfrac{1}{6}K'''(T_0)(T - T_0)^3 + \cdots \\ &= \tfrac{1}{2}z^2 + \tfrac{1}{6}\lambda_3(T_0)z^3 + \tfrac{1}{24}\cdot\lambda_4(T_0)z^4 + \cdots, \end{aligned} \tag{3.2}$$

where $z = (T - T_0)[K''(T_0)]^{1/2}$, and w is chosen to have the same sign as $\mathscr{I}(z)$ on the contour. Inversion of the series yields an expansion

$$z = iw + \tfrac{1}{6}\lambda_3(T_0)w^2 + \{\tfrac{1}{24}\cdot\lambda_4(T_0) - \tfrac{5}{72}\lambda_3^2(T_0)\}iw^3 + \cdots$$

* The proof given in [13] contains an error which will be corrected in the forthcoming new edition.

convergent in some neighbourhood of $w = 0$. The contribution to (2.2) from this part of the contour is then

$$\frac{n}{2\pi i}\frac{e^{n[K(T_0)-T_0\bar{x}]}}{[K''(T_0)]^{1/2}}\int_{-A}^{B} e^{-nw^2/2}\frac{dz}{dw}\,dw,$$

to which Watson's lemma can be applied. Contributions to the integral from the rest of the contour are negligible since for $T = T_0 + iy$ with y outside $(-\alpha_1, \alpha_2)$ we have

$$|M(T)e^{-T\bar{x}}| \leqq \rho|M(T_0)e^{-T_0\bar{x}}|$$

for some $\rho < 1$, so that the extra terms contain the factor ρ^n and may be neglected. We thus obtain the asymptotic expansion

$$f_n(\bar{x}) \sim \left[\frac{n}{2\pi K''(T_0)}\right]^{1/2} e^{n[K(T_0)-T_0\bar{x}]}\left\{a_0 + \frac{a_1}{n} + \frac{a_2}{n^2} + \cdots\right\}. \tag{3.3}$$

From the Lagrange expansion of dz/dw we find

$$a_\tau = \frac{1}{2^\tau r!}\frac{d^{2\tau}}{dz^{2\tau}}\left\{\frac{z}{iw(z)}\right\}^{2\tau+1}\Bigg|_{z=0}. \tag{3.4}$$

The coefficients of this series can be shown to be identical with those obtained by the method of Section 2 (see Appendix).

4. A Generalisation of the Edgeworth Expansion

We now show how the work of Cramér [3], [4] on the Edgeworth series can also be employed to establish the asymptotic nature of (2.6), using a technique similar to that adopted by Cramér [5] and Khinchin [14].

It has been proved that on any admissible path of the form $T = \tau + iy$ the integrand attains its maximum modulus only at $T = \tau$. Consequently (2.6) is only one of a family of series for $f_n(\bar{x})$ which can be derived in a similar way by integrating along $T = \tau + iy$, τ taking any value in $(-c_1, c_2)$. In particular, $\tau = 0$ gives the Edgeworth series, whose asymptotic character was demonstrated by Cramér [3].

We have

$$e^{K(T)-T\bar{x}} = \int_{-\infty}^{\infty} e^{T(x-\bar{x})}f(x)\,dx = \int_{-\infty}^{\infty} e^{Tu}f(u+\bar{x})\,du. \tag{4.1}$$

On the path $T = \tau + iy$ we can put

$$e^{K(T)-T\bar{x}} = e^{K(t)-\tau x}\phi(y),$$

where

$$\phi(y) = \frac{\int_{-\infty}^{\infty} e^{iyu} \cdot e^{\tau u} f(u + \bar{x})\, du}{\int_{-\infty}^{\infty} e^{\tau u} f(u + \bar{x})\, du} \tag{4.2}$$

is the characteristic function for a random variable u having the density function $h(u) \propto e^{\tau u} f(u + \bar{x})$. The inversion formula (2.2) then becomes

$$\begin{aligned} f_n(\bar{x}) &= e^{n[K(\tau) - t\bar{x}]} \cdot (n/2\pi) \int_{-\infty}^{\infty} \phi^n(y)\, dy \\ &= e^{n[K(\tau) - \tau\bar{x}]} h_n(0), \end{aligned}$$

where $h_n(\bar{u})$ is the density function for the mean $\bar{u}$ of n independent u's. Using the fact that

$$\log \phi = [K'(\tau) - \bar{x}]iy + \sum_{j \geqq 2} K^{(j)}(\tau) \frac{(iy)^j}{j!}$$

we may replace $h_n(0)$ by its Edgeworth series and obtain the family of asymptotic expansions

$$\begin{aligned} f_n(\bar{x}) \sim \exp n\{K(\tau) - \tau\bar{x} - [K'(\tau) - \bar{x}]^2 / 2K''(\tau)\} \\ \cdot [n/2\pi K''(\tau)]^{1/2} \{1 + A_1/n^{1/2} + A_2/n + \cdots\} \end{aligned} \tag{4.3}$$

where

$$\begin{aligned} A_1 &= (1/3!)\lambda_3(\tau) H_3([K'(\tau) - \bar{x}][n/K''(\tau)]^{1/2}), \\ A_2 &= (1/4!)\lambda_4(\tau) H_4([K'(\tau) - \bar{x}][n/K''(\tau)]^{1/2}) \\ &\quad + (10/6!)\lambda_3^2(\tau) H_6([K'(\tau) - \bar{x}][n/K''(\tau)]^{1/2}), \end{aligned}$$

etc., the H's being Hermite polynomials.

When $\tau = 0$ this reduces to the Edgeworth series for $f_n(\bar{x})$. (Since c_1 or c_2 can be zero it may not be possible to take the expansion beyond a certain number of terms in this case). On the other hand when $\tau = T_0$, so that $K'(T_0) = \bar{x}$, all the odd powers of $n^{-1/2}$ vanish and we get (2.6), which is an expansion in powers of n^{-1}. In particular the dominant term $g_n(\bar{x})$ has the same accuracy as the first two terms of the Edgeworth series. Unlike the latter, however, $g_n(\bar{x})$ can never be negative, and is shown in Section 7 to have a further important advantage over the other approximations.

5. Examples

The method is applied to three examples.

EXAMPLE 5.1.

$$f(x) = \frac{1}{\sigma\sqrt{2\pi}} e^{-(x-m)^2/2\sigma^2}, \qquad -\infty \leqq x \leqq \infty.$$

$$K(T) = mT + \tfrac{1}{2}\sigma^2 T^2, \qquad K'(T) = m + \sigma^2 T_0 = \bar{x},$$

$$T_0 = (\bar{x} - m)/\sigma^2, \qquad K''(T_0) = \sigma^2,$$

$$g_n(\bar{x}) = \frac{1}{\sigma}\left(\frac{n}{2\pi}\right)^{1/2} e^{-n(\bar{x}-m)^2/2\sigma^2}.$$

In this case $g_n(\bar{x}) = f_n(\bar{x})$ for every value of n.

EXAMPLE 5.2.

$$f(x) = (c^\alpha/\Gamma(\alpha))x^{\alpha-1}e^{-cx}, \qquad 0 \le x \le \infty.$$

$$K(T) = -\alpha \log(1 - T/c), \qquad K'(T_0) = \alpha/(c - T_0) = \bar{x},$$

$$K''(T_0) = \alpha/(c - T_0)^2 = \bar{x}^2/\alpha,$$

$$g_n(\bar{x}) = (n\alpha/2\pi)^{1/2} e^{n\alpha}(c/\alpha)^{n\alpha}\bar{x}^{n\alpha-1}e^{-nc\bar{x}}.$$

The exact result is

$$f_n(\bar{x}) = [(nc)^{n\alpha}/\Gamma(n\alpha)]\bar{x}^{n\alpha-1}e^{-nc\bar{x}}$$

which differs from $g_n(\bar{x})$ only in that $\Gamma(n\alpha)$ is replaced by Stirling's approximation in the normalising factor. As this can always be readjusted ultimately to make the total probability unity, we can regard $g_n(\bar{x})$ as being in this sense "exact" for all n.

EXAMPLE 5.3.

$$f(x) = \tfrac{1}{2}, \qquad -1 \leqq x \leqq 1.$$

The density function for the mean of n independent rectangular variables in $(-1, 1)$ is known to be

$$f_n(\bar{x}) = \frac{n^n}{2^n(n-1)!}\sum_{s=0}^{n}(-1)^s\binom{n}{s}\left\langle 1 - \bar{x} - \frac{2s}{n}\right\rangle^{n-1}, \qquad |x| \leqq 1$$

where $\langle z \rangle = z$ for $z \geqq 0$ and $= 0$ for $z < 0$. (Seal [16] gives a historical note on this result.) We have

$$K(T) = \log\left(\frac{\sinh T}{T}\right), \qquad K'(T_0) = \coth T_0 - \frac{1}{T_0} = \bar{x},$$

$$K''(T_0) = \frac{1}{T_0^2} - \operatorname{cosech}^2 T_0,$$

$$g_n(\bar{x}) = \left(\frac{n}{2\pi}\right)^{1/2}\left\{\frac{1}{T_0} - \operatorname{cosech}^2 T_0\right\}^{-1/2}\left(\frac{\sinh T_0}{T_0}\right)^n e^{-T_0\bar{x}}.$$

When T_0 is large and positive, $\bar{x} \sim 1 - 1/T_0$ and

$$K(T_0) \sim \log(e^{T_0}/2T_0), \qquad K''(T_0) \sim 1/T_0^2.$$

So for small $1 - \bar{x}$,

$$g_n(\bar{x}) \sim (n/2\pi)^{1/2}(\tfrac{1}{2}e)^n(1-\bar{x})^{n-1}$$

which agrees with $f_n(\bar{x}) = [n^n/2^n(n-1)!](1-\bar{x})^{n-1}$ when $\bar{x} > 1 - 2/n$ except for the normalising constant, and there is similar agreement for $\bar{x}$ near -1. Actually $\log_e g_n(\bar{x})$ is remarkably close to $\log_e f_n(\bar{x})$ for quite moderate values of n over the whole range of x. Table 1 shows the agreement for $n = 6$, which could be improved by adjusting the normalising constant. With n as low as 6, $g_n(\bar{x})$ never differs from $f_n(\bar{x})$ by as much as 4 percent. This example leads one to enquire under what conditions $f_n(\bar{x})/g_n(\bar{x}) \to 1$ *uniformly for all* $\bar{x}$ as $n \to \infty$, so that the relative accuracy of the approximation is maintained up to the ends of the range of $\bar{x}$. In Section 7 we show that the result is true for a wide class of density functions.

6. The Real Roots of $K'(T) = \xi$

In this section we discuss the existence and properties of the real roots of the equation $K'(T) = \xi$, upon which the approximation $g_n(\bar{x})$ depends. The conditions are here relaxed so that the distribution need not have a density function. The moment generating function is still assumed to satisfy the conditions of Section 2, namely that

$$M(T) = e^{K(T)} = \int_{-\infty}^{\infty} e^{Tx}\, dF(x)$$

converges for real T in $-c_1 < T < c_2$ where $0 \leqq c_1 \leqq \infty$ and $0 \leqq c_2 \leqq \infty$ but $c_1 + c_2 > 0$. Throughout this section T is supposed to take real values only.

The distribution may extend from $-\infty$ to ∞, or it may be limited at either or both ends. We shall write

$$\begin{aligned} F(x) &= 0, & x &< a, \\ 0 < F(x) &< 1, & a &< x < b, \\ F(x) &= 1, & b &< x, \end{aligned}$$

where if desired $a = -\infty$ or $b = \infty$, or both. Note that $b < \infty$ implies $c_2 = \infty$ so that $c_2 < \infty$ implies $b = \infty$, and similarly for a and c_1. The converse is not true since b and c_2 (or a and c_1) can both be infinite.

We now establish the conditions under which $K'(T) = \xi$ has no real root when ξ lies outside the interval (a, b), and has a unique simple root T_0 for

Table 1

$\bar{x}$	.1	.2	.3	.4	.5	.6	.7	.8	.9
$\log_e f_6(\bar{x})$	0.419	0.172	−0.249	−0.860	−1.687	−2.778	−4.216	−6.243	−9.709
$\log_e g_6(\bar{x})$	0.445	0.199	−0.221	−0.829	−1.653	−2.742	−4.188	−6.228	−9.695
Difference	0.026	0.027	0.028	0.031	0.034	0.036	0.028	0.015	0.014

every ξ in (a, b). It is convenient to consider first the case where both a and b are finite.

Theorem 6.1. *$F(x) = 0$ for $x < a$, and $F(x) = 1$ for $x > b$ if and only if $K(T)$ exists for all real T and $K'(T) = \xi$ has no real root whenever $\xi < a$ or $\xi > b$.*

PROOF. Write

$$M(T, \xi) = e^{K(T)-T\xi} = \int_{-\infty}^{\infty} e^{T(x-\xi)}\, dF(x).$$

If $dF(x) = 0$ outside (a, b) then $M(T, \xi)$ exists for all real T and

$$M'(T, \xi) = \int_a^b (x - \xi)e^{T(x-\xi)}\, dF(x)$$

exists and has constant sign for all T when $\xi < a$ or $\xi > b$, and $K'(T) = \xi$ has then no real root.

Conversely, suppose $K(T)$ exists for all T and $K'(T) = \xi$ has no real root when $\xi < a$ or $\xi > b$. Then $M'(T, \xi)$ has constant sign in the domains $\xi < a$, $-\infty < T < \infty$ and $\xi > b$, $-\infty < T < \infty$ so that $M(T, \xi)$ is monotonic in T for these values of ξ.

Moreover $M(T, \xi)$ must increase with T for all $\xi < a$, and decrease with T for all $\xi > b$. For if $M(T, \xi)$ increases with T, then $dF(x) = 0$ for every $x < \xi$, otherwise $M(-\infty, \xi) = \infty$ and if this were true for all $\xi > b$ we should have $dF(x) = 0$ for all x. Similarly $M(T, \xi)$ cannot decrease with T for $\xi < a$.

Hence when $\xi < a$, $dF(x) = 0$ for all $x < \xi$, that is $F(x) = 0$ for all $x < a$. In the same way $F(x) = 1$ for all $x > b$.

Theorem 6.2. *Let $F(x) = 0$ for $x < a$, $0 < F(x) < 1$ for $a < x < b$, $F(x) = 1$ for $b < x$, where $-\infty < a < b < \infty$. Then for every ξ in $a < \xi < b$ there is a unique simple rool T_0 of $K'(T) = \xi$. As T increases from $-\infty$ to ∞, $K'(T)$ increases continuously from $\xi = a$ to $\xi = b$.*

PROOF. When $a < \xi < b$, $M'(-\infty, \xi) = -\infty$ and $M'(\infty, \xi) = \infty$, and $M'(T, \xi)$ is strictly increasing with T since $M''(T) > 0$. So for each ξ in $a < \xi < b$ there is a unique root T_0 of $M'(T, \xi) = 0$ and hence of $K'(T) = \xi$. Also $K''(T_0) = M''(T_0, \xi)/M(T_0, \xi)$ so that $0 < K''(T_0) < \infty$, and T_0 is a simple root and $K'(T_0)$ is a strictly increasing function of T_0.

For all $T, M'(T, b) < 0$ and so $K'(T) < b$, but $M'(T, b - \varepsilon) \to \infty$ as $T \to \infty$ for every $\varepsilon > 0$ so that $K'(T) > b - \varepsilon$ for all sufficiently large T. Hence $K'(T) \to b$ as $T \to \infty$. Similarly $K'(T) \to a$ as $T \to -\infty$. This also implies $K''(T) \to 0$ as $T \to \pm\infty$.

The theorem has an obvious interpretation in terms of the family of conjugate distributions (the term is due to Khinchin [14])

$$dF(x,T) = Ce^{Tx}\,dF(x)$$

which have mean $K'(T)$ and variance $K''(T)$.

A complication arises when a and b are allowed to be infinite. Suppose for example that a is finite but $b = \infty$, so that $K(T)$ exists in $-\infty < T < c_2$ where $0 \leqq c_2 \leqq \infty$. If $c_2 = \infty$, then $K'(T) \to \infty$ as $T \to \infty$ and the theorems still hold, for however large ξ is, $M'(T,\xi) \to \infty$ as $T \to \infty$ and so $K'(T) > \xi$ for all sufficiently large T.

But if $c_2 < \infty$ the corresponding theorems do not hold without a further condition, for it is not necessarily true that $K'(T) \to \infty$ as $T \to c_2$. Consider the class of distributions

$$dF(x) = e^{-c_2 x}\,dG(x)$$

where $\int_a^\infty dG(x) = m_0 < \infty$ and $\int_a^\infty x\,dG(x) = m_1 < \infty$, but $\int_a^\infty e^{\varepsilon x}\,dG(x) = \infty$, for all $\varepsilon > 0$. Here $K'(T)$ increases from $-\infty$ to m_1/m_0 as T increases from $-\infty$ to c_2, but $K'(T) = \infty$ for all $T > c_2$. So for $\xi > m_1/m_0$, $K'(T) = \xi$ has no real root though the distribution may extend to ∞.

The case $a = -\infty$ can be discussed similarly. In the general case where $K(T)$ exists in $-c_1 < T < c_2$ and a and b may be infinite, the conditions

$$\lim_{T \to -c_2} K'(T) = b, \qquad \lim_{T \to -c_1} K'(T) = a, \tag{6.1}$$

are required for every ξ in (a,b) to have a corresponding T_0 in $(-c_1, c_2)$. They will be automatically satisfied except when a or b is infinite and the corresponding c_1 or c_2 is finite, in which case the appropriate condition has to be stated explicitly. But even when (6.1) is not satisfied the approximation $g_n(\bar{x})$ and the expansion (2.6) can still be used whenever $\bar{x}$ lies within the restricted range of values assumed by $K'(T)$.

7. Accuracy at the Ends of the Range of $\bar{x}$

We return to the distributions having a density function, and examine the accuracy of $g_n(\bar{x})$ and the expansion (2.6) for values of $\bar{x}$ near the ends of its admissible range (a,b), where the approximation might be expected to fail. It is assumed that the appropriate conditions hold for $K'(T) = \bar{x}$ to have a unique real root T_0 for every $\bar{x}$ in (a,b).

It has been proved that

$$|f_n(\bar{x})/g_n(\bar{x}) - 1| < A(\bar{x})/n, \tag{7.1}$$

where $A(\bar{x})$ may depend on $\bar{x}$ since it is a function of T_0. The family of expansions (4.3) provides similar inequalities, and in particular an inequality of type (7.1) holds for symmetrical distributions when $g_n(\bar{x})$ is replaced by the limiting normal approximation to $f_n(\bar{x})$. But it is well known that the relative accuracy of the normal approximation, and of the Edgeworth series generally, deteriorates in most cases as $\bar{x}$ approaches the ends of its range. For example, if the interval (a, b) is finite and $f_n(\bar{x}) \to 0$ as $\bar{x} \to a$ or b, what corresponds to $A(\bar{x})$ in (7.1) becomes intolerably large as x approaches a or b, since the normal approximation can never be zero.

We now show that for a wide class of distributions $g_n(\bar{x})$ satisfies (7.1) with $A(\bar{x}) = A$, independent of $\bar{x}$, as $\bar{x}$ approaches a or b. In fact, for such distributions the asymptotic expansion of $f_n(\bar{x})/g_n(\bar{x})$ given by (2.6) is valid uniformly as $\bar{x} \to a$ or b. This will be so if $\lambda_j(T)$ remains bounded as $T \to -c_1$ or c_2 for every fixed j, so we examine the behaviour of $\lambda_j(T)$ near the ends of the interval. Equivalently, we study the conjugate distributions with density function

$$f(x, T) = Ce^{Tx} f(x) \tag{7.2}$$

whose jth cumulant is $K^{(j)}(T)$. The form of $f(x, T)$ as T approaches $-c_1$ or c_2 depends on the behaviour of $f(x)$ as x approaches a or b. For the commonest end conditions on $f(x)$, it will appear that $f(x, T)$ approximates either to the gamma form of Example 5.2 or to the normal form as $T \to -c_1$ or c_2. In the first case $\lambda_j(T)$ is bounded for given j; in the second case $\lambda_j(T) \to 0$ so that $g_n(\bar{x})$, for any n, becomes progressively more accurate as $\bar{x} \to b$, its relative error tending to a limiting value which is of smaller order than any power of n^{-1}.

We begin by discussing distributions with $b = \infty$ and first consider asymptotic forms of $f(x)$ when x is large for which $f(x, T)$ approximates to the gamma form.

EXAMPLE 7.1.

$$f(x) \sim Ax^{\alpha-1}e^{-cx}, \qquad \alpha > 0,\ c > 0.$$

Let X be large. Then

$$M^{(j)}(T) = \int_{-\infty}^{\infty} x^j e^{Tx} f(x)\, dx = I_1 + I_2,$$

where $I_1 = \int_{-\infty}^{X} x^j e^{Tx} f(x)\, dx$ is bounded as $T \to c$, and for small $c - T$,

$$\begin{aligned} I_2 &\sim \int_X^{\infty} x^{j+\alpha-1} e^{-(c-T)x}\, dx = \frac{A}{(c-T)^{j+\alpha}} \int_{X(c-T)}^{\infty} w^{j+\alpha-1} e^{-w}\, dw \\ &\sim A\Gamma(j+\alpha)/(c-T)^{j+\alpha}. \end{aligned}$$

Thus

$$K^{(j)}(T) \sim \frac{\alpha}{(c-T)^j}; \qquad \lambda_j(T) \sim \alpha^{1-j/2}, \tag{7.3}$$

for every j. In this case $f(x, T)$ tends to the gamma form as $T \to c$. The result is in fact a familiar Abelian theorem for Laplace transforms, and a more general form of it (Doetsch [7, p. 460]) can be restated for our purpose as follows.

Theorem 7.1. *Let $f(x) \sim Ax^{\alpha-1}l(x)e^{-cx}$ for $\alpha > 0$ and $c > 0$, where $l(x)$ is continuous and $l(kx)/l(x) \to 1$ as $x \to \infty$ for every $k > 0$. Then, as $T \to c$,*

$$M^{(j)}(T) \sim A \frac{\Gamma(j+\alpha)}{(c-T)^{j+\alpha}}\, l\left(\frac{1}{c-T}\right) \qquad \textit{and} \qquad \lambda_j(T) \sim \alpha^{1-j/2}.$$

This enables us to include end conditions of the form $Ax^{\alpha-1} \log x\, e^{-cx}$ or $Ax^{\alpha-1} \log\log x\, e^{-cx}$, etc. In all such cases $f(x, T)$ tends to the gamma form as $T \to c$.

In the second class of end conditions $f(x, T)$ approximates to the normal form for limiting values of T. We first consider heuristically some typical examples, again with $b = \infty$.

Example 7.2.

$$f(x) \sim A \exp(\beta x^\alpha - cx), \qquad \beta > 0,\ c > 0,\ 0 < x < 1.$$

Here we might expect $\lambda_j(T) \to 0$ as $T \to c$, for when $c - T$ is small the dominant part of $f(x, T)$ lies in the region of large x where

$$f(x, T) \sim CA \exp(\beta x^\alpha - (c-T)x).$$

This has a unique maximum at $x_0 = [\alpha\beta/(c-T)]^{1/(1-\alpha)}$ which is large for small $c - T$. If we put $x = x_0 y$ the corresponding density for y is $c' \exp[\beta x_0^\alpha(y^\alpha - \alpha y)]$ which has a sharp maximum at $y = 1$, near which it approximates to the normal form $c'' \exp[-\frac{1}{2}\beta\alpha(1-\alpha)x_0^\alpha(y-1)^2]$; it is relatively negligible elsewhere.

Example 7.3.

$$f(x) \sim A \exp(-\beta x^\alpha), \qquad \beta > 0,\ \alpha > 1.$$

In this case T can be indefinitely large. We again expect $\lambda_j(T) \to 0$ as $T \to \infty$, for

$$f(x, T) \sim CA \exp[-\beta x^\alpha + Tx]$$

has a unique maximum at $x_0 = (T/\alpha\beta)^{1/(\alpha-1)}$ which tends to infinity with T; with $x = z_0 y$ the density for y becomes $c' \exp[\beta x_0^\alpha(y^\alpha - \alpha y)]$, which approximates to $c'' \exp[-\frac{1}{2}\beta\alpha(\alpha-1)x_0^\alpha(y-1)^2]$ as before.

These examples are included in the following general theorem concerning end conditions of the type $f(x) \sim e^{-h(x)}$, where $x^2 h''(x) \to \infty$ as $x \to \infty$. Subject to a restriction on the variation of $h''(x)$ it is shown that $\lambda_j(T) \to 0$ in such cases as T tends to its upper limit.

Theorem 7.2. *Let $f(x) \sim e^{-h(x)}$ for large x, where $h(x) > 0$ and $0 < h''(x) < \infty$. Let $v(x)$ and $w(x)$ exist such that:*

$$\text{(i)}\quad [v(x)]^2 h''(x) \to \infty; \qquad \text{(ii)}\quad e^{-w(x)} h''(x) \to 0;$$

monotonically as $x \to \infty$, where

$$v(x) > 0, \qquad |v'(x)| \leqq \alpha < \infty, \qquad w(x) = \int (1/v(x))\, dx.$$

Then $\lambda_j(T) \to 0$ as T tends to its upper limiting value.

Examples 7.2 and 7.3 are covered by $v(x) = x/\gamma$ for some $\gamma > 0$, conditions (i) and (ii) reducing to $x^2 h''(x) \to \infty$, and $x^{-\gamma} h''(x) \to 0$. For $h(x) = e^x$ one can take $v(x) = \frac{1}{2}$, for $h(x) = e^{e^x}$ take $v(x) = \frac{1}{2}e^{-x}$, and so on. In all cases $v(x)/x$ is bounded and $w(x)$ increases at least as fast as $\log x$.

Since $0 < h''(x) < \infty$, $h'(x)$ is strictly increasing and $h'(x) \to c \leqq \infty$ as $x \to \infty$. Thus for large x, $f(x, T) \sim Ce^{Tx - h(x)}$ has a single maximum at the unique root x_0 of $h'(x_0) = T$, where $x_0 \to \infty$ as $T \to c \leqq \infty$.

The jth moment of $f(x, T)$ about x_0 is

$$\mu_j(T) = C \int_{-\infty}^{\infty} (x - x_0)^j f(x) e^{Tx}\, dx.$$

It will be shown that as $T \to c$ the major contribution to the integral comes from within a range $x_0 \pm \varepsilon v(x_0)$ where ε is arbitrarily small. Consider first the behaviour of $v(x)$ and $w(x)$ in this interval. Since $|v'(x)| < \alpha$ as $x \to \infty$ we have for large x_0 and $|x - x_0| \leqq \varepsilon v(x_0)$,

$$|v(x) - v(x_0)| \leqq \alpha |x - x_0| < \alpha\varepsilon v(x_0),$$

that is,

$$|v(x)/v(x_0) - 1| < \alpha\varepsilon. \tag{7.4}$$

Also for some x_1 in (x, x_0),

$$w(x) - w(x_0) = (x - x_0) w'(x_1) = (x - x_0)/v(x_1)$$

so that for $|x - x_0| < \varepsilon v(x_0)$,

$$|w(x) - w(x_0)| \leqq \varepsilon \frac{v(x_0)}{v(x_1)} \leqq \frac{\varepsilon}{1 - \alpha\varepsilon}. \tag{7.5}$$

Let X be large, but choose T so that $x_0 > X + j/T$. Then

$$\mu_j(T) \sim C \int_{-\infty}^{X} (x - x_0)^j f(x) e^{Tx}\, dx$$

$$+ C \left\{ \int_X^{x_0 - \varepsilon v(x_0)} + \int_{x_0 - \varepsilon v(x_0)}^{x_0 + \varepsilon v(x_0)} + \int_{x_0 + \varepsilon v(x_0)}^{\infty} \right\} [(x - x_0)^j \exp[x h'(x_0) - h(x)]\, dx$$

$$= I_1 + I_2 + I_3 + I_4$$

say. We examine the magnitude of each term as $T \to c$.

Since $(x_0 - x)^j e^{Tx}$ has its maximum at $(x_0 - j/T) > X$,

$$|I_1| < C(x_0 - x)^j e^{TX} F(X) < Cx_0^j e^{Xh'(x_0)}$$

For I_2,

$$|I_2| = Ce^{x_0 h'(x_0) - h(x_0)} \int_x^{x_0 - \varepsilon v(x_0)} (x_0 - x)^j e^{-\psi(x, x_0)}\, dx, \tag{7.6}$$

where we write $\psi(x, x_0) = h(x) - h(x_0) - (x - x_0)h'(x_0)$. By condition (ii) of the theorem, when $x \leqq x_0$,

$$h''(x) \geqq e^{w(x) - w(x_0)} h''(x_0) > 0$$

and for $x \leqq x_0 - \varepsilon v(x_0)$,

$$\begin{aligned} h'(x_0) - h'(x) &> h''(x_0) \int_{x_0 - \varepsilon v(x_0)}^{x_0} e^{w(x) - w(x_0)}\, dx \\ &\geqq \eta v(x_0) h''(x_0), \end{aligned}$$

where $\eta = \varepsilon e^{-\varepsilon/(1-\alpha\varepsilon)}$, by (7.5). So $\psi(x, x_0) > \eta v(x_0) h''(x_0)(x_0 - x)$, and from (7.6),

$$|I_2| < Ce^{x_0 h'(x_0) - h(x_0)} / [\eta v(x_0) h''(x_0)]^{j+1}.$$

For I_3,

$$\begin{aligned} I_3 &= Ce^{x_0 h'(x_0) - h(x_0)} \left\{ \int_{x_0 - \varepsilon v(x_0)}^{x_0} + \int_{x_0}^{x_0 \varepsilon v(x_0)} \right\} (x - x_0)^j e^{-\psi((x, x_0)}\, dx \\ &= Ce^{x_0 h'(x_0) - h(x_0)} \{J_1 + J_2\}, \end{aligned}$$

say. When $x_0 - \varepsilon v(x_0) \leqq x \leqq x_0$ we have from (i) and (ii),

$$e^{w(x) - w(x_0)} \leqq h''(x)/h''(x_0) \leqq [v(x_0)/v(x)]^2 \tag{7.7}$$

and so from (7.4) and (7.5),

$$\tfrac{1}{2} h''(x_0)(x - x_0)^2 e^{-\varepsilon/(1-\alpha\varepsilon)} \leqq \psi(x, x_0) \leqq \tfrac{1}{2} h''(x_0)(x - x_0)^2 (1 + \alpha\varepsilon)^2.$$

Putting $u = (x - x_0)[h''(x_0)]^{1/2}$ in J_1 makes the lower limit of integration become $-\varepsilon v(x_0)[h''(x_0)]^{1/2}$, which tends to $-\infty$ as $x_0 \to \infty$ for fixed ε, by (i). Hence

$$J_1 \sim (-)^j \frac{2^{(j-1)/2} \Gamma[(j+1)/2]}{[h''(x_0)]^{(j+1)/2}} \{1 + O(\varepsilon)\}.$$

In the range $x_0 \leqq x \leqq x_0 + \varepsilon v(x_0)$ the inequalities (7.7) are reversed and

$$\tfrac{1}{2} h''(x_0)(x - x_0)^2 (1 - \alpha\varepsilon)^2 \leqq \psi(x, x_0) \leqq \tfrac{1}{2} h''(x_0)(x - x_0)^2 e^{\varepsilon/(1-\alpha\varepsilon)}$$

so that

$$J_2 \sim \frac{2^{(j-1)/2} \Gamma[(j+1)/2]}{[h''(x_0)]^{(j+1)/2}} \{1 + O(\varepsilon)\}.$$

Hence if j is even

$$I_3 \sim C\frac{e^{x_0h'(x_0)-h(x_0)}}{[h''(x_0)]^{(j+1)/2}}\frac{(2j)!}{2^jj!}(2\pi)^{1/2}\{(1+O(\varepsilon)\},$$

while if j is odd

$$I_3 \sim C\frac{e^{x_0h'(x_0)-h(x_0)}}{[h''(x_0)]^{(j+1)/2}}\cdot O(\varepsilon).$$

For I_4,

$$I_4 = e^{x_0h'(x_0)-h(x_0)}\int_{x_0+\varepsilon v(x_0)}^{\infty}(x-x_0)^je^{-\psi(x,x_0)}\,dx.$$

The inequality $h''(x) \geqq [v(x_0)/v(x)]^2h''(x_0) > 0$ shows, as with I_2, that

$$I_4 < \frac{Ce^{x_0h'(x_0)-h(x_0)}j!}{[\varepsilon(1-\alpha\varepsilon)v(x_0)h''(x_0)]^{j+1}}.$$

We now show that I_3 is the dominant term. First let j be even. As $T \to c$, both I_2/I_3 and I_4/I_3 are $O\{[v^2(x_0)h''(x_0)]^{-(j+1)/2}\}$ and so $\to 0$ for fixed ε. Further,

$$|I_1|/I_3 < x_0^j[h''(x_0)]^{(j+1)/2}e^{h(X)-\psi(X,x_0)}.$$

From (ii), $h''(x_0) < e^{w(x_0)}$ as x_0 becomes large. Also since $v(x)/x$ is bounded, (i) implies that $(x-X)h''(x)v(x) \to \infty$ and so for all large enough x_0,

$$\psi(X,x_0) = \int_X^{x_0}(x-X)h''(x)\,dx > A\int_X^{x_0}(1/v(x))\,dx = A\{w(x_0)-w(X)\}$$

whatever $A > 0$. Thus

$$|I_1|/I_3 = O\{\exp[j\log x_0 - [A-\tfrac{1}{2}(j+1)]w(x_0)]\},$$

which tends to zero as $T \to c$ for fixed X if A is large enough, since $w(x)$ increases at least as fast as $\log x$.

It follows that for even j,

$$\begin{aligned}\mu_j(T) &\sim \frac{Ce^{x_0h'(x_0)-h(x_0)}}{[h''(x_0)]^{(j+1)/2}}\cdot\frac{(2j)!}{2^jj!}(2\pi)^{1/2}\\ &\sim [h''(x_0)]^{-j/2}\frac{(2j)!}{2^jj!},\end{aligned}$$

since $\mu_0(T) = 1$. Similarly when j is odd,

$$\mu_j(T) \sim [h''(x_0)]^{-j/2}O(\varepsilon)$$

as $T \to c$, so the odd moments can be made relatively negligible for arbitrarily small ε. Thus the moments tend to those of the normal distribution and $\lambda_j(T) \to 0$ as $T \to c$.

Turning now to the case where $x \leqq b < \infty$ we consider forms of $f(x)$ when $b - x$ is small. Again there are found to be two classes of end conditions for which $\lambda_j(T)$ is bounded as $T \to \infty$, where $f(x, T)$ tends respectively to the gamma and to the normal form. It is convenient to put $u = b - x$ and regard $(-)^j K^{(j)}(T)$ as the jth cumulant of the distribution of u with density $f(b-u, T) = Be^{-Tu}f(b-u)$ for $u \geqq 0$.

Example 7.4.

$$f(x) \sim A(b-x)^{\alpha-1}, \qquad \alpha > 0.$$

The jth moment of u about its origin is

$$B\int_0^\infty u^j e^{-Tu} f(b-u)\,du \sim BA\int_0^\delta u^{j+\alpha-1}e^{-Tu}du + B\int_\delta^\infty u^j e^{-Tu}f(b-u)\,du$$

$$\sim BA\,\frac{\Gamma(\alpha+j)}{T^{\alpha+j}}, \qquad T \to \infty,$$

where δ is small, the remainder being $O(e^{-T\delta})$ for large T. It follows that $\lambda_j(T) \sim (-)^j\alpha^{1-j/2}$. As in Example 7.1 this is a well-known result on Laplace transforms, and its more general form (Doetsch [7, p. 476]) yields the following theorem.

Theorem 7.3. *Let $f(x) \sim A(b-x)^{\alpha-1}l(b-x)$ for $\alpha > 0$, where $l(u)$ is continuous and $[l(ku)]/l(u) \to 1$ as $u \to 0$ for every $k > 0$. Then $\lambda_j(T) \sim (-)^j\alpha^{1-j/2}$.*

For example, $l(u)$ could be $\log(1/u)$ or $\log\log(1/u)$, etc.

The second class of end conditions is typified by the following example.

Example 7.5.

$$f(x) \sim A\exp[-\beta/(b-x)^\gamma], \qquad \beta > 0,\ \gamma > 0.$$

As in Example 7.2 we expect $\lambda_j(T) \to 0$ as $T \to \infty$, for

$$Ce^{-Tu}f(b-u) \sim CA\exp[-Tu - \beta/u^\gamma]$$

has a unique maximum at $u_0 = (\beta\gamma/T)^{1/(\gamma+1)}$, and the density function for $y = u/u_0$ is

$$C'\exp[-\beta u_0^{-\gamma}(\gamma y + y^{-\gamma})] \sim C''\exp[-\tfrac{1}{2}\beta\gamma(\gamma+1)u_0^{-\gamma}(y-1)^2].$$

The general theorem analogous to Theorem 7.2 is:

Theorem 7.4. *Let $f(x) \sim e^{-h(x)}$ for small $b - x$, where $h(x) > 0$ and $0 < h''(x) < \infty$. Let $v(u)$ and $w(u)$ exist such that:*

$$\text{(i)}\quad [v(b-x)]^2 h''(x) \to \infty; \qquad \text{(ii)}\quad e^{w(b-x)}h''(x) \to 0;$$

monotonically as $x \to b$, *where* $v(0) = 0$ *and* $w(u) = \int[1/v(u)]\,du$, *and* $0 < v'(u) \leqq \alpha < \infty$ *for* $u > 0$. *Then* $\lambda_j(T) \to 0$ *as* $T \to \infty$.

As before $h'(x)$ is strictly increasing, and $h'(x) \to \infty$ as $x \to b$ since (i) implies $(b-x)^2 h''(x) \to \infty$, and $h'(x_0) = T$ has a unique root x_0 where $x_0 \to b$ as $T \to \infty$. Thus $f(b-u, T)$ has a unique maximum at $u_0 = b - x_0$ for large T, and $u_0 \to 0$ as $T \to \infty$. The jth moment of u about u_0 is

$$\mu_j(T) = B \int_0^\infty (u-u_0)^j e^{-Tu} f(b-u)\,du.$$

We write

$$\int_0^\infty = \int_0^{u_0 - \varepsilon v(u_0)} + \int_{u_0-\varepsilon v(u_0)}^{u_0+\varepsilon v(u_0)} + \int_{u_0+\varepsilon v(u_0)}^{\delta} + \int_\delta^\infty,$$

where ε and δ are small. The proof then proceeds with appropriate modifications as in Theorem 7.2.

8. Discrete Variables

The discussion has so far been concerned with approximations to probability densities, but the saddlepoint method provides similar approximations to probabilities when the variable is discrete, and indeed it is typically used for this purpose in statistical mechanics. Consider, for example, a variable x which takes only integral values $x = r$ with nonzero probabilities $p(r)$. The moment generating function,

$$M(T) = e^{K(T)} = \sum_r p(r) e^{Tr} \tag{8.1}$$

is assumed to satisfy conditions (6.1).

The mean $\bar{x}$ of n independent x's can take only values $\bar{x} = r/n$, for which the probabilities are

$$p_n(\bar{x}) = \frac{1}{2\pi i} \int_{\tau-i\pi}^{\tau'+i\pi} e^{n[K(T) - T\bar{x}]}\,dT \tag{8.2}$$

analogous to (2.2). The contour is again chosen to be the line $T = T_0 + iy$ passing through the unique real saddle point T_0, but it now terminates at $T_0 \pm i\pi$. This ensures that the integrand attains its maximum modulus at T_0 but nowhere else on the contour, provided we exclude cases where $p(r) = 0$ except at multiples of an integer greater than unity. The discussion of Section 2 shows that the maximum modulus is attained when y satisfies $\cos(ry - \alpha) = 1$ for some α and all integral r, and $y = 0$ is the only possible value in $(-\pi, \pi)$. The argument then proceeds as before and leads to the approximation

$$p_n(\bar{x}) \sim \frac{e^{n[K(T_0)-T_0\bar{x}]}}{[2\pi n K''(T_0)]^{1/2}}\{1+O(n^{-1})\}, \tag{8.3}$$

where $\bar{x} = r/n$ and r is an integer.

As an example, consider the binomial distribution

$$p(r) = \binom{N}{r}(1-p)^{\tau}p^{n-\tau}.$$

Here

$$K(T) = N\log\{1+p(e^T-1)\}, \qquad K'(T_0) = Npe^{T_0}/[1+p(e^{T_0}-1)] = \bar{x},$$

$$e^{T_0} = [\bar{x}/(N-\bar{x})]\cdot[(1-p)/p], \qquad K''(T_0) = \bar{x}(N-\bar{x})/N,$$

$$p_n(\bar{x}) \sim \frac{N^{nN+1/2}}{(2\pi n)^{1/2}}\frac{(1-p)^{n(N-\bar{x})}p^{n\bar{x}}}{(N-\bar{x})^{n(N-\bar{x})+1/2}\bar{x}^{n\bar{x}+1/2}}\{1+O(n^{-1})\}.$$

This is the familiar intermediate form obtained on replacing the factorials by Stirling's approximation before passing to the normal limit.

9. Ratio of Sums of Random Variables

The saddlepoint technique can also be applied to the distribution of ratios. Cramér [4] has shown that if x and y are two independent random variables with densities $f_1(x)$ and $f_2(y)$ and characteristic functions $\phi_1(t)$ and $\phi_2(u)$, and if $y \geqq 0$, the density function for $r = x/y$ is given by

$$f(r) = \frac{1}{2\pi i}\int_{-\infty}^{\infty}\phi_1(t)\phi_2'(-rt)\,dt \tag{9.1}$$

provided y has a finite mean. (Gurland [11] relaxes this condition by introducing principal values. Cramér states the condition differently and appears to require unnecessarily that x shall have a finite mean also.) Cramér deduced the result from the distribution of $x - ry$ for fixed r, but it also follows on applying Parseval's theorem to the formula

$$f(r) = \int_0^{\infty} f_1(ry)f_2(y)y\,dy, \tag{9.2}$$

where y must have a finite mean to make $\phi_2'(-rt)$ the Fourier transform of $iyf_2(y)$. In terms of cumulant generating functions (9.1) takes the form

$$f(r) = \frac{1}{2\pi i}\int_{\tau-i\infty}^{\tau'+i\infty} e^{K_1(T)+K_2(-\tau T)}K_2'(-rT)\,dT.$$

Let $x_1, x_2, \ldots, x_n$ and $y_1, y_2, \ldots, y_n$ be independent random samples from these distributions, their sums being X and Y. The density for $R = x/y$ is then

$$f_{n_1,n_2}(R) = \frac{n_2}{2\pi i}\int_{\tau-i\infty}^{\tau'+i\infty} e^{n_1K_1(T)+n_2K_2(-RT)}K_2'(-RT)\,dT.$$

When n_1 and n_2 are large, an approximation is found by passing the path of integration through a saddlepoint T_0 of the exponential part of the integrand, given by

$$n_1K_1'(T_0) - n_2RK_2'(-RT) = 0. \tag{9.3}$$

Assuming conditions (6.1) to be satisfied, both $K_1'(T)$ and $K_2'(T)$ are increasing functions of T taking every admissible value of X and Y respectively as T varies over its appropriate interval, so that to every R there is a single real root T_0 of (9.3). (However, it is possible for the same T_0 to correspond to more than one value of R, since $TK_2'(T)$ is not necessarily monotonic and so dT_0/dR may change sign). Proceeding as before, expanding $K_2'(-RT)$ also, we obtain an asymptotic expansion whose dominant term is

$$g_{n_1,n_2}(R) = \frac{n_2K_2'(-RT_0)e^{n_1K_1(T_0)+n_2K_2(-RT_0)}}{\{2\pi[n_1K_1''(T_0)+n_2R^2K_2''(-RT_0)]\}^{1/2}},$$

the remainder being relatively $O(n^{-1})$ where $n = \min(n_1, n_2)$.

Example 9.1.

$$f_1(x) = A_1x^{\alpha_1-1}e^{-\beta_1x}, \qquad f_2(y) = A_2y^{\alpha_2-1}e^{-\beta_2y},$$

where $x, y, \alpha_1, \beta_1, \alpha_2, \beta_2$ are all positive. In this case

$$T_0 = \frac{1}{R}\frac{(n_2\alpha_2\beta_1R - n_1\alpha_1\beta_2)}{(n_1\alpha_1 + n_2\alpha_2)}.$$

The approximation is found to be

$$g_{n_1,n_2}(R) = \frac{\beta_1^{n_1\alpha_1}\beta_2^{n_2\alpha_2}(n_1\alpha_1+n_2\alpha_2)^{n_1\alpha_1+n_2\alpha_2-1/2}}{(2\pi)^{1/2}(n_1\alpha_1)^{n_1\alpha_1-1/2}(n_2\alpha_2)^{n_2\alpha_2-1/2}}\frac{R^{n_1\alpha_1-1}}{(\beta_1R+\beta_2)^{n_1\alpha_1+n_2\alpha_2}},$$

which differs from the exact density function only in the normalising constant, and so is "exact" in the sense of Example 5.2. This suggests that there may again be a class of distributions for which the relative error is bounded uniformly over the whole range of R for every n.

An extension of (9.1) is available when the variables are not independent (Cramér [6, p. 317, Ex. 6]; Geary [10]). If (x, y) has a bivariate density function $f(x, y)$ everywhere and characteristic function $\phi(t, u)$, and if $y \geqq 0$, the density function for $r = x/y$ is

$$f(r) = \frac{1}{2\pi i}\int_{-\infty}^{\infty}\left[\frac{\partial\phi(t,u)}{\partial u}\right]_{u=-rt} dt \tag{9.4}$$

provided the integral is absolutely convergent, which requires y to have a finite mean. The following proof of (9.4) shows the integrand to be proportional to a characteristic function which attains its maximum modulus only at $t = 0$, so that the previous methods are applicable. Corresponding to (9.2) we have

$$f(r) = \int_0^\infty f(ry, y)y\,dy. \tag{9.5}$$

Write $\eta = E(y)$ and define a new distribution with density and characteristic function

$$h(x, y) = \frac{1}{\eta}\,yf(x, y), \qquad \phi(x, y) = \frac{1}{\eta}\frac{\partial\phi(t, u)}{i\partial u}. \tag{9.6}$$

From (9.5) it is seen that $(1/\eta)f(r)$ can be regarded as the probability density at zero of the variable $w = x - ry$, where (x, y) has the distribution (9.6). The result then follows from the fact that w has the characteristic function

$$\frac{1}{\eta}\left[\frac{\partial\phi(t, u)}{i\partial u}\right]_{u=-rt}.$$

For a random sample of n, the ratio R of the sums X and Y has density

$$f_n(R) = \frac{n}{2\pi i}\int_{\tau-i\infty}^{\tau'+i\infty} e^{nK(T,-RT)}\left[-\frac{1}{T}\frac{\partial K(T, -RT)}{\partial R}\right]\partial T$$

in terms of the bivariate cumulant generating function. The saddlepoint approximation is

$$g_n(R) = \left\{\frac{n}{2\pi K''(T_0, -RT_0)}\right\}^{1/2} e^{nK(T_0,-RT_0)}\left[\frac{-1}{T_0}\frac{\partial K(T_0, -RT_0)}{\partial R}\right],$$

where

$$\frac{\partial K(T_0, -RT_0)}{\partial T_0} = 0.$$

Example 9.2. Let $x = \frac{1}{2}u^2$ and $y = \frac{1}{2}v^2$, where u and v have a bivariate normal distribution with unit variances and correlation coefficient ρ. Thus $R = X/Y$ is a "variance ratio" calculated from two equal correlated samples. The exact distribution of R has been given by Bose [1] and Finney [8]. We find

$$K(T, -RT) = \tfrac{1}{2}\log\{1 + (R-1)T - RT^2(1-\rho^2)\},$$

$$T_0 = \frac{(R-1)}{2R(1-\rho^2)}, \qquad K''(T_0, -RT_0) = \frac{4R(1-\rho^2)^2}{[(1+R)^2 - 4\rho^2 R]},$$

$$\frac{-1}{T_0}\frac{\partial K(T_0, -RT_0)}{\partial R} = \frac{(R+1)(1-\rho^2)}{[(1+R)^2 - 4\rho^2 R]},$$

$$g_n(R) = 2^{n-1}\left(\frac{n}{2\pi}\right)^{1/2}\frac{(1-\rho^2)^{n/2}R^{(n/2)-1}(1+R)}{[(1+R)^2 - 4\rho^2 R]^{(n+1)/2}},$$

which again agrees with the exact distribution except for the normalising constant.

In the most general situation where the sample members are themselves correlated, the saddlepoint method can still be applied. In each particular case the contribution to the integral from parts of the contour outside a neighbourhood of the saddlepoint must be established as negligible. One can obtain, in this way an approximation to the distribution of the sample serial correlation coefficient of lag 1 from a linear Markoff population. With the usual "circular" definitions it turns out to be the approximation given by Leipnik [15], but a similar approximation can also be found for the noncircular case. A detailed account of this work will appear elsewhere.

10. Acknowledgments

I am much indebted to D.V. Lindley and Sir Harold Jeffreys for many stimulating discussions and useful suggestions, to a referee for comments which led to improvements in the paper, and to D.A. East for computing Table 1.

11. Appendix

The identity of the series (2.6) and (3.3) may be established as follows. For the contour $T = T_0 + iy$ the inversion formula is

$$f_n(\bar{x}) = \frac{n}{2\pi} e^{n[K(T_0)-T_0\bar{x}]} \int_{-\infty}^{\infty} e^{-nw^2/2}\, dy,$$

where w^2 is defined by (3.2). With $v = y[nK''(T_0)]^{1/2}$ and $s = n^{-1/2}$ this becomes

$$f_n(\bar{x}) = \frac{1}{2\pi}\left[\frac{n}{K''(T_0)}\right]^{1/2} e^{n[K(T_0)-T_0\bar{x}]} \int_{-\infty}^{\infty} e^{-w^2(ivs)/2s^2}\, dv$$

with $z = ivs$ in (3.2). To get (2.5) the integrand is expanded as a power series in s. Term-by-term integration gives (2.6). Thus

$$\exp\left[-\frac{w^2(ivs)}{2s^2}\right] = \sum_{m=0}^{\infty} b_m(v)s^m,$$

where

$$b_m(v) = \frac{1}{m!}\frac{\partial^m}{\partial s^m}\exp\left[-\frac{w^2(ivs)}{2s^2}\right]\Bigg|_{s=0}$$

$$= \frac{1}{m!}v^m\frac{\partial^m}{\partial x^m}\exp\left[-\frac{v^2w^2(ix)}{2x^2}\right]\Bigg|_{x=0}.$$

Since $w^2(ix)/x^2 \sim 1 + O(x)$, for small x we can interchange the order of differentiation with respect to x and integration with respect to v. Only the even terms survive and

$$\int_{-\infty}^{\infty} b_{2r}(v)\,dv = \frac{1}{(2r)!}\frac{d^{2r}}{dx^{2r}}\int_{-\infty}^{\infty} v^{2r}\exp\left[-\frac{v^2w^2(ix)}{2x^2}\right]dv\Bigg|_{x=0}$$

$$= \frac{(2\pi)^{1/2}}{2^r r!}\frac{d^{2r}}{dx^{2r}}\left[\frac{x}{w(ix)}\right]^{2r+1}\Bigg|_{x=0} = (2\pi)^{1/2}a_r,$$

putting $z = ix$ in (3.4).

References

[1] S. Bose. On the distribution of the ratio of variance of two samples drawn from a given normal bivariate correlated population. *Sankhya*, Vol. 2 (1935), pp. 65–72.

[2] D.R. Cox. A note on the asymptotic distribution of the range. *Biometrika*, Vol. 35 (1948), pp. 311–315.

[3] H. Cramér. On the composition of elementary errors. *Skand. Aktuarietids.*, Vol. 11 (1928), pp. 13–74, 141–180.

[4] H. Cramér. *Random Variables and Probability Distributions*. Cambridge University Press, Cambridge, 1937.

[5] H. Cramér. Sur un nouveau théorème-limite de la théorie des probabilités. *Actualités Scientifiques et Industrielles*, No. 736. Hermann, Paris, 1938.

[6] H. Cramér. *Mathematical Methods of Statistics*. Princeton University Press, Princeton, NJ, 1946.

[7] G. Doetsch. *Handbuch der Laplace Transformation*, Band I. Birkhäuser, Basel, 1950.

[8] D.J. Finney. The distribution of the ratio of estimates of two variances in a sample from a normal bivariate population. *Biometrika*, Vol. 30 (1938), pp. 190–192.

[9] R.H. Fowler. *Statistical Mechanics*. Cambridge University Press, Cambridge, 1936.

[10] R.C. Geary. Extension of a theorem by Harald Cramér on the frequency distribution of a quotient of two variables. *J. Roy. Statist. Soc.*, Vol. 17 (1944), pp. 56–57.

[11] J. Gurland. Inversion formulae for the distribution of ratios. *Ann. Math. Statist*, Vol. 19 (1948), pp. 228–237.

[12] H. Jeffreys. *Theory of Probability*. Oxford University Press, Oxford, 1948.

[13] H. Jeffreys and B.S. Jeffreys. *Methods of Mathematical Physics*. Cambridge University Press, Cambridge, 1950.

[14] A.I. Khinchin. *Mathematical Foundations of Statistical Mechanics*. Dover, New York, 1949.
[15] R.B. Leipnik. Distribution of the serial correlation in a circularly correlated universe. *Ann. Math. Statist.*, Vol. 18 (1947), pp. 80–87.
[16] H.L. Seal. Spot the prior reference. *J. Inst. Actuaries, Students' Soc.*, Vol. 10 (1951), pp. 255–258.
[17] G.N. Watson. *Theory of Bessel Functions*. Cambridge University Press, Cambridge, 1948.
[18] A. Wintner. *The Fourier Transforms of Probability Distributions*. Johns Hopkins University, Baltimore, 1947.

Introduction to Cooley and Tukey (1965) An Algorithm for the Machine Calculation of Complex Fourier Series

I.J. Good

History is not merely what happened, it is what happened in the context of what might have happened.

H.R. Trevor-Roper (Valedictory Lecture, Oxford, 1980).

The fast Fourier transform (FFT) ... has changed ... science and engineering so much so that ... *life as we know it would be very different without the FFT. ... life as we know it would be considerably different if ... the FFT community had made systematic and heavy use of matrix-vector notation!*

Van Loan (1992, Preface).

Fourier (frequency) analysis is important in statistics, mathematics, and all the hard sciences (as in x-ray crystallography), and in signal processing, so Fast Fourier Transforms (FFTs) are widely applicable. For example, Costain (1974) said the FFTs are regarded as an economic breakthrough in geophysics, especially in seismology.

The purpose of a Fast Fourier Transform is to cut down *greatly* on the work of calculating a Discrete Fourier Transform (DFT), defined and discussed below. But small savings *in addition* are also worthwhile and special-purpose equipment can be valuable. Hence a vast "literature" on various FFTs and their implementations has been generated. In 1984, a Bibliography of over 2000 items had already been compiled (Heideman and Burrus, 1984), and the later books on FFTs give many more references, so the present survey can be only a condensed introduction. It includes some early personal material that might otherwise have remained unknown to historians.

Fourier analysis is valuable because of:

(i) the ubiquity of frequency (cycles per second);
(ii) the relation to convolution and autocorrelation;

(iii) orthogonality of the sinusoidal system;
(iv) "completeness" of the system; and
(v) periodicity.

Let us first discuss *frequency*.

Frequencies are fundamental for communication because they are invariant in a system describable by linear differential equations, though the amplitudes and phases may be changed. So Nature invented the cochlea, an analogue device in the inner ear which does a frequency (or Fourier) analysis of sound. It is shaped like a cone, wrapped in a spiral to save space, somewhat like a French horn. Different parts of the cochlea resonate to distinct frequencies and neurons lead to the brain from these parts. The *quality* of a subjective sensation depends on what part of the body the neurons come from, in this case what part of the cochlea, whereas the *intensities*, the Fourier coefficients, depend on the frequencies of the activities of the neurons! Perhaps snails do a Fourier analysis with their shells but I have not seen this mentioned anywhere.

Another breakthrough was made by *humans*, the ancient Greeks, when they invented trigonometry, mainly spherical (for applications to astronomy). Trigonometrical series were widely used in the eighteenth century largely in the theory of vibrations and, by Euler, for interpolation when raw data is given, using finite trigonometric series. Euler showed that rational functions can be expanded as infinite trigonometrical series in say $(0, 2\pi)$. Fourier in 1811 (see Kline (1972, p. 672)) intuited that any function with only a finite number of discontinuities could be represented by a convergent trigonometrical series in a finite interval (Fourier, 1822). (That would make the Ptolemaic system of epicycles bound to seem superficially like a good theory if there is no penalty for complexity. Compare Jeffreys (1957, pp. 130–132).) Laplace, Lagrange, and Legendre objected to Fourier's lack of rigor but Fourier's intuition was right if the function is absolutely integrable, and wherever the function is differentiable (Tolstov, 1962/76, p. 75). The coefficients can be obtained by the familiar integration formula which depends on the orthogonality of the Fourier system.

For a very short biography of Fourier, see Bracewell (1986, Chap. 24) who cites Herivel (1975).

The Fourier integral or transform is attributed by Kline (1972, Chap. 28) to Laplace, Fourier, Cauchy, and Poisson. But the characteristic function in probability theory, which is of course a Fourier or Fourier–Stieltjes integral, was introduced, according to Cramér (1937, p. 23n), by Lagrange (1770–1773) and used systematically by Laplace (1812/1820).

When writing down the values of raw data there can be only a finite number of observations and, for interpolation, it is convenient for these to be equally spaced. The data need have no periodicity, but periodicity with any sufficiently large period can be artificially enforced by adding zeros outside the range of the observations. This process may be called *circularization* (just as the function x has a Fourier series in $(-\pi, \pi)$ and over the real line is replaced by a saw-tooth function). Circularization was used by Euler and by Gauss, leading soon to what is now called the Discrete Fourier Transform or DFT. This is a transformation from a sequence or vector of say t numbers a_r $(r = 0, 1, \ldots, t-1)$ to another sequence a_s^* $(s = 0, 1, \ldots, t-1)$ by means of the formula

$$a_s^* = \sum_{r=0}^{t-1} a_r \omega^{rs} \qquad (\omega = e^{2\pi i/t}), \tag{1}$$

or, in matrix/vector notation,

$$\mathbf{a}^* = F\mathbf{a} \tag{2}$$

where F, the DFT matrix, is (ω^{rs}). Its trace is the Gaussian sum which is important for the theory of quadratic residues in the theory of numbers.

For the case where t is not a prime, Gauss (1876, posthumous), "while considering the problem of determining the orbit of certain asteroids from sample locations" (Heideman et al. 1985, p. 269), found a way to compute $\mathbf{a}^*$ by means of some short-cuts. This was presumably the first FFT if Euler missed it. Gauss's work was translated by Goldstine (1977, pp. 249–258) who says that it was neglected and was rediscovered by Cooley and Tukey (1965). The work is described by Heideman et al. whose account is better than Gauss's, who did not submit his account for publication. But they erred when they said that Good (1979) called the DFT the "discrete Gauss transform." I used that expression, in honor of Gauss, for the transform with matrix $(\rho^{(r-s)^2})$, although Gauss might not have been directly concerned with that matrix.

Cooley and Tukey give two algorithms known as the Mixed Radix and its special case the Radix Two method (MRFFT and R2FFT). Gauss's method was the MRFFT, (See also Singleton (1969).)

Another anticipation of Cooley and Tukey (1965) was in Runge and König (1924), according to Cooley et al. (1967, p. 77). Then Danielson and Lanczos (1942) (cited by Whittaker and Robinson (1944/1967, Chap. X)) again anticipated the R2FFT, attributing it to Lanczos. They cite two papers by Runge, of 1903 and 1905, which *missed* using Lanczos's method. They didn't know the relevant passage in Runge and König (1924). Lanczos used a method that reduced an analysis for $t2^n$ points to 2^n analyses for t points. This is the essential basis of the R2FFT.

Consider now the DFT as such rather than methods for calculating it.

In Good (1950a) there is a relevant exercise on page 58. Briefly, a sequence of digits each have independent *chances* $a_0, a_1, \ldots, a_9$ of being $0, 1, \ldots, 9$. These digits are added modulo 10 in blocks of N, thus producing a new sequence with chances $a'_0, a'_1, \ldots, a'_9$. Then

$$a'_r = \frac{1}{10} \sum_{s=0}^{9} (a_s^*)^N \omega^{-rs}. \tag{3}$$

The method has since been used by RAND Corporation, by SWAC and by ERNIE (British Government lottery) and presumably in many other places.

Good (1950b), mentioned that the eigenvalues of a circulix (circulant matrix; the ending *ant* is traditionally used, for example, by Muir (1933/60), for various kinds of *determinant*), are given by the DFT of the top row, and that this leads to an easy way to invert a circulix. This paper was written because the inversion had been required by Quenouille (1949) for trend elimination.

In Good (1951) the name Discrete Fourier Transform was introduced and a multidimensional DFT was defined, by natural analogy with a multidimensional Fourier transform, as

$$f^*(\mathbf{s}) = \sum_{\mathbf{r}} f(\mathbf{r})\omega_1^{r_1 s_1}\omega_2^{r_2 s_2}\cdots\omega_n^{r_n s_n} \qquad (\omega_1 = e^{2\pi i/t_1}, \text{etc.}). \tag{4}$$

(The bold **r** and **s** have the natural meanings and both run through $t_1 t_2 \ldots t_n$ values.) Statements were given for the inversion formula, a Rayleigh–Parseval formula, and for the theorem that the multidimensional DFT of the sum of independent random variables on a torus (anchor ring) is the product of the separate DFTs. The paper shows that, if $t_1, t_2, \ldots, t_n$ *are mutually prime in pairs, then the multidimensional DFT can be expressed as a one-dimensional DFT, and conversely, by making use of the Chinese Remainder Theorem* (Sun Tsu, first century: see Hardy and Wright (1938, pp. 93 and 105)). This correspondence has since been used by many writers. It is often called the Sino correspondence. Good (1951) also provided a method, using the DFT and toroidalization, for approximating problems involving continuous random motion in a bounded simply-connected domain surrounded by an absorbing barrier.

In Good (1955) the DFT is used for a simple evaluation of Gauss's sum $\sum \omega^{r^2}$ and to prove that the Legendre polynomial is given by the series (new at the time)

$$P_N(x) = \frac{1}{t}\sum_{r=0}^{t-1}\{x + (x^2 - 1)^{1/2}\cos(2\pi r/t)\}^N \qquad \text{for all} \quad t > N. \tag{5}$$

By letting $t \to \infty$ we get Laplace's integral. The method (related to an analogue of Poisson's summation formula), which is widely applicable, depends on a device, using roots of unity, due to Thomas Simpson (1757/1758), for summing every tth coefficient in a power series whose sum is otherwise neatly expressible. The method provides an application for FFTs (Good, 1969). It should much increase the usefulness of generating functions. When the power series is a polynomial the method extracts a single coefficient for all sufficiently large t (compare circularization with the "circumference" large enough), and does much the same for an infinite series if one term dominates. This idea was used, with a two-variable generating function, for finding the "exact" distribution of Pearson's χ^2 for an "equiprobable" multinomial; see Good (1957b, p. 880; 1981, 1982, 1987) and Holtzman and Good (1986). Finding the distribution, to five d.p., when there are 1001 objects in 1001 cells, took only 6 minutes in 1981 on an IBM 3032 (programmer: Luen Fure Lee) although the number of Hardy–Ramanujan partitions of 1001 is about 2×10^{31}. Pearson's chi-squared test was selected by *Science 84* as one of the "20 discoveries [in this century] that changed our lives."

Good (1962) gave an elementary proof for a discrete multidimensional form of Poisson's summation formula, and it might have an application to confounded factorial experiments.

Further interesting applications of the DFT are given in Good (1953a, 1957a, 1958/1960, 1959). A table of 29 DFT pairs was compiled by Conolly and Good (1977). Apart from the mathematical interest, they could be used for checking FFT programs. One pair at least, due to Gauss, has an application to concert hall acoustics and to spread spectrum communication (Schroeder, 1980).

To discuss Good (1958), which was the only paper cited by Cooley and Tukey (1965), some work of Frank Yates must first be mentioned.

Yates (1937, pp. 11–16, 27–30) defined the main effects and interactions for a 2^n factorial experiment by means of a simple adding-and-subtracting algorithm. He emphasized (p. 13) that the expressions "are really a matter of definition." Yates told me at an Open Day of the Atlas Computer Laboratory on May 21 or 22, 1966 that he did not know that the main effects and interactions could be regarded as the components of a mod $2n$-dimensional DFT. (This is a Walsh–Hadamard transform on which there have been several symposia at the Naval Research Laboratory.) This way of defining the interactions had been pointed out in Good (1953b, 1967) who said also that the inversion formula (which Yates did not mention) was useful for checking the calculations. It was independently used by Hunter (1966) for detecting aberrant observations. See also Kempthorne (1952, p. 259).

Yates's little algorithm (or definition) was generalized in Good (1958) to apply to any multidimensional DFT and much more generally. This point of view shows that the equation of high-order interactions to zero is analogous to filtering out high frequencies. Then, by applying the inverse transform, we can smooth the original observations. We return to this topic soon in more general terms.

The main theorem in Good (1958) uses Kronecker or direct products of matrices (McDuffee, 1946/1956, Chap. VII). For its relevance see equations (11) and (12) below. (The calculation of the DFT reduces to repeated summations: see Good (1971).) It is proved that, if M is any t by t matrix, then the nth Kronecker power of M, denoted by $M^{[n]}$, is the ordinary nth power of a "sparse" matrix G (a matrix containing many zeros),

$$M^{[n]} = G^n, \tag{6}$$

where

$$G = \{m_{r_1,s_n}\delta_{r_2}^{s_1}\delta_{r_3}^{s_2}\cdots\delta_{r_n}^{s_{n-1}}\}. \tag{7}$$

This indicates succintly (perhaps too much so!) that an element of G vanishes unless the last $(n-1)$ components of **r** match the first $(n-1)$ components of **s**.

For example, when $t = 3$ and $n = 2$, the theorem states that $M^{[2]} = G^2$ where

$$G = \{m_{r_1,s_2}\delta_{r_2}^{s_1}\} \qquad (r_1, r_2, s_1, s_2 = 0, 1, 2), \tag{8}$$

$$M = \begin{bmatrix} m_{00} & m_{01} & m_{02} \\ m_{10} & m_{11} & m_{12} \\ m_{20} & m_{21} & m_{22} \end{bmatrix} \qquad (\text{with } t^n = 3^2 = 9 \text{ elements}), \tag{9}$$

and G, when unpacked, is the t^n by t^n (= 9 by 9) matrix

$$\begin{array}{c|ccccccccc} r_1r_2\backslash s_1s_2 & 00 & 01 & 02 & 10 & 11 & 12 & 20 & 21 & 22 \\ \hline 00 & m_{00} & m_{01} & m_{02} & 0 & 0 & 0 & 0 & 0 & 0 \\ 01 & 0 & 0 & 0 & m_{00} & m_{01} & m_{02} & 0 & 0 & 0 \\ 02 & 0 & 0 & 0 & 0 & 0 & 0 & m_{00} & m_{01} & m_{02} \\ 10 & m_{10} & m_{11} & m_{12} & 0 & 0 & 0 & 0 & 0 & 0 \\ 11 & 0 & 0 & 0 & m_{10} & m_{11} & m_{12} & 0 & 0 & 0 \\ 12 & 0 & 0 & 0 & 0 & 0 & 0 & m_{10} & m_{11} & m_{12} \\ 20 & m_{20} & m_{21} & m_{22} & 0 & 0 & 0 & 0 & 0 & 0 \\ 21 & 0 & 0 & 0 & m_{20} & m_{21} & m_{22} & 0 & 0 & 0 \\ 22 & 0 & 0 & 0 & 0 & 0 & 0 & m_{20} & m_{21} & m_{22} \end{array}. \tag{10}$$

The first t^{n-1} rows of G are obtained by regarding the first row of M (of length t) as a "stair" and using it to construct a flight of t^{n-1} stairs. The next t^{n-1} rows of G are similarly obtained from the second row of M, and so on, and the whole matrix G consists of a staircase of t flights of stairs. Thus each element of M occurs t^{n-1} times in G. Of the t^{2n} elements of G, at most t^{n+1} are nonzero.

When a vector is multiplied n times by G, the number of products of numbers (real or complex) is at most nt^{n+1} whereas t^{2n} would be required if the matrix $M^{[n]}$ were used once instead of G being used n times. The natural generalization, to the case where the moduli are $t_1, t_2, \ldots, t_n$, follows by a direct generalization of the proof of the special case on page 364 of Good (1958). The generalization is stated explicitly, and independently, by Andrews and Caspari (1970b, p. 83) and by Good (1971). The number of products of numbers, real or complex, is then reduced from τ^2, where $\tau = t_1t_2\ldots t_n$, to $(t_1 + t_2 + \cdots t_n)\tau$. This reduction was stated for a somewhat different reason in Good (1958, p. 371).

The multidimensional DFT, in matrix/vector notation, can be written as

$$a_{\mathbf{s}}^{*} = \Omega a_{\mathbf{r}}^{*}, \tag{11}$$

where Ω is the τ by τ matrix with components $\omega^{\mathbf{rs}} = \omega_1^{r_1 s_1} \ldots \omega_n^{r_n s_n}(r_1,\ s_1 = 0, 1, \ldots, t_1 - 1; \ldots; r_n,\ s_n = 0, 1, \ldots, t_n - 1)$. The rows (and columns) can be ordered lexicographically. Thus Ω is the Kronecker product of matrices (t_1 by $t_1, \ldots, t_n$ by t_n),

$$\Omega = \left(\omega_1^{r_1 s_1}\right) \otimes \cdots \otimes \left(\omega^{r_n s_n}\right), \tag{12}$$

and the generalization of (6), just mentioned, can be applied in order to calculate a multidimensional DFT. For this purpose the numbers $t_1, t_2, \ldots, t_n$ do not have to be distinct or relatively prime.

Now, as mentioned before, a one-dimensional DFT can be expressed as a multidimensional DFT, when $t_1, t_2, \ldots, t_n$ are mutually prime, by making use of the Chinese Remainder Theorem. For example, if $\tau = 7 \cdot 8 \cdot 9 \cdot 11 = 5544$, the numerical work would be cut by a factor of 158 (not allowing for subsidary tricks). So this algorithm is certainly an FFT. It is often called the prime factor algorithm or PFA but *mutually prime factor FFT* is a fuller description.

In the MRFFT (and R2FFT) a correspondence between the unidimensional and multidimensional DFT is defined, even if $t_1, t_2, \ldots, t_n$ are not mutually prime. The method depends on expressing r and s in mixed radix or reversed mixed radix form. But the PFA has been found to have other advantages in later developments. It helped to inspire Winograd (1978) (as he acknowledges) to produce a novel FFT (the WFFT). He used a polynomial version of the Chinese Remainder Theorem. For various values of τ, the WFFT requires appreciably fewer multiplications than the MRFFT.

For more details concerning the relationship between the MRFFT and PFA see Good (1971). That paper also introduces a Fourier–Galois transform (from one integer sequence to another one) for its mathematical or coding interest. It was invented in a conference in 1956 but not then published.

The Cooley–Tukey paper became immediately in big demand and has been very heavily cited. This was largely due to the efforts of Richard L. Garwin who said he acted as a "missionary," "knows a good thing when he sees it," and "... much more than an idea or even a publication is required...." (Garwin, 1969). His great impact was acknowledged by Cooley (1969).

In 1956, December 6, John Tukey came to dinner. I told him I had a fast way to do practical Fourier analysis, generalizing Yates's definition of interactions in 2^n factorial experiments. My work was submitted for publication 8 months later (Good, 1958) and it was the only paper cited in the Cooley–Tukey paper. At first the three of us thought the two methods were identical but later we saw they were not though they are closely related.

On 1957, September 10, Richard Garwin came to dinner. He told me about one of the experiments in which parity in physics was found to break down in weak interactions (Garwin et al., 1957). I had intended to tell him of my FFT work but his justified enthusiasm about the physics experiment was so great that I thought it would be rude to change the subject entirely. He wrote to me on February 9, 1976, and said "Had we talked about it [an FFT] in 1957, I could have stolen it from you then instead of from John Tudkey in 1963 or thereabouts." If I had been less polite, the history of the FFT would probably have been entirely different!

For important later developments, up to 1978, the book by McClellan and Rader (1979) is strongly recommended. It gives first the background on number theory, groups, rings, and fields. Then there are excellent introductions to the reprints of thirteen well selected articles by Kolba and Parks; Rader (who found an ingenious FFT for prime t by using a primitive root of t); Agarwal and Cooley; Winograd (2); Good; Pollard; Agarwal and Burrus (2); McClellan; Leibowitz; Reed, and Truong; and Nussbaumer. The book concludes with a Fortran program for the WFFT.

Elementary number theory has become a branch of applied mathematics. In that respect see also Burr (ed.) (1991) and Schroeder (1984/86). The more recent book by Van Loan (1992) unifies most of the FFT literature in terms of the factorization of the DFT matrix mentioned above. There are many other good books on the FFT.

In image processing, as distinct from signal analysis, other orthogonal systems can be as useful as the Fourier system. The factorization of Kronecker products [generalizing equation (6)] is again relevant and is used by Andrews and Kane (1970) or Andrews and Caspari (1970a) for the discussion of generalized spectral analysis. See also Fino and Algazi (1977).

For multidimensional transforms in general, let us ask what corresponds to a low-pass filter, the equating to zero of higher-order coefficients, and then using the inverse transform, in the hope of smoothing the data or image. (Note that there is more to image processing than just smoothing the image. One example is "edge enhancement"; see, for example, the indexes of Andrews (1970), and Nadler and Smith (1993).) For example, we could equate to zero all coefficients having "complicated" definitions (e.g., high-order interactions, as in the analysis of multidimensional contingency tables by maximum entropy (Good, 1963)) or they could be merely *squashed toward zero, the squashing being greater for the more complicated terms.* Or we could see which of the complicated terms in the DFT, when equated to zero (or squashed), followed by an inverse DFT, have the least effect on some measure of roughness of the adjusted values of $a_{\mathbf{r}}$. One interpretation of degree of complexity would be *the size of the denominator of* $(s_1/t_1) + (s_2/t_2) + \cdots + (s_n/t_n)$ *when it is reduced to its lowest terms.* This resembles the concept of "sequency" (number of zero crossings), a term attributed to H.F. Harmuth by Andrews and Caspari (1970b, p. 93; or 1970a, p. 22). That concept was introduced for a somewhat different purpose.

Conclusion

An invention can be the mother of necessity as in supply-side economics. The FFTs collectively provide an example. Their great value is tied to modern electronics so it is perhaps fair to call them collectively the algorithm of the century in spite of Gauss (1876).

References

Andrews, H.C. (1970). *Computer Techniques in Image Processing*. Academic Press, New York.

Andrews, H.C. and Caspari, K.L. (1970a). A generalized technique for spectral analysis. *IEEE Trans. Comput.*, **C19**, 16–25.

Andrews, H.C. and Caspari, K.L. (1970b). Orthogonal transformations. In Andrews (1970, pp. 73–103).

Andrews, H.C. and Kane, J. (1970). Kronecker matrices, computer implementation, and generalized spectra. *J. Assoc. Comput. Mach.*, **17**, 260–268.

Bracewell, R.N. (1986). *The Fourier Transform and Its Applications*, 2nd edn. revised. McGraw-Hill, New York.

Burr, S.A., ed. (1991). *The Unreasonable Effectiveness of Number Theory*. American Mathematical Society, Providence, RI.

Conolly, B. and Good, I.J. (1977). A table of discrete Fourier transform pairs. *SIAM J. Appl. Math.* **32**, 810–822. Errata, **33**, 534.

Cooley, J.W. (1969). The impact of the Fast Fourier Transform. *IEEE Trans. Audio Electroacoustics*, **AU-17**, 66–68.

Cooley, J.W. Lewis, P.A.W., and Welch, P.D. (1967). Historical notes on the Fast Fourier Transform. *IEEE Trans. Audio Electroacoustics*, **AU-15**, 76–79.

Cooley, J.W. and Tukey, J.W. (1965). An algorithm for the machine calculation of complex Fourier series. *Mathe. Comput.*, **19**, 297–301.

Costain, J.K. (1974). Personal communication.

Cramér, H. (1937). *Random Variables and Probability Distributions*. Cambridge University Press, London.

Danielson, G.C. and Lanczos, C. (1942). Some improvements in practical Fourier analysis and their application to X-ray scattering from liquids. *J. Franklin Inst.*, **233**, 365–380, 435–452.

Fino, B.T. and Algazi, V.R. (1977). A unified treatment of discrete fast unitary transforms. *SIAM J. Comput.*, **6**, 700–717.

Fourier, J.B.J. (1822). *Théorie Analytique de la Chaleur*. Also in his *Oeuvres*, **1**.

Garwin, R.L. (1969). The Fast Fourier Transform as an example of the difficulty in gaining wide use for a new technique. *IEEE Trans. Audio Electroacoustics*, **AU-17**, 68–72.

Garwin, R.L., Lederman, L.M., and Weinrich, M. (1957). Observation of the failure of parity and charge conjugation in meson decays. *Phys. Rev.*, **105**, 1415–1417.

Gauss, K.F. (1876). Theoria interpolationes methodo nova tractata. *Nachlass* [literary remains] *Werke*, **III**, 265–327, esp. pp. 303ff. (Cited by Goldstine (1977, p. 249).)

Goldstine, H.H. (1977). *A History of Numerical Analysis*. Springer-Verlag, New York.

Good, I.J. (1950a). *Probability and the Weighing of Evidence*. Griffin, London; Hafner, New York.

Good, I.J. (1950b). On the inversion of circulant matrices. *Biometrika*, **37**, 185–186.

Good, I.J. (1951). Random motion on a finite Abelian group. *Proc. Cambridge Philos. Soc.*, **47**, 756–762. Corrigenda, **48** (1952), 368.

Good, I.J. (1953a). The serial test for sampling numbers and other tests for randomness. *Proc. Cambridge Philos. Soc.*, **49**, 276–284.

Good, I.J. (1953b). Review of M.H. Quenouille. Design and Analysis of Experiment. *Ann. Eugenics*, **18**, 263–266.

Good, I.J. (1955). A new finite series for Legendre polynomials. *Proc. Cambridge Philos. Soc.*, **51**, 385–388.

Good, I.J. (1957a). On the serial test for random sequences. *Ann. Math. Statist.*, **28**, 262–264.

Good, I.J. (1957b). Saddle-point methods for the multinomial distribution. *Ann. Math. Statist.*, **28**, 861–881.

Good, I.J. (1958). The interaction algorithm and practical Fouriers analysis. *J. Roy. Statist. Soc. Ser. B*, **20**, 361–372; **22** (1960), 372–375.

Good, I.J. (1959). Contribution to the discussion of a paper by W.E. Thomson. ERNIE–A mathematical and statistical analysis. *J. Roy. Statist. Soc. Ser. A*, **122**, 326–328.

Good, I.J. (1962). Analogues of Poisson's summation formula. *Amer. Math. Monthly*, **69**, 259–266.

Good, I.J. (1963). Maximum entropy for hypothesis formulation, especially for multidimensional contingency tables. *Ann. Math. Statist.*, **34**, 911–934.

Good, I.J. (1967). Checks on Yates's algorithm. *Biometrics*, **23**, 573.

Good, I.J. (1969). Polynomial algebra: An application of the fast Fourier transform. *Nature*, **222**, 1302.

Good, I.J. (1971). The relationship between two Fast Fourier Transforms. *IEEE Trans. Comput.*, **C20**, 310–317. Reprinted in McClellan and Rader (1970).

Good, I.J. (1979). The inversion of the discrete Gauss transform. *Appl. Anal.*, **9**, 205–218.

Good, I.J. (1981). The fast calculation of the exact distribution of Pearson's chi-squared and of the number of repeats within the cells of a multinomial by using a Fast Fourier Transform. C119 in *J. Statist. Comput. Simulation*, **14**, 71–78; **16**, 319; **15** (1982), 336–337.

Good, I.J. (1987). A survey of the use of the Fast Fourier Transform for computing distributions. *J. Statist. Comput. Simulation*, **28**, 87–93.

Good, I.J. and K. Caj. Doog, (1958/60). A paradox concerning rate of information. *Inform. and Control*, **1**, 113–126; **2** (1959), 195–197; Effective sampling rates for signal detection: Or can the Gaussian model be salvaged? *Inform. and Control*, **3** (1960), 116–140.

Hardy, G.H. and Wright, E.M. (1938). *An Introduction to the Theory of Numbers*. Clarendon Press, Oxford.

Heideman, M.T. and Burrus, C.S. (1984). A bibliography of fast transform and convolution algorithm. Department of Electrical Engineering, Technical Report 8402, Rice University, Houston, TX. (Cited by Van Loan (1992, p. xi).)

Heideman, M.T., Johnson, D.H., and Burrus, C.S. (1985). Gauss and the history of the Fast Fourier Transform. *Arch. Hist. Exact Sci.*, **34**, 265–277.

Herivel, J. (1975). *Joseph Fourier, the Man and the Physicist*. Clarendon Press, Oxford.

Holtzman, G.I. and Good, I.J. (1986). The Poisson and chi-squared approximations as compared with the true upper-tail probability of Pearson's χ^2 for equiprobable multinomials. *J. Statist. Plann. Inference*, **13**, 283–295.

Hunter, J.S. (1966). The inverse Yates algorithm. *Technometrics*, **8**, 177–183.

Jeffreys, Harold (1957). *Scientific Inference*, 2nd edn. Cambridge University Press, Cambridge.

Kempthorne, O. (1952). *Design and Analysis of Experiments*. Wiley, New York.
Kline, M. (1972). *Mathematical Thought from Ancient to Modern Times*. Oxford University Press, New York.
Lagrange, J.L. (1770–73). Mémoire sur l'utilité de la methode de prendre le milieu entre les résultats de plusieurs observations. *Misc. Taurinensia*, **5**, 167–232. *Oeuvres*, **2**. Paris (1868).
Laplace, P.S. (1812/14/20). *Théorie Analytique des Probabilités*. Courcier, Paris.
MacDuffee, C.C. (1946/56). *The Theory of Matrices*. Chelsea, New York.
McClellan, J.H. and Rader, C.M. (1979). *Number Theory in Digital Signal Processing*. Prentice-Hall, Englewood Cliffs, NJ.
Muir, Thomas (1933/60). *A Treatise on the Theory of Determinants*. Longman Green, London; Dover, New York, 1960.
Nadler, M. and Smith, E.P. (1993). *Pattern Recognition Engineering*. Wiley, New York.
Quenouille, M.H. (1949). On a method of trend elimination. *Biometrika*, **36**, 75–91.
Runge, C. and König, H. (1924). *Die Grundlehren der Matematischen Wissenschaften, Vorlesungen über Numerische Rechnen*, vol. **11**. Springer-Verlag, Berlin.
Schroeder, M.R. (1980). Constant-amplitude antenna arrays with beam patterns whose lobes have equal magnitudes. *Archiv für Electronic und Uebertragungstechnik* (Electronics and communication), **34**, 165–168.
Schroeder, M.R. (1984/86). *Number Theory in Science and Communication*. Springer-Verlag, New York.
Simpson, Thomas (1757/58). The invention of a general method for determining the sum of every 2nd, 3rd, 4th, or 5th, etc., term of a series, taken in order, the sum of the whole series being known. *Philos. Trans. Roy. Soc. London*, **50**, 757–769.
Singleton, R.C. (1969). An algorithm for computing the mixed-radix Fast Fourier Transform. *IEEE Trans. Audio Electroacoustics*, **AU-17**, 93–103.
Tolstov, G.P. (1962/76). *Fourier Series*. Translated from the Russian. Prentice-Hall, Englewood Cliffs, NJ; Dover, New York.
Van Loan, C. (1992). *Computational Frameworks for the Fast Fourier Transform*. SIAM, Philadelphia, PA.
Whittaker, E.T. and Robinson, G. (1944/1967). *The Calculus of Observations*, 4th edn. Blackie, London; Dover, New York.
Winograd, S. (1978). On computing the discrete Fourier transform. *Math. Comput.*, **32**, 175–199. Reprinted in McClellan and Rader (1979).
Yates, F. (1937). *The Design and Analysis of Factorial Experiments*. Imperial Bureau of Soil Science, Harpenden, England.

An Algorithm for the Machine Calculation of Complex Fourier Series*

James W. Cooley and John W. Tukey

An efficient method for the calculation of the interactions of a 2^m factorial experiment was introduced by Yates and is widely known by his name. The generalization to 3^m was given by Box et al. [1]. Good [2] generalized these methods and gave elegant algorithms for which one class of applications is the calculation of Fourier series. In their full generality, Good's methods are applicable to certain problems in which one must multiply an N-vector by an $N \times N$ matrix which can be factored into m sparse matrices, where m is proportional to $\log N$. This results in a procedure requiring a number of operations proportional to $N \log N$ rather than N^2. These methods are applied here to the calculation of complex Fourier series. They are highly composite numbers. The algorithm is here derived and presented in a rather different form. Attention is given to the choice of N. It is also shown how special advantage can be obtained in the use of a binary computer with $N = 2^m$ and how the entire calculation can be performed within the array of N data storage locations used for the given Fourier coefficients.

Consider the problem of calculating the complex Fourier series

$$X(j) = \sum_{k=0}^{N-1} A(k) \cdot W^{jk}, \qquad j = 0, 1, \ldots, N-1, \tag{1}$$

where the given Fourier coefficients $A(k)$ are complex and W is the principal Nth root of unity,

$$W = e^{2\pi i/N}. \tag{2}$$

* Received August 17, 1964. Research in part at Princeton University under the sponsorship of the Army Research Office (Durham). The authors wish to thank Richard Garwin for his essential role in communication and encouragement.

A straightforward calculation using (1) would require N^2 operations where "operation" means, as it will throughout this note, a complex multiplication followed by a complex addition.

The algorithm described here iterates on the array of given complex Fourier amplitudes and yields the result in less than $2N \log_2 N$ operations without requiring more data storage than is required for the given array A. To derive the algorithm, suppose N is a composite, i.e., $N = r_1 \cdot r_2$. Then let the indices in (1) be expressed

$$\begin{aligned} j &= j_1 r_1 + j_0, & j_0 &= 0, 1, \ldots, r_1 - 1, & j_1 &= 0, 1, \ldots, r_2 - 1, \\ k &= k_1 r_2 + k_0, & k_0 &= 0, 1, \ldots, r_2 - 1, & k_1 &= 0, 1, \ldots, r_1 - 1. \end{aligned} \tag{3}$$

Then, one can write

$$X(j_1, j_0) = \sum_{k_0} \sum_{k_1} A(k_1, k_0) \cdot W^{jk_1 r_2} W^{jk_0}. \tag{4}$$

Since

$$W^{jk_1 r_2} = W^{j_0 k_1 r_2}, \tag{5}$$

the inner sum, over k_1, depends only on j_0 and k_0 and can be defined as a new array,

$$A_1(j_0, k_0) = \sum_{k_1} A(k_1, k_0) \cdot W^{j_0 k_1 r_2}. \tag{6}$$

The result can then be written

$$X(j_1, j_0) = \sum_{k_0} A_1(j_0, k_0) \cdot W^{(j_1 r_1 + j_0) k_0}. \tag{7}$$

There are N elements in the array A_1, each requiring r_1 operations, giving a total of Nr_1 operations to obtain A_1. Similarly, it takes Nr_2 operations to calculate X from A_1. Therefore, this two-step algorithm, given by (6) and (7), requires a total of

$$T = N(r_1 + r_2) \tag{8}$$

operations.

It is easy to see how successive applications of the above procedure, starting with its application to (6), give an m-step algorithm requiring

$$T = N(r_1 + r_2 + \cdots + r_m) \tag{9}$$

operations, where

$$N = r_1 \cdot r_2 \cdots r_m. \tag{10}$$

If $r_j = s_j t_j$ with $s_j, t_j > 1$, then $s_j + t_j < r_j$ unless $s_j = t_j = 2$, when $s_j + t_j = r_j$. In general, then, using as many factors as possible provides a minimum to (9), but factors of 2 can be combined in pairs without loss. If we are able to choose N to be highly composite, we may make very real gains. If all r_j are equal to r, then, from (10) we have

$$m = \log_r N \tag{11}$$

and the total number of operations is

$$T(r) = rN \log_r N. \tag{12}$$

If $N = r^m s^n t^p \cdots$, then we find that

$$\begin{aligned} \frac{T}{N} &= m \cdot r + n \cdot s + p \cdot t + \cdots, \\ \log_2 N &= m \cdot \log_2 r + n \cdot \log_2 s + p \cdot \log_2 t + \cdots, \end{aligned} \tag{13}$$

so that

$$\frac{T}{N \log_2 N}$$

is a weighted mean of the quantities

$$\frac{r}{\log_2 r}, \quad \frac{s}{\log_2 s}, \quad \frac{t}{\log_2 t}, \quad \ldots,$$

whose values run as follows:

r	$\dfrac{r}{\log_2 r}$
2	2.00
3	1.88
4	2.00
5	2.15
6	2.31
7	2.49
8	2.67
9	2.82
10	3.01.

The use of $r_j = 3$ is formally most efficient, but the gain is only about 6% over the use of 2 or 4, which have other advantages. If necessary, the use of r_j up to 10 can increase the number of computations by no more than 50%. Accordingly, we can find "highly composite" values of N whithin a few percent of any given large number.

Whenever possible, the use of $N = r^m$ with $r = 2$ or 4 offers important advantages for computers with binary arithmetic, both in addressing and in multiplication economy.

The algorithm with $r = 2$ is derived by expressing the indices in the form

$$\begin{aligned} j &= j_{m-1} \cdot 2^{m-1} + \cdots + j_1 \cdot 2 + j_0, \\ k &= k_{m-1} \cdot 2^{m-1} + \cdots + k_1 \cdot 2 + k_0, \end{aligned} \tag{14}$$

where j_v and k_v are equal to 0 or 1 and are the contents of the respective bit positions in the binary representation of j and k. All arrays will now be written as functions of the bits of their indices. With this convention (1) is written

$$X(j_{m-1},\ldots,j_0)=\sum_{k_0}\sum_{k_1}\cdots\sum_{k_{m-1}}A(k_{m-1},\ldots,k_0)\cdot W^{jk_{m-1}\cdot 2^{m-1}+\cdots+jk_0}, \quad (15)$$

where the sums are over $k_v = 0, 1$. Since

$$W^{jk_{m-1}\cdot 2^{m-1}}=W^{j_0k_{m-1}\cdot 2^{m-1}}, \quad (16)$$

the innermost sum of (15), over k_{m-1}, depends only on $j_0, k_{m-2}, \ldots, k_0$ and can be written

$$A_1(j_0,k_{m-2},\ldots,k_0)=\sum_{k_{m-1}}A(k_{m-1},\ldots,k_0)\cdot W^{j_0k_{m-1}\cdot 2^{m-1}}. \quad (17)$$

Proceeding to the next innermost sum, over k_{m-2}, and so on, and using

$$W^{j\cdot k_{m-l}\cdot 2^{m-l}}=W^{(j_{l-1}\cdot 2^{l-1}+\cdots+j_0)k_{m-l}\cdot 2^{m-l}}, \quad (18)$$

one obtains successive arrays,

$$A_l(j_0,\ldots,j_{l-1},k_{m-l-1},\ldots,k_0)$$
$$=\sum_{k_{m-l}}A_{l-1}(j_0,\ldots,j_{l-2},k_{m-1},\ldots,k_0)\cdots W^{(j_{l-1}\cdot 2^{l-1}+\cdots+j_0)\cdot k_{m-l}\cdot 2^{m-l}} \quad (19)$$

for $l = 1, 2, \ldots, m$.

Writing out the sum this appears as

$$\begin{aligned}A_l(j_0,&\ldots,j_{l-1},k_{m-l-1},\ldots,k_0)\\ &=A_{l-1}(j_0,\ldots,j_{l-2},0,k_{m-l-1},\ldots,k_0)\\ &\quad+(-1)^{j_{l-1}}i^{j_{l-2}}A_{l-1}(j_0,\ldots,j_{l-2},1,k_{m-l-1},\ldots,k_0)\\ &\quad\cdot W^{(j_{l-3}\cdot 2^{l-3}+\cdots+j_0)\cdot 2^{m-l}}, \qquad j_{l-1}=0,1.\end{aligned} \quad (20)$$

According to the indexing convention, this is stored in a location whose index is

$$j_0\cdot 2^{m-1}+\cdots+j_{l-1}\cdot 2^{m-l}+k_{m-l-1}\cdot 2^{m-l-1}+\cdots+k_0. \quad (21)$$

It can be seen in (20) that only the two storage locations with indices having 0 and 1 in the 2^{m-l} bit position are involved in the computation. Parallel computation is permitted since the operation described by (20) can be carried out with all values of $j_0, \ldots, j_{l-2}$, and $k_0, \ldots, k_{m-l-1}$ simultaneously. In some applications* it is convenient to use (20) to express A_l in terms of A_{l-2}, giving what is equivalent to an algorithm with $r = 4$.

* A multiple-processing circuit using this algorithm was designed by R.E. Miller and S. Winograd of the IBM Watson Research Center. In this case $r = 4$ was found to be most practical.

The last array calculated gives the desired Fourier sums,

$$X(j_{m-1},\ldots,j_0)=A_m(j_0,\ldots,j_{m-1}) \tag{22}$$

in such an order that the index of an X must have its binary bits put in reverse order to yield its index in the array A_m.

In some applications, where Fourier sums are to be evaluated twice, the above procedure could be programmed so that no bit-inversion is necessary. For example, consider the solution of the difference equation,

$$aX(j+1)+bX(j)+cX(j-1)=F(j). \tag{23}$$

The present method could be first applied to calculate the Fourier amplitudes of $F(j)$ from the formula

$$B(k)=\frac{1}{N}\sum_j F(j)W^{-jk}. \tag{24}$$

The Fourier amplitudes of the solution are, then,

$$A(k)=\frac{B(k)}{aW^k+b+cW^{-k}}. \tag{25}$$

The $B(k)$ and $A(k)$ arrays are in bit-inverted order, but with an obvious modification of (20), $A(k)$ can be used to yield the solution with correct indexing.

A computer program for the IBM 7094 has been written which calculates three-dimensional Fourier sums by the above method. The computing time taken for computing three-dimensional $2^a\times 2^b\times 2^c$ arrays of data points was as follows:

a	b	c	No. Pts.	Time (minutes)
4	4	3	2^{11}	0.02
11	0	0	2^{11}	0.02
4	4	4	2^{12}	0.04
12	0	0	2^{12}	0.07
5	4	4	2^{13}	0.10
5	5	3	2^{13}	0.12
13	0	0	2^{13}	0.13

References

[1] G.E.P. Box, L.R. Connor, W.R. Cousins, O.L. Davies (ed.), F.R. Hirnsworth, and G.P. Silitto. *The Design and Analysis of Industrial Experiments*. Oliver & Boyd, Edinburgh, 1954.

[2] I.J. Good. The interaction algorithm and practical Fourier series. *J. Roy. Statist. Soc. Ser. B*, **20**, 1958, pp. 361–372; Addendum, **22**, 1960, pp. 372–375. MR **21**, #1674; MR **23**, #A4231.

Introduction to Hájek (1970) A Characterization of Limiting Distributions of Regular Estimates

Rudolf J. Beran
University of California, Berkeley

Hájek's paper combined disparate ideas to reach surprising and forceful conclusions for estimation in large samples. In the paper itself, the convolution theorem emerged as structure hidden behind the asymptotic information bound. In recent work, the convolution theorem has re-emerged as structure underlying correct convergence of bootstrap distributions. Both bootstrapping and the information inequality deceive naive users at the superefficiency points characteristic of sophisticated estimators.

One set of ideas behind the paper was the research, from the mid-1950s onward, that sought to clarify the validity of the asymptotic information inequality. Two developments motivated this work. In small sample theory for the multivariate normal model, the discovery that the sample mean is inadmissible under quadratic loss (Stein, 1956; James and Stein, 1961) cast doubt on unbiased or equivariant estimation and on the method of maximum likelihood. In large sample theory, the discovery of superefficient estimators such as the Hodges estimator (Le Cam, 1953)—or as was noticed later, the James–Stein estimator—reinforced the need to correct ideas about the information inequality. In response, Bahadur (1964), Wolfowitz (1965), Kaufman (1966), Schmetterer (1966), and Roussas (1968) characterized classes of estimators for which asymptotic forms of the information inequality hold. The mathematical hypotheses used in these papers were effective, yet seemed too strong to have a cogent statistical interpretation.

This work in the 1960s demonstrably contained the seeds of Hájek's convolution theorem. Indeed, Inagaki (1970) independently proved a version of the convolution theorem under assumptions similar to those of Kaufman. However, it was a second set of ideas, expressed in Le Cam's (1960) formulation and study of local asymptotic normality, that enabled Hájek to prove his result under virtually minimal hypotheses. Le Cam's work was in turn influenced by Wald's (1943) pioneering paper on asymptotic optimality of tests. Among theoretical statisticians, the scope of Le Cam's ideas on local asymptotic normality first became apparent through Hájek's work. Notable stages were the investigation of the asymptotic power of rank tests [see Hájek and Šidák (1967)], the present paper, and Hájek's (1972) paper on the local asymptotic minimax bound.

Jaroslav Hájek was born in 1926 Poděbrady, on the river Labe (Elbe) about 45 kilometers east of Prague, in the state of Czechoslovakia created at the end of World War I. Attending the Technical University in Prague after the next war, he earned the degree of Statistical Engineer in 1949. The requirements for this degree went beyond mathematics to include probability and theoretical statistics, insurance and financial mathematics, sample surveys, bookkeeping, computational methods, economics, law, and statistical methods for transport, insurance, agriculture, and industry.* Having completed his postgraduate training with the C.Sc. degree, he worked from 1954 to 1964 as a researcher in the Mathematical Institute of the Czechoslovak Academy of Sciences. He became Head of the Department of Mathematical Statistics at Charles University in 1964 and was appointed Professor in 1966. The growth of statistics at Charles University owes a great deal to Hájek's efforts. His research was often hampered by lack of access to books and journals published in the West. On several occasions, he discovered results that turned out to be known in the unavailable literature. Visiting Professor appointments at Berkeley in 1965–1966 and at Florida State in 1969–1970 made a breakthrough in his scientific contacts with Western statisticians.

Too ill to travel, Hájek convened the Prague Symposium on Asymptotic Statistics in September 1973. Both the list of participants and the papers contributed to the Proceedings are noteworthy. In June 1974, at the age of 48, Hájek died from chronic kidney disease. A review article by Dupač (1975) gives further biographical information. A memorial volume, edited by Jurečková, was created in the months after Hájek's death. Nonscholarly causes delayed its publication until 1979. The Prague Symposia have recurred at 5-year intervals, the Fifth Symposium taking place in September 1993. A one-day symposium commemorating the personality and work of Hájek was held in Prague in June 1994, 20 years after his death. Papers presented on this occasion appeared in a special issue of *Kybernetika*, Number 3 of Volume 31 in 1995.

With the passage of time, the statistical significance of the convolution theorem, of Hájek's regularity assumption on estimators, and of Le Cam's LAN property for models have all become sharper. For simplicity, let us consider the case where the sample consists of n independent, identically distributed (i.i.d.) random variables whose joint distribution is $P_{\theta_0,n}$; the value of θ_0 is unknown; and θ_0 is restricted to the parameter space Θ, which

* I am grateful to Professor Vaclav Dupač for this information.

is an open subset of R^k. Suppose that the problem is to estimate the R^m-valued parametric function $\tau = \tau(\theta)$, where $m \leq k$ and τ is differentiable at every θ in Θ, the derivative matrix $\nabla\tau(\theta)$ having full rank.

Let $P^c_{\theta,n}$ denote the part of $P_{\theta,n}$ that is absolutely continuous with respect to $P_{\theta_0,n}$. For $h \in R^k$, let $L_n(h, \theta_0)$ designate the log-likelihood ratio of $P^c_{\theta_0+n^{-1/2}h,n}$ with respect to $P_{\theta_0,n}$. The model $P_{\theta,n}$ is *locally asymptotically* normal (LAN) at θ_0 if there exist a column vector $Y_n(\theta_0)$ and a nonsingular matrix $I(\theta_0)$ such that, under $P_{\theta_0,n}$,

$$L_n(h_n, \theta_0) = h'Y_n(\theta_0) - 2^{-1}h'I(\theta_0)h + o_p(1) \tag{1}$$

for every $h \in R^k$ and for every sequence $\{h_n \in R^k\}$ converging to h; and if the distribution of $Y_n(\theta_0)$ under $P_{\theta_0,n}$ converges weakly to $N(0, I(\theta_0))$ as $n \to \infty$.

Formulated by Le Cam (1960), the LAN property characterizes parametric models for which the asymptotic information bound is pertinent. The matrix $I(\theta_0)$ in the definition plays the role of the classical information matrix, without requiring the existence of second derivatives. The statistical interpretation of LAN is apparent: For a LAN model, the log-likelihood ratio behaves asymptotically like the log-likelihood ratio of $N(h, I^{-1}(\theta_0))$ with respect to $N(0, I^{-1}(\theta_0))$. What we know about estimation or testing in the normal location model has analogues for estimation in any LAN model. Smoothly parametrized exponential families are the most familiar LAN models. However, some models where maximum likelihood estimators fail, such as the three-parameter log-normal [see Hill (1963)], are nevertheless LAN at every θ_0 and help motivate the development of estimation theory for LAN models.

Let $\{T_n: n \geq 1\}$ be any sequence of estimators of $\tau(\theta)$ and let $H_n(\theta)$ denote the distribution of $n^{1/2}(T_n - \theta)$ under $P_{\theta,n}$. Suppose that $H_n(\theta_0)$ converges weakly to a limit distribution $H(\theta_0)$ as $n \to \infty$. The estimators $\{T_n\}$ are *locally asymptotically equivariant* (LAE) at θ_0 if, for every $h \in R^k$ and for every sequence $\{h_n \in R^k\}$ that converges to h, the distributions $\{H_n(\theta_0 + n^{1/2}h_n)\}$ converge weakly to $H(\theta_0)$. The LAE property at θ_0 is equivalent to asserting that the maximum Prohorov distance between $H_n(\theta)$ and $H(\theta_0)$, evaluated over all θ such that $n^{1/2}|\theta - \theta_0| \leq b$, converges to zero for every finite, positive choice of b as $n \to \infty$. Local asymptotic robustness, though only under small perturbations within the parametric model, is one statistical interpretation of the LAE property.

The LAE property turns out to be very common. Suppose that $H_n(\theta)$ converges weakly to a limit distribution $H(\theta)$ for every $\theta \in \Theta$. Then, under conditions on the model weaker than LAN at every θ, there exists a Lebesgue null set E such that the $\{T_n\}$ are LAE at every $\theta \in \Theta - E$ [Le Cam (1973); Le Cam and Yang (1990, pp. 76–77)]. In particular, standard estimators obtained from maximum likelihood or method-of-moments or minimum distance criteria are typically LAE at every θ. On the other hand, model selection and shrinkage estimators lack the LAE property at key points in the parameter space.

Model Selection Example. Suppose that the observations are i.i.d. random variables, each having $N(\theta, 1)$ distribution, where $\theta \in R$. Let $\bar{X}_n$ denote the sample mean. Consider the Hodges estimator $T_{n,H}$ of θ that equals 0 whenever $|\bar{X}_n| \le n^{-1/4}$ and equals $\bar{X}_n$ otherwise (Le Cam, 1953). This estimator selects between fitting the $N(0, 1)$ model and the $N(\theta, 1)$ model to the data. The Hodges estimator is LAE at every $\theta \neq 0$ but is not LAE at $\theta = 0$.

Shrinkage Example. Let I_k denote the k-dimensional identity matrix. Suppose that the observations are i.i.d. random k-vectors, each distributed according to $N(\theta, I_k)$, where $\theta \in R^k$. Let $[\cdot]^+$ represent the positive-part function on the real line. For $k \ge 3$, the James–Stein (1961) estimator of θ is $T_{n,JS} = [1 - (k-2)/(n|\bar{X}_n|^2)]^+ \bar{X}_n$. While this estimator operates more smoothly on the data than the Hodges estimator, it too is LAE at every $\theta \neq 0$ and is not LAE at $\theta = 0$.

The convolution theorem characterizes asymptotic behavior of estimators at LAE points. In LAN models, preferred estimation methods such as maximum likelihood, or one-step maximum likelihood when maximum likelihood fails, generate estimators $T_{n,E}$ of $\tau(\theta)$ whose asymptotic structure under $P_{\theta_0,n}$ is

$$T_{n,E} = \tau(\theta_0) + n^{-1/2} \nabla\tau(\theta_0) I^{-1}(\theta_0) Y_n(\theta_0) + o_p(n^{-1/2}) \tag{2}$$

for every $\theta_0 \in \Theta$. Estimators with this property are LAE at θ_0 and are usually called *asymptotically efficient* at θ_0. Information bounds stemming from the convolution theorem (see below) justify this terminology, but only within the class of all estimators that are LAE at θ_0.

Let $K_n(\theta)$ denote the joint distribution of $n^{1/2}(T_n - T_{n,E})$ and $Y_n(\theta)$ under $P_{\theta,n}$. Let $\sum_\tau(\theta_0) = \nabla\tau(\theta_0) I^{-1}(\theta_0) \nabla'\tau(\theta_0)$. A current statement of the convolution theorem is the following.

Convolution Theorem. *Suppose that the model is LAN at θ_0 and that the distribution $H_n(\theta_0)$ of $n^{1/2}(T_n - \theta_0)$ under $P_{\theta_0,n}$ converges weakly to $H(\theta_0)$. Then the following two statements are equivalent:*

(a) *The estimators $\{T_n\}$ are LAE at θ_0 with limit distribution $H(\theta_0)$.*
(b) *For every $h \in R^k$ and every sequence $\{h_n \in R^k\}$ that converges to h,*

$$K_n(\theta_0 + n^{1/2} h_n) \Rightarrow D(\theta_0) \times N(0, I(\theta_0)) \tag{3}$$

for some distribution $D(\theta_0)$ such that $H(\theta_0) = D(\theta_0) * N(0, \sum_\tau(\theta_0))$.

The distribution $D(\theta_0)$ is a point mass at the origin if and only if $\{T_n\}$ has the asymptotic structure (2) *of $\{T_{n,E}\}$.*

Display (3) asserts that $n^{1/2}(T_n - T_{n,E})$ and $Y_n(\theta_0)$ are asymptotically independent under $P_{\theta_0,n}$, with marginal limit distributions $D(\theta_0)$ and $N(0, I(\theta_0))$. In view of (2), the limit distribution $H(\theta_0)$ of $n^{1/2}(T_n - T_{n,E}) +$

$n^{1/2}(T_{n,E} - \tau(\theta_0))$ is the convolution of $D(\theta_0)$ with $N(0, \sum_\tau(\theta_0))$. It is this structure in $H(\theta_0)$ that gives the result its name. If the hypotheses of the theorem are satisfied for every θ_0, then the LAE property and its equivalent hold almost everywhere in the parameter space.

Hájek's paper treated models for independent nonidentically distributed random variables. Specialized to the i.i.d. case, his theorem in Section 2 of the paper differs from the above statement in several ways. He used a definition of LAN weaker than Le Cam's, in which (1) is required to hold only when $h_n = h$ for every n. Instead of LAE, he used the weaker hypothesis, called *regularity*, that $H_n(\theta_0 + n^{1/2}h) \Rightarrow H(\theta_0)$ for every $h \in R^k$. From these two hypotheses, he deduced the convolution structure of $H(\theta_0)$ and the condition under which $D(\theta_0)$ is degenerate at the origin.

What is the significance of these differences? On the one hand, Hájek's formulation is nearly minimal for the immediate purpose of proving the convolution structure of $H(\theta_0)$. On the other hand, Le Cam's definition of LAN seems to have wider statistical implications. If the model is LAn in Le Cam's sense, then Hájek regularity of estimators is equivalent to the LAE property. Moreover, the connection between the convolution theorem and bootstrap convergence, discussed below, relies on Le Cam's form of the LAN property. These points do not overshadow Hájek's insight in linking the convolution structure to regular estimators in LAN models.

Soon after Hájek's announcement of the convolution theorem, P.J. Bickel sent him a letter that sketched a short characteristic function proof of the theorem. This argument entered the literature through the monograph of Roussas (1972) and has been used frequently since. By different reasoning, Le Cam (1972) generalized the convolution theorem to models that have limit experiments but are not necessarily LAN. More recently, van der Vaart (1988, 1989) gave another proof for the convolution theorem and investigated what happens when Hájek regularity is weakened. The form of the convolution theorem stated above draws on ideas in Ehm and Müller (1983), Pfanzagl (1994), and Beran (1996). The large literature that has grown from the theorem is surveyed by Beran (1995).

A real-valued function l on R^k is *subconvex* if the set $\{x \in R^k : l(x) \leq c\}$ is convex for every real c. The function is *symmetric* if $l(x) = l(-x)$ for every $x \in R^k$. The essence of Section 3 in Hájek's paper is the following result.

Asymptotic Information Bound. *Suppose that l is any real-valued, symmetric, subconvex, nonnegative function defined on R^k. Suppose that the model is LAN at θ_0 and that the estimators $\{T_n\}$ are LAE at θ_0. Then*

$$\liminf_{n\to\infty} \mathrm{E}_{\theta_0} l[n^{1/2}(T_n - \tau(\theta_0))] \geq \mathrm{E}l\left[\sum_\tau^{-1/2}(\theta_0)Z_k\right], \tag{4}$$

where Z_k has a standard normal distribution on R^k. If l is continuous a.e. and *bounded, then*

$$\lim_{n\to\infty} \mathrm{E}_{\theta_0} l[n^{1/2}(T_{n,E} - \tau(\theta_0))] = \mathrm{E}l\left[\sum_{\tau}^{-1/2}(\theta_0) Z_k\right]. \tag{5}$$

Indeed, inequality (4) is immediate from the second part of the convolution theorem, Fatou's lemma, and Anderson's lemma as developed by Ibragimov and Has'minskii (1981, p. 155). Equality (5) justifies the claim that the estimators $\{T_{n,E}\}$ of $\tau(\theta)$ are asymptotically efficient among all estimators that are LAE at θ_0. Moreover, no estimators $\{T_n\}$ can achieve smaller asymptotic risk than $\{T_{n,E}\}$ except at non-LAE points in the parameter space, a set of Lebesgue measure zero.

If an estimator sequence $\{T_n\}$ is such that

$$\limsup_{n\to\infty} \mathrm{E}_{\theta_0} l[n^{1/2}(T_n - \tau(\theta_0))] < \mathrm{E}l\left[\sum_{\tau}^{-1/2}(\theta_0) Z_k\right], \tag{6}$$

then θ_0 is called a *superefficiency* point of $\{T_n\}$. The discussion in the previous paragraph shows that points of superefficiency must be non-LAE points, thereby reproving the result of Le Cam (1953) and Bahadur (1964) that superefficiency points form a Lebesgue null set. Examples show that the set of non-LAE points is, in general, larger than the set of superefficiency points.

Is superefficiency desirable in estimation? At non-LAE points, limiting risks computed pointwise in θ do not provide enough information to answer this question. For the Hodges estimator, $\lim_{n\to\infty} n\mathrm{E}_\theta(T_{n,H} - \theta)^2$ is 0 when $\theta = 0$ and is 1 otherwise. The estimator is superefficient at the origin under quadratic loss and is best LAE at all other values of θ. However, for each n, the normalized risk $n\mathrm{E}_\theta(T_{n,H} - \theta)^2$ is less than the information bound 1 in a neighborhood of $\theta = 0$, then rises quickly above 1, and finally drops slowly toward 1 as $|\theta|$ increases [see Lehmann (1983, Chap. 6)]. The neighborhood of the origin where risk is less than 1 narrows as n increases. Such poor behavior in the maximum risk of superefficient estimators is intrinsic when the dimension k of the parameter space is very small [see Le Cam (1953) and Hájek (1972) for $k = 1$; Le Cam (1974, pp. 187–188) for $k \leq 2$].

On the other hand, when $k \geq 3$, superefficient estimators in LAN models can have desirable risk. For the James–Stein estimator, $\lim_{n\to\infty} n\mathrm{E}_\theta|T_{n,JS} - \theta|^2$ is 2 when $\theta = 0$ and is k otherwise. The estimator is superefficient at the origin and is best LAE at every other value of θ. Unlike the risk of the Hodges estimator, the normalized risk $n\mathrm{E}_\theta|T_{n,JS} - \theta|^2$ is strictly less than the information bound for every value of n and θ. When the dimension k is not too small, the risk of a well-designed superefficient estimator can dominate the information bound over the entire parameter space, especially near the points of superefficiency.

A close connection exists between the convolution theorem and conditions under which natural bootstrap distributions converge correctly. In the present setting, let $\{\hat{\theta}_n\}$ be a sequence of estimators for θ. Efron's (1979) parametric bootstrap estimator for the distribution $H_n(\theta)$ of $n^{1/2}(T_n - \tau(\theta))$ is $H_n(\hat{\theta}_n)$. Let $J_n(\theta)$ denote the distribution of $n^{1/2}(\hat{\theta}_n - \theta)$ under $P_{\theta,n}$.

Bootstrap Consistency Theorem. *Suppose that the model is LAN at θ_0, that $H_n(\theta_0)$ converges weakly to $H(\theta_0)$, and that $J_n(\theta_0)$ converges weakly to a limit distribution $J(\theta_0)$ with full support on R^k. Then the following three statements are equivalent:*

(a) *$H_n(\hat{\theta}_n)$ converges weakly to $H(\theta_0)$ in $P_{\theta_0,n}$-probability.*
(b) *The estimators $\{T_n\}$ are LAE at θ_0.*
(c) *$K_n(\hat{\theta}_n) \Rightarrow D(\theta_0) \times N(0, I(\theta_0))$ in $P_{\theta_0,n}$-probability for some distribution $D(\theta_0)$ such that $H(\theta_0) = D(\theta_0) * N(0, \sum_\tau(\theta_0))$.*

The full support assumption on $J(\theta_0)$ is not strong. It holds at every θ_0 for asymptotically efficient and other standard estimators of θ. More generally, if the LAN assumption and the weak convergence of $J_n(\theta_0)$ both hold at every θ_0, then the estimators $\{\hat{\theta}_n\}$ must be LAE almost everywhere, as discussed earlier. In that event, by the convolution theorem, $J(\theta_0)$ has full support for almost all θ_0.

Parts (a) and (b) of this theorem tell us that standard parametric bootstrap distributions for $n^{1/2}(T_n - \tau(\theta))$ converge correctly at θ_0 if and only if θ_0 is a point where $\{T_n\}$ is LAE. It is unwise, therefore, to trust ordinary bootstrap distributions for model selection estimators, or shrinkage estimators, or signal recovery estimators. These bootstrap distributions lack consistency at superefficiency points, the parameter values of greatest interest. Part (c) of the theorem provides a logical basis for bootstrap diagnostics akin to regression diagnostics. Modified bootstrap distributions can be made to converge correctly at non-LAE points. Beran (1996) gives details of the bootstrap consistency theorem and further references.

The LAE property describes circumstances under which the asymptotic risk of T_n, computed at θ_0, approximates the actual risk, computed near or at θ_0. Hájek's convolution theorem, a brilliant melody in estimation theory, expresses the unforeseen implications of LAE and reveals the hidden story of that pragmatic technique, the bootstrap.

References

Bahadur, R. (1964). A note on Fisher's bound on asymptotic variance. *Ann. Math. Statist.*, **35**, 1545–1552.

Beran, R. (1995). The role of Hájek's convolution theorem in statistics. *Kybernetika*, **31**, 221–237.

Beran, R. (1996). Diagnosing bootstrap success. Revised preprint.

Dupač, V. (1975). Jaroslav Hájek, 1926–1974. *Ann. Statist.*, **3**, 1031–1037.

Efron, B. (1979). Bootstrap methods: Another look at the jackknife. *Ann. Statist.*, **7**, 1–26.

Ehm, W. and Müller, D.W. (1983). Factorizing the information contained in an experiment. *Z. Wahrsch. Verw. Gebiete*, **65**, 121–134.

Hájek, J. (1970). A characterization of limiting distributions of regular estimators. *Z. Wahrsch. Verw. Gebiete*, **14**, 323–330.

Hájek, J. (1972). Local asymptotic minimax and admissibility in estimation. In *Proc. Fifth Berkeley Symp. Math. Statist. Prob.* (L.M. Le Cam, J. Neyman, and E.M. Scott, eds.), Vol. 1. University of California Press, Berkeley, pp. 175–194.

Hájek, J. and Šidák, Z. (1967). *Theory of Rank Tests*. Academic Press, New York.

Hill, B.M. (1963). The three-parameter lognormal distribution and Bayesian analysis of a point source epidemic. *J. Amer. Statist. Assoc.*, **58**, 72–84.

Ibragimov, I.A. and Has'minskii, R.Z. (1981). *Statistical Estimation: Asymptotic Theory*. Springer-Verlag, New York.

Inagaki, N. (1970). On the limiting distribution of a sequence of estimators with uniformity property. *Ann. Inst. Statist. Math.*, **22**, 1–13.

James, W. and Stein, C. (1961). Estimation with quadratic loss. In *Proc. Fourth Berkeley Symp. Math. Statist. Prob.* (J. Neyman, ed.), Vol. 1. University of California Press, Berkeley, pp. 361–380.

Kaufman, S. (1966). Asymptotic efficiency of the maximum likelihood estimator. *Ann. Inst. Statist. Math.*, **18**, 155–178.

Le Cam, L. (1953). On some asymptotic properties of maximum likelihood estimates and related Bayes estimates. *Univ. California Publ. Statist.*, **1**, 125–142.

Le Cam, L. (1960). Locally asymptotically normal families of distributions. *Univ. California Publ. Statist.*, **3**, 37–98.

Le Cam, L. (1972). Limits of experiments. In *Proc. Fifth Berkeley Symp. Math. Statist. Prob.* (L.L. Le Cam, J. Neyman, and E.M. Scott, eds.), Vol. 1. University of California Press, Berkeley, pp. 245–261.

Le Cam, L. (1973). Sur les contraintes imposées par les passages à limite usuels en statistique. *Bull. Internat. Statist. Inst.*, **45**, 169–180.

Le Cam, L. (1974). *Notes on Asymptotic Methods in Statistical Decision Theory*. Centre de Recherches Mathématiques, Université de Montréal.

Le Cam, L. and Yang, G.L. (1990). *Asymptotics in Statistics*. Springer-Verlag, New York.

Lehmann, E.L. (1983). *Theory of Point Estimation*. Wiley, New York.

Pfanzagl, J. (1994). *Parametric Statistical Theory*. de Gruyter, Berlin.

Roussas, G.G. (1968). Some applications of the asymptotic distribution of likelihood functions to the asymptotic efficiency of estimates. *Z. Wahrsch. Verw. Gebiete*, **10**, 252–260.

Roussas, G.G. (1972). *Contiguous Probability Measures: Some Applications in Statistics*. Cambridge University Press, Cambridge.

Schmetterer, L. (1966). On asymptotic efficiency of estimates. In *Research Papers in Statistics: Festschrift for J. Neyman* (F.N. David, ed.). Wiley, New York, pp. 310–317.

Stein, C. (1956). Inadmissibility of the usual estimator for the mean of a multivariate normal distribution. In *Proc. Third Berkeley Symp. Math. Statist. Prob.* (J. Neyman, ed.), Vol. 1. University of California Press, Berkeley, pp. 197–206.

van der Vaart, A. (1988). *Statistical Estimation in Large Parameter Spaces*. C.W.I. Tract 44. Cen trum voor Wiskunde en Informatica, Amsterdam.

van der Vaart, A. (1989). On the asymptotic information bound. *Ann. Statist.*, **17**, 1487–1500.

Wald, A. (1943). Tests of statistical hypotheses concerning several parameters when the number of observations is large. *Trans. Amer. Math. Soc.*, **54**, 426–482.

Wolfowitz, J. (1965). Asymptotic efficiency of the maximum likelihood estimator. *Theory Probab. Appl.*, **10**, 247–260.

A Characterization of Limiting Distributions of Regular Estimates

Jaroslav Hájek

Summary

We consider a sequence of estimates in a sequence of general estimation problems with a k-dimensional parameter. Under certain very general conditions we prove that the limiting distribution of the estimates, if properly normed, is a convolution of a certain normal distribution, which depends only of the underlying distributions, and of a further distribution, which depends on the choice of the estimate. As corollaries we obtain inequalities for asymptotic variances and for asymptotic probabilities of certain sets, generalizing some results of J. Wolfowitz (1965), S. Kaufman (1966), L. Schmetterer (1966) and G.G. Roussas (1968).

1. Introduction

In several papers there were established asymptotic lower bounds for variances and for probabilities of some sets provided the estimates are regular enough to avoid superefficiency. See Rao (1963), Bahadur (1964), Wolfowitz (1965), Kaufman (1966), Schmetterer (1966), Roussas (1968), for example. In the present paper we obtain some of their results as corollaries of a representation theorem describing the class of all possible limiting laws. The present result refers to a general sequence of experiments with general norming matrices K_n. Our condition (2) below is implied by conditions imposed on the families of distributions in the above mentioned papers. The same is true about regularity condition (3) concerning the limiting distribution of estimates. A comparison with Kaufman (1966) is given in Remark 1 below, and the same comparison could be made with Wolfowitz (1965). Schmetterer's (1966) condition, namely that the distribution law of normed estimates converges continuously in θ, also entails (3). See also Remark 2 for Bahadur's (1964) condition, which is weaker than (3).

Roughly speaking, condition (2) means that the likelihood functions may be locally asymptotically approximed by a family of normal densities differing in location only. Similarly, condition (3) expresses that the family of distributions of an estimate T_n may be approximated, in a local asymptotic sense, by a family of distributions differing again in location only.

The idea of the proof is based on considering the parameter as a random vector that has uniform distribution over certain cubes. In this respect the spirit of the proof is Bayesian, similarly as in Weiss and Wolfowitz (1967). The mathematical technique of the present paper is borrowed from LeCam (1960).

2. The Theorem

Consider a sequence of statistical problems $(\mathscr{X}_n, \mathscr{A}_n, P_n(\cdot, \theta))$, $n \geqq 1$, where θ runs through an open subset Θ of R^k. The nature of sets $\mathscr{X}_n$ may be completely arbitrary, and the only connection between individual statistical problems will be given by conditions (2) and (3). In particular, the nth problem may deal with n observations which may be independent, or may form a Markov chain, or whatsoever. A point from $\mathscr{X}_n$ will be denoted by x_n, and X_n will denote the abstract random variable defined by the identity function $X_n(x_n) = x_n$, $x_n \in \mathscr{X}_n$. The norm of a point $h = (h_1, \ldots, h_k)$ from R^k will be defined by $|h| = \max_{1 \leqq i \leqq k} |h_i|$; $h \leqq r$ will mean that $h_i \leqq r_i$, $1 \leqq i \leqq k$, and $h'r$ will denote the scalar product of two vectors $h, r \in R^k$.

Take a point $t \in \Theta$ and assume it is the true value of the parameter θ. We shall abbreviate $P_n = P_n(\cdot, t)$ and $P_{nh} = P_n(\cdot, t + K_n^{-1}h)$ where K_n^{-1} are inverses of some norming regular $(k \times k)$-matrices. Most usually the theorem below is applicable with K_n diagonal and having $n^{1/2}$ for its diagonal elements; then simply $K_n^{-1}h = n^{-1/2}h$. We shall need that $\{K_n^{-1}\}$ be a sequence of contracting transforms in the sense that for every $\varepsilon > 0$ and $a > 0$ there exists an n_0 such that $|h| < a$ entails $|K_n^{-1}h| < \varepsilon$ for all $n > n_0$. This property will be insured by

$$\lim_{n\to\infty} (\text{the minimal eigenvalue of } K_n'K_n) = \infty, \tag{1}$$

where K_n' denotes the transpose of K_n. Since Θ is open, (1) ensures that for all $h \in R^k$ the points $t + K_n^{-1}h$ belong to if n is sufficiently large. If $\Theta = R^k$, the theorem below makes sense and is true without condition (1).

Given two probability measures P and Q, denote by $dQ\ dP$ the Radon–Nikodym derivative of the *absolutely continuous part* of Q with respect to P. Thus, generally, we do not obtain $\int_A (dQ/dP)\, dP = Q(A)$, but only $\leqq Q(A)$.

Introduce the family of likelihood ratios

$$r_n(h, x_n) = \frac{dP_{nh}}{dP_n}(x_n), \qquad h \in R^k, \quad n \geqq n_h, \quad x_n \in \mathscr{X}_n,$$

where n_h denotes the smallest n_0 such that $n \geqq n_0$ entails $t + K_n^{-1}h \in \Theta$. In what follows the argument x_n will be usually omitted in order to stress the other arguments.

The distribution law of a random vector $Y_n = Y_n(x_n)$ under P_n will be denoted by $\mathscr{L}(Y_n|P_n)$ and its weak convergence to L by $\mathscr{L}(Y_n|P_n) \to L$. The k-dimensional normal distribution with zero expectation vector and covariance matrix Γ will be denoted by $\phi(\cdot|\Gamma)$, and the corresponding law by $\mathscr{N}(0, \Gamma)$. The expectation with respect P_n will be denoted by E_n, i.e. $E_n(\cdot) = \int(\cdot)\, dP_n$.

Let $T_n = T_n(x_n)$ be a sequence of estimates of θ, i.e. for every n T_n is a measurable transform from $\mathscr{X}_n$ to R^k. (We shall say that the estimates T_n are regular, if they satisfy condition (3) below.) Now we are prepared to formulate the following

Theorem. *In the above notations, let us assume that* (1) *holds and that*

$$r_n(h) = \exp\{h'\Delta_n - \tfrac{1}{2}h'\Gamma h + Z_n(h)\}, \qquad h \in R^k, \quad n \geqq n_h, \tag{2}$$

where $\mathscr{L}(\Delta_n|P_n) \to \mathscr{N}(0, \Gamma)$ *and* $Z_n(h) \to 0$ *in* P_n*-probability for every* $h \in R^k$. *Further consider a sequence of estimates satisfying*

$$P_{nh}(K_n(T_n - t) - h \leqq v) \to L(v), \qquad \text{for every} \quad h \in R^k, \tag{3}$$

in continuity points of some distribution function $L(v)$, $v \in R^k$.

Then, if the matrix Γ *is regular, we have*

$$L(v) = \int \Phi(v - u|\Gamma^{-1})\, dG(u), \tag{4}$$

where $G(u)$ *is a certain distribution function in* R^k.

Remark. The dependence of Δ_n, Γ and $Z_n(h)$ on t, and of Δ_n and $Z_n(h)$ on x_n, is suppressed in our notation.

PROOF. By Lemma 1 below we may also write

$$r_n(h) = \exp\{h'\Delta_n^* - \tfrac{1}{2}h'\Gamma h + Z_n^*(h)\}, \tag{5}$$

where the properties $\mathscr{L}(\Delta_n^*|P_n) \to \mathscr{N}(0, \Gamma)$ and $Z_n^*(h) \to 0$ in P_n-probability are preserved, and, in addition

$$B_{nh} = [E_n \exp(h'\Delta_n^* - \tfrac{1}{2}h'\Gamma h)]^{-1} \tag{6}$$

satisfies

$$\sup_{|h|<j} |B_{nh} - 1| \to 0 \qquad \text{for every} \quad j > 0, \quad \text{as } n \to \infty. \tag{7}$$

Put

$$r_n^*(h) = B_{nh} \exp(h'\Delta_n^* - \tfrac{1}{2}h'\Gamma h). \tag{8}$$

From our assumptions and from (7) it follows that $N(-\frac{1}{2}h'\Gamma h, h'\Gamma h)$ is a limit for $\mathscr{L}(\log r_n(h)|P_n)$ as well as for $\mathscr{L}(\log r_n^*(h)|P_n)$. Let Y be a random variable such that $\mathscr{L}(Y) = \mathscr{N}(-\frac{1}{2}h'\Gamma h, h'\Gamma h)$. Then $Ee^Y = 1$. Further, by (8) and (6),

$$E_n r_n^*(h) = 1.$$

On the other hand, since $r_n(h)$ is a the Radon–Nikodym derivative of the absolutely continuous part of P_{nh} relative to P_n,

$$\limsup_n E_n r_n(h) \leqq 1.$$

Since $r_n(h)$ is nonnegative and convergent in distribution to e^Y, we also have

$$\liminf_n E_n r_n(h) \geqq e^Y = 1.$$

Consequently,

$$\lim_{n\to\infty} E_n r_n(h) = 1.$$

Since, furthermore, $[r_n(h) - r_n^*(h)] \to 0$ in P_n-probability, Lemma 2 below may be applied to the effect that

$$\int |r_n(h) - r_n^*(h)| dP_n \to 0, \qquad h \in R^k. \tag{9}$$

Put

$$Q_{nh}(A) = \int_A r_n^*(h)\, dP_n, \qquad A \in \mathscr{A}_n. \tag{10}$$

Then (3) and (9) entail that

$$Q_{nh}(K_n(T_n - t) - h \leqq v) \to L(v). \tag{11}$$

From (8) it follow that $Q_{nh}(K_n(T_n - t) - h \leqq v)$ is a measurable function of h for every $v \in R^k$.

Let $\lambda_j(dh)$ be the uniform distribution over the cube $|h| \leqq j$. It is easy to see that (11) entails for every natural j

$$L_{nj}(v) \overset{\text{Df}}{=} \int Q_{nh}(K_n(T_n - t) - h \leqq v)\, d\lambda_j(h) \to L(v). \tag{12}$$

By a LeCam's lemma (LeCam, 1960 or Hájek and Šidák, 1967, VI.1.4) we deduce

$$\mathscr{L}(\Gamma^{-1}\Delta_n^*|Q_{nh}) \to \mathscr{N}(h, \Gamma^{-1}). \tag{13}$$

Put $a_j = (-j, \ldots, -j)$, $b_j = (j, \ldots, j)$ and $c_j = (\sqrt{j}, \ldots, \sqrt{j})$. Let U be some random k-vector possessing normal distribution with zero expectation and covariance matrix Γ^{-1}, i.e. $\mathscr{L}(U) = \mathscr{N}(0, \Gamma^{-1})$. Then, by (13),

$$\begin{aligned}\int Q_{nh}(a_j + c_j \leqq \Gamma^{-1}\Delta_n^* \leqq b_j - c_j)\, d\lambda_j(h) \\ \rightarrow \int P(a_j + c_j \leqq U + h \leqq b_j - c_j)\, d\lambda_j(h) \\ \geqq \int P(a_j + 2c_j \leqq h \leqq b_j - 2c_j)\, d\lambda_j(h) - P(|U| > \sqrt{j}) \\ = (1 - 2/\sqrt{j})^k - P(|U| > \sqrt{j}).\end{aligned} \tag{14}$$

For every natural j denote by m_j some integer such that

$$\sup_{|h|<j} |B_{nh} - 1| < \frac{1}{j}, \qquad n > m_j, \tag{15}$$

$$\rho(\bar{L}_{nj}, L) < \frac{1}{j}, \qquad n > m_j, \tag{16}$$

(where $\bar{L}_{nj}$ is given by (12) and ρ denotes the Lévy distance), and such that

$$\begin{aligned}\int Q_{nh}(a_j + c_j \leqq \Gamma^{-1}\Delta_n^* \leqq b_j - c_j)\, d\lambda_j(h) \\ > (1 - 2/\sqrt{j})^k - P(|U| > \sqrt{j}) - 1/j, \qquad n > m_j.\end{aligned} \tag{17}$$

All this may be satisfied in view of (7), (12) and (14). We may assume that $m_1 < m_2 < \cdots$. Let $j(n)$ be defined by

$$m_{j(n)} \leqq n < m_{j(n)+1},$$

and set

$$\bar{L}_n = \bar{L}_{nj(n)}, \qquad \bar{Q}_n = \int Q_{nh}\, d\lambda_{j(n)}(h). \tag{18}$$

From (16) it follows that

$$\bar{L}_n \rightarrow L. \tag{19}$$

On the other hand $\bar{L}_n(v)$ may also be interpreted as the probability of $\{K_n(T_n - t) - h \leqq v\}$, if the joint distribution of (X_n, h) is given by the prior distribution $\lambda_{j(n)}$ and by the family Q_{nh} of conditional distributions of X_n given h. Denoting the posterior distribution function of h by $D_n(y|X_n)$, we may therefore write

$$\bar{L}_n(v) = \int [1 - D_n(K_n(T_n - t) - v - 0|X_n)]\, d\bar{Q}_n. \tag{20}$$

Now, in view of (8), for $a_{j(n)} \leqq y \leqq b_{j(n)}$,

$$D_n(y|X_n) = c(X_n) \int_{a_{j(n)}}^{y} B_{nh} \exp(h'\Delta_n^* - \tfrac{1}{2}h'\Gamma h)\, dh$$

$$= c'(X_n) \int_{a_{j(n)}}^{y} B_{nh} \exp\{(h - \Gamma^{-1}\Delta_n^*)'\Gamma(h - \Gamma^{-1}\Delta_n^*)\}\, dh, \tag{21}$$

where $c(X_n)$ and $c'(X_n)$ are some constants depending on X_n only. Now, in view of (15),

$$\sup_{|h|<j(n)} |B_{nh} - 1| \to 1, \tag{22}$$

and in view of (17),

$$-\infty \leftarrow a_{j(n)} - \Gamma^{-1}\Delta_n^* \leqq h - \Gamma^{-1}\Delta_n^* \leqq b_{j(n)} - \Gamma^{-1}\Delta_n^* \to \infty, \tag{23}$$

where $-\infty = (-\infty, \ldots, -\infty)$, $\infty = (\infty, \ldots, \infty)$ and the convergence is in $\bar{Q}_n$-probability. Consequently

$$D_n(y|X_n) \to \Phi(y - \Gamma^{-1}\Delta_n^* | \Gamma^{-1}) \tag{24}$$

in $\bar{Q}_n$-probability, and, in turn

$$|\bar{L}_n(v) - \int \Phi(v - K_n(T_n - t) + \Gamma^{-1}\Delta_n^* | \Gamma^{-1})\, d\bar{Q}_n| \to 0. \tag{25}$$

Denoting

$$G_n(u) = \bar{Q}_n(K_n(T_n - t) - \Gamma^{-1}\Delta_n^* \leqq u) \tag{26}$$

we may also write

$$|L_n(v) - \int \Phi(v - u|\Gamma^{-1})\, dG_n(u)| \to 0. \tag{27}$$

Taking a subsequence $\{m\} \subset \{n\}$ such that $G_m \to G$, we obtain from (19) and (27)

$$L(v) \leftarrow \bar{L}_m(v) \to \int \Phi(v - u|\Gamma^{-1})\, dG(u).$$

Since $L(v)$ is a distribution function, $G(u)$ has to be a distribution function, too. This completes the proof.

Lemma 1. *Let $\mathscr{L}(\Delta_n/P_n) \to \mathscr{N}(0, \Gamma)$ hold. Then there exists a truncated version Δ_n^* of Δ_n such that $(\Delta_n - \Delta_n^*) \to 0$ in P_n-probability and*

$$\sup_{|h|<j} |E_n \exp(h'\Delta_n^* - \tfrac{1}{2}h'\Gamma h) - 1| \to 0, \qquad j > 0. \tag{28}$$

Proof. For every natural i let us put

$$\Delta_{ni} = \Delta_n, \qquad \text{if } |\Delta_n| \leqq i,$$
$$= 0, \qquad \text{otherwise.}$$

Let Y be some random vector such that $\mathscr{L}(Y) = \mathscr{N}(0, \Gamma)$ and let $Y_i = Y$, if $|Y| \leqq i$, and 0 otherwise. Let m_i be an integer such that

$$\sup_{|h| \leqq i} |E_n \exp(h'\Delta_{ni} - \tfrac{1}{2}h'\Gamma h) - E \exp(h'Y_i - \tfrac{1}{2}h'\Gamma h)| < \frac{1}{i}, \qquad n > m_i.$$

Such an m_i exists, since $\mathscr{L}(\Delta_{ni}|P_n) \to \mathscr{L}(Y_i)$ and the distributions are concentrated on a compact subset on which the system of functions $\exp(h'y - \frac{1}{2}h\Gamma h)$ is compact in the supremum metric. Note that further

$$\lim_{i \to \infty} \sup_{|h|<j} |E \exp(h'Y_i - \tfrac{1}{2}h'\Gamma h) - 1| = 0.$$

Assume that $m_1 < m_2 < \cdots$, define $i(n)$ by $m_{i(n)} \leqq n < m_{i(n)+1}$, and put $\Delta_n^* = \Delta_{ni(n)}$.

Then the conclusion easily follows.

Lemma 2. *Let $\{(U_n, V_n), n \geqq 1\}$ be a sequence of pairs of nonnegative random variables such that:*

(a) *U_n converges in distribution to U, $EU < \infty$.*
(b) *$(U_n - V_n) \to 0$ in probability.*
(c) *$E_n U_n \to EU$, $E_n V_n \to EU$, where E_n and E denote expectations.*

Then

$$E_n|U_n - V_n| \to 0 \qquad as \quad n \to \infty. \tag{29}$$

Proof. The proof follows from Loève (1963, Theorem 11.4.A).

3. Corollaries

Corollary 1. *Let C be a convex symmetric set in R^k, and let Y be a random variable such that $\mathscr{L}(Y) = \mathscr{N}(0, \Gamma^{-1})$. Then, if the assumptions of the above theorem are satisfied for some $t \in \Theta$, we have for T_n from the same theorem*

$$\limsup_{n \to \infty} P_n[K_n(T_n - t) \in C|t] \leqq P(Y \in C). \tag{30}$$

Proof. Since the boundary of C has zero Lebesgue measure and since the limiting law of $K_n(T_n - t)$ is absolutely continuous, in view of (4), we have

$$\lim_{n \to \infty} P_n[K_n(T_n - t) \in C|t] = P(Z \in C),$$

where $\mathscr{L}(Z) = L$. Furthermore, (4) with the well-known lemma of Anderson (1955) entail

$$P(Z \in C) \leqq P(Y \in C).$$

Remark 1. Corollary 1 provides the essence of the main result by Kaufman (1966). His conditions, namely that the distribution of $K_n(T_n - \theta)$ converges uniformly in $R^k \times C$ where C is any compact subset of Θ, and his Lemma 5.1 entail our assumption (3). His regularity conditions for densities entail assumption (2). Contrary to Kaufman, we do not claim that the maximum likelihood estimate is after norming asymptotically normal $\mathcal{N}(0, \Gamma^{-1})$. For this conclusion we would need some global conditions concerning the distributions.

Under LeCam's *DN* conditions in [4], there is an estimate, not necessarily the maximum likelihood one, which is after norming asymptotically normal $\mathcal{N}(0, \Gamma^{-1})$.

Remark 2. The theorem also entails

$$\liminf_{n\to\infty} E_n(h'K_n(T_n - t))^2 \geqq h'\Gamma^{-1}h. \tag{31}$$

If we assume that $K_n(T_n - \theta)$ is asymptotically normal with zero variance, then by Bahadur (1964), we could derive (31) from weaker assumptions than (3). Actually, then (3) may be replaced by

$$P_{nh}(h'K_n(T_n - t) \leqq h'h) \to \tfrac{1}{2}, \qquad h \in R^k. \tag{32}$$

Remark 3. We have introduced K_n instead of $n^{1/2}I$ in order to cover such instances as

$$p_n(\theta) = \prod_{i=1}^{n} f\left(x_i - \sum_{j=1}^{k} \theta_j c_{nij}\right) \tag{33}$$

where we may put $K_n = \{\sum_{i=1}^{n} c_{nij}c_{nij'}\}_{j,j'=1}^{k}$. Alternatively, we may take for K_n a diagonal matrix coinciding on the diagonal with the previous one. Instances of this character occur in asymptotic theory of linear regression.

Corollary 2. *Put $A_v = \{y\colon a'y \leqq v\}$ where $a \in R^k$ and $v \in R$. Assume that*

$$\int_{A_v} dL(y) = \int_{A_{v^*}} d\Phi(y|\Gamma^{-1}). \tag{34}$$

Then, under the assumptions of the above theorem

$$\int_{A_{v+s} \div A_v} dL(y) \leqq \int_{A_{v^*+s} \div A_{v^*}} d\Phi(y|\Gamma^{-1}), \qquad s \in R, \tag{35}$$

where $\div$ denotes the symmetric difference.

Proof. Consider two distributions of the pair $(Y, U) \in R^k \times R^k$, namely

$$\Phi(\cdot|\Gamma^{-1}) \times G \qquad \text{and} \qquad \Phi(\cdot + \lambda\Gamma^{-1}a|\Gamma^{-1}) \times G, \qquad \lambda > 0.$$

Then (34) entails that the tests rejecting $\Phi(\cdot|\Gamma^{-1}) \times G$ if $a'Y \leqq v^*$ and $a'(Y+U) \leqq v$, respectively, have the same significance level, and (35) simply means that the former test has for $\lambda a'\Gamma^{-1}a = s > 0$ a larger power than the latter test, which easily follows from the Neyman–Pearson lemma. For $s < 0$ we proceed similarly.

Remark 4. Corollary 2 generalizes a result by Wolfowitz (1965) and Roussas (1968), if we note that

$$P_n(a'K_n(T_n - t) \leqq v) \to \int_{A_v} dL(v),$$

which follows again from the fact the A_v has zero boundary.

Remark 5. If the loss incurred by T_n if θ is true equals $l(K_n(T_n - \theta))$, where $l(\cdot)$ may be represented as a mixture of indicators of complements of convex symmetric sets, then Corollary 1 entails that

$$\liminf_{n\to\infty} E_n l_n[K_n(T_n - t)] \geqq El(Y),$$

where $\mathscr{L}(Y) = \mathscr{N}(0, \Gamma^{-1})$. In this sense $\mathscr{N}(0, \Gamma^{-1})$ may be regarded as a best possible limiting distribution. From (26) it is apparent that $\mathscr{L}(K_n(T_n - t)|P_n) \to \mathscr{N}(0, \Gamma^{-1})$ if the only if $[K_n(T_n - t) - \Gamma^{-1}\Delta_n^*] \to 0$ in $\bar{Q}_n$-probability.

References

[1] Anderson, T.W. The integral of a symmetric unimodal function. *Proc. Amer. Math. Soc.*, **6**, 170–176 (1955).

[2] Bahadur, R.R. A note on Fisher's bound for asymptotic variance. *Ann. Math. Statist.*, **35**, 1545–1552 (1964).

[3] Hájek, J. and Šidák, Z. *Theory of Rank Tests.* Academia, Prague; Academic Press, London, 1967.

[4] LeCam, L. Locally asymptotically normal families of distributions. *Univ. California Publ. Statist.*, 1960.

[5] Kaufman, S. Asymptotic efficiency of the maximum likelihood estimator. *Ann. Inst. Statist. Math.*, **18**, 155–178 (1966).

[6] Loève, M. Probability Theory, 3rd ed. D. Van Nostrand, Princeton, NJ, 1963.

[7] Rao, C.R. Criteria of estimation in large samples. *Sankhya*, **25**a, 189–206 (1963).

[8] Roussas, G.G. Some applications of the asymptotic distribution of likelihood functions to the Asymptotic efficiency of estimates. *Z. Wahrsch. Verw. Gebiete*, **10**, 252–260 (1968).

[9] Schmetterer, L. *On Asymptotic Efficiency of Estimates.* Research Papers in Statistics. (F.N. David, ed.). Wiley, New York, 1966.

[10] Weiss, L. and Wolfowitz, J. Generalized maximum likelihood estimators. *Teor. Veroyatnost. Primenen*, **11**, 68–93 (1966).

[11] Weiss, L. and Wolfowitz, J. Maximum probability estimators. *Ann. Inst. Statist. Math.*, **19**, 103–206 (1967).

[12] Wolfowitz, J. Asymptotic efficiency of the maximum likelihood estimator. *Teor. Veroyatnost. Primenen*, **10**, 267–281 (1965).

Introduction to Hastings (1970) Monte Carlo Sampling Methods Using Markov Chains and Their Applications

T.L. Lai

Consider the problem of generating a distribution π on a high-dimensional space, with π specified via a function $\tilde{\pi}$ to which it is proportional, where the integral (or sum) of $\tilde{\pi}$ over the support of π may not be 1. Such problems are of common occurrence in many statistical applications. For example, the posterior density $p(\theta|\mathbf{x})$ of an unknown parameter θ given the observed sample $\mathbf{x}$ is proportional to the product of the prior density of θ and the density function of $\mathbf{x}$ given θ, but the normalizing constant is often difficult to compute. The Metropolis–Hastings algorithms, introduced by Metropolis et al. (1953) and by Hastings in the present paper, provide powerful practical methods for these important but seemingly intractable problems. While Metropolis and his collaborators considered exclusively the case where π is the equilibrium distribution of some physical system in statistical mechanics, Hastings considered general π and provided a far-reaching generalization of Metropolis's method and subsequent variants thereof in the statistical mechanics literature.

Statistical mechanics studies aggregate properties of a large number of components, such as atoms or molecules, in liquid or solid matter or gases. Because of the very large number of atoms or molecules, only the most probable behavior of the system in thermal equilibrium at a given temperature is observed in experiments. This can be characterized by the average taken over the ensemble of identical systems and small fluctuations about this average. The equilibrium distribution of the system at absolute temperature T is Boltzmann's distribution with probability density proportional to $\exp(-E(x)/kT)$, where k is Boltzmann's constant and $E(x)$ is the energy at state x of the system. According to ergodic theory, the proportion of time that the system spends in state x is also proportional to $\exp(-E(x)/kT)$. To evaluate the expectation, usually denoted by $\langle f \rangle$ in statistical mechanics, of a function $f(X)$ such that the random variable X has the equilibrium distribution, Metropolis et al. (1953) introduced the following Markov Chain Monte Carlo method. Let Q be the transition probability function of an irreducible, aperiodic Markov chain on the state space of the system such that $Q(x,y) = Q(y,x)$. At stage $t+1$, generate X'_{t+1} given the value of X_t using the transition matrix Q. Let $\Delta E_{t+1} = E(X'_{t+1}) - E(X_t)$ denote the change in energy. If $\Delta E_{t+1} \leq 0$, define $X_{t+1} = X'_{t+1}$. If $\Delta E_{t+1} > 0$, let $X_{t+1} = X'_{t+1}$ with probability $\exp(-\Delta E_{t+1}/kT)$ and let $X_{t+1} = X_t$ with probability $1 - \exp(-\Delta E_{t+1}/kT)$. The Markov chain $\{X_t, t \geq 0\}$ thus generated has Boltzmann's distribution as its stationary distribution, and therefore $n^{-1}\sum_{t=1}^{n} f(X_t)$ with sufficiently large n yields a good approximation to $\langle f \rangle$.

Barker (1965) subsequently proposed an alternative randomization method to generate X_t having the Boltzmann distribution as its stationary distribution. Generate X'_{t+1} given X_t from Q as in Metropolis's method, but use the following randomization scheme to generate X_{t+1}. With probability $p_t = E(X'_{t+1})/\{E(X_t) + E(X'_{t+1})\}$ let $X_{t+1} = X'_{t+1}$, and with probability $1 - p_t$ let $X_{t+1} = X_t$. Which of Metropolis's or Barker's method is better? Using a general representation of both methods, Hastings suggested in the present paper that Metropolis's method might be preferable. A rigorous proof was subsequently provided by Hastings's doctoral student Peskun (1973).

Hastings's paper given a far-reaching generalization of these Markov Chain Monte Carlo methods in statistical mechanics. First, Boltzmann's distribution is replaced by a general distribution π and the transition matrix Q need not be symmetric. Secondly, the randomization schemes of Metropolis and Barker are replaced by the following general form: Let $X_{t+1} = X'_{t+1}$ with probability $\alpha(X_t, X'_{t+1})$ and let $X_{t+1} = X_t$ with probability $1 - \alpha(X_t, X'_{t+1})$, where

$$\alpha(x,y) = s(x,y)/\{1 + \pi(x)Q(x,y)/\pi(y)Q(y,x)\}$$

and $s(x,y)$ is a symmetric function of x, y so chosen that $0 \leq \alpha(x,y) \leq 1$ for all x and y. Thus, Hastings's general algorithm is as follows: Suppose $X_t = x$. Then generate X'_{t+1} according to the probability distribution $Q(x,\cdot)$, in which Q need not be symmetric. Suppose $X'_{t+1} = y$. Then take $X_{t+1} = y$ with probability $\alpha(x,y)$ and $X_{t+1} = x$ with probability $1 - \alpha(x,y)$. Metropolis's and Barker's methods correspond to the symmetric case $Q(x,y) = Q(y,x)$ and $s(x,y) = 1$ for Barker's method but

$$s(x,y) = 1 + \min\{\pi(x)/\pi(y), (\pi(y)/\pi(x)\}$$

for Metropolis's method. Hastings's paper further points out that more generally we can choose

$$s(x,y) = g(\min\{\pi(x)Q(x,y)/\pi(y)Q(y,x), \pi(y)Q(y,x)/\pi(x)Q(x,y)\}),$$

in which $0 \leq g(t) \leq 1 + t$ for all $0 \leq t \leq 1$. In particular, taking $g(t) = 1 + t$ yields

$$\alpha(x,y) = \min\{1, \pi(y)Q(y,x)/\pi(x)Q(x,y)\},$$

which has become the commonly used choice in applications of Hastings's algorithm, and Peskun (1973) showed this choice to be optimal in some sense.

Let $P(x, y) = \alpha(x, y)Q(x, y)$ for $x \neq y$ and let $P(x, x) = 1 - \sum_{y \neq x} P(x, y)$. From the symmetry property $s(x, y) = s(y, x)$, it follows that $\pi(x)P(x, y) = \pi(y)P(y, x)$, and therefore the Markov chain $\{X_t, t \geq 0\}$ that has transition probability function P is reversible and π is its stationary distribution, as noted in Subsection 2.2 of the paper. The sample mean $\bar{f}_n = n^{-1} \sum_{t=1}^{n} f(X_t)$ therefore provides a Monte Carlo estimate of the expected value $E_\pi(f) \sum_x f(x)\pi(x)$ with respect to the stationary distribution π of the chain, and Subsection 2.1 uses standard results in time series to give an estimate of the standard error of $\bar{f}_n$. Subsection 2.6 provides some practical guidelines for the choice of the transition matrix Q and the initial state X_0, and for assessing the adequacy of the standard deviation to measure error and of the length n of the realization. In particular, concerning the choice of Q, it is suggested that the rejection rate, which is defined as the proportion of times t for which $X_{t+1} = X_t$, should be kept low and that the sample point in one step should be able to move as large a distance as possible in the sample space.

Subsection 2.4 of the paper considers the multivariate case where π is d-dimensional and the simulated chain is $X_t = (X_{1,t}, \ldots, X_{d,t})$. It mentions the following three possibilities in the transition from time $t - 1$ to t:

(i) All coordinates of X_t may be changed.
(ii) Only one coordinate of X_t may be changed, with coordinate i chosen according to a probability distribution on $\{1, \ldots, d\}$.
(iii) Only one coordinate may be changed, the coordinates being selected in a fixed rather than random sequence (e.g., $1, \ldots, d, 1, \ldots, d, \ldots$).

A special case of Hastings's Markov Chain Monte Carlo methods in the multivariate case is the Gibbs sampler introduced by Geman and Geman (1984), which adopts (iii) or (ii) above and chooses the rejection probability $\alpha(x, y)$ to be 0. The past decade has witnessed major advances in spatial statistics, image restoration, genetic linkage analysis, hierarchical models and Bayesian computations by making use of the Gibbs sampler and other Markov Chain Monte Carlo methods in the multivariate case considered by Hastings. The discussion papers by Besag and Green (1993), Besag et al. (1995), Gelman and Rubin (1992), Geyer (1992), Geyer and Thompson (1992), Smith and Roberts (1993), and Tierney (1994) and the monographs by Geman (1991) and Tanner (1991) highlight some of these important developments and give useful pointers to the rapidly developing literature on statistical applications of Markov Chain Monte Carlo methods.

Subsection 2.5 of the paper discusses importance sampling for variance reduction. The mean $E_\pi(f) = \sum_x f(x)\tilde{\pi}(x) / \sum_x \tilde{\pi}(x)$ of a distribution π

that is specified via a function $\tilde{\pi}$ to which it is proportional can be expressed as

$$E_\pi(f) = \left\{\sum_x (f(x)\tilde{\pi}(x)/p(x))p(x)\right\} \Big/ \left\{\sum_x (\tilde{\pi}(x)/p(x))p(x)\right\}$$
$$= E_p(f\tilde{\pi}/p)E_p(\tilde{\pi}/p),$$

where p is another distribution. Therefore, to evaluate $E_\pi(f)$ by Markov Chain Monte Carlo methods, we can set up the Markov chain X_t using p (which needs only be specified up to a constant of proportionality) instead of $\tilde{\pi}$. For some recent developments of this powerful idea, called *Metropolized importance sampling*, see Liu (1996).

Markov Chain Monte Carlo has also emerged as an active area of research in probability. It is easy to extend the Metropolis–Hastings algorithms from finite to general state spaces, but convergence to the corresponding stationary distributions becomes a much more difficult problem. Moreover, rates of convergence of the algorithms are closely related to some deep geometric and mixing properties and eigenvalue bounds for Markov chains. The recent papers by Amit (1996), Athreya et al. (1996), Diaconis and Saloff-Coste (1996), Mengersen and Tweedie (1996), Rosenthal (1995), and Sinclair and Jerrum (1989) address some of these issues. Another area of active current research is Markov Chain Monte Carlo methods for generating random elements of large finite groups in computational group theory, see Finkelstein and Kantor (1993) and Celler et al. (1995). It is interesting to note that Hastings envisioned such applications of Markov Chain Monte Carol in Section 3 of his paper.

W. Keith Hastings was born on July 21, 1930, in Toronto. He received his B.A. degree in applied mathematics in 1953 and the Ph.D. degree in mathematics/statistics in 1962 from the University of Toronto. From 1955 to 1959, he worked as a consultant in computer applications at H.S. Gellman & Company in Toronto. After finishing his Ph.D. thesis on invariant fiducial distributions, he spent 2 years at the University of Canterbury in New Zealand as Senior Lecturer in Mathematics and another 2 years at Bell Laboratories as a member of the Technical Staff in the Statistics Department. He returned to the University of Toronto in 1966 as Associate Professor of Mathematics. In 1971 he moved to the University of Victoria, British Columbia.

References

Amit, Y. (1996). Convergence properties of the Gibbs sampler for perturbations of Gaussians. *Ann. Statist.*, **24**, 122–140.

Athreya, K.B., Doss, H., and Sethuraman, J. (1996). On the convergence of the Markov chain simulation method. *Ann. Statist.*, **24**, 69–100.

Barker, A.A. (1965). Monte Carlo calculations of the radial distribution functions for a proton–electron plasma. *Austral. J. Phys.*, **18**, 119–133.

Besag, J.E. and Green, P.J. (1993). Spatial statistics and Bayesian computation (with discussion). *J. Roy. Statist. Soc. Ser. B*, **55**, 25–37.

Besag, J., Green, P., Higdon, D., and Mengersen, K. (1995). Bayesian computation and stochastic systems (with discussion). *Statist. Sci.*, **10**, 3–66.

Celler, F., Leedham-Green, C., Murray, S., Niemeyer, A., and O'Brien, E. (1995). Generating random elements of a finite group. *Comm. Algebra*, **23**, 4931–4948.

Diaconis, P. and Saloff-Coste, L. (1996). Walks on generating sets of Abelian groups. *Probab. Theory Related Fields*, **105**, 393–421.

Finkelstein, L. and Kantor, W. (1993). *Groups and Computations*. American Mathematical Society, Providence, RI.

Gelman, A. and Rubin, D. (1992). Inference from iterative simulation using multiple sequences (with discussion). *Statist. Sci.*, **7**, 457–472, 503–511.

Geman, D. (1991). *Random Fields and Inverse Problems in Imaging*. Lecture Notes in Mathematics, vol. **1427**. Springer-Verlag, New York.

Geman, S. and Geman, D. (1984). Stochastic relaxation, Gibbs distributions and Bayesian restoration of images. *IEEE Trans. Pattern Anal. Machine Intelligence*, **6**, 721–741.

Geyer, C.J. (1992). Practical Markov chain Monte Carlo (with discussion). *Statist. Sci.*, **7**, 473–503.

Geyer, C.J. and Thompson, E.A. (1992). Constrained Monte Carlo maximum likelihood for dependent data (with discussion). *J. Roy. Statist. Soc. Ser. B*, **54**, 657–699.

Liu, J. (1996). Metropolized independent sampling with comparisons to rejection sampling and importance sampling. *Statist. Comput.*, **6**, 113–119.

Mengersen, K.L. and Tweedie, R.L. (1996). Rates of convergence of the Hastings and Metropolis algorithms. *Ann. Statist.*, **24**, 101–121.

Metropolis, N., Rosenbluth, A.W., Rosenbluth, M.N., Teller, A.H., and Teller, E. (1953). Equations of state calculations by fast computing machines. *J. Chem. Phys.*, **21**, 1087–1092.

Peskun, P.H. (1973). Optimum Monte Carlo sampling using Markov chains. *Biometrika*, **60**, 607–612.

Rosenthal, J.S. (1995). Minorization conditions and convergence rates for Markov chain Monte Carlo. *J. Amer. Statist. Assoc.*, **90**, 558–566.

Sinclair, A.J. and Jerrum, M.R. (1989). Approximate counting, uniform generation and rapidly mixing Markov chains. *Inform. and Comput.*, **82**, 93–133.

Smith, A.F.M. and Roberts, G.O. (1993). Bayesian computation via the Gibbs sampler and related Markov chain Monte Carlo methods (with discussion). *J. Roy. Statist. Soc. Ser. B*, **55**, 3–24.

Tanner, M.A. (1991). *Tools for Statistical Inference: Observed Data and Data Augmentation Methods*. Lecture Notes in Statistics, vol. **67**. Springer-Verlag, New York.

Tierney, L. (1994). Markov chains for exploring posterior distributions (with discussion). *Ann. Statist.*, **22**, 1701–1762.

Monte Carlo Sampling Methods Using Markov Chains and Their Applications

W.K. Hastings
University of Toronto

Summary

A generalization of the sampling method introduced by Metropolis et al. (1953) is presented along with an exposition of the relevant theory, techniques of application and methods and difficulties of assessing the error in Monte Carlo estimates. Examples of the methods, including the generation of random orthogonal matrices and potential applications of the methods to numerical problems arising in statistics, are discussed.

1. Introduction

For numerical problems in a large number of dimensions, Monte Carlo methods are often more efficient than conventional numerical methods. However, implementation of the Monte Carlo methods requires sampling from high-dimensional probability distributions and this may be very difficult and expensive in analysis and computer time. General methods for sampling from, or estimating expectations with respect to, such distributions are as follows:

(i) If possible, factorize the distribution into the product of one-dimensional conditional distributions from which samples may be obtained.

(ii) Use importance sampling, which may also be used for variance reduction. That is, in order to evaluate the integral

$$J = \int f(x)p(x)\,dx = E_p(f),$$

where $p(x)$ is a probability density function, instead of obtaining independent samples $x_1, \ldots, x_y$ from $p(x)$ and using the estimate $\hat{J}_1 = \sum f(x_i)/N$, we instead obtain the sample from a distribution with density $q(x)$ and use the estimate $\hat{J}_2 = \sum\{f(x_i)p(x_i)\}/\{q(x_i)N\}$. This may be advantageous if it is easier to sample from $q(x)$ than $p(x)$, but it is a difficult method to use in a large number of dimensions, since the values of the weights $w(x_i) = p(x_i)/q(x_i)$ for reasonable values of N may all be extremely small, or a few may be extremely large. In estimating the probability of an event A, however, these difficulties may not be as serious since the only values of $w(x)$ which are important are those for which $x \in A$. Since the methods proposed by Trotter and Tukey (1956) for the estimation of conditional expectations require the use of importance sampling, the same difficulties may be encountered in their use.

(iii) Use a simulation technique; that is, if it is difficult to sample directly from $p(x)$ or if $p(x)$ is unknown, sample from some distribution $q(y)$ and obtain the sample x values as some function of the corresponding y values. If we want samples from the conditional distribution of $x = g(y)$, given $h(y) = h_0$, then the simulation technique will not be satisfactory if $\text{pr}\{h(y) = h_0\}$ is small, since the condition $h(y) = h_0$ will be rarely if ever satisfied even when the sample size from $q(y)$ is large.

In this paper, we shall consider Markov chain methods of sampling that are generalizations of a method proposed by Metropolis et al. (1953), which has been used extensively for numerical problems in statistical mechanics. An introduction to Metropolis's method and its applications in statistical mechanics is given by Hammersley and Handscomb (1964, p. 117). The main features of these mothods for sampling from a distribution with density $p(x)$ are:

(a) The computations depend on $p(x)$ only through ratios of the form $p(x')/p(x)$, where x' and x are sample points. Thus, the normalizing constant need not be known, no factorization of $p(x)$ is necessary, and the methods are very easily implemented on a computer. Also, conditional distributions do not require special treatment and therefore the methods provide a convenient means for obtaining correlated samples from, for example, the conditional distributions given by Fraser (1968) or the distributions of the elements in a multiway table given the marginal totals.

(b) A sequence of samples is obtained by simulating a Markov chain. The resulting samples are therefore correlation and estimation of the standard deviation of an estimate and assessment of the error of an estimate may require more care than with independent samples.

2. Dependent Samples Using Markov Chains

2.1. Basic Formulation of the Method

Let $\mathbf{P} = \{p_{ij}\}$ be the transition matrix of an irreducible Markov chain with states $0, 1, \ldots, S$. Then, if $X(t)$ denotes the state occupied by the process at time t, we have

$$\mathrm{pr}\{X(t+1) = j | X(t) = i\} = p_{ij}.$$

If $\boldsymbol{\pi} = (\pi_0, \pi_1, \ldots, \pi_S)$ is a probability distribution with $\pi_i > 0$ for all i, and if $f(\cdot)$ is a function defined on the states, and we wish to estimate

$$I = E_\pi(f) = \sum_{i=0}^{S} f(i)\pi_i,$$

we may do this in the following way. Choose $\mathbf{P}$ so that $\boldsymbol{\pi}$ is its unique stationary distribution, i.e. $\boldsymbol{\pi} = \boldsymbol{\pi}\mathbf{P}$. Simulate this Markov chain for times $t = 1, \ldots, N$ and use the estimate

$$\hat{I} = \sum_{i=1}^{N} f\{X(t)\}/N.$$

For finite irreducible Markov chains we know that $\hat{I}$ is asymptotically normally distributed and that $\hat{I} \to I$ in mean square as $N \to \infty$ (Chung, 1960, p. 99).

In order to estimate the variance of $\hat{I}$, we observe that the process $X(t)$ is asymptotically stationary and hence so is the process $Y(t) = f\{X(t)\}$. The asymptotic variance of the mean of such a process is independent of the initial distribution of $X(0)$, which may, for example, attach probability 1 to a single state, or may be $\boldsymbol{\pi}$ itself, in which case the process is stationary. Thus, if N is large enough, we may estimate var $(\hat{I})$, using results appropriate for estimating the variance of the mean of a stationary process.

Let ρ_j be the correlation of $Y(t)$ and $Y(t+j)$ and let $\sigma^2 = \mathrm{var}\{Y(t)\}$. It is well known (Bartlett, 1966, p. 284) that for a stationary process

$$\mathrm{var}(\bar{Y}) = \frac{\sigma^2}{N} \sum_{j=-N+1}^{N-1} \left(1 - \frac{|j|}{N}\right) \rho_j$$

and that as $N \to \infty$

$$\operatorname{var}(\bar{Y}) \simeq 2\pi g(0)/N,$$

where $g(\omega)$ is the spectral density function at frequency ω. If the ρ_j are negligible for $j \geq j_0$, then we may use Hannan's (1957) modification of an estimate of $\operatorname{var}(\bar{Y})$ proposed by Jowett (1955), namely

$$v_{\bar{Y}}^2 = \frac{N}{(N-j_0)(N-j_0+1)}\left\{\sum_{j=-j_0+1}^{j_0-1}\left(1-\frac{|j|}{N}\right)(c_j - \bar{Y}^2)\right\}, \tag{1}$$

where

$$c_j = \sum_{t=1}^{N-j} Y(t)Y(t+j)/(N-j) \qquad \text{for} \quad j \geq 0 \quad \text{and} \quad c_{-j} = c_j.$$

A satisfactory alternative which is less expensive to compute is obtained by making use of the pilot estimate, corrected for the mean, for the spectral density function at zero frequency suggested by Blackman and Tukey (1958, p. 136) and Blackman (1965). We divide our observations into L groups of K consecutive observations each. Denoting the mean of the ith block by

$$\bar{Y}_i = \sum_{t=1}^{K} Y\{(i-1)K+t\}/K,$$

we use the estimate

$$s_{\bar{Y}}^2 = \sum_{i=1}^{L}(\bar{Y}_i - \bar{Y})^2/\{L(L-1)\}. \tag{2}$$

This estimate has approximately the stability of a chi-squared distribution on $(L-1)$ degrees of freedom. Similarly, the covariance of the means of two jointly stationary processes $Y(t)$ and $Z(t)$ may be estimated by

$$s_{\bar{Y}\bar{Z}} = \sum_{i=1}^{L}(\bar{Y}_i - \bar{Y})(\bar{Z}_i - \bar{Z})/\{L(L-1)\}. \tag{3}$$

2.2. Construction of the Transition Matrix

In order to use this method for a given distribution $\boldsymbol{\pi}$, we must construct a Markov chain $\mathbf{P}$ with $\boldsymbol{\pi}$ as its stationary distribution. We now describe a general procedure for doing this which contains as special cases the methods which have been used for problems in statistical mechanics, in those cases where the matrix $\mathbf{P}$ was made to satisfy the reversibility condition that for all i and j

$$\pi_i p_{ij} = \pi_j p_{ji}. \tag{4}$$

The property ensures that $\sum \pi_i p_{ij} = \pi_j$, for all j, and hence that $\boldsymbol{\pi}$ is a stationary distribution of $\mathbf{P}$. The irreducibility of $\mathbf{P}$ must be checked in each specific application. It is only necessary to check that there is a positive probability of going from state i to state j in some finite number of transitions, for all pairs of states i and j.

We assume that p_{ij} has the form

$$p_{ij} = q_{ij}\alpha_{ij} \qquad (i \neq j), \tag{5}$$

with

$$p_{ij} = 1 - \sum_{j \neq i} p_{ij},$$

where $\mathbf{Q} = \{q_{ij}\}$ is the transition matrix of an arbitrary Markov chain on the states $0, 1, \ldots, S$ and α_{ij} is given by

$$\alpha_{ij} = \frac{s_{ij}}{1 + \dfrac{\pi_i}{\pi_j}\dfrac{q_{ij}}{q_{ji}}}, \tag{6}$$

where s_{ij} is a symmetric function of i and j chosen so that $0 \leq \alpha_{ij} \leq 1$ for all i and j. With this form for p_{ij} it is readily verified that $\pi_i p_{ij} = \pi_j p_{ji}$, as required. In order to simulate this process we carry out the following steps for each time t:

(i) assume that $X(t) = i$ and select a state j using the distribution given by the ith row of $\mathbf{Q}$; and
(ii) take $X(t+1) = j$ with probability α_{ij} and $X(t+1) = i$ with probability $1 - \alpha_{ij}$.

For the choices of s_{ij} we will consider, only the quantity $(\pi_j q_{ji})/(\pi_i q_{ij})$ enters into the simulation and we will henceforth refer to it as the test ratio.

Two simple choices for s_{ij} are given for all i and j by

$$s_{ij}^{(M)} = \begin{cases} 1 + \dfrac{\pi_i q_{ij}}{\pi_j q_{ji}} & \left(\dfrac{\pi_j q_{ji}}{\pi_i q_{ij}} \geq 1\right), \\[2ex] 1 + \dfrac{\pi_j q_{ji}}{\pi_i q_{ij}} & \left(\dfrac{\pi_j q_{ji}}{\pi_i q_{ij}} \leq 1\right), \end{cases}$$

$$s_{ij}^{(B)} = 1.$$

With $q_{ij} = q_{ji}$ and $s_{ij} = s_{ij}^{(M)}$ we have the method devised by Metropolis et al. (1953) and with $q_{ij} = q_{ji}$ and $s_{ij} = s_{ij}^{(B)}$ we have Barker's (1965) method.

Little is known about the relative merits of these two choices for s_{ij}, but when $q_{ij} = q_{ji}$, we have

$$\alpha_{ij}^{(M)} = \begin{cases} 1 & (\pi_j/\pi_i \geq 1), \\ \pi_j/\pi_i & (\pi_j/\pi_i < 1), \end{cases}$$

$$a_{ij}^{(B)} = \pi_j/(\pi_i + \pi_j).$$

Thus we see that if $\pi_j = \pi_i$, we will take $X(t+1) = j$ with probability 1 with Metropolis's method and with probability $\frac{1}{2}$ with Barker's method. This suggests that Metropolis's method may be preferable since it seems to encourage a better sampling of the states.

More generally, we may choose

$$s_{ij} = g[\min\{(\pi_i q_{ij})/(\pi_j q_{ji}), (\pi_j q_{ji})/(\pi_i q_{ij})\}],$$

where the function $g(x)$ is chosen so that $0 \le g(x) \le 1 + x$ for $0 \le x \le 1$, and $g(x)$ may itself be symmetric in i and j. For example, we may choose $g(x) = 1 + 2(\frac{1}{2}x)^\gamma$ with the constant $\gamma \ge 1$, obtaining $s_{ij}^{(M)}$ with $\gamma = 1$ and $s_{ij}^{(B)}$ with $\gamma = \infty$.

We may define a *rejection rate* as the proportion of times t for which $X(t+1) = X(t)$. Clearly, in choosing $\mathbf{Q}$, high rejection rates are to be avoided. For example, if $X(t) = i$ and i is near the mode of an unimodal distribution, then $\mathbf{Q}$ should be chosen so that j is not too far from i, otherwise, π_j/π_i will be small and it is likely that $X(t+1) = i$. For each simulation it is useful to record the rejection rate since a high rejection rate may be indicative of a poor choice of initial state or transition matrix, or of a "bug" in the computer program.

We shall apply these methods to distributions $\boldsymbol{\pi}$ defined mathematically on an infinite sample space although, when we actually simulate the Markov chains on a digital computer, we will have a large but finite number of states. When $\boldsymbol{\pi}$ is continuous, we will have a discrete approximation to $\boldsymbol{\pi}$. Let $\pi(x)\,d\mu(x), p(x,x')\,d\mu(x')$ and $q(x,x')\,d\mu(x')$ be the probability elements for the distribution $\boldsymbol{\pi}$ and the Markov processes analogous to $\mathbf{P}$ and $\mathbf{Q}$, respectively. Let the possible values for x and x' on the computer be $x_0, \ldots, x_S$. These values depend on the word length and on the representation in the computer, float-point, fixed-point, etc. The probability elements may be approximated by $\pi(x_i)\delta\mu(x_i)$, $p(x_i, x_j)\delta\mu(x_j)$ and $q(x_i, x_j)\delta\mu(x_j)$, respectively. Substituting $\pi(x_i)\delta\mu(x_i)$ for π_i etc., we have

$$p(x_i, x_j) = q(x_i, x_j)\alpha_{ij}$$

and $(\pi_i q_{ij})/(\pi_j q_{ji})$ is replaced by $\{\pi(x_i)q(x_i, x_j)\}/\{\pi(x_j)q(x_j, x_i)\}$. Therefore, so long as α_{ij} is chosen so that it depends on $\boldsymbol{\pi}$ and $\mathbf{Q}$ only through the quantity $(\pi_i q_{ij})/(\pi_j q_{ji})$, we may use the densities in place of the corresponding probability mass functions, and we may think in terms of continuous distributions ignoring the underlying approximation which will be more than adequate in most applications.

For simplicity we have considered reversible Markov chains only. An example of an irreversible $\mathbf{P}$ is given in Handscomb (1962), where the states may be subdivided into finite subsets of states, where the states within a given subset are equally probable. Transitions amongst states within such a subset are made in a cyclic fashion; all other transitions are reversible.

2.3. Elementary Examples

EXAMPLE 1. Let $\boldsymbol{\pi}$ be the Poisson distribution with $\pi_i = \lambda^i e^{-\lambda}/i!(i = 0, 1, \ldots)$. For λ small we may use the following choice of $\mathbf{Q}$ to generate samples from $\boldsymbol{\pi}$:

$$q_{ij} = \tfrac{1}{2} \quad (j = i-1, i+1; i \neq 0) \qquad q_{00} = q_{01} = \tfrac{1}{2}.$$

Note that $\pi_{i+1}/\pi_i = \lambda/(i+1)$ and $\pi_{i-1}/\pi_i = i/\lambda$ so that the computations may be performed rapidly. For λ large $\mathbf{Q}$ must be chosen so that step sizes greater than unity are permitted or else very long realizations will be required in order to ensure adequate coverage of the sample space.

EXAMPLE 2. To sample from the standard normal distribution we define $\mathbf{Q}_k (k = 1, 2)$ in the following ways. Let $X(t)$ be the state at time t and choose $X'(t)$ to be uniformly distributed on the interval $[\varepsilon_k X(t) - \Delta, \varepsilon_k X(t) + \Delta]$, where $\Delta > 0$ is a constant, $\varepsilon_1 = +1$ and $\varepsilon_2 = -1$. To use Metropolis's method with either of these choices of $\mathbf{Q}$, both of which are symmetric, at each time t compute the test ratio $\beta(t) = \exp(\frac{1}{2}[X^2(t) - \{X'(t)\}^2])$. If $\beta(t) \geq 1$, set $X(t+1) = X'(t)$; if $\beta(t) < 1$, $X(t+1) = X'(t)$ with probability $\beta(t)$ and $X(t+1) = X(t)$ with probability $1 - \beta(t)$. Computing time can be saved if we compare $X^2(t)$ with $\{X'(t)\}^2$ instead of $\beta(t)$ with 1 and compute the exponential only when $\{X'(t)\}^2 > X^2(t)$.

Simulations carried out on an IBM7094II using the above transition matrices yielded the following estimates $\bar{X}$ of the mean of the distribution where, in all cases, we chose $N = 1000$ and $L = 25$: with $X(0) = 0$ and $\Delta = 1$, we obtained $\bar{X} = -0.12$ and $s_{\bar{x}} = 0.11$ with $\varepsilon_k = +1$, and $\bar{X} = -0.013$ and $s_{\bar{x}} = 0.02$ with $\varepsilon_k = -1$. The estimated standard deviation of the estimate is smaller with $\varepsilon_k = -1$, and is comparable to the theoretical standard deviation of the mean with 1000 independent observations, 0.031. Here we have an ideal situation with a symmetric distribution and known mean, but a similar device for variance reduction may sometimes be applicable in other problems of more practical interest. Other results indicated that there is little to choose between moderate values of Δ in the range 0.2–1.8, but extreme values of Δ led to poor estimates. For example, with $X(0) = 1.0$ and $\Delta = 0.001$ we obtained $\bar{X} = 1.001$ and $s_{\bar{x}} = 0.002$.

2.4. Multidimensional Distributions

If the distribution $\boldsymbol{\pi}$ is d-dimensional and the simulated process is $\mathbf{X}(t) = \{X_1(t), \ldots, X_d(t)\}$, there are many additional techniques which may be used to construct $\mathbf{P}$:

(1) In the transition from time t to $(t+1)$ all co-ordinates of $\mathbf{X}(t)$ may be changed.
(2) In the transition from time t to $t+1$ only one co-ordinate of $\mathbf{X}(t)$ may be changed, that selection being made at random from amongst the d co-ordinates.
(3) Only one co-ordinate may change in each transition, the co-ordinates being selected in a fixed rather than a random sequence.

Method (3) was used by Ehrman, Fosdick, and Handscomb (1960), who justified the method for their particular application. A general justification for the method may be obtained in the following way. Let the transition matrix when co-ordinate k is to be moved by $\mathbf{P}_k (k = 1, \ldots, d)$. Assume that the co-ordinates are moved in the order $1, 2, \ldots$ and that the process is observed only at times $0, d, \ldots$. The resulting process is a Markov process with transition matrix $\mathbf{P} = \mathbf{P}_1 \ldots \mathbf{P}_d$. If, for each k, $\mathbf{P}_k$ is constructed so that $\boldsymbol{\pi}\mathbf{P}_k = \boldsymbol{\pi}$, then $\boldsymbol{\pi}$ will be a stationary distribution of $\mathbf{P}$ since $\boldsymbol{\pi}P = \boldsymbol{\pi}\mathbf{P}_1 \ldots \mathbf{P}_d = \boldsymbol{\pi}\mathbf{P}_2 \ldots \mathbf{P}_d = \cdots = \boldsymbol{\pi}$. In practice, we must check the irreducibility of $\mathbf{P}$ to ensure uniqueness of the stationary distribution. Note that for the validity of this proof it is not necessary that each $\mathbf{P}_k$ satisfy the reversibility conditions (4). Also, in our estimates we may average observed values of the function $f(\cdot)$ at all times t of the original process although it would not be desirable to do so if the function values only change every d steps.

Example 3. To illustrate method (3) we consider sampling from a distribution with probability density function $p(\mathbf{x}) = p(x_1, \ldots, x_d)$ defined over the domain

$$0 \leq x_1 \leq \cdots \leq x_d < \infty.$$

Let the co-ordinates at time t be $\mathbf{x}(t) = \{x_1(t), \ldots, x_d(t)\}$ and assume co-ordinate $k (k \neq d)$ is to be changed. Let $x'_k(t)$ be chosen from the uniform distribution on the interval $\{x_{k-1}(t), x_{k+1}(t)\}$, where we assume that $x_0(t) \equiv 0$, and define $x'_i(t) = x_i(t)(i \neq k)$. Using Metropolis's method, we find that the test ratio is $p\{\mathbf{x}'(t)\}/p\{\mathbf{x}(t)\}$. When $k = d$, we may choose $x'_d(t)$ from the uniform distribution on the interval

$$\left[\tfrac{1}{2}\{x_{d-1}(t) + x_d(t)\}, 2x_d(t) - x_{d-1}(t)\right]$$

and use the test ratio

$$[p\{x'(t)\}\{x_d(t) - x_{d-1}(t)\}]/[p\{x(t)\}\{x'_d(t) - x_{d-1}(t)\}].$$

A few computer runs using this form of transition matrix have yielded satisfactory results for (i) sampling from the distribution of eigenvalues of a Wishart matrix, and (ii) estimating the conditional expected values of order statistics needed for the probability plotting method proposed by Wilk and Gnandesikan (1968). The latter problem requires estimation of

$E_p(x_k)(k = 1, \ldots, d)$, where

$$p(\mathbf{x}) = c\prod_{i=1}^{d}(\beta_i)^{\beta_i} x_i^{\beta_i - 1} \exp(-\beta_i x_i)/\Gamma(\beta_i) \qquad (0 \le x_i \le \cdots \le x_d),$$

the β_i's are constants and c is a normalizing constant.

Generalizations of these methods are readily obtained by noting that, for example, the co-ordinates need not be moved equally often and that the co-ordinates may be moved in groups instead of one at a time. For example, in statistical mechanics applications, one particle and hence three co-ordinates are moved at a time.

When only a few of the co-ordinates are moved at one time, computing effort can usually be saved by employing a recurrence formula for obtaining $f\{\mathbf{X}(t+1)\}$ from $f\{\mathbf{X}(t)\}$. However it is then necessary to guard against excessive build-up of error. One way to accomplish this is periodically to calculate $f\{\mathbf{X}(t)\}$ without the aid of the recursion, although an error analysis in some cases may indicate that this is unnecssary. If, however, it is expensive to compute either of π_i or $f(i)$, it may be preferable to attempt to move all of the co-ordinates a small distance each at time t.

2.5. Importance Sampling

Usually, **P** is constructed so that it depends only on the rations π_j/π_i as in the methods of Metropolis and Barker. In this case we need only know the π_i's up to a constant of proportionality, and if $\pi_0 + \cdots + \pi_S \neq 1$, we are estimating

$$E(f) = \sum_{i=0}^{S} f(i)\pi_i \Big/ \sum_{i=0}^{S} \pi_i.$$

This expression may be rewritten in the form

$$\frac{\sum_{i=0}^{S}\{f(i)\pi_i/\pi_i'\}\pi_i' \Big/ \sum_{i=0}^{S}\pi_i'}{\sum_{i=0}^{S}(\pi_i/\pi_i')\pi_i' \Big/ \sum_{i=0}^{S}\pi_i'},$$

which we recognize as $E_{\pi'}\{f(i)\pi_i/\pi_i'\}/E_{\pi'}(\pi_i/\pi_i')$. Hence, if we set up the Markov chain using the distribution $\boldsymbol{\pi}'$ instead of $\boldsymbol{\pi}$ we can estimate $I = E_\pi(f)$ by the ratio

$$\hat{I} = \frac{\sum_{t=1}^{N}[f\{X(t)\}\pi_{X(t)}/\pi'_{X(t)}]/N}{\sum_{t=1}^{N}\{\pi_{X(t)}/\pi'_{X(t)}\}/N}. \qquad (7)$$

This device permits us to use importance sampling with the Markov chain methods as suggested by Fosdick (1963).

If $\pi_0 + \cdots + \pi_S = 1$ and $\pi'_0 + \cdots + \pi'_S = 1$, then we may replace the denominator in $\hat{I}$ by 1, simplifying the estimate. Otherwise we have a ratio estimate and we can estimate its variance using the usual approximation. Thus, if we denote $\hat{I}$ by $\bar{Y}/\bar{Z}$, the variance is given approximately by

$$\operatorname{var}(\bar{Y}/\bar{Z}) = \{\operatorname{var}(\bar{Y}) - 2I \operatorname{cov}(\bar{Y}, \bar{Z}) + I^2 \operatorname{var}(Z)\}/\{E(\bar{Z})\}^2.$$

This may be estimated by

$$(s_{\bar{Y}}^2 - 2\hat{I} s_{\bar{Y}\bar{Z}} + \hat{I}^2 s_{\bar{Z}})/\bar{Z}^2,$$

where $s_{\bar{Y}}^2$, $s_{\bar{Y}\bar{Z}}$ and $s_{\bar{Z}}^2$ are obtained as in (2) and (3). For a discussion of the validity of this approximation, see Hansen, Hurwitz and Madow (1953, p. 164).

Importance sampling for variance reduction is more easily implemented with the Markov chain methods than with methods using independent samples, since with the Markov chain methods it is not necessary to construct the distribution $\boldsymbol{\pi}'$ so that independent samples can be obtained from it. If, however, we can obtain independent samples from the distribution $\boldsymbol{\pi}'$, we may obtain estimates of $E_\pi(f)$ using the Markov chain methods, which have the advantage that they do not involve the weights $\pi_{x(t)}/\pi'_{x(t)}$; see the discussion of importance sampling in §1. To accomplish this we set up the Markov chain $\mathbf{P}$ so that it has $\boldsymbol{\pi}$, not $\boldsymbol{\pi}'$, as its stationary distribution and choose $q_{ij} = \pi'_j$ for all i and j.

2.6. Assessment of the Error

In using Monte Carlo methods to estimate some quantity we usually use the standard deviation of the estimate to obtain some indication of the magnitude of the error of the estimate. There are, of course, many sources of error common to all Monte Carlo methods whose magnitude cannot be assessed using the standard deviation alone. These include: (i) the source of uniform random numbers used to generate the sample, which should be of as high a quality as possible; (ii) the nonnormality of the distribution of the estimate; (iii) computational errors which arise in computing the estimate; (iv) computation errors which arise in generating the samples (including discretization and truncation of the distribution); and (v) errors induced because the sample size is too small, which are often best overcome by methods other than increasing the sample size; for example, by the use of importance sampling. In what follows we shall concentrate upon categories (iv) and (v) above.

In generating successive samples using the Markov chain methods, errors will arise in computing the new state $X(t+1)$ and in computing the test ratio. An error analysis may sometimes be useful for the computation of $X(t+1)$ (see, for example, §3), but it is difficult to assess the effects of using inaccurate test ratios. The situation is also difficult to analyze, in general, when successive samples are independent and are generated using a factorization of the probability density function. To see this, let

$$p(\mathbf{x}) = p(x_1, x_2, \ldots x_d) = \prod_{i=1}^{d} p_i(x_i)$$

be the joint density function, where $p_i(x_i)$ is the conditional density function for x_i given $x_1, \ldots, x_{i-1}$. When we attempt to sample from each of the one-dimensional distributions $p_i(x_i)$ in turn, errors will be introduced and we will instead be sampling from some distribution $\bar{p}_i(x_i) = p_i(x_i)(1 + y_i)$, where y_i is a function of $x_1, \ldots, x_i$ and will not, in general, be small for all sample points. Consequently, we will be generating samples from a distribution

$$\bar{p}(\mathbf{x}) = p(\mathbf{x}) \prod_{i=1}^{d} (1 + y_i)$$

and, especially when d is large, it will be difficult to assess the error in our estimate induced by the y_i's. A similar, but more involved, analysis might also be applied to the Markov chain methods if we consider a single realization of length N as a single sample from a distribution of dimension Nd; sampling at each step of the Markov chain would correspond to sampling from a single factor above.

If the sample size is not large enough, important regions of the sample space may be inadequately represented. For example, if we are estimating the integral $\int f(x)p(x)\,dx$ by sampling from $p(x)$ and if the major contribution to the value of the integral comes from a small region of low probability in which $f(x)$ has very large values, then we may obtain a very poor estimate and a deceptively small standard deviation even with seemingly large sample sizes. This difficulty may be encountered with any method of numerical quadrature and there is no substitute for a thorough study of the integrand and a consequent adjustment of the method if gross errors are to be avoided. With the Markov chain methods the influence on the result of the choice of initial state and the correlation of the samples, which may be considerable if the sample size is small, may be minimized by adopting the following procedures:

(i) Choose a transition matrix $\mathbf{Q}$ so that the sample point in one step may move as large a distance as possible in the sample space, consistent with a low rejection rate. This offers some protection against the possibility that the sample points for the whole realization remain near one mode of the distribution which is separated from other modes by a deep trough, or that the initial state is in a region of low probability.

(ii) If possible, choose the initial state so that it is in a region of high probability, by sampling from $\boldsymbol{\pi}$ if this is possible, or set the co-ordinates of the initial state equal to their expected values, or asymptotic approximations of these.

In view of the many sources of error one or more of the following techniques, depending upon the application, should be used in practice to aid in assessing the suitability of the Markov chain methods, the magnitude of the error and the adequacy of the length of the realization and of the standard deviation as a measure of error:

(a) Test the method on problems which bear a close resemblance to the problem under study and for which the results are known analytically.
(b) If the expected value of some function with respect to $\boldsymbol{\pi}$ is known, this may be estimated for checking purposes, and possible for variance reduction, while other aspects of $\boldsymbol{\pi}$ are under study.
(c) Compare estimates obtained from different segments of the same realization to see if there is evidence of nonconvergence.
(d) Compare results obtained with and without the use of importance sampling, or using different choices of $\mathbf{P}$, or using different random numbers, or using the Markov chain method and some other numerical method, for which adequate assessment of the error may also be difficult as is often the case with asymptotic results, for example.

The illustrations given by Fox and Mayers (1968), for example, show how even the simplest of numerical methods may yield spurious results if insufficient care is taken in their use, and how difficult it often is to assess the magnitude of the errors. The discussion above indicates that the situation is certainly no better for the Markov chain methods and that they should be used with appropriate caution.

3. Random Orthogonal Matrices

We now consider methods for generating random orthogonal and unitary matrices and their application to the evaluation of averages over the orthogonal group with respect to invariant measure, a problem considered analytically by James (1955), and to other related problems.

Let $\mathbf{H}$ be an orthogonal $m \times m$ matrix with $|\mathbf{H}| = 1$ and let $\mathbf{E}_{ij}(\theta)$ be an elementary orthogonal matrix with elements given by

$$e_{ii} = \cos\theta, \qquad e_{ij} = \sin\theta, \qquad e_{ji} = -\sin\theta, \qquad e_{jj} = \cos\theta,$$
$$e_{\alpha\alpha} = 1 \quad (\alpha \neq i,j), \qquad e_{\alpha\beta} = 0 \quad \text{otherwise},$$

for some angle θ. We may generate a sequence of orthogonal matrices $\mathbf{H}(t) (t = 1, 2, \ldots)$ using Metropolis's method, which is suitable for estimating averages over the orthogonal group with respect to invariant measure $\{d\mathbf{H}\}$, in the following way:

(i) Let $\mathbf{H}(0) = \mathbf{H}_0$, where $\mathbf{H}_0$ is an orthogonal matrix.
(ii) For each t, select i and j at random from the set $\{1, \ldots, m\}$ with $i \neq j$. Select θ from the uniform distribution on $[0, 2\pi]$.
(iii) Let $\mathbf{H}'(t) = \mathbf{E}_{ij}(\theta)\mathbf{H}(t)$.

Since the measure $\{d\mathbf{H}\}$ is left invariant, $\{d\mathbf{H}'(t)\}$ and $\{d\mathbf{H}(t)\}$ are equal and hence, in Metropolis's method, the new state is always the newly computed value, i.e. $\mathbf{H}(t+1) = \mathbf{H}'(t)$. To estimate the integral

$$J = \int_{O(m)} f(\mathbf{H})\{d\mathbf{H}\},$$

where $O(m)$ denotes the group of orthogonal matrices, we use the estimate

$$J = \sum_{t=1}^{N} f\{\mathbf{H}(t)\}/N.$$

To remove the restriction that $|\mathbf{H}| = 1$ we may use the following procedure. After i and j are selected as above, select one of i and j at random, each with probability $\frac{1}{2}$, and call the result i'. Multiply the i'th row of $\mathbf{E}_{ij}(\theta)$ by ± 1, the sign being selected at random, to form a new matrix which we denote by $\mathbf{E}'_{ij}(\theta)$; $\mathbf{E}'_{ij}(\theta)$ is then used in step (iii) above, in place of $\mathbf{E}_{ij}(\theta)$.

To show that the Markov chain is irreducible we need only show that every orthogonal matrix $\mathbf{H}$ may be represented as the product of matrices of the form $\mathbf{E}'_{ij}(\theta)$. This, in turn, may be done by a simple annihilation argument. Choose θ so that $\mathbf{E}_{12}(\theta)\mathbf{H} = \mathbf{H}^{(1)}$ has the elements $h_{21}^{(1)} = 0$. Choose θ so that $\mathbf{E}_{13}(\theta)\mathbf{H}^{(1)} = \mathbf{H}^{(2)}$ has elements $h_{31}^{(2)} = h_{21}^{(2)} = 0$. Continue in this way until we have the matrix $\mathbf{H}^{(m-1)}$ with the first column annihilated. Since $\mathbf{H}^{(m-1)}$ is orthogonal, we must have $h_{12}^{(m-1)} = \cdots = h_{1m}^{(m-1)} = 0$ and $h_{11}^{(m-1)} = \pm 1$. Continuing this procedure we may annihilate the remaining off diagonal elements and obtain a diagonal matrix with elements ± 1 on the diagonal. The desired factorization is easily deduced from this.

We now show that the process, by which the sequence of orthogonal matrices is generated, is numerically stable. Denote the computed values of $\mathbf{H}(t)$ and $\mathbf{E}_{ij}(\theta)$ by $\mathbf{H}_c(t)$ and $\mathbf{E}_c(t)$. Let $\mathbf{H}_c(t) = \mathbf{H}(t) + \mathbf{H}_e(t)$, $\mathbf{E}_c(t) = \mathbf{E}_{ij}(\theta) + \mathbf{E}_e(t)$ and $\mathbf{H}_c(t+1) = \mathbf{E}_c(t)\mathbf{H}_c(t) + \mathbf{F}(t)$. We are interested in how far $\mathbf{H}_e(t)$ is from zero, which we measure by $\|\mathbf{H}_e(t)\|$, where the Euclidean norm $\|\mathbf{A}\|$ of a matrix $\mathbf{A}$ is defined by

$$\|\mathbf{A}\| = \left(\sum_{i=1}^{m}\sum_{j=1}^{m} a_{ij}^2\right)^{1/2}.$$

Using the fact that $\|\mathbf{A}\|$ is preserved under orthogonal transformations and the inequalities $\|\mathbf{AB}\| \leq \|\mathbf{A}\|\|\mathbf{B}\|$ and $\|\mathbf{A}+\mathbf{B}\| \leq \|\mathbf{A}\| + \|\mathbf{B}\|$, we can easily show that

$$\|\mathbf{H}_e(t+1)\| \leq (1+k)\|\mathbf{H}_e(t)\| + k,$$

where k is chosen such that $\|\mathbf{E}_c(t)\| + \|\mathbf{F}(t)\| \leq k$ for all t, and by induction that $\|\mathbf{H}_c(t)\| \leq U_t$, where U_t is the solution of the difference equation $U_{t+1} = (1+k)U_t + k$ with $U_0 = \|\mathbf{H}_c(0)\|$. Solving the difference equation, we arrive at the bound

$$\|\mathbf{H}_e(t)\| \leq (1+k)^t - 1 + \|\mathbf{H}_e(0)\|.$$

Therefore, unless the word length of the computer is short or t is very large, the accumulation of error may be neglected since, for all t, $\|\mathbf{F}(t)\|$ will be small in view of the few arithmetic operations involved in forming the elements of the product $\mathbf{E}_c(t)\mathbf{H}_c(t)$, and $\|\mathbf{E}_e(t)\|$ will also be small if the method below is used for the generation of the four nontrivial elements of $\mathbf{E}_c(t)$. For short computer word length or very large t, orthonormalization of the columns of $\mathbf{H}_c(t)$ at regular intervals will prevent excessive accumulation of error.

One method for computing $\cos\theta$ and $\sin\theta$ required in $\mathbf{E}_{ij}(\theta)$ is that of von Neumann. Let U_1 and U_2 be independent and uniformly distributed on $[0, 1]$. If $U_1^2 + U_2^2 \leq 1$, compute

$$\cos\theta = (U_1^2 - U_2^2)/(U_1^2 + U_2^2), \qquad \sin\theta = \pm 2U_1U_2/(U_1^2 + U_2^2),$$

where the sign of $\sin\theta$ is chosen at random. If $U_1^2 + U_2^2 > 1$, obtain new values of U_1 and U_2 until $U_1^2 + U_2^2 \leq 1$.

EXAMPLE 4. When $f(\mathbf{H}) = h_{11}^2 + \cdots + h_{mm}^2$, then $J = 1$. Using $N = 1000$, $L = 25, m = 50$ and $\mathbf{H}(0) = \mathbf{I}$, we obtained the estimate $\hat{J} = 3.5$ with standard deviation 1.5. This estimate is poor because of the very poor choice for $\mathbf{H}(0)$ but the estimated standard deviation gives us adequate warning. It would be natural to increase the sample size N in order to obtain greater precision but this may also be achieved by a better choice of $\mathbf{H}(0)$ with a saving in computer time. For problems of the kind being considered here the following choice of $\mathbf{H}(0)$ is generally useful and better than that considered above. For $j = 1, \ldots, m$ and for m even, set

$$h_{1j} = 1/\sqrt{m}, h_{2j} = (1/\sqrt{m})\cos\{(j-1)\pi\},$$

$$h_{rj} = \sqrt{(2/m)}\cos\{(j-1)(r-2)2\pi/m\} \qquad (r = 3, 4, \ldots, \tfrac{1}{2}m+1),$$

$$h_{sj} = \sqrt{(2/\pi)}\sin\{(j-1)(s-\tfrac{1}{2}m-1)2\pi/m\} \qquad (s = \tfrac{1}{2}m+2, \ldots, m).$$

For m odd a similar choice for $\mathbf{H}(0)$ may be made. Using this matrix for $\mathbf{H}(0)$, we obtained the estimate $\hat{J} = 0.96$ with standard deviation 0.03; the computer time required was 0.08 minutes.

The techniques discussed above may be applied to the following problems:

(i) To generate random permutation matrices we set $\theta = \frac{1}{2}\pi$. To generate random permutations, of m elements, at each time t, we interchange a randomly selected pair of elements in the latest permutation in order to generate a new permutation.

(ii) To generate k orthogonal vectors in m dimensions we replace $\mathbf{H}(t)$ by an $m \times k$ matrix where columns will be the desired vectors.

(iii) To generate random unitary matrices the above procedure for orthogonal matrices is modified as follows:

In place of $\mathbf{E}_{ij}(\theta)$ we use an elementary unitary matrix $\mathbf{U}_{ij}(\theta, \phi)$ with elements given by

$$U_{ij} = \cos\theta, \qquad U_{ij} = e^{-i\phi}\sin\theta, \qquad U_{ji} = -e^{\,i\phi}\sin\theta, \qquad U_{jj} = \cos\theta,$$
$$U_{\alpha\alpha} = 1 \quad (\alpha \neq i, j), \qquad U_{\alpha\beta} = 0 \quad \text{otherwise.}$$

The matrix $\mathbf{U}'_{ij}(\theta, \phi)$ is obtained by multiplying the ith row of $\mathbf{U}'_{ij}(\theta, \phi)$ by $e^{i\gamma}$. Here, θ, ϕ and γ are chosen to be independent and uniformly distributed on $[0, 2\pi]$. Irreducibility may again be established by an annihilation argument and the proof of numerical stability given above requires only minor modifications.

(iv) To sample from a distribution defined on a group $G = \{g\}$, the transition matrix $\mathbf{Q}$ must generate $g'(t)$ from $g(t)$, and this may be done by arranging that $g'(t) = hg(t)$, where $h \in G$ and h is chosen from some distribution on G. If the probability element for g is $p(g)\{d\mu(g)\}$, where μ is the left invariant measure on G, then the ratio corresponding to π_j/π_i is $p\{g'(t)\}/p\{g(t)\}$. At above we must ensure that $g(t)$ is close to being a group element.

(v) To generate samples from a distribution with probability element

$$p(\mathbf{x})\,d\mathbf{x} = \int_{O(m)} q(\mathbf{x}, \mathbf{H})\{d\mathbf{H}\}\,d\mathbf{x}$$

we may remove the integral and sample from $q(\mathbf{x}, \mathbf{H})\{d\mathbf{H}\}d\mathbf{x}$ with, unfortunately, a consequent increase in the dimension of the space being sampled. The sampling may be carried out by combining the methods of §2.4, alternating the changes of $\mathbf{x}$ and $\mathbf{H}$, with the techniques given above for orthogonal matrices. Note that the value of the test ratio for changing $\mathbf{H}$ will no longer be unity. More generally, sampling from any distribution whose density may be expressed as an average may be accomplished in a similar way. Many of the normal theory distributions of multivariate analysis given by James (1964) have this form.

In the method described above for orthogonal matrices the successive matrices are statistically dependent and we now consider a method in which successive matrices are independent. Let x_{ij} $(i = 1, \ldots, k;\ j = 1, \ldots, m)$ be independent standard normal variates and form the k vectors

$$x_i = (x_{i1}, \ldots, x_{im}) \qquad (i = 1, 2, \ldots, k).$$

If we now apply a Gram–Schmidt orthonormalization to these vectors, we will obtain k vectors y_i, with the desired properties, since it is easy to see that the joint distribution of they y_i's will be unchanged if they are subjected to an arbitrary orthogonal transformation. For the case $k = 1$, see Tocher (1963). For each set of k orthogonal vectors generated by this procedure we require mk standard normal deviates, k square roots and, approximately, $2mk^2$ arithmetic operations. The Markov chain method, on the other hand, requires only $6k$ arithmetic operations and the cosine and sine of a uniformly distributed angle and, if the function $f(\cdot)$ being averaged is evaluated only at times $T, 2T, \ldots$, we must have T as large as $\frac{1}{3}mk$ before the computing times of the two procedures are comparable. The degree of correlation amongst the sample values for the Markov chain method will depend upon the particular function $f(\cdot)$ and will determine, ignoring error analysis considerations, which of the two methods is better.

Acknowledgements

It is a pleasure to thank Messrs. Damon Card, Stuart Whittington and Professor John P. Valleau, who generously shared their considerable knowledge of and experience with the Markov chain methods in statistical mechanics applications, Professor John C. Ogilvie, who made several helpful suggestions and Mr. Ross D. MacBride, who capably prepared the computer programs. I am grateful to the referees whose comments were very helpful in the revision of this paper. The work was supported by the National Research Council of Canada.

References

Barker, A.A. (1965). Monte Carlo calculations of the radial distribution functions for a proton–electron plasma. *Austral. J. Phys.*, **18**, 119–133.

Bartlett, M.S. (1966). *An Introduction to Stochastic Processes*, 2nd ed. Cambridge University Press, Cambridge.

Blackman, R.B. (1965). *Data Smoothing and Prediction.* Addison-Wesley, Reading, MA.

Blackman, R.B. and Tukey, J.W. (1958). *The Measurement of Power Spectra.* Dover, New York.

Chung, K.L. (1960). *Markov Processes with Stationary Transition Probabilities.* Springer-Verlag, Heidelberg.

Erhman, J.R., Fosdick, L.D., and Handscomb, D.C. (1960). Computation of order parameters in an Ising lattice by the Monte Carlo method. *J. Math. Phys.*, **1**, 547–558.

Fosdick, L.D. (1963). Monte Carlo calculations on the Ising lattice. *Methods Comput. Phys.*, **1**, 245–280.

Fox, L. and Mayers, D.F. (1968). *Computing Methods for Scientists and Engineers.* Clarendon Press, Oxford.

Fraser, D.A.S. (1968). *The Structure of Inference*. Wiley, New York.

Hammersley, J.M. and Handscomb, D.C. (1964). *Monte Carlo Methods*. Methuen, London.

Handscomb, D.C. (1962). The Monte Carlo method in quantum statistical mechanics. *Proc. Cambridge Philos. Soc.*, **58**, 594–598.

Hannan, E.J. (1957). The variance of the mean of a stationary process. *J. Roy. Statist. Soc. Ser. B*, **19**, 282–285.

Hansen, M.H., Hurwitz, W.N., and Madow, W.G. (1953). *Sample Survey Methods and Theory*. Wiley, New York.

James, A.T. (1955). A generating function for averages over the orthogonal group. *Proc. Roy. Soc. A*, **229**, 367–375.

James, A.T. (1964). Distribution of matrix variates and latent roots derived from normal samples. *Ann. Math. Statist.*, **35**, 475–501.

Jowett, G.H. (1955). The comparison of means of sets of observations from sections of independent stochastic series. *J. Roy. Statist. Soc. Ser. B*, **17**, 208–227.

Metropolis, N., Rosenbluth, A.W., Rosenbluth, M.N., Teller, A.H. and Teller, E. (1953). Equations of state calculations by fast computing machines. *J. Chem. Phys.*, **21**, 1087–1092.

Tocher, K.D. (1963). *The Art of Simulation*, English Universities Press, London.

Trotter, H.F. and Tukey, J.W. (1956). Conditional Monte Carlo for normal samples. In *Symposium on Monte Carlo Methods*, ed. H.A. Meyer, pp. 64–79. Wiley, New York.

Wilk, M.B. and Gnanadesikan, R. (1968). Probability plotting methods for the analysis of data. *Biometrika*, **55**, 1–17.

Introduction to Lindley and Smith (1972) Bayes Estimates for the Linear Model

John J. Deely

Introduction

In 1972 there were no MCMC methods readily available to the statistical community in general and to the Bayesian statisticians in particular. Realistic formulations of models for practical situations were dismissed if solutions were neither available in closed form nor amenable to numerical calculation. Thus much effort was devoted to either approximating solutions to these practically desirable models or formulating other models that were both realistic and solvable. It was also the case the Bayesian modeling did not find general acceptance unless it could be shown that such models had desirable frequentist properties. Thus, when viewed in this context, this contribution by Lindley and Smith (hereafter LS) must surely qualify for the accolade, "breakthrough."

They started by considering the normal linear model and, as good Bayesians, noted that often in practical applications of this model there is some form of prior information available about the "parameters" as a group. This approach is to be contrasted to situations in which there exists prior information about parameters individually. It is intuitively reasonable (later experimentally verified) that using such prior information should produce better statistical inferences and thus we should try to incorporate such information into the mathematical model. To model this "group" idea in the Bayesian paradigm, they further noted that choosing a prior selected via the notion of exchangeability could quite naturally accommodate this group dependence. They called their model "multistage" but then adopted the description "hierarchy" since it is more descriptive of exactly what is happening. By making efficient use of matrix algebra these "hierarchical" Bayesian estimates became totally transparent. As a result their proposed estimates could then be easily compared to frequentist least squares estimates and show them up as wanting. This latter point of course was required in those days in order to even get a hearing in the first instance. Thus, in light of all this, the word "breakthrough" seems completely appropriate.

Of course, the concepts of parameter dependence and exchangeability had previously appeared in the literature. The pioneering work of Stein (1956) and James and Stein (1960) had fostered the frequentist's interests and contributions to "shrinkage," a concept which related the individual parameters in a multivariate situation to one another by devising ad hoc methods which caused the estimates of individual parameters to be "shrunk" from their individual sample means toward the overall mean.

Concerning the concept of exchangeability, there had been the early seminal work of de Finetti (1937, 1964) which, although dealing with more elementary probability models, fostered interest amongst the larger statistical community in the more complicated mathematical structures. In particular, there was considerable interest in mixtures of distributions and their relationship to exchangeability (see, e.g., Hill (1969)).

All of this may now seem completely obvious and as is often the case, after you once see the obvious you wonder why you did not see it was obvious before someone pointed it out to you. Who but the presumptuous would have the temerity to claim that they would have stated the obvious if only they had been asked first? The fact that this paper was the first to make this connection is borne out by the comments of de Finetti, who was one of the discussants. It also appears that this paper really popularized the words "hierarchy" and "hyperparameters," the former a more definitive description of the multistage model, the latter describing the statistical variables in the later stages. However, it is generally acknowledged that I.J. Good was the first to use the word to describe statistical structures. As one of the discussants for this paper, Barnett also uses the expression, "hierarchical normal."

Further evidence supporting the claim that in fact this paper represented a significant breakthrough is provided by noting the enormous amount of research that ensued in the area of hierarchical Bayesian models. A casual perusal of the literature will show how often it is listed in the references. LS had proposed a model which in its simplicity was elegant; they then demonstrated that simple and straightforward inferences could be obtained while leaving the more complicated for later development. Because it provided this kind of valuable insight into multistage Bayesian modeling, it encouraged researchers to dig deeper and opened up their horizons to see why the Bayesian model really got closer to reality than classical least squares. Nelder (one of the discussants) even comments than the techniques seem so reasonable that you don't have to be a Bayesian to use them.

Brief Review

LS propose the multistage normal linear model as

$$\mathbf{Y} \sim N(\mathbf{A}_1\boldsymbol{\theta}_1, \mathbf{C}_1), \qquad \theta_1 \sim N(\mathbf{A}_2\boldsymbol{\theta}_2, \mathbf{C}_2), \qquad \theta_2 \sim N(\mathbf{A}_3\boldsymbol{\theta}_3, \mathbf{C}_3),$$

where the $\mathbf{A}_i$ are known design matrices and $\mathbf{C}_i$ are nonsingular dispersion matrices. The paper contains an Introduction and five sections: (1) Exchangeability; (2) General Bayesian Linear Model; (3) Examples; (4) Estimation with Unknown Covariance Structure; and (5) Examples with Unknown Covariance Structure. The Introduction succinctly provides the motivation by pointing out the deficiencies that are known to exist with least squares in many practical situations. It also shows how in that time that it was necessary to convince frequentists in their language of the value of the Bayesian paradigm. Section 1 sets out in elegant simplicity the valuable role exchangeability can play in modeling the prior distribution in the context of the multistage or hierarchical model. Mathematical notation and key matrix identities essential to general Bayesian estimates are developed in Section 2, which is followed in the next section by three well-known practical examples illustrating the methodology:

(i) a typical AOV two-way layout;
(ii) a regression problem over several groups (say schools) with n_i observations within a group, with each group having its own regression equation but the respective group coefficients assumed to be exchangeable; and
(iii) a regression problem with n observations from a single regression equation with p coefficients which are assumed to be exchangeable.

These examples in their simplicity and practicality are extremely helpful in understanding the application of exchangeability in modeling prior information. For these examples the vector $\boldsymbol{\theta}_1$ constitutes the vector of the "parameters" (i.e., unobserved quantities) which are of primary interest. In each case it is quite reasonable to assume that knowing the value of one quantity (i.e., one component of the vector $\boldsymbol{\theta}_1$) is helpful in estimating the value of another and that some prior information about these collective values as a group is likely to be available. One easy way of modeling this concept is via exchangeability. This of course simply amounts to a more convenient way to arrive at the prior distribution of $\boldsymbol{\theta}_1$ which will express the kind of prior information available. When this type of information leads to a kind of dependence between the components of $\boldsymbol{\theta}_1$ then the prior should express this possibility. The device of exchangeability says that the prior can the written as a special mixture of the form

$$p(\vartheta_1) = \int \prod_{j=1}^{p} p(\vartheta_{1j}|\vartheta_2)h(\vartheta_2)\,d\vartheta_2$$

and hence is one way of modeling the dependence of the parameters. It is clear that other types of mixtures as well as other types of multivariate distributions could also be candidates to model the available prior information. The authors are very careful to point this out in a statement which has become a classic: "In any application the particular form of the prior distribution has to be carefully considered." (see Introduction, last line, third last paragraph). Their approach in this paper was to show that a fairly simple model could describe various types of prior information that we might expect to arise in practical scenarios.

Section 4 describes the situation as it existed then when "nuisance" parameters (i.e., covariances $\mathbf{C}_i$) were unknown. The first paragraph beautifully describes the Bayesian paradigm and indicates why at that time they had to seek approximations as indicated in this section. These approximations and intuitively reasonable estimation procedures are followed by the same three examples as given is Section 3 but with the added assumption that the matrices $\mathbf{C}_1$ and $\mathbf{C}_2$ are unknown. In the present day of readily available MCMC computational methods, these proposed procedures have surely been superseded.

Some Weaknesses

This of course is the normal model and there is now no need to assume the hyperprior is normal. This points out how their model is quite often not realistic in that the designation of the hyperprior should be free to express opinions about the parameters (unobserved but practically defined important measurable quantities) as a group and what relationship they have to each other. To be fair, the authors are quite aware of this important concept as they point out in the Introduction: "We argue that it is typically true that there is available prior information about the parameters ... we explore a particular form of prior information based on de Finetti's important concept of exchangeability."

Another weakness perhaps is their preoccupation with the noninformative case at the third stage. This artefact essentially eliminates the third stage. Specifically LS develop a general formula for the posterior [see (12) and (13) in LS] which says that the posterior distribution of $\boldsymbol{\theta}_1$ is a multivariate normal with dispersion matrix $\mathbf{D}$ where

$$\mathbf{D}^{-1} = \mathbf{A}_1^T\mathbf{C}_1^{-1}\mathbf{A}_1 + (\mathbf{C}_2 + \mathbf{A}_2\mathbf{C}_3\mathbf{A}_2^T)^{-1}.$$

They apply the matrix lemma (10) to this expression to obtain (15) as

$$\mathbf{A}_1^T\mathbf{C}_1^{-1}\mathbf{A}_1 + \mathbf{C}_2^{-1} - \mathbf{C}_2^{-1}\mathbf{A}_2(\mathbf{A}_2^T\mathbf{C}_2^{-1}\mathbf{A}_2 + \mathbf{C}_3^{-1})^{-1}\mathbf{A}_2^T\mathbf{C}_2^{-1}.$$

Then "... by supposing the third stage dispersion matrix $\mathbf{C}_3$ to be large, or to let its inverse, the precision matrix, be zero," that is, set $\mathbf{C}_3^{-1} = 0$ in (15), they obtain an expression for the inverse of the posterior dispersion matrix $\mathbf{D}_0^{-1}$ as given in (16)

$$\mathbf{D}_0^{-1} = \mathbf{A}_1^T\mathbf{C}_1^{-1}\mathbf{A}_1 + \mathbf{C}_2^{-1} - \mathbf{C}_2^{-1}\mathbf{A}_2(\mathbf{A}_2^T\mathbf{C}_2^{-1}\mathbf{A}_2)^{-1}\mathbf{A}_2^T\mathbf{C}_2^{-1}.$$

There are two technical problems here which do not arise for the specific situations in LS, but which can be important in a general context. First, the only way in which the substitution of $\mathbf{C}_3^{-1} = 0$ is justified is to use a limit argument for $\mathbf{C}_3 \to \infty$. This has to be done carefully since all paths for which all kinds of $\mathbf{C}_3$'s become unbounded may not lead to $\mathbf{C}_3^{-1} = 0$ in the limit. Second, in order that the posterior under the vague prior on the third stage has a properly defined dispersion matrix, the matrix $\mathbf{D}_0^{-1}$ must be shown to be nonsingular, a fact that is not always true.

In addition to these technical problems, LS were not able to adopt a proper hierarchical Bayesian approach to the other covariances $\mathbf{C}_1$ and $\mathbf{C}_2$ when they were assumed to be unknown as indicated in their Sections 4 and 5. This problem of course is now numerically solvable and has fostered considerable research over the last 5 years in particular.

Finally, the important area of predictive inference was not discussed due no doubt to the fact that it would have introduced another level of complexity. Again this approach is now possible and has appeared in many forms in the literature.

Discussants

When this paper was presented at a meeting of the Royal Statistical Society, twelve discussants added their comments and, after the meeting, six others made contributions. Observing their respective career paths over the intervening years cannot but impress us with their credentials. Their comments are well worth reading and make a valuable contribution to the overall effect of the paper. As LS point out in their "Rejoinder," all but two of the eighteen were quite complimentary with their comments, "... Most of them have been prepared to look at the (our) estimates ... and not get involved in continual arguments on philosophy. ... " Perhaps a flavor of the feeling engendered by their comments can be gained by quoting one small phrase contributed by Fienberg: "It is more than faint praise for me to remark that I wish I had written this paper."

Summary

Twenty-five years ago these two authors entered the arena to do battle with the "least squares" giant and it is now clear that they were supremely victorious. They have succeeded in showing how the multistage or hierarchical structure can accommodate a wide variety of applications with their correspondent prior information and thus completely overpower least squares. By showing that prior information about the ***group*** of parameters (their $\boldsymbol{\theta}_1$), as opposed to ***individual*** members of the group (i.e., components of the vector $\boldsymbol{\theta}_1$), could be conveniently modeled by exchangeability, they pioneered a new area of thought. With elegant exposition and beautiful simplicity their paper gave clear direction to future research.

It is perhaps somewhat disconcerting that LS did not really develop the informative approach to prior information, but of course this is understandable, given the state of numerical techniques at that time. They do acknowledge the real possibilities of prior information and that it should be incorporated into the analysis. Yet on the other hand, they are forced to adopt a noninformative model at the third stage. It is clear that in 1972 their noninformative approach arose because of computational difficulties. It is worth pondering what form their model would have taken if MCMC methods had been well known at that time. The same could also be said of the many other less attractive "ad hockeries" which were also suggested during that period. With regard to the development of present-day MCMC methods, it should be pointed out that it is a historical fact that the second author, over the interim, has indeed pioneered application of such MCMC methods, not only to the linear model but to many other areas as well.

Finally, when we think of the overall contribution of this paper, it is apparent from the volume of research that has followed in hierarchical Bayesian methods, applied not only to the normal linear model but to far wider areas as well, that the material presented by LS was so lucid and vital that others were inspired to further develop both their strengths and weaknesses. Indeed, in the light of history it can safely be said that LS made a breakthrough.

References

de Finetti, B. (1937). Foresight: Its logical laws, its subjective sources. Translated and reprinted in *Studies in Subjective Probability* (H. Kyburg and H. Smokler, eds.), 1964, pp. 93–158. Wiley, New York.

Hill, B.M. (1969). Foundations for the theory of least squares. *J. Roy. Statist. Soc. Ser. B*, **31**, 89–97.

James, W. and Stein, C. (1960). Estimation with quadratic loss. *Proc. Fourth Berkeley Sympos.*, **1**, 361–380. University of California Press, Berkeley.

Stein, C. (1956). Inadmissibility of the usual estimator for the mean of a multivariate normal distribution. *Proc. Third Berkeley Sympos.*, **1**, 197–206. University of California Press, Berkeley.

Bayes Estimates for the Linear Model*

D.V. Lindley and A.F.M. Smith
University College, London

Summary

The usual linear statistical model is reanalyzed using Bayesian methods and the concept of exchangeability. The general method is illustrated by applications to two-factor experimental designs and multiple regression.

Introduction

Attention is confined in this paper to the linear model, $E(\mathbf{y}) = \mathbf{A}\boldsymbol{\theta}$, where $\mathbf{y}$ is a vector of observations, $\mathbf{A}$ a known design matrix and $\boldsymbol{\theta}$ a vector of unknown parameters. The usual estimate of $\boldsymbol{\theta}$ employed in this situation is that derived by the method of least squares. We argue that it is typically true that there is available prior information about the parameters and that this may be exploited to find improved, and sometimes substantially improved, estimates. In this paper we explore a particular form of prior information based on de Finetti's (1964) important idea of exchangeability.

The argument is entirely within the Bayesian framework. Recently there has been much discussion of the respective merits of Bayesian and non-Bayesian approaches to statistics: we cite, for example, the paper by Cornfield (1969) and its ensuing discussion. We do not feel that it is necessary or desirable to add to this type of literature, and since we know of no reasoned argument against the Bayesian position we have adopted it here. Never-

* Read before the Royal Statistical Society at a meeting organized by the Research Section on Wednesday, December 8th, 1971, Mr. M.J.R. Healy in the Chair.

theless the reader not committed to this approach may like to be reminded that many techniques of the sampling-theory school are basically unsound: see the review by Lindley (1971b). In particular the least-squares estimates are typically unsatisfactory: or, in the language of that school, are inadmissible in dimensions greater than two. This follows since, by a well-known device in least-squares theory (see, for example, Plackett, 1960, p. 59), we may write the linear model after transformation in the form $E(z_i) = \xi_i$ for $i \leq p$ and $E(z_i) = 0$ for $i > p$. Here the z's are transforms of the data, and the ξ's of the parameters. Adding the assumption of normality, we can appeal to the results of Brown (1966), generalizing those of Stein (1956), which show that for a very wide class of loss functions the estimate of ξ_i by z_i, for $i \leq p$ is inadmissible. In Section 1 of this paper we do comment on the admissibility of the Bayesian estimates and try to show, in a way that might appeal to an adherent of orthodox ideas, that they are likely to be superior, at least in some situations, to the least-squares estimates.

1. Exchangeability

We begin with a simple example. Suppose, in the general linear model, that the design matrix is the unit matrix so that $E(y_i) = \theta_i$ for $i = 1, 2, \ldots, n$, and that $y_1, y_2, \ldots, y_n$ are independent, normally distributed with known variance σ^2. Such a simple model might arise if y_i was the observation on the ith variety in a field trial, of average yield θ_i. In considering the prior knowledge of the θ_i it may often be reasonable to assume their distribution *exchangeable*. That is, that it would be unaltered by any permutation of the suffixes: so that, in particular, the prior opinion of θ_7 is the same as that of θ_4, or any other θ_i; and similarly for pairs, triplets and so on. Now one way of obtaining an exchangeable distribution $p(\boldsymbol{\theta})$ is to suppose

$$p(\boldsymbol{\theta}) = \int \prod_{i=1}^{n} p(\theta_i|\mu)\, dQ(\mu),$$

where $p(\theta_i|\mu)$, for each μ, and $Q(\mu)$ describe arbitrary probability distributions. In other words, $p(\boldsymbol{\theta})$ is a *mixture*, by $Q(\mu)$, of independent and identical distributions, given μ. Indeed, Hewitt and Savage (1955), in generalization of de Finetti's original result, have shown that if exchangeability is assumed for *every n*, then a mixture is the *only* way to generate an exchangeable distribution.

In the present paper we study situations where we have exchangeable prior knowledge and assume this exchangeability described by a mixture. In the example this implies $E(\theta_i) = \mu$, say, a common value for each i. In other words there is a linear structure to the *parameters* analogous to the linear structure supposed for the observations $\mathbf{y}$. If we add the premise that the distribution from which the θ_i appear as a random sample is normal, the parallelism between the two stages, for $\mathbf{y}$ and $\boldsymbol{\theta}$, becomes closer. In this paper we study the situation in which the parameters of the general linear model themselves have a general linear structure in terms of other quantities which we call *hyperparameters.** In this simple example there is just one hyperparameter, μ.

Indeed, we shall find it necessary to go further and let the hyperparameters also have a linear structure. This will be termed a *three-stage model* and is analysed in detail in the next section. There are straightforward extensions to any number of stages.

Returning to the simple example with $E(y_i) = \theta_i$, $E(\theta_i) = \mu$ and respective variances σ^2 and τ^2, say, the situation will be completely specified once a prior distribution has been given for μ. (Effectively this is the third stage just mentioned.) Supposing μ to have a uniform distribution over the real line—a situation usually described by saying there is vague prior knowledge of μ—Lindley (1971a) has obtained the posterior distribution of θ_i and found its mean to be

$$E(\theta_i|\mathbf{y}) = \frac{y_i/\sigma^2 + y./\tau^2}{1/\sigma^2 + 1/\tau^2}, \tag{1}$$

where $y. = \sum y_i/n$. The detailed analysis has been given in the reference just cited, so we content ourselves with a brief discussion to serve as an introduction to the general theory in the next section.

The estimates, (1), will be referred to as *Bayes* estimates, and it is these that we propose as substitutes for the usual least-squares estimates. We denote them by θ_i^*, and reserve the usual notation, $\hat{\theta}_i$, for the ordinary estimates. Notice that θ_i^* is a weighted average of $y_i = \hat{\theta}_i$ and the overall mean, $y.$, with weights inversely proportional to the variances of y_i and θ_i. Hence the natural estimates are pulled towards a central value $y.$, the extreme values experiencing most shift. We shall find the weighted average phenomenon will persist even within the general model. Of course the estimate (1) depends on σ^2 and τ^2, which will typically be unknown, but their estimation presents no serious difficulties. If, for each i, there is replication of they y_i then σ^2 may be estimated as the usual within variance. Since we have replication (from the distribution $N(\mu, \tau^2)$ underlying the exchangeability assumption) for the θ_i, τ^2 may be estimated. For example $\sum(\theta_i^* - \theta^*)^2/(n-1)$ might be a reasonable estimate of τ^2, although in fact the reference just cited shows this can be improved upon. These estimates of σ^2 and τ^2 can be used in place of the known values used in (1) and the cycle repeated.

Let us now digress from the Bayesian viewpoint and try to persuade an orthodox statistician that (1) is a sensible estimate for him to consider, and indeed is better than the least-squares estimate. Of course, θ_i^* is a biased estimate of θ_i, so its merit cannot be judged by its variance. We use instead

* We believe we have borrowed this terminology from I.J. Good but are unable to trace the reference.

the mean-square error $E(\theta_i^* - \theta_i)^2$. This is just a criterion for judging the merit of one of the n estimates, so let us look at the average mean-square error over the n values. Simple, but tedious, calculations enable this to be found and compared with the corresponding quantity for $\hat{\theta}_i$, namely σ^2. The condition for the average m.s.e. for θ_i^* to be less than that for $\hat{\theta}_i$ is that

$$\sum(\theta_i - \theta)^2/(n-1) < 2\tau^2 + \sigma^2. \tag{2}$$

The m.s.e. for θ_i^* depends on θ_i and hence this condition does also. Consequently the Bayes estimates are not always superior to least-squares. But consider when (2) obtains. The θ_i are, by supposition, given μ, τ^2, a random sample from $N(\mu, \tau^2)$ so that the left-hand side of (2) is the usual estimate of τ^2, had the θ_i been known. Hence the condition is that the estimate of τ^2 be less than $2\tau^2 + \sigma^2$. The distribution of the estimate is a multiple of χ^2 and simple calculations show that the chance—according to the $N(\mu, \tau^2)$ distribution—of (2) being satisfied is high for n as low as 4 and rapidly tends to 1 as n increases. But τ^2, as we have seen, can itself be estimated, so with this in (1) we are almost certain to have a smaller m.s.e. for θ_i^* than for $\hat{\theta}_i$. In particular the expectation (over the θ-distribution) is always in favour of the Bayes estimate.

That argument is heuristic. Our estimates are similar to those proposed by Stein (1956), which he rigorously showed to be superior (in the average m.s.e. sense) to the least-squares estimates. It has been pointed out to us by L. Brown (personal communication) that (1), with known σ^2, τ^2, is an admissible estimate. Essentially this is because the impropriety in our prior distribution is confined to one dimension—in μ. We digress to amplify this statement.

If a *proper* prior distribution (that is, one whose integral over the whole space is unity) and a *bounded* utility function are used, then the estimate obtained by using as an estimate that value which maximizes the expected (over the parameter distribution) utility is always admissible. This is easy to demonstrate since, under the two conditions stated, all the usual mathematical operations, such as reversals of order of integration, are valid. Difficulties arise if either of the italicized conditions above are violated. Quadratic loss, leading to m.s.e. is unbounded, but can conveniently be replaced by

$$1 - \exp\{-(\boldsymbol{\theta} - \mathbf{e})^{\mathrm{T}}\boldsymbol{\Lambda}(\boldsymbol{\theta} - \mathbf{e})\} \tag{3}$$

for estimate $\mathbf{e}$, where $\boldsymbol{\Lambda}$ is positive semi-definite and, in particular, a unit matrix. The use of vague prior knowledge, with a uniform, and therefore improper, prior distribution does cause difficulties and it is this feature, at least in dimensions higher than two, that gives rise to inadmissible estimates, as Stein was the first to show. In the general theory of the next section all our estimates will be admissible in terms of the bounded loss function (3) provided the prior distribution is proper; we conjecture admissibility if the impropriety is confined to at most two dimensions.

Returning, then, to the inequality (2), we see that there is good reason within the orthodox framework for preferring the new estimates to the old. Further justification may be found in papers by Hoerl and Kennard (1970a, b) who discuss a special case of the estimates that we shall develop in Section 5.3. We do not take these justifications very seriously, feeling that the Bayesian viewpoint is supported by so many general considerations in which criteria, like mean-square error, play little or no part, that the additional validation they provide is of small consequence.

Before proceeding to the general discussion one point must be emphasized. In the example we have assumed an exchangeable prior distribution. The estimates (1) are therefore only suggested when this assumption is practically realistic. It is the greatest strength of the Bayesian argument that is provides a formal system within which an inference or decision problem can be described. In passing from the real-world problem to its mathematical formulation it becomes necessary to make, and to expose, the assumptions. (This applies to any formalism, Euclidean geometry, for example, and not just to Bayesian statistics.) Here exchangeability is one such assumption, and its practical relevance must be assessed before the estimates based on it are used. For example, if, as suggested above, our model described the observed yields of n varieties in an agricultural field trial, the exchangeability assumption would be inappropriate if one or more varieties were controls and the remainder were experimental. However, the assumption might be modified to one of exchangeability within controls and separately within experimental varieties. Similarly with a two-way classification into rows and columns, it might be reasonable to assume separately that the rows and the columns were exchangeable. In any application the particular form of the prior distribution has to be carefully considered.

It should be noted that in assigning a prior distribution to the θ_i of the above form, whilst we are effectively regarding them as a random sample from $N(\mu, \tau^2)$, we are not thereby passing to a Model II, random effects, situation such as has been discussed by Fisk (1967) and Nelder (1968). We are interested in the estimation of the *fixed* effects. One of us (A.F.M.S.) has studied the genuine Model II situation and obtained estimates for μ ($\boldsymbol{\theta}_2$ in the general model below) but this will be reported separately.

We now turn to the general theory. The mathematics is not difficult for someone familiar with matrix algebra, and the main result is stated as a theorem with corollaries. The results in Section 2 all assume *known* variances. The extensions to unknown variances will be described later.

2. General Bayesian Linear Model

The notation $\mathbf{y} \sim N(\boldsymbol{\mu}, \mathbf{D})$ means that the column vector $\mathbf{y}$ has a (multivariate) normal distribution with mean $\boldsymbol{\mu}$, a column vector, and dispersion $\mathbf{D}$, a positive semi-definite matrix.

Lemma. *Suppose, given* $\boldsymbol{\theta}_1$, *a vector of* p_1 *parameters,*

$$\mathbf{y} \sim N(\mathbf{A}_1\boldsymbol{\theta}_1, \mathbf{C}_1) \tag{4}$$

and that, given $\boldsymbol{\theta}_2$, *a vector of* p_2 *hyperparameters,*

$$\boldsymbol{\theta}_1 \sim N(\mathbf{A}_2\boldsymbol{\theta}_2, \mathbf{C}_2). \tag{5}$$

Then (a) *the marginal distribution of* $\mathbf{y}$ *is*

$$N(\mathbf{A}_1\mathbf{A}_2\boldsymbol{\theta}_2, \mathbf{C}_1 + \mathbf{A}_1\mathbf{C}_2\mathbf{A}_1^{\mathrm{T}}), \tag{6}$$

and (b) *the distribution of* $\boldsymbol{\theta}_1$, *given* $\mathbf{y}$, *is* $N(\mathbf{Bb}, \mathbf{B})$ *with*

$$\mathbf{B}^{-1} = \mathbf{A}_1^{\mathrm{T}}\mathbf{C}_1^{-1}\mathbf{A}_1 + \mathbf{C}_2^{-1} \tag{7}$$

and

$$\mathbf{b} = \mathbf{A}_1^{\mathrm{T}}\mathbf{C}_1^{-1}\mathbf{y} + \mathbf{C}_2^{-1}\mathbf{A}_2\boldsymbol{\theta}_2. \tag{8}$$

(*Here* $\mathbf{y}$ *is a vector of n elements and* $\mathbf{A}_1, \mathbf{A}_2, \mathbf{C}_1$ *and* $\mathbf{C}_2$ *are known positive-definite matrices of obvious dimensions.*)

The lemma is well known but we prove it here, both for completeness and because the proof has an unexpected byproduct.

To prove (a) we write (4) in the form $\mathbf{y} = \mathbf{A}_1\boldsymbol{\theta}_1 + \mathbf{u}$, where $\mathbf{u} \sim N(\mathbf{0}, \mathbf{C}_1)$ and (5) as $\boldsymbol{\theta}_1 = \mathbf{A}_2\boldsymbol{\theta}_2 + \mathbf{v}$ where $\mathbf{v} \sim N(\mathbf{0}, \mathbf{C}_2)$. Hence, putting these two equalities together, we have $\mathbf{y} = \mathbf{A}_1\mathbf{A}_2\boldsymbol{\theta}_2 + \mathbf{A}_1\mathbf{v} + \mathbf{u}$. But, by the standard properties of normal distributions, $\mathbf{A}_1\mathbf{v} + \mathbf{u}$, a linear function of independent normal random variables, is $N(\mathbf{0}, \mathbf{C}_1 + \mathbf{A}_1\mathbf{C}_2\mathbf{A}_1^{\mathrm{T}})$ and the result follows.

To prove (b) we use Bayes's theorem,

$$p(\boldsymbol{\theta}_1|\mathbf{y}) \propto p(\mathbf{y}|\boldsymbol{\theta}_1)p(\boldsymbol{\theta}_1).$$

The product on the right-hand side is $e^{-1/2Q}$ where Q is given by

$$\begin{aligned}(\mathbf{y} - \mathbf{A}_1\boldsymbol{\theta}_1)^{\mathrm{T}}\mathbf{C}_1^{-1}(\mathbf{y} - \mathbf{A}_1\boldsymbol{\theta}_1) &+ (\boldsymbol{\theta}_1 - \mathbf{A}_2\boldsymbol{\theta}_2)^{\mathrm{T}}\mathbf{C}_2^{-1}(\boldsymbol{\theta}_1 - \mathbf{A}_2\boldsymbol{\theta}_2) \\ &= \boldsymbol{\theta}_1^{\mathrm{T}}\mathbf{B}^{-1}\boldsymbol{\theta}_1 - 2\mathbf{b}^{\mathrm{T}}\boldsymbol{\theta}_1 + \{\mathbf{y}^{\mathrm{T}}\mathbf{C}_1^{-1}\mathbf{y} + \boldsymbol{\theta}_2^{\mathrm{T}}\mathbf{A}_2^{\mathrm{T}}\mathbf{C}_2^{-1}\mathbf{A}_2\boldsymbol{\theta}_2\}\end{aligned}$$

on collecting the quadratic and linear terms in $\boldsymbol{\theta}_1$ together, and using the expressions (7) and (8) for $\mathbf{b}$ and $\mathbf{B}$. Completing the square in $\boldsymbol{\theta}_1$, Q may finally be written

$$(\boldsymbol{\theta}_1 - \mathbf{Bb})^{\mathrm{T}}\mathbf{B}^{-1}(\boldsymbol{\theta}_1 - \mathbf{Bb}) + \{\mathbf{y}^{\mathrm{T}}\mathbf{C}_1^{-1}\mathbf{y} + \boldsymbol{\theta}_2^{\mathrm{T}}\mathbf{A}_2^{\mathrm{T}}\mathbf{C}_2^{-1}\mathbf{A}_2\boldsymbol{\theta}_2 - \mathbf{b}^{\mathrm{T}}\mathbf{Bb}\}. \tag{9}$$

The term in braces is a constant as far as the distribution of $\boldsymbol{\theta}_1$ is concerned, and the remainder of the expression demonstrates the truth of (b).

The proof of the lemma is complete, but by combining the separate proofs of (a) and (b) an interesting result can be obtained. On integrating $e^{-1/2,Q}$, with Q given by (9), with respect to $\boldsymbol{\theta}_1$, the result is proportional to the density of $\mathbf{y}$, already obtained in (a). The integration does not affect the term in braces in (9) so that, in particular, the quadratic term in $\mathbf{y}$ in (9)—

remembering that $\mathbf{b}$ contains $\mathbf{y}$—may be equated to the quadratic term obtained directly from (6), with the result that

$$\mathbf{C}_1^{-1} - \mathbf{C}_1^{-1}\mathbf{A}_1\mathbf{B}\mathbf{A}_1^{\mathrm{T}}\mathbf{C}_1^{-1} = \{\mathbf{C}_1 + \mathbf{A}_1\mathbf{C}_2\mathbf{A}_1^{\mathrm{T}}\}^{-1}.$$

We therefore have the

Matrix Lemma. *For any matrices $\mathbf{A}_1, \mathbf{A}_2, \mathbf{C}_1$ and $\mathbf{C}_2$ of appropriate dimensions and for which the inverses stated in the result exist, we have*

$$\mathbf{C}_1^{-1} - \mathbf{C}_1^{-1}\mathbf{A}_1(\mathbf{A}_1^{\mathrm{T}}\mathbf{C}_1^{-1}\mathbf{A}_1 + \mathbf{C}_2^{-1})^{-1}\mathbf{A}_1^{\mathrm{T}}\mathbf{C}_1^{-1} = \{\mathbf{C}_1 + \mathbf{A}_1\mathbf{C}_2\mathbf{A}_1^{\mathrm{T}}\}^{-1}. \qquad (10)$$

The result follows from the last equation on inserting the form for $\mathbf{B}$, equation (7). It is, of course, easy to prove the result (10) directly once its truth has been conjectured: furthermore $\mathbf{C}_1$ and $\mathbf{C}_2$ do not have to be positive definite. It suffices to multiply the left-hand side of (10) by $\mathbf{C}_1 + \mathbf{A}_1\mathbf{C}_2\mathbf{A}_1^{\mathrm{T}}$ and verify that the result is a unit matrix. The above proof is interesting because it does not require an initial conjecture and because it uses a probabilistic argument to derive a purely algebraic result. The matrix lemma is important to us since it provides simpler forms than would otherwise be available for our estimates. This result has been given by Rao (1965, Exercise 2.9, p. 29).

We next proceed to the main result. As explained in Section 1, we are dealing with the linear model, which is now written in the form $E(\mathbf{y}) = \mathbf{A}_1\boldsymbol{\theta}_1$, the suffixes indicating that this is the first stage in the model. We generalize to an arbitrary dispersion matrix, $\mathbf{C}_1$, for $\mathbf{y}$. The prior distribution of $\boldsymbol{\theta}_1$ is expressed in terms of hyperparameters $\boldsymbol{\theta}_2$ as another linear model, $E(\boldsymbol{\theta}_1) = \mathbf{A}_2\boldsymbol{\theta}_2$ with dispersion matrix $\mathbf{C}_2$. This can proceed for as many stages as one finds convenient: it will be enough for us to go to three, supposing the mean, as well as the dispersion, known at the final stage. For our inferences, and in particular for estimation, we require the posterior distribution of $\boldsymbol{\theta}_1$. This is provided by the following result.

Theorem. *Suppose that, given $\boldsymbol{\theta}_1$,*

$$\mathbf{y} \sim N(\mathbf{A}_1\boldsymbol{\theta}_1, \mathbf{C}_1), \qquad (11.1)$$

given $\boldsymbol{\theta}_2$,

$$\boldsymbol{\theta}_1 \sim N(\mathbf{A}_2\boldsymbol{\theta}_2, \mathbf{C}_2) \qquad (11.2)$$

and given $\boldsymbol{\theta}_3$,

$$\boldsymbol{\theta}_2 \sim N(\mathbf{A}_3\theta_3, \mathbf{C}_3). \qquad (11.3)$$

Then the posterior distribution of $\boldsymbol{\theta}_1$, given $\{\mathbf{A}_i\}, \{\mathbf{C}_i\}, \boldsymbol{\theta}_3$ and $\mathbf{y}$ is $n(\mathbf{Dd}, \mathbf{D})$ with

$$\mathbf{D}^{-1} = \mathbf{A}_1^{\mathrm{T}}\mathbf{C}_1^{-1}\mathbf{A}_1 + \{\mathbf{C}_2 + \mathbf{A}_2\mathbf{C}_3\mathbf{A}_2^{\mathrm{T}}\}^{-1} \qquad (12)$$

and

$$\mathbf{d} = \mathbf{A}_1^{\mathrm{T}}\mathbf{C}_1^{-1}\mathbf{y} + \{\mathbf{C}_2 + \mathbf{A}_2\mathbf{C}_3\mathbf{A}_2{}^{\mathrm{T}}\}^{-1}\mathbf{A}_2\mathbf{A}_3\boldsymbol{\theta}_3. \tag{13}$$

(Here $\boldsymbol{\theta}_i$ is a vector of p_i elements and the dispersion matrices, $\mathbf{C}_i$, are all supposed non-singular.)

The joint distribution of $\boldsymbol{\theta}_1$ and $\boldsymbol{\theta}_2$ is described in (11.2) and (11.3). The use of part (a) of the lemma enables the marginal distribution of $\boldsymbol{\theta}_1$ to be written down as

$$\boldsymbol{\theta}_1 \sim N(\mathbf{A}_2\mathbf{A}_3\boldsymbol{\theta}_3, \mathbf{C}_2 + \mathbf{A}_2\mathbf{C}_3\mathbf{A}_2^{\mathrm{T}}). \tag{14}$$

(Notice that this is the prior distribution of $\boldsymbol{\theta}_1$ free of the hyperparameters $\boldsymbol{\theta}_2$. We could have expressed the prior in this way but in applications we find the hierarchical form more convenient.)

Then, with (14) as prior, (11.1) as likelihood, part (b) of the lemma shows that the posterior distribution of $\boldsymbol{\theta}_1$ is as stated.

In particular the mean of the posterior distribution may be regarded as a point estimate of $\boldsymbol{\theta}_1$ to replace the usual least-squares estimate. The form of this estimate is a generalization of the form noted in the example of Section 1; namely, it is a weighted average of the least-squares estimate $(\mathbf{A}_1^{\mathrm{T}}\mathbf{C}_1^{-1}\mathbf{A}_1)^{-1}$ $\mathbf{A}_1^{\mathrm{T}}\mathbf{C}_1^{-1}\mathbf{y}$ and the prior mean $\mathbf{A}_2\mathbf{A}_3\boldsymbol{\theta}_3$ (equation (14)) with weights equal to the inverses of the corresponding dispersion matrices, $\mathbf{A}_1^{\mathrm{T}}\mathbf{C}_1^{-1}\mathbf{A}_1$ for the least-squares values, $\mathbf{C}_2 + \mathbf{A}_2\mathbf{C}_3\mathbf{A}_2^{\mathrm{T}}$ for the prior distribution (14). For the simple example considered in Section 1 we produced an heuristic argument to show that, with respect to our prior distribution, we were confident of satisfying inequality (2) and thus achieving smaller mean square error than with the least-squares estimate. This result can be shown to hold generally for Bayes's estimates derived from hierarchical prior structures, as in (11.1)–(11.3), and will be presented in a future paper.

The matrix lemma enables us to obtain several alternative forms for the term in braces in (12), and hence for the posterior mean and variance, both of which involve this expression. These alternatives look more complicated than those already stated but are often useful in applications. Notice that a computational advantage of the matrix lemma is that its use reduces the order of the matrices to be inverted. The matrix on the right-hand side of (10) is of order n, whereas on the left-hand side, apart from $\mathbf{C}_1$ which is usually of a simple structure (often $\mathbf{C}_1 = \sigma^2\mathbf{I}$), the matrix to be inverted is of order p_1, typically much less than n.

Corollary 1. *An alternative expression for* $\mathbf{D}^{-1}$ (equation (12)) *is*

$$\mathbf{A}_1^{\mathrm{T}}\mathbf{C}_1^{-1}\mathbf{A}_1 + \mathbf{C}_2^{-1} - \mathbf{C}_2^{-1}\mathbf{A}_2(\mathbf{A}_2^{\mathrm{T}}\mathbf{C}_2^{-1}\mathbf{A}_2 + \mathbf{C}_3^{-1})^{-1}\mathbf{A}_2^{\mathrm{T}}\mathbf{C}_2^{-1}. \tag{15}$$

This is immediate on applying (10), with the suffixes all increased by one, to the second term in (12).

In most applications of these results the design of the experiment rather naturally suggests the second stage, (11.2), in the hierarchy but at the third stage we find ourselves in a position where the prior knowledge is weak.

(Least-squares results apply when the second-stage prior knowledge is weak.) It is natural to express this by supposing the third-stage dispersion matrix $\mathbf{C}_3$ to be large, or to let its inverse, the precision matrix, be zero. In the original form of (12) and (13) it is not easy to see what happens when $\mathbf{C}_3^{-1} = \mathbf{0}$, but (15) enables the form to be seen easily.

Corollary 2. *If* $\mathbf{C}_3^{-1} = \mathbf{0}$, *the posterior distribution of* $\boldsymbol{\theta}_1$ *is* $N(\mathbf{D}_0\mathbf{d}_0, \mathbf{D}_0)$ *with*

$$\mathbf{D}_0^{-1} = \mathbf{A}_1^T\mathbf{C}_1^{-1}\mathbf{A}_1 + \mathbf{C}_2^{-1} - \mathbf{C}_2^{-1}\mathbf{A}_2(\mathbf{A}_2^T\mathbf{C}_2^{-1}\mathbf{A}_2)^{-1}\mathbf{A}_2^T\mathbf{C}_2^{-1} \tag{16}$$

and

$$\mathbf{d}_0 = \mathbf{A}_1^T\mathbf{C}_1^{-1}y. \tag{17}$$

The form for $\mathbf{D}_0^{-1}$ follows by direct substitution of $\mathbf{C}_3^{-1} = \mathbf{0}$ in (15). That for $\mathbf{d}_0$ follows by remarking that if the second and third terms in (15) are postmultiplied by $\mathbf{A}_2$ the result is zero, but such postmultiplication takes place in the original expression for $\mathbf{d}$, equation (13).

This corollary is the form we shall most often use in applications.

It is possible to extend the theorem to cases where some or all of the dispersion matrices $\mathbf{C}_i$ are singular. This can be accomplished using generalized inverses and will be the subject of a separate paper. Notice that we have not assumed, as in the usual least-squares theory, that $\mathbf{A}_1^T\mathbf{C}_1^{-1}\mathbf{A}_1$ is non-singular. (The case $\mathbf{C}_1 = \sigma^2\mathbf{I}$ will be more familiar.) In the standard exposition it is usual to constrain the individual parameters in the vector $\boldsymbol{\theta}_1$ to preserve identifiability in the likelihood function. Identifiability problems do not arise in the Bayesian formulation since, provided the prior distribution is proper, so is the posterior, whether or not the parameters referred to in these two distributions are identifiable or not in the likelihood function. An example below will help to make this clear.

The situation described in Section 1 has already been discussed in detail by Lindley (1971a), though not within the general framework which was briefly described in Lindley (1969). The interested reader can easily fit the example into the argument of this section. Corollary 2 is relevant and it is an easy matter to perform the necessary matrix calculations. We proceed to the discussion of other examples.

3. Examples

3.1. Two-Factor Experimental Designs

Consider t "treatments" assigned to n experimental units arranged in b "blocks." If the ith treatment is applied within the jth block and yields an observation y_{ij}, the usual model is

$$E(y_{ij}) = \mu + \alpha_i + \beta_j \qquad (1 \le i \le t, 1 \le j \le b)$$

with the errors independent $N(0, \sigma^2)$. In the general notation of (11.1)

$$\boldsymbol{\theta}_1^{\mathrm{T}} = (\mu, \alpha_1, \alpha_2, \ldots, \alpha_t, \beta_1, \beta_2, \ldots, \beta_b)$$

and $\mathbf{A}_1$ describes the design used.

For the second stage we argue as follows. It might be reasonable to assume that our prior knowledge of the treatment constants $\{\alpha_i\}$ was exchangeable, and similarly that of the block constants $\{\beta_j\}$, but that these were independent. We emphasize the word "might" in the last sentence. In repetition of the point made in Section 1, we remind the reader that this *assumption* is not always appropriate and our recipes below are not necessarily sensible when this form of exchangeability is unreasonable. For example, it may be known that the treatments are ordered, say $\alpha_1 \leq \alpha_2 \leq \cdots \leq \alpha_t$. In this case other forms of prior information are available and alternative estimates are sensible: these will be reported on in a separate paper.

Adding the assumptions of normality we therefore describe the second stage (11.2) by

$$\alpha_i \sim N(0, \sigma_\alpha^2), \qquad \beta_j \sim N(0, \sigma_\beta^2), \qquad \mu \sim N(\omega, \sigma_\mu^2),$$

these distributions being independent. The means of α_i and β_j have been chosen to be zero. Any other value would do since the likelihood provides no information about them, but the choice of zero mean is convenient, since it leads to straightforward comparisons of the Bayes and (constrained) least-squares estimates as deviations from an average level. We shall consider the case where the prior knowledge of μ is vague, so that $\sigma_\mu^2 \to \infty$; ω will then be irrelevant. A third stage is not necessary. We proceed to calculate expressions (12) and (13) for the posterior distribution of $\boldsymbol{\theta}_1$.

The matrix $\mathbf{C}_2$ is diagonal, so the same is true of $\mathbf{C}_2^{-1}$ and its leading diagonal is easily seen to be

$$(\sigma_\mu^{-2}, \sigma_\alpha^{-2}, \ldots, \sigma_\alpha^{-2}, \sigma_\beta^{-2}, \ldots, \sigma_\beta^{-2})$$

and as $\sigma_\mu^2 \to \infty$, the first element tends to zero. $\mathbf{C}_1$ is the unit matrix times σ^2. We can therefore substitute these values into (12) and (13), remembering that $\mathbf{C}_3 = 0$ and $(\mathbf{A}_3\mu)^{\mathrm{T}} = (\omega, 0, \ldots, 0)$ and easily obtain

$$\mathbf{D}^{-1} = \sigma^{-2}\mathbf{A}_1^{\mathrm{T}}\mathbf{A}_1 + \mathbf{C}_2^{-1}$$

and

$$\mathbf{d} = \sigma^{-2}\mathbf{A}_1^{\mathrm{T}}\mathbf{y}.$$

Hence $\boldsymbol{\theta}_1^*$, the Bayes estimate $\mathbf{Dd}$, satisfies the equations

$$(\mathbf{A}_1^{\mathrm{T}}\mathbf{A}_1 + \sigma^2\mathbf{C}_2^{-1})\boldsymbol{\theta}_1^* = \mathbf{A}_1^{\mathrm{T}}\mathbf{y}. \tag{18}$$

These differ from the least-squares equations only in the inclusion of the extra term $\sigma^2\mathbf{C}_2^{-1}$.

In the case of a complete randomized-block design where each treatment occurs exactly once in each block we have, on arranging the elements of $\mathbf{y}$ in lexicographical order,

$$(\mathbf{A}_1^{\mathrm{T}}\mathbf{A}_1 + \sigma^2\mathbf{C}_2^{-1}) = \begin{pmatrix} bt & b\mathbf{1}_t^{\mathrm{T}} & t\mathbf{1}_b^{\mathrm{T}} \\ b\mathbf{1}_t & (b+\sigma^2/\sigma_\alpha^2)\mathbf{I}_t & \mathbf{J}_{t,b} \\ t\mathbf{1}_b & \mathbf{J}_{b,t} & (t+\sigma^2/\sigma_\beta^2)\mathbf{I}_b \end{pmatrix}, \tag{19}$$

where $\mathbf{1}_m$ is a vector of m 1's, $\mathbf{I}_m$ is the unit matrix of order m and $\mathbf{J}_{m,n}$ is a matrix of order $m \times n$ all of whose elements are 1. As usual

$$(\mathbf{A}_1^{\mathrm{T}}\mathbf{y})^{\mathrm{T}} = (bty_{..}, by_{1.}, \ldots, by_{t.}, ty_{.1}, \cdots, ty_{.b}).$$

Notice that the matrix (19) is non-singular and the solution to (18) is easily seen to be

$$\mu^* = y_{..}, \quad \alpha_i^* = (b\sigma_\alpha^2 + \sigma^2)^{-1}b\sigma_\alpha^2(y_{i.} - y_{..}), \quad \beta_j^* = (t\sigma_\beta^2 + \sigma^2)^{-1}t\sigma_\beta^2(y_{.j} - y_{..}). \tag{20}$$

Consequently the estimators of the treatment and block effects (on being measured from the overall mean) are shrunk towards zero by a factor depending on the ratio of σ^2 to σ_α^2 or σ_β^2 respectively. This is in agreement with the result, equation (1), quoted above. Because this is an orthogonal design the magnitude of the "shrinkage" of the treatment effect does not depend on the exchangeability for the blocks, and vice versa. With a non-orthogonal design, such as balanced incomplete blocks, the same remark is not true.

3.2. Exchangeability Between Multiple Regression Equations

The following practical example stimulated our extension from the example of Section 1 to the general model, and we shall report on its use in Section 5.2. The context was educational measurement where variables x and y were related with the usual linear regression structure. However the values of the regression parameters depended on the school the student had attended. Novick (personal communication) suggested to us that improved estimates might be obtained for any one school by combining the data for all schools. This is just what the Bayes estimates do, and would seem to be appropriate whenever exchangeability *between* regressions (schools) is a sensible assumption. The mathematics for p regressor variables goes as follows.

Suppose

$$\mathbf{y}_j \sim N(\mathbf{X}_j\boldsymbol{\beta}_j, \mathbf{I}_{nj}\sigma_j^2) \tag{21}$$

for $j = 1, 2, \ldots, m$ and $\boldsymbol{\beta}_j$ a vector of p parameters: that is m linear, multiple regressions on p variables. In the notation of the Theorem, $\mathbf{A}_1$, expressed in terms of submatrices, is diagonal with $\mathbf{X}_j$ as the jth diagonal submatrix; $\boldsymbol{\theta}_1^{\mathrm{T}}$ is $(\boldsymbol{\beta}_1^{\mathrm{T}}, \boldsymbol{\beta}_2^{\mathrm{T}}, \ldots, \boldsymbol{\beta}_m^{\mathrm{T}})$ of mp elements. The exchangeability of the individual $\boldsymbol{\beta}_j$ added to normality gives us the second stage as

$$\boldsymbol{\beta}_j \sim N(\boldsymbol{\xi}, \boldsymbol{\Sigma}) \tag{22}$$

say. Here $\mathbf{A}_2$ is a matrix of order $mp \times p$, all of whose $p \times p$ submatrices are unit matrices, and $\boldsymbol{\theta}_2 = \xi$. We shall suppose vague prior knowledge of ξ and use the special form of Corollary 2.

Simple calculations show that $(\mathbf{A}_2^{\mathrm{T}}\mathbf{C}_2^{-1}\mathbf{A}_2)^{-1} = m^{-1}\boldsymbol{\Sigma}$ and then that

$$\mathbf{C}_2^{-1}\mathbf{A}_2(\mathbf{A}_2^{\mathrm{T}}\mathbf{C}_2^{-1}\mathbf{A}_2)^{-1}\mathbf{A}_2^{\mathrm{T}}\mathbf{C}_2^{-1}$$

is a matrix of order mp *all* of whose $p \times p$ submatrices are $m^{-1}\boldsymbol{\Sigma}^{-1}$. In the usual way $\mathbf{A}_1^{\mathrm{T}}\mathbf{C}_1^{-1}\mathbf{A}_1$, expressed in terms of submatrices, is diagonal with $\sigma_j^{-2}\mathbf{X}_j^{\mathrm{T}}\mathbf{X}_j$ as the jth diagonal submatrix. The equations for the Bayes estimates $\boldsymbol{\beta}_j^*$ are then found to be

$$\begin{pmatrix} \sigma_1^{-2}\mathbf{X}_1^{\mathrm{T}}\mathbf{X}_1 + \boldsymbol{\Sigma}^{-1} & \cdots & \mathbf{0} \\ \cdots & \sigma_2^{-2}\mathbf{X}_2\mathrm{T}\mathbf{X}_2 + \boldsymbol{\Sigma}^{-1} & \cdots \\ \vdots & \vdots & \vdots \\ 0 & \cdots & \sigma_m^{-2}\mathbf{X}_m^{\mathrm{T}}\mathbf{X}_m + \boldsymbol{\Sigma}^{-1} \end{pmatrix}$$

$$\times \begin{pmatrix} \boldsymbol{\beta}_1^* \\ \boldsymbol{\beta}_2^* \\ \vdots \\ \boldsymbol{\beta}_m^* \end{pmatrix} - \boldsymbol{\Sigma}^{-1} \begin{pmatrix} \boldsymbol{\beta}_{\cdot}^* \\ \boldsymbol{\beta}_{\cdot}^* \\ \vdots \\ \boldsymbol{\beta}_{\cdot}^* \end{pmatrix} = \begin{pmatrix} \sigma_1^{-2}\mathbf{X}_1^{\mathrm{T}}\mathbf{y} \\ \sigma_2^{-2}\mathbf{X}_2^{\mathrm{T}}\mathbf{y} \\ \vdots \\ \sigma_m^{-2}\mathbf{X}_m^{\mathrm{T}}\mathbf{y} \end{pmatrix}, \tag{23}$$

where $\boldsymbol{\beta}_{\cdot}^* = \sum \boldsymbol{\beta}_i^*/m$. These equations are easily solved for $\boldsymbol{\beta}_{\cdot}^*$ and then, in terms of $\boldsymbol{\beta}_{\cdot}^*$, the solution is

$$\boldsymbol{\beta}_j^* = (\sigma_j^{-2}\mathbf{X}_j^{\mathrm{T}}\mathbf{X}_j + \boldsymbol{\Sigma}^{-1})^{-1}(\sigma_j^{-2}\mathbf{X}_j^{\mathrm{T}}\mathbf{y} + \boldsymbol{\Sigma}^{-1}\boldsymbol{\beta}_{\cdot}^*), \tag{24}$$

a compromise between the least-squares estimate and an average of the various estimates. The example of Section 1 is a special case with $p = 1$.

Noting that $\mathbf{D}_0^{-1}$, given in Corollary 2 (16), may, for this application, be written in the form,

$$\begin{pmatrix} \sigma_1^{-2}\mathbf{X}_1^{\mathrm{T}}\mathbf{X}_1 + \boldsymbol{\Sigma}^{-1} & 0 \\ \vdots & \vdots \\ 0 & \sigma_m^{-2}\mathbf{X}_m^{\mathrm{T}}\mathbf{X}_m + \boldsymbol{\Sigma}^{-1} \end{pmatrix} - m^{-1} \begin{pmatrix} \boldsymbol{\Sigma}^{-1} \\ \vdots \\ \boldsymbol{\Sigma}^{-1} \end{pmatrix}$$

$$\times \begin{pmatrix} \boldsymbol{\Sigma} & \cdots & \mathbf{0} \\ \vdots & \vdots & \vdots \\ \mathbf{0} & \cdots & \boldsymbol{\Sigma} \end{pmatrix} (\boldsymbol{\Sigma}^{-1} \cdots \boldsymbol{\Sigma}^{-1})$$

and thus may be inverted by the matrix Lemma (10), we can obtain an explicit form for $\boldsymbol{\beta}_j^*$. After some algebra we obtain the weighted form of (24) with $\boldsymbol{\beta}_\cdot^*$ replaced by $\sum \mathbf{W}_i\hat{\boldsymbol{\beta}}_i$ where,

$$\mathbf{W}_i = \left[\sum_{j=1}^{m}(\mathbf{X}_j^{\mathrm{T}}\mathbf{X}_j\sigma_j^{-2} + \boldsymbol{\Sigma}^{-1})^{-1}\mathbf{X}_j^{\mathrm{T}}\mathbf{X}_j\sigma_j^{-2}\right]^{-1}(\mathbf{X}_i^{\mathrm{T}}\mathbf{X}_i\sigma_i^{-2} + \boldsymbol{\Sigma}^{-1})^{-1}\mathbf{X}_i^{\mathrm{T}}\mathbf{X}_i\sigma_i^{-2}.$$

This shows explicitly how the information from the ith regression equation is combined with the information from all equations.

3.3. Exchangeability Within Multiple Regression Equations

In contrast to the last section suppose that we have a single multiple regression situation

$$\mathbf{y} \sim N(\mathbf{X}\boldsymbol{\beta}, \mathbf{I}_n\sigma^2). \tag{25}$$

In the educational context, the p regressor variables might be the results of p tests applied to students and the dependent variable, y, a measure of the students' performance after training. We are interested in the case where the individual regression coefficients in $\boldsymbol{\beta}^{\mathrm{T}} = (\beta_1, \beta_2, \ldots, \beta_p)$ are exchangeable. To achieve this it may be necessary to rescale the regressor variables: for example, to write (25) in correlation form in which the diagonal elements of $\mathbf{X}^{\mathrm{T}}\mathbf{X}$ are unity and the off-diagonals are the sample correlations. (Again we emphasize the point that this is an assumption and may not be appropriate). If the assumption is sensible then we may fit it into our general model by supposing

$$\beta_j \sim N(\xi, \sigma_\beta^2). \tag{26}$$

There are at least two useful possibilities: (i) to suppose vague prior knowledge for ξ (Corollary 2), (ii) to put $\xi = 0$, reflecting a feeling that the β_i are small.

In (i) simple but tedious calculations analogous to those of Section 3.2 show that

$$\boldsymbol{\beta}^* = \{\mathbf{I}_p + k(\mathbf{X}^{\mathrm{T}}\mathbf{X})^{-1}(\mathbf{I}_p - p^{-1}\mathbf{J}_p)\}^{-1}\hat{\boldsymbol{\beta}}, \tag{27}$$

where $k = \sigma^2/\sigma_\beta^2$. Similar calculations in (ii), using only a two-stage model, give

$$\boldsymbol{\beta}^* = \{\mathbf{I}_p + k(\mathbf{X}^{\mathrm{T}}\mathbf{X})^{-1}\}^{-1}\hat{\boldsymbol{\beta}}. \tag{28}$$

The estimates (27) and (28) are very similar to those prposed by Hoerl and Kennard (1970a). The main difference is that k in their argument is a constant introduced for various intuitively sensible reasons, whereas here it is a variance ratio. Also the derivation is different: Hoerl and Kennard argue within the orthodox, sampling theory framework, whereas we use the formal theory. We do not attempt to reproduce their most convincing argument against the least-squares estimates and in favour of (27) and (28), merely referring the sampling-theorist to it and saying that we agree with its conclusions with the reservation that we feel that the estimates may not be so sensible if the exchangeability within the regression equation is inappropriate. We return to this example in Section 5.3 where the estimation of k is discussed.

Examples 3.2 and 3.3 may be combined when there is exchangeability between *and* within regressions. We omit the details of this and many other extensions and instead consider how we might remove the major impediment to the application of the general theory, namely the assumption that all the variances are known. In the next section we show that the simple device of replacing the known variances by estimated values in the Bayes estimates is satisfactory.

4. Estimation with Unknown Covariance Structure

For the purpose of the immediate exposition denote by θ the parameters of interest in the general model and by ϕ the nuisance parameters. The latter will include the dispersion matrices $\mathbf{C}_i$ when these are unknown. Consider how the Bayesian treatment proceeds. We first assign a joint prior distribution to θ and ϕ—instead of just to θ—and combine this with the likelihood function to provide the joint posterior distribution $p(\theta, \phi|\mathbf{y})$. This distribution then has to be integrated with respect to ϕ, thus removing the nuisance parameters and leaving the posterior for θ. Finally, if we are using quadratic loss or generally one of the forms given by (3), we shall require the mean of this distribution, necessitating another integration. The calculation of the mean will also require the constant of proportionality in Bayes's formula to be evaluated, involving yet another integration. Any reasonable prior distributions for ϕ that we have considered lad to integrals which cannot all be expressed in closed form and, as a result, the above argument is technically most complex to execute. We therefore consider an approximation to it which is technically much simpler and yet yields the bulk, though not unfortunately all, of the information required for the estimation.

The first approximation consists in using the *mode* of the posterior distribution in place of the *mean*. Secondly, we mostly use the mode of the *joint* distribution rather than that of the *θ-margin*. The modal values satisfy the equations

$$\frac{\partial}{\partial \theta} p(\theta, \phi|\mathbf{y}) = \frac{\partial}{\partial \phi} p(\theta, \phi|\mathbf{y}) = 0.$$

These equations may be re-written in terms of conditional and marginal distributions. In particular that for θ may be expressed as

$$\frac{\partial}{\partial \theta} p(\theta|\phi, \mathbf{y}) p(\phi|\mathbf{y}) = 0$$

or, assuming $p(\phi|\mathbf{y}) \neq 0$, as

$$\frac{\partial}{\partial\theta} p(\theta|\phi, \mathbf{y}) = 0. \tag{29}$$

But the conditional density $p(\theta|\phi, \mathbf{y})$ in (29) is exactly what has been found in the general theory of Section 2, where it was shown to be normal, with mode consequently equal to the mean. Hence we have the result that the θ-value of the posterior mode of the joint distribution of θ and ϕ is equal to the mode of the conditional distribution of θ evaluated at the modal value of ϕ. Consequently all we have to do is to take the estimates derived in Section 2 and replace the unknown values of the nuisance parameters by their modal estimates. For example, the simple estimate (1) is replaced by

$$\frac{y_i/s^2 + y/t^2}{1/s^2 + 1/t^2},$$

where s^2 and t^2 are respectively modal estimates of σ^2 and τ^2. This approach avoids the integrations referred to above. The modal estimates of ϕ may, analogous to (29), be found by supposing θ known, and then replacing θ in the result by their modes.

It is reasonably clear that the approximations are only likely to be good if the samples are fairly large and the resulting posterior distributions approximately normal. Also the approach does not provide information about the precision of the estimates, such as a standard error (of the posterior, not the sampling-theoretic distribution!) would provide. But as a first step on the way to a satisfactory description of the posterior distribution, it seems to go a long way and has the added merit of being intuitively sensible. In practice we shall find it convenient to proceed as follows. For an assumed ϕ calculate the mode $\theta^{(1)}$, say. Treating $\theta^{(1)}$ as known we can find the mode for ϕ, $\phi^{(1)}$ say. This may be used to find $\theta^{(2)}$, and so on. This sequence of iterations typically converges and only involves equations for the modes of one parameter, knowing the value of the other.

We now proceed to apply these ideas to the situations discussed in Section 3. At the moment we have no general theory to parallel that of Section 2. The reason for this is essentially that we do not have an entirely satisfactory procedure for estimating the dispersion matrix of a multivariate normal distribution. This might appear an odd statement to make when there are numerous texts on multivariate analysis available that discuss this problem. But just as the usual estimates of the means are inadmissible, so are those of the variances and covariances (Brown, 1968), and are, in any case, obtained from unrealistic priors. We hope to report separately on this problem and defer discussion of a general theory.

5. Examples with Unknown Covariance Structure

5.1. Two-Factor Experimental Designs

We saw in Section 3.1 that there were three variances in this situation: σ^2 the usual residual variance contributing to the likelihood function, and $\sigma_\alpha^2, \sigma_\beta^2$ being respectively the variances of the treatment and block effects. ($\sigma_\mu^2 \to \infty$ so does not enter.) It is first necessary to specify prior distributions for these and this we do through the appropriate conjugate family, which is here inverse-χ^2, assuming the three variances independent. This conjugate family involves two parameters and is sufficiently flexible for most applications. Specifically we suppose

$$\frac{\nu\lambda}{\sigma^2} \sim \chi_\nu^2, \qquad \frac{\nu_\alpha\lambda_\alpha}{\sigma_\alpha^2} \sim \chi_{\nu_\alpha}^2 \qquad \text{and} \qquad \frac{\nu_\beta\lambda\beta}{\sigma_\beta^2} \sim \chi_{\nu_\beta}^2. \tag{30}$$

The joint distribution of all quantities involved can then be written down as proportional to

$$\begin{aligned}(\sigma^2)^{-1/2(n+\nu+2)} &\exp\left[-\frac{1}{2\sigma^2}\{\nu\lambda + S^2(\mu, \alpha, \beta)\}\right] \\ &\times (\sigma_\alpha^2)^{-1/2(t+\nu_\alpha+2)} \exp\left[-\frac{1}{2\sigma_\alpha^2}\{\nu_\alpha\lambda_\alpha + \sum \alpha_i^2\}\right] \\ &\times (\sigma_\beta^2)^{-1/2(b+\nu\beta+2)} \exp\left[-\frac{1}{2\sigma_\beta^2}\{\nu_\beta\lambda_\beta + \sum \beta_j^2\}\right],\end{aligned} \tag{31}$$

where $S^2(\mu, \alpha, \beta)$ is the sum of squares $\sum(y_{ij} - \mu - \alpha_i - \beta_j)^2$.

If σ^2, σ_α^2 and σ_β^2 are known, the mode of this distribution has been found—equation (18), or in the balanced case, equation (20). We have only to substitute the modal estimates of the three variances into these expressions. To find these modal estimates we can, reversing the roles of θ and ϕ in the general argument of the previous paragraph, suppose μ, α and β known. Using the corresponding Roman letters for these modes, we easily see them to be, from (31),

$$\left.\begin{aligned} s^2 &= \{\nu\lambda + S^2(\mu^*, \alpha^*, \beta^*)\}/(n + \nu + 2), \\ s_\alpha^2 &= \{\nu_\alpha\lambda_\alpha + \sum \alpha_i^{*2}\}/(t + \nu_\alpha + 2), \\ s_\beta^2 &= \{\nu_\beta\lambda_\beta + \sum \beta_j^{*2}\}/(b + \nu_\beta + 2). \end{aligned}\right\} \tag{32}$$

These equations, together with (18) (or (20)), can now be solved iteratively. With trial values of $\sigma^2, \sigma_\alpha^2$ and σ_β^2, (18) can be solved for μ^*, α^* and β^*. These values can be inserted into (32) to give revised values for s^2, s_α^2 and

s_β^2, which can again be used in (18). The cycle can be repeated until the values converge.

A few points about these solutions are worth noting. Firstly, the value of S^2 that occurs is not the usual residual sum of squares, which is evaluated about the least-squares value, but the sum about the Bayes estimates. Since the former minimizes the sum of squares, our S^2 is necessarily greater than the residual: s^2 could therefore be larger than the usual estimate. Secondly, whilst it would be perfectly possible to put $\nu = 0$ (referring to σ^2), so avoiding the specification of a value for λ and thereby taking the usual vague prior for a variance, one cannot put ν_α and ν_β zero. If this is done the modal estimates for the treatment and block effects are all zero. The point is discussed in detail in connection with the example of Section 1 in Lindley (1971a). Essentially the estimation of σ_α^2 and σ_β^2 is difficult, in the sense that the data contain little information about them, when they are small in comparison with σ^2: the residual "noise" is too loud. In the contrary case where σ_α^2 and σ_β^2 are large in comparison with σ^2, the actual values of ν_α, λ_α, ν_β and λ_β do not matter much provided the ν's are both small.

5.2. Exchangeability Between Multiple Regression Equations

We continue the discussion of Section 3.2 but mainly confine our attention to the homoscedastic case where $\sigma_j^2 = \sigma^2$, say, for all j. It is only necessary to specify prior distributions for σ^2 and $\boldsymbol{\Sigma}$, the dispersion matrix of the regression coefficients (equation (22)). As in the last example we suppose $\nu\lambda/\sigma^2 \sim \chi_\nu^2$. The conjugate distribution for $\boldsymbol{\Sigma}$ is to suppose $\boldsymbol{\Sigma}^{-1}$ has a Wishart distribution with ρ, say, degrees of freedom and matrix $\mathbf{R}$. We are not too happy with this assumption but at least it provides a large-sample solution (see the remarks at the end of Section 4). $\boldsymbol{\Sigma}$ and σ^2 are supposed independent.

The joint distribution of all the quantities is now

$$(\sigma^2)^{-1/2n} \exp\left\{-\frac{1}{2\sigma^2}\sum_{j=1}^{m}(\mathbf{y}_j - \mathbf{X}_j\boldsymbol{\beta}_j)^{\mathrm{T}}(\mathbf{y}_j - \mathbf{X}_j\boldsymbol{\beta}_j)\right\}$$
$$\times |\boldsymbol{\Sigma}|^{-1/2m} \exp\left\{-\frac{1}{2}\sum_{j=1}^{m}(\boldsymbol{\beta}_j - \boldsymbol{\xi})^{\mathrm{T}}\boldsymbol{\Sigma}^{-1}(\boldsymbol{\beta}_j - \boldsymbol{\xi})\right\}$$
$$\times |\boldsymbol{\Sigma}|^{-1/2(\rho-p-1)} \exp\left\{-\frac{1}{2}\operatorname{tr}\boldsymbol{\Sigma}^{-1}\mathbf{R}\right\}$$
$$\times (\sigma^2)^{-1/2(\nu+2)} \exp\{-\nu\lambda/2\sigma^2\}, \tag{33}$$

assuming $\boldsymbol{\xi}$ to have a uniform distribution over p-space. (The four lines of (33) come respectively from the likelihood, the distribution of $\boldsymbol{\beta}$, (22), the

Wishart distribution for $\boldsymbol{\Sigma}^{-1}$ and the inverse-χ^2 for σ^2.) The integration with respect to $\boldsymbol{\xi}$ is straightforward and effectively results in the usual loss of one degree of freedom. The joint posterior density for $\boldsymbol{\beta}$, σ^2 and $\boldsymbol{\Sigma}^{-1}$ is then proportional to

$$(\sigma^2)^{-1/2(n+v+2)} \exp\left[-\frac{1}{2\sigma^2}\sum_{j=1}^{m}\{m^{-1}v\lambda + (\mathbf{y}_j - \mathbf{X}_j\boldsymbol{\beta}_j)^{\mathrm{T}}(\mathbf{y}_j - \mathbf{X}_j\boldsymbol{\beta}_j)\}\right]$$

$$\times\, |\boldsymbol{\Sigma}|^{-1/2(m+\rho-p-2)} \exp\left[-\frac{1}{2}\operatorname{tr}\boldsymbol{\Sigma}^{-1}\left\{\mathbf{R} + \sum_{j=1}^{m}(\boldsymbol{\beta}_j - \boldsymbol{\beta}.)(\boldsymbol{\beta}_j - \boldsymbol{\beta}.)^{\mathrm{T}}\right\}\right], \quad (34)$$

where

$$\boldsymbol{\beta}. = m^{-1}\sum_{j=1}^{m}\boldsymbol{\beta}_j.$$

The modal estimates are then easily obtained. Those for $\boldsymbol{\beta}_j$ are as before, equation (24), with $\boldsymbol{\Sigma}$ and σ^2 replaced by modal values. The latter are seen to satisfy

$$s^2 = \sum_{j=1}^{m}\{m^{-1}v\lambda + (\mathbf{y}_j - \mathbf{X}_j\boldsymbol{\beta}_1^*)^{\mathrm{T}}(\mathbf{y}_j - \mathbf{X}_j\boldsymbol{\beta}_j^*)\}/(n+v+2),$$

and

$$\boldsymbol{\Sigma}^* = \left\{\mathbf{R} + \sum_{j=1}^{m}(\boldsymbol{\beta}_j^* - \boldsymbol{\beta}_\cdot^*)(\boldsymbol{\beta}_j^* - \boldsymbol{\beta}_\cdot^*)^{\mathrm{T}}\right\}\Big/(m+\rho-p-2). \quad (35)$$

It is possible in this case, as in Sections 5.1 and 5.3, to proceed a little differently and obtain the posterior distribution of the $\boldsymbol{\beta}_j$'s, free of σ^2 and $\boldsymbol{\Sigma}$ and consider the modes of this. This is because the integration of (34) with respect to $\boldsymbol{\Sigma}^{-1}$ and σ^2 is possible in closed form. The result is

$$\left[\sum_{j=1}^{n}\{m^{-1}v\lambda + (\mathbf{y}_j - \mathbf{X}_j\boldsymbol{\beta}_j)^{\mathrm{T}}(\mathbf{y}_j - \mathbf{X}_j\boldsymbol{\beta}_j)\}\right]^{-1/2(n+v)}$$

$$\times\left|\mathbf{R} + \sum_{j=1}^{m}(\boldsymbol{\beta}_j - \boldsymbol{\beta})(\boldsymbol{\beta}_j - \boldsymbol{\beta})^{\mathrm{T}}\right|^{-1/2(m+\rho-1)}. \quad (36)$$

The mode of this distribution can be used in place of the modal values for the wider distribution. The differentiation is facilitated by using the result that, with $\mathbf{V}$ equal to the matrix whose determinant appears in (36),

$$\frac{\partial}{\partial\boldsymbol{\beta}_i}\log|\mathbf{V}| = 2\mathbf{V}^{-1}(\boldsymbol{\beta}_i - \boldsymbol{\beta}).$$

It is then possible to verify that the modes for $\boldsymbol{\beta}_j$ satisfy the same equations as before, (24), with $\boldsymbol{\Sigma}$ and σ^2 replaced by values given by (35) except that the divisors on the right-hand sides are $(n+\nu)$ and $(m+\rho-1)$ rather than $(n+\nu+2)$ and $(m+\rho-p-2)$.

It is possible to extend this model significantly by reverting to the heteroscedastic case as originally considered, (21). Here we have to specify a joint distribution for the σ_j^2. A possible device is to suppose that, like the means, the σ_j^2 are exchangeable. A convenient distribution to generate the exchangeability is to suppose $\nu\lambda/\sigma_j^2 \sim \chi_\nu^2$. In the context of several means (Section 1) Lindley (1971a) has shown how the estimates of the variances get pulled towards a central value. The details are so similar here that we do not repeat them.

As explained in Section 3.2, it was Novick's suggestion to consider this problem in an educational context, and we conclude this section by briefly reporting on an application that he, in conjunction with Jackson, Thayer and Cole (1972), have made of these results. We are most grateful to them for permission to include the details here. Their analysis used data from the American College Testing Program on the prediction of grade-point average at 22 colleges from the results of 4 tests; namely, English, Mathematics, Social Studies and Natural Sciences. We therefore have the situation studied in this section with $p = 5$ (one variable corresponding to the mean), $m = 22$, and n_j varying from 105 to 739. They used the heteroscedastic model but the basic equations (24) and (35) are essentially as here described. With the substantial amounts of data available the prior constants, ν, λ, ρ and $\mathbf{R}$ scarcely affect the analysis: the first three were taken to be small and changes of origin of the regressor variables effected to make the prior judgment that $\mathbf{R}$ was diagonal. With ρ small the diagonal elements again play little role. We omit details of how the calculations were performed and refer the interested reader to their paper.

Data were available for 1968 and 1969. The approach was to use the 1968 data to estimate the regressions, to use these estimated equations on the 1969 x-data to estimate the y's, the grade-point averages, and then to compare these predictions with the actual 1969 y-values, using as a criterion of prediction the mean of the squares of the differences. This operation was done twice; once with the full 1968 data, and once with a random 25 per cent sample from each College. The results are summarized in Table 1.

The first row refers to the analysis of the whole data and shows that the Bayesian method only reduces the error by under 2 per cent. With such

Table 1. Comparison of Predictive Efficiency.

	Average mean-square error	
	Least-squares	Bayes
All data	0.5596	0.5502
25% sample	0.6208	0.5603

large samples there is little room for improvement. With the quarter sample, however, in the second row of the table, the reduction is up to 9 per cent and most strikingly the error is almost down to the value reached with the least-squares estimates for all the data. In other words, 25 per cent of the data and Bayes are as good as all the data and least squares: or the Bayesian method provides a possible 75 per cent saving in sample size. They also provide details of the comparisons between the two estimates of the regression coefficients. These tend to be "shrunk" towards a common value (for each regressor variable) and in some cases with the quarter sample the shrinkage is substantial.

It would be dangerous to draw strong conclusions from one numerical study but the analysis should do something to answer the criticism of those who have said that Bayesian methods are not "testable." We favour the method because of its coherence, but the pragmatists may like to extend the method of Novick *et al.* to other data sets, remembering, of course, that we have made an assumption of exchangeability, and the method cannot be expected to work when this is unreasonable.

5.3. Exchangeability Within Multiple Regression Equations

In this section we briefly indicate how the analysis of Section 3.3 proceeds when σ^2, the residual regression variance, and σ_β^2, the variance of the regression coefficients, are both unknown. As before, we assume that independently

$$\nu\lambda/\sigma^2 \sim \chi_\nu^2, \qquad \nu_\beta\lambda_\beta/\sigma_\beta^2 \sim \chi_{\nu_\beta}^2.$$

As in Section 5.2 the integration with respect to ξ, the mean of the β_j's, may be performed and the result is that the posterior distribution of $\boldsymbol{\beta}$, σ^2 and σ_β^2 is proportional to

$$(\sigma^2)^{-1/2(n+\nu+2)} \exp\left[-\frac{1}{2\sigma^2}\{\nu\lambda + (\mathbf{y}-\mathbf{X}\boldsymbol{\beta})^{\mathrm{T}}(\mathbf{y}-\mathbf{X}\boldsymbol{\beta})\}\right]$$

$$\times\, (\sigma_\beta^2)^{-1/2(p+\nu\beta+1)} \exp\left[-\frac{1}{2\sigma_\beta^2}\left\{\nu_\beta\lambda_\beta + \sum_{j=1}^{p}(\beta_j - \beta_.)^2\right\}\right], \quad (37)$$

where

$$\beta_. = p^{-1}\sum_{j=1}^{p}\beta_j.$$

The modal equations are then easily seen to be (the first coming from (23))

$$\left.\begin{aligned}
\boldsymbol{\beta}^* &= \{\mathbf{I}_p + k^*(\mathbf{X}^{\mathrm{T}}\mathbf{X})^{-1}(\mathbf{I}_p - p^{-1}\mathbf{J}_p)\}^{-1}\hat{\boldsymbol{\beta}},\\
s^2 &= \{\nu\lambda + (\mathbf{y}-\mathbf{X}\boldsymbol{\beta}^*)^{\mathrm{T}}(\mathbf{y}-\mathbf{X}\boldsymbol{\beta}^*)\}/(n+\nu+2),\\
s_\beta^2 &= \left\{\nu_\beta\lambda_\beta + \sum_{j=1}^{p}(\beta_j^* - \beta_.^*)^2\right\}\Big/(p+\nu_\beta+1).
\end{aligned}\right\} \quad (38)$$

Table 2. 10-Factor Multiple Regression Example.

Estimate	β_1	β_2	β_3	β_4	β_5	
Least-squares	−0.185	−0.221	−0.359	−0.105	−0.469	
Bayes	−0.256	−0.178	−0.326	−0.086	−0.289	
Ridge	−0.295	−0.110	−0.245	−0.050	−0.040	
Estimate	β_6	β_7	β_8	β_9	β_{10}	k
Least-squares	0.813	0.285	0.383	0.092	0.094	0.000
Bayes	0.592	0.195	0.349	0.117	0.116	0.039
Ridge	0.325	0.050	0.240	0.125	0.125	0.250

The value of k^* is of course s^2/s_β^2. The marginal posterior distribution of $\boldsymbol{\beta}$ can be obtained in a manner similar to that described in the last section.

We are now in a position to compare our method with that of Hoerl and Kennard (1970b). We have taken the example of a 10-factor, non-orthogonal multiple regression summarized in Gorman and Toman (1966) and re-analysed by Hoerl and Kennard using their ridge regression method. The results are summarized in Table 2.

As already explained, the main difference between ridge regression and the Bayes approach lies in the choice of $k(=\sigma^2/\sigma_\beta^2)$ in equation (23). This has the value zero for least-squares, is chosen subjectively in the ridge method by selecting it so large that the regression estimates stabilize, and is estimated from the data in the Bayes method. In applying the Bayes method we started with $k^* = 0$ in (38), obtained estimates $\boldsymbol{\beta}^*$, which were then used in the other equations in (38) to obtain s^2 and s_β^*. It was found that 10 iterations were needed until the cycle converged. The solution is fairly insensitive to changes in the small, positive values of ν and ν_β and these were set to zero.

In the case of non-orthogonal data, the least-squares procedure has a tendency to produce regression estimates which are too large in absolute value, of incorrect sign and unstable with respect to small changes in the data. The ridge method attempts to avoid some of these undesirable features. The Bayesian method reaches the same conclusion but has the added advantage of dispensing with the rather arbitrary choice of k and allows the data to estimate it. It will be seen from Table 2 that except for β_1, β_9 and β_{10}, all the estimates are pulled towards zero, the effect being greater with the ridge method than with Bayes, the latter choosing a considerably larger value of k than the data suggest.

Acknowledgements

We are very grateful to Melvin R. Novick who first stimulated our interest in these problems, has continually made fruitful suggestions and, with his colleagues, has allowed us to include the example in Section 5.2. We are indebted to the referees for constructive comment on a first draft of this paper. The second author would like to thank the Science Research Council and the Central Electricity Generating Board for financial support during the course of this research.

References

Brown, L.D. (1966). On the admissibility of invariant estimators of one or more location parameters. *Ann. Math. Statist.*, **37**, 1087–1136.

Brown, L.D. (1968). Inadmissibility of the usual estimators of scale parameters in problems with unknown location and scale parameters. *Ann. Math. Statist.*, **39**, 29–48.

Cornfield, J. (1969). The Bayesian outlook and its application. *Biometrics*, **25**, 617–657.

De Finetti, B. (1964). Foresight: its logical laws, its subjective sources. In *Studies in Subjective Probability* (H.E. Kyburg, Jr. and H.E. Smokler eds.), pp. 93–158. Wiley, New York.

Fisk, P.R. (1967). Models of the second kind in regression analysis. *J. Roy. Statist. Soc. Ser. B*, **29**, 266–281.

Gorman, J.W. and Toman, R.J. (1966). Selection of variables for fitting equations to data. *Technometrics*, **8**, 27–51.

Hewitt, E. and Savage, L.J. (1955). Symmetric measures on Cartesian products. *Trans. Amer. Math. Soc.*, **80**, 470–501.

Hoerl, A.E. and Kennard, R.W. (1970a). Ridge regression: Biased estimation for nonorthogonal problems. *Technometrics*, **12**, 55–67.

Hoerl, A.E. and Kennard, R.W. (1970b). Ridge regression: Applications to nonorthogonal problems. *Technometrics*, **12**, 69–82.

Lindley, D.V. (1969). Bayesian least squares. *Bull. Inst. Internat. Statist.*, **43**(2), 152–153.

Lindley, D.V. (1971a). The estimation of many parameters. In *Foundations of Statistical Inference* (V.P. Godambe and D.A. Sprott, eds.), pp. 435–455. Holt, Rinehart and Winston, Toronto.

Lindley, D.V. (1971b). *Bayesian Statistics, A Review*. SIAM, Philadelphia, PA.

Nelder, J.A. (1968). Regression, model-building and invariance. *J. Roy. Statist. Soc. Ser. A*, **131**, 303–315.

Novick, M.R., Jackson, P.H., Thayer, D.T., and Cole, N.S. (1972). Estimating multiple regressions in *m*-groups; a cross-validation study. *British. J. Math. Statist. Psych.*

Plackett, R.L. (1960). *Principles of Regression Analysis.* Clarendon Press, Oxford.

Rao, C.R. (1965). *Linear Statistical Inference and its Applications*. Wiley, New York.

Stein, C. (1956). Inadmissibility of the usual estimator for the mean of a multivariate normal distribution. *Proc. Third Berkeley Sympos.*, **1**, 197–206. University of California Press, Berkeley.

Introduction to Besag (1974) Spatial Interaction and the Statistical Analysis of Lattice Systems

Richard L. Smith
University of North Carolina

Julian Besag has made a career of identifying emerging areas of statistics and writing important papers about them, just as they were beginning to attract serious attention. Thus Besag (1986) was the second major paper [after Geman and Geman (1984), featured elsewhere in this volume] on the Markov random fields approach to image analysis, while in the 1990s he made a number of contributions to Markov chain Monte Carlo (MCMC) sampling [e.g., Besag and Green (1993); Besag et al. (1995)]. His first major paper, however, laid out the foundations for statistical inference in lattice systems, and so provided the background for much later work, both his own and that of many others.

There are some papers of which we can say that the whole is greater than the sum of its parts, and I believe this to be a case in point. No single part of the paper represents a unique contribution in its own right. Its proof of the Hammersley–Clifford theorem, though simple and elegant, was by no means the final word on this result. The inferential techniques in Section 6 were just the first steps that were to lead, a decade or more later, to MCMC methods of inference. When the paper is considered as a whole, however, we see a very big picture: from the consistency relations that must be satisfied by a family of conditional probabilities, through the Hammersley–Clifford theorem as a general characterization of the resulting probability models, to the development of several classes of concrete models, statistical inference based on those models, and finally some by no means elementary examples. It is this all-encompassing nature of the paper which qualifies it, in my view, to be considered a "breakthrough in statistics."

For the remainder of this Introduction I discuss a number of aspects of the paper itself, then I summarize the discussion which was published at the time, and finally give a very brief overview of the main developments in spatial statistics which may be considered to have been directly influenced by Besag's paper.

The Background

Besag was not, of course, the first person to think about statistics for lattice systems. The need to take account of spatial variation in analyzing agricultural field trials was an important motivation for R.A. Fisher's work on randomization, and more explicit methods of modeling spatial effects also go back as far as the 1930s (e.g., the Papadakis method for estimating treatment effects in an analysis of variance). More recently, Bartlett and Whittle had developed stochastic models for spatial processes. Nevertheless, I think if we contrast the state of spatial statistics in the early 1970s with that of (linear) time-series analysis, one can see that the latter had an extremely rich and powerful theory which was essentially complete at that time, whereas the former consisted of a handful of isolated techniques with no coherent structure. By no means the least of Besag's achievements was to provide a framework that others could build upon.

Summary of the Paper

After an introductory section about the kinds of problems to which the paper is addressed (most of these have an agricultural flavor; no hint here of the subsequent applications to digital image processing!), the author discusses in Section 2 the fundamential difficulties associated with defining a field in terms of its one-dimensional conditional probabilities. The key formula here is (2.2), which specifies the relationship between the joint and the conditional probabilities, and which also clarifies the consistency conditions which the latter have to satisfy (essentially, there are many different ways of changing a vector $\mathbf{x}$ into a vector $\mathbf{y}$ one component at a time, but they must all lead to the same ratio $P(\mathbf{x})/P(\mathbf{y})$).

Section 3 is about the Hammersley–Clifford theorem which states, loosely, that if a random field is Markov with respect to some specified system of neighbors, then the function $Q(\mathbf{x})$, defined by (3.1), is expressible as a sum over "cliques," i.e., subsets of lattice sites which are mutual neighbors. The history of this result is of some interest, since the original version had been obtained by Hammersley and Clifford in 1971 but was not published. Besag had an early preprint of this work and succeeded in deriving a much simpler proof, the one published here. In the meantime, a number of other authors published proofs of the main result at about the same time. Further information about the history of the result, along with the original Hammersley–Clifford proof, is in Clifford (1990). One feature that may have held up publication of the original proof was its apparently awkward dependence on a "positivity condition"—existence of a product state space in which there are no states of zero probability. However, this condition is needed for Besag's proof as well, and as subsequently emerged, is needed for the result. Further information about this was given in Hammersley's contribution to the discussion which followed Besag's paper.

Sections 4 and 5 provide numerous special cases of models derived from this general theory. In particular, Subsections 4.1 and 4.2 define classes of "auto-models" which, for all their alleged disadvantages (p. 200), remain the most widely used classes of models some 20 years later. See, for example, Cox and Wermuth (1994).

Section 6 brings us to the subject of statistical inference, and here Besag offers three approaches: the "coding" approach, where an exact likelihood is constructed *conditionally* on some subset of lattice sites, the "unilateral approximations" approach, where the model is approximated by one allowing the likelihood function to be written as a product of one-sided conditional probabilities, and approximate maximum likelihood approaches for Gaussian models—this is closely related to Whittle's approach which, in its turn, is a generalization of Whittle's methods for constructing likelihood approximations in time series analysis.

However, perhaps the paper's most famous contribution to statistical inference is one that it does not contain! The idea of inference based on the pseudo-likelihood function

$$\prod_i P\{x_i|x_j,\, j \neq i\}$$

is widely cited [by Winkler (1995), to give just one example] as a product of the present paper, whereas in fact it was introduced for the first time in Besag (1975). To be fair, it is clear that there is a common circle of ideas here, and the method of pseudo-likelihood is very closely related to the coding methods which are in this paper (one interpretation of pseudo-likelihood, in the case of a regular lattice, is that it is the product of the individual coding-based likelihoods corresponding to different coding schemes). For further discussion about pseudo-likelihood and related concepts in both spatial statistics and survival data analysis, see Barndorff-Nielsen and Cox (1994, Chap. 8).

The final substantive section of the paper contains two real data examples. The second of these is the famous Mercer–Hall wheat yields dataset, much analyzed both previously and subsequently [by Cressie (1993), for instance]. Besag fits two Markov random field models by each of the three approximate maximum likelihood schemes, and compares the results. With present-day computing speeds, it is not too difficult to compute the exact maximum likelihood estimators, but it is understandable that Besag did not attempt to do this in 1974. A more substantive difficulty with this dataset, as pointed out by Besag and by a number of authors who have analyzed it, is that the data does not provide a very good fit to a stationary Markov random field, so it would probably be wiser to regard any attempted estimation of the parameters as being primarily for illustrative purposes.

The Discussion

This being a Royal Statistical Society read paper, it was followed by a published discussion (not reproduced here). Royal Statistical Society discussions have a reputation, which sometimes tends to get rather exaggerated, of being very sharp and critical; this one, however, was highly complimentary to the author. D.R. Cox, as proposer of the formal Vote of Thanks, called the paper "original, lucid and comprehensive," praising it for its "emphasis ... on models for the analysis of data." A.G. Hawkes, as seconder of the Vote of Thanks, likewise felt that the main contribution was its "methods which [are] readily available for use by practising statisticians [and which] will become used quite widely." Peter Clifford's preference for the term "Markov meadows" instead of "Markov fields" (and "Markov streams" instead of "Markov chains") appears not to have caught on. Perhaps the most intriguing contribution was by Hammersley, who reviewed the history of the Hammersley–Clifford theorem and the unsuccessful attempts to free it from the "positivity condition." The punchline is that the result is apparently false without such condition, one of Hammersley's students having constructed a counterexample!

Subsequent Developments in Spatial Statistics

The field is much too vast for anyone to give, in a short overview, a comprehensive summary of all the subsequent developments in spatial statistics (including image analysis) which might be said to have stemmed, directly or indirectly, from Besag's paper. So I will confine myself to a few highlights and personal remarks.

Methodology for statistical inference in Markov random fields has continued to develop; the paper by Besag (1975) has already been mentioned; others were by Besag and Moran (1975) and Besag (1977). In recent years, of course, the possibilities have expanded greatly with the popularization of MCMC methods. For instance, Penttinen (1984) presented a method for simulation-based approximations of maximum likelihood estimates, extended by Geyer and Thompson (1992); this method is directly applicable to inference from Markov random fields. Such methods might lead us to suppose that simple methods such as pseudo-likelihood are now obsolete, as indeed appears to be Besag's own view, though personally I am not so convinced of this. Pseudo-likelihood-based estimators have the advantage of being easy to compute without worrying about all the convergence and computational efficiency issues associated with MCMC, and we still have only an incomplete understanding of the relative properties of the two methods (see, e.g., Comets (1992), Winkler (1995), and Ji and Seymour (1996)).

Image analysis is so large a subject that it seems hardly appropriate to give any survey of the current state of knowledge; suffice to say that Geman and Geman (1984) and Besag (1986) both made strong use of Markov random field ideas and this has continued to be one of the main themes in subsequent research.

The interplay between Markov random fields and simulation methodology has also played a very large role in the development of MCMC methods themselves. One facet of this has been the exchange of ideas between the statistics and statistical physics literatures; for example, in recent years Besag has been an enthusiastic advocate of methods derived from the Swendsen–Wang (1987) algorithm and multigrid methods; see, for example, Besag and Green (1993).

The one area where I feel that the development has been rather slow is in the more traditional spatial statistics problems, especially if we go beyond the agricultural problems with which Besag illustrated his own paper. Certainly there have been recent developments, and Cressie (1993) gave a number of examples in different fields, but we do not get the impression of fundamental advances in either the models or the methodology. However, the development of new computational methodology, particularly in the MCMC area, surely creates many possibilities for new models and new analysis techniques. It seems to me that this is an area where we may expect to see a lot of new development in the next few years.

Biographical Information

Julian Besag was born in 1945 in the central English town of Loughborough. After an abortive 2 years studying engineering as an undergraduate in Cambridge, he transferred to Birmingham to study mathematics. Birmingham University at that time had a very strong statistics group (Henry Daniels, Frank Downton, Vic Barnett, David Wishart, John Nelder) and they influenced him to obtain his degree in statistics. John Nelder and Stephan Vajda were particularly strong influences on him at this time. An undergraduate research project on sequential analysis brought him into contact with Maurice Bartlett, who offered him a Research Assistant position at Oxford. It was during this period (1968–1969), and under Bartlett's influence, that he began his work on spatial statistics, but after a year he was offered a Lectureship at Liverpool University, so he continued his work there. British academia in the 1970s was less concerned with formal academic qualifications than are most institutions today; under the rules that were then in force, his position at Oxford, technically contract employee rather than student, prevented him registering for a higher degree, and as a consequence he never obtained a doctorate.

The 1974 paper was originally submitted to, and accepted by, *Biometrika*, but Professor D.R. Cox, as Editor of *Biometrika*, suggested that he resubmit it to the Royal Statistical Society as a Read Paper.

In 1975 Besag moved to the University of Durham as Reader in Mathematical Statistics, and in 1985 he became that university's first Professor of Statistics. Since 1989 he has been Professor and Director of the Center for Spatial Statistics, University of Washington. His formal honors include being awarded the Guy Medal in Silver of the Royal Statistical Society, the citation for which made particular reference to his 1974 paper.

Acknowledgments

I am grateful to Julian Besag and Peter Green for comments on a preliminary draft of this commentary. I would especially like to thank Julian Besag for filling in a considerable amount of background information. The opinions expressed are mine alone, as of course is the responsibility for any errors of fact or interpretation that remain.

References

Barndorff-Nielsen, O.E. and Cox, D.R. (1994). *Inference and Asymptotics*. Chapman and Hall, London.

Besag, J. (1975). Statistical analysis of non-lattice data. *The Statistician*, **24**, 179–195.

Besag, J. (1977). Efficiency of pseudolikelihood estimation for simple Gaussian field. *Biometrika*, **64**, 616–618.

Besag, J.E. (1986). On the statistical analysis of dirty pictures (with discussion). *J. Roy. Statist. Soc. Ser. B*, **48**, 259–302.

Besag, J.E. and Green, P.J. (1993). Spatial statistics and Bayesian computation. *J. Roy. Statist. Soc. Ser. B*, **55**, 25–37.

Besag, J., Green, P.J., Higdon, D., and Mengersen, K. (1995). Bayesian computations and stochastic systems. *Statist. Sci.*, **10**, 1–66.

Besag, J. and Moran, P.A.P. (1975). On the estimation and testing of spatial interaction in Gaussian lattice processes. *Biometrika*, **62**, 555–562.

Clifford, P. (1990). Markov random fields in statistics. In *Disorder in Physical Systems: A Volume in Honour of John M. Hammersley* (G.R. Grimmett and D.J.A. Welsh, eds.). Oxford University Press, Oxford.

Comets, F. (1992). On consistency of a class of estimators for exponential families of Markov random fields on a lattice. *Ann. Statist.*, **20**, 455–468.

Cox, D.R. and Wermuth, N. (1994). A note on the quadratic exponential binary distribution. *Biometrika*, **81**, 403–408.

Cressie, N. (1993). *Statistics for Spatial Data*, 2nd ed. Wiley, New York.

Geman, S. and Geman, D. (1984). Stochastic relaxation, Gibbs distributions and the Bayesian restoration of images. *IEEE Trans. Pattern Anal. Machine Intelligence*, **6**, 721–741.

Geyer, C.J. and Thompson, E.A. (1992). Constrained Monte Carlo maximum likelihood for dependent data (with discussion). *J. Roy. Statist. Soc. Ser. B*, **54**, 657–699.

Ji, C. and Seymour, L. (1996), A consistent model selection procedure for Markov random fields based on the pseudolikelihood. *Ann. Appl. Probab.*, **6**, 423–443.

Penttinen, A. (1984). Modelling interaction in spatial point patterns: parametric estimation by the maximum likelihood method. *J. Stud. Comput. Sci. Econ. Statist.*, **7**.

Swendsen, R.H. and Wang, J.-S. (1987). Nonuniversal critical dynamics in Monte Carlo simulations. *Phys. Rev. Lett.*, **58**, 86–88.

Winkler, G. (1995). *Image Analysis, Random Fields and Dynamic Monte Carlo Analysis: A Mathematical Introduction*. Springer-Verlag, Berlin.

Spatial Interaction and the Statistical Analysis of Lattice Systems*

Julian Besag
University of Liverpool

Summary

The formulation of conditional probability models for finite systems of spatially interacting random variables is examined. A simple alternative proof of the Hammersley–Clifford theorem is presented and the theorem is then used to construct specific spatial schemes on and off the lattice. Particular emphasis is placed upon practical applications of the models in plant ecology when the variates are binary or Gaussian. Some aspects of infinite lattice Gaussian processes are discussed. Methods of statistical analysis for lattice schemes are proposed, including a very flexible coding technique. The methods are illustrated by two numerical examples. It is maintained throughout that the conditional probability approach to the specification and analysis of spatial interaction is more attractive than the alternative joint probability approach.

1. Introduction

In this paper, we examine some stochastic models which may be used to describe certain types of spatial processes. Potential applications of the models occur in plant ecology and the paper concludes with two detailed numerical examples in this area. At a formal level, we shall largely be concerned with a rather arbitrary system, consisting of a finite set of sites, each

* Read before the Royal Statistical Society at a meeting organized by the Research Section on Wednesday, March 13th, 1974, Professor J. Durbin in the Chair.

site having associated with it a univariate random variable. In most ecological applications, the sites will represent points or regions in the Euclidean plane and will often be subject to a rigid lattice structure. For example, Cochran (1936) discusses the incidence of spotted wilt over a rectangular array of tomato plants. The disease is transmitted by insects and, after an initial period of time, we should clearly expect to observe clusters of infected plants. The formulation of spatial stochastic models will be considered in Sections 2–5 of the paper. Once having set up a model to describe a particular situation, we should then hope to be able to estimate any unknown parameters and to test the goodness-of-fit of the model on the basis of observation. We shall discuss the statistical analysis of lattice schemes in Sections 6 and 7.

We begin by making some general comments on the types of spatial systems which we shall, and shall not, be discussing. Firstly, we shall not be concerned here with any random distribution which may be associated with the locations of the sites themselves. Indeed, when setting up models in practice, we shall require quite specific information on the relative positions of sites, in order to assess the likely interdependence between the associated random variables. Secondly, although, as in Cochran's example above, the system may, in reality, have developed continuously through time, we shall always assume that observation on it is only available at an isolated instant; hence, we shall not be concerned here with the setting up of spatial–temporal schemes. This has the important consequence that our models will not be mechanistic and must be seen as merely attempts at describing the "here and now" of a wider process. In many practical situations, this is a reasonable standpoint, since we can only observe the variables at a single point in time (for example, the yields of fruit trees in an orchard) but, in other cases, a spatial–temporal approach may be more appropriate. In fact, the states of the tomato plants, in Cochran's example, were observed at three separate points in time and it is probably most profitable to use a classical temporal autoregression to analyse the system. A similar comment applies to the hop plants data of Freeman (1953). Ideally, even when dealing with a process at a single instant of time, we should first set up an intuitively plausible spatial–temporal model and then derive the resulting instantaneous spatial structure. This can sometimes be done if we are prepared to assume stationarity in both time and space (see Bartlett, 1971a) but, unfortunately, such an assumption is unlikely to be realistic in our context. However, when this approach is justifiable, it is of course helpful to check that our spatial models are consistent with it; for simple examples, see Besag (1972a). Otherwise, regarding the transient spatial structure of a spatial–temporal process, this is almost always intractable and hence there exists a need to set up and examine purely spatial schemes without recourse to temporal considerations.

The following examples are intended as typical illustrations of the spatial situations we shall have in mind. They are classified according to the nature of:

(a) the system of sites (regular or irregular);
(b) the individual sites (points or regions); and
(c) the associated random variables (discrete or continuous).

1.1. A *regular lattice of point sites with discrete variables* commonly occurs under experimental conditions in plant ecology. Examples include the pattern of infection in an array of plants (Cochran, 1936, as described above; Freeman, 1953, on the incidence of nettlehead virus in hop plants) and the presence or absence of mature plants seeded on a lattice and subsequently under severe competition for survival (data kindly supplied by Dr. E.D. Ford, Institute of Tree Biology, Edinburgh, relates to dwarf French marigolds on a triangular lattice of side 2 cm). Often, as above, the data are binary.

1.2. A *regular lattice of point sites with continuous variables* commonly occurs in agricultural experiments, where individual plant yields are measured (Mead, 1966, 1967, 1968, on competition models; Batchelor and Reed, 1918, on fruit trees). It is often reasonable to assume that the variates have a multivariate normal distribution.

1.3. A *regular lattice of regions with discrete variables* arises in sampling an irregularly distributed population when a rectangular grid is placed over an area and counts are made of the number of individuals in each quadrat (Professor P. Greig-Smith on *Carex arenaria*, in Bartlett, 1971b; Gleaves, 1973, on *Plantago lanceolata*; Clarke, 1946, and Feller, 1957, p. 150, on flying-bomb hits in South London during World War II; Matui, 1968, on the locations of farms and villages in an area of Japan). In plant ecology, the quadrats are often so small that few contain more than a single plant and it is then reasonable to reduce the data to a binary (presence/absence) form.

1.4. A *regular lattice of regions with continuous variables* typically occurs in field trials where aggregate yields are measured (Mercer and Hall, 1911, on wheat plots). Multivariate normality is often a reasonable assumption.

1.5. *Irregular point sites with discrete variables* arise in sampling natural plant populations. Examples include the presence or absence of infection in individuals and the variety of plant at each site in a multi-species community.

1.6. *Irregular point sites with continuous variables* again occur in sampling natural plant populations (Brown, 1965, on tree diameters in pine forests; Mead, 1971, on competition models).

1.7. *Irregular regions with discrete or continuous variables* have applications particularly in a geographical context, with regions defined by administrative boundaries (O'Sullivan, 1969, and Ord, 1974, on aspects of the economy of Eire).

It has previously been stated that in the practical construction of spatial models, we shall require precise information concerning the relative positions of the various sites. Where the sites are regions, rather than points, the data are by definition, aggregate data and the assumption of single, uniquely defined locations for each of the associated variables is clearly open to criticism. For example, quadrat counts (Section 1.3) are usually used to examine spatial pattern rather than spatial interaction. Further comments will appear in Section 5.

Combinations of the above situations may occur. For example, in competition experiments where yields are measured, "missing observations" may be due to intense competition and should then be specifically accounted for by the introduction of mixed distributions. We shall not contemplate such situations here.

2. Conditional Probability Approach to Spatial Processes

There appear to be two main approaches to the specification of spatial stochastic processes. These stem from the non-equivalent definitions of a "nearest-neighbour" system, originally due to Whittle (1963) and Bartlett (1955, Section 2.2, 1967, 1968), respectively. Suppose, for definiteness, that we temporarily restrict attention to a rectangular lattice with sites labelled by integer pairs (i, j) and with an associated set of random variables $\{X_{i,j}\}$. For the moment, we ignore any problems concerning the finiteness or otherwise of the lattice. Then Whittle's basic definition requires that the *joint* probability distribution of the variates should be of the product form

$$\prod_{i,j} Q_{i,j}(x_{i,j}; x_{i-1,j}, x_{i+1,j}, x_{i,j-1}, x_{i,j+1}), \tag{2.1}$$

where $x_{i,j}$ is a value of the random variable, $X_{i,j}$. On the other hand, Bartlett's definition requires that the *conditional* probability distribution of $X_{i,j}$, given all other site values, should depend only upon the values at the four nearest sites to (i, j), namely $x_{i-1,j}$, $x_{i+1,j}$, $x_{i,j-1}$, and $x_{i,j+1}$. Whilst the conditional probability formulation may be said to have rather more intuitive appeal, this is marred by a number of disadvantages.

Firstly, there is no obvious method of deducing the joint probability structure associated with a conditional probability model. Secondly, the conditional probability structure itself is subject to some unobvious and highly restrictive consistency conditions. When these are enforced, it can be shown (Brook, 1964) that the conditional probability formulation is degenerate with respect to (2.1). Thirdly, it has been remarked by Whittle (1963) that the natural specification of an equilibrium process in statistical mechanics is in terms of the joint distribution rather than the conditional distribution of the variables.

These problems were partially investigated in a previous paper (Besag, 1972a). The constraints on the conditional probability structure were identified for homogeneous systems and found to be so severe that they actually generated particular spatial models, given the nature of the variables. Had these models failed to retain any practical appeal, then there would have been little further scope for discussion. However, this is not the case. For example, with binary variables, the conditional probability formulation *necessarily* generates just that basic model (the Ising model of ferromagnetism) which has been at the centre of so much work in statistical mechanics. Thus, although this model may *classically* be formulated in terms of joint probabilities, it is generated in a natural way through basic *conditional* probability assumptions. This fact may also be related to the problem of degeneracy. There is surely no indignity in studying a subclass of schemes provided that subclass is of interest in its own right. However, we go further. Suppose we consider wider classes of conditional probability models in which the conditional distribution of $X_{i,j}$ is allowed to depend upon the values at more remote sites. We can build up a hierarchy of models, more and more general, which eventually will include the scheme (2.1) and any particular generalization of it. That is, we extend the concept of first-, second- and higher-order Markov chains in one dimension to the realm of spatial processes. There is then no longer any degeneracy associated with the conditional probability models. This is the approach taken in the present paper. It has been made possible by the advent of the celebrated Hammersley–Clifford theorem which, sadly, has remained unpublished by its authors.

Finally, in this section, we examine the problems and implications of deriving the joint probability structure associated with the site variables, given their individual conditional distributions. We no longer restrict attention to "nearest-neighbour" models nor even to lattice schemes but instead consider a fairly arbitrary system of sites. Suppose then that we are concerned with a *finite* collection of random variables. $X_1, \ldots, X_n$, which are associated with sites labelled $1, \ldots, n$, respectively. For each site, $P(x_i|x_1, \ldots, x_{i-1}, x_{i-1}, \ldots, x_n)$, the conditional distribution of X_i, given all other site values, is specified and we require the joint probability distribution of the variables. Our terminology will be appropriate to discrete variables but the arguments equally extend to the continuous case.

We make the following important assumption: if $x_1, \ldots, x_n$ can individually occur at the sites $1, \ldots, n$, respectively, then they can occur together. Formally, if $P(x_i) > 0$ for each i, then $P(x_1, \ldots, x_n) > 0$. This is called the *positivity* condition by Hammersley and Clifford (1971) and will be assumed throughout the present paper. It is usually satisfied in practice. We define the sample space Ω to be the set of all possible realizations $\mathbf{x} = (x_1, \ldots, x_n)$ of the system. That is, $\Omega = \{\mathbf{x}: P(\mathbf{x}) > 0\}$. It then follows that for any two given realizations $\mathbf{x}$ and $\mathbf{y} \in \Omega$,

$$\frac{P(\mathbf{x})}{P(\mathbf{y})} = \prod_{i=1}^{n} \frac{P(x_i|x_1, \ldots, x_{i-1}, y_{i+1}, \ldots, y_n)}{P(y_i|x_1, \ldots, x_{i-1}, y_{i+1}, \ldots, y_n)}. \tag{2.2}$$

The proof of this result resembles that of equation (6) in Besag (1972a). Clearly, we may write

$$P(\mathbf{x}) = P(x_n|x_1,\ldots,x_{n-1})P(x_1,\ldots,x_{n-1}); \tag{2.3}$$

however, $P(x_1,\ldots,x_{n-1})$ cannot be factorized in a useful way since, for example, $P(x_{n-1}|x_1,\ldots,x_{n-2})$ is not easily obtained from the given conditional distributions. Nevertheless, we can introduce y_n, write

$$P(\mathbf{x}) = \frac{P(x_n|x_1,\ldots,x_{n-1})}{P(y_n|x_1,\ldots,x_{n-1})}P(x_1,\ldots,x_{n-1},y_n)$$

and now operate on x_{n-1} in $P(x_1,\ldots,x_{n-1},y_n)$. This yields

$$P(x_1,\ldots,x_{n-1},y_n) = \frac{P(x_{n-1}|x_1,\ldots,x_{n-2},y_n)}{P(y_{n-1}|x_1,\ldots,x_{n-2},y_n)}P(x_1,\ldots,x_{n-2},y_{n-1},y_n),$$

after the similar introduction of y_{n-1}. Continuing the reduction process, we eventually arrive at equation (2.2) which clearly determines the joint probability structure of the system in terms of the given conditional probability distributions. We require the positivity condition merely to ensure that each term in the denominator of (2.2) is non-zero.

Equation (2.2) highlights the two fundamental difficulties concerning the specification of a system through its conditional probability structure. Firstly, the labeling of individual sites in the system being arbitrary implies that many factorizations of $P(\mathbf{x})/P(\mathbf{y})$ are possible. All of these must, of course, be equivalent and this, in turn, implies the existence of severe restrictions on the available functional forms of the conditional probability distributions in order to achieve a mathematically consistent joint probability structure. This problem has been investigated by Lévy (1948), Brook (1964), Spitzer (1971), Hammersley and Clifford (1971) and Besag (1972a) and we discuss it in detail in the next section. Seconly, whilst expressions for the relative probabilities of two realizations may be fairly straightforward, those for absolute probabilities, in general, involve an extremely awkward normalizing function with the consequence that direct approaches to statistical inference through the likelihood function are rarely possible. We shall have to negotiate this problem in Section 6 of the paper.

3. Markov Fields and the Hammersley–Clifford Theorem

In this section, we examine the constraints on the functional form of the conditional probability distribution available at each of the sites. We restate a theorem of Hammersley and Clifford (1971) and give a simple alternative proof. This theorem, which has received considerable attention recently, is essential to the construction of valid spatial schemes through the conditional probability approach. We begin by describing the problem more precisely. Our definitions will closely follow those of Hammersley and Clifford.

The first definition determines the *set of neighbours* for each site. Thus, site j ($\neq i$) is said to be a neighbour of site i if and only if the functional form of $P(x_i|x_1, \ldots, x_{i-1}, x_{i+1}, \ldots, x_n)$ is dependent upon the variable x_j. As the simplest example, suppose that $X_1, \ldots, X_n$ is a Markov chain. Then it is easily shown that site i $(2 \leq i \leq n-1)$ has neighbours $i-1$ and $i+1$ whilst the sites 1 and n have the single neighbours 2 and $n-1$, respectively. For a more interesting spatial example, suppose the sites form a finite rectangular lattice and are now conveniently labelled by integer pairs (i, j). Then, if $P(x_{i,j}\,|\,\text{all other site values})$ depends only upon $x_{i,j}$, $x_{i-1,j}$, $x_{i+1,j}$, $x_{i,j-1}$ and $x_{i,j+1}$ for each internal site (i, j), we have a so-called "nearest-neighbour" lattice scheme. In such a case, each internal site (i, j) has four neighbours, namely $(i-1, j)$, $(i+1, j)$, $(i, j-1)$, and $(i, j+1)$. (There is a slight inconsistency in the usage of the word "neighbour" here: this will be resolved in later sections by introducing the term "first-order" scheme.) Any system of n sites, each with specified neighbours, clearly generates a class of valid stochastic schemes. We call any member of this class a *Markov field*. Our aim is to be able to identify the class in any given situation.

Any set of sites which either consists of a single site or else in which every site is a neighbour of every other site in the set is called a *clique*. Thus, in the "nearest-neighbour" situation described above, there are cliques of the form $\{(i, j)\}$, $\{(i-1, j), (i, j)\}$ and $\{(i, j-1), (i, j)\}$ over the entire lattice, possibly with adjustments at the boundary. The definition of a clique is crucial to the construction of valid Markov fields.

We now make two assumptions, again following Hammersley and Clifford. Firstly, we suppose that there are only a finite number of values available at each site, although we shall relax this condition later in the section. Secondly, we assume that the value zero is available at each site. If this is originally untrue, it can always be subsequently brought about by re-indexing the values taken at the offending sites, a procedure which will be illustrated in Section 4.3. This second assumption, which is therefore made for purely technical reasons, ensures that, under the positivity condition, an entire realization of zeros is possible. That is, $P(\mathbf{0}) > 0$ and we may legitimately define

$$Q(\mathbf{x}) \equiv \ln\{P(\mathbf{x})/P(\mathbf{0})\} \tag{3.1}$$

for any $\mathbf{x} \in \Omega$. Lastly given any $\mathbf{x} \in \Omega$, we write $\mathbf{x}_i$ for the realization

$$(x_1, \ldots, x_{i-1}, 0, x_{i+1}, \ldots, x_n).$$

The problem to which Hammersley and Clifford addressed themselves may now be stated as follows: *given the neighbours of each site, what is the most general form which $Q(\mathbf{x})$ may take in order to give a valid probability structure to the system*? Since

$$\begin{aligned}\exp\{Q(\mathbf{x}) - Q(\mathbf{x}_i)\} \\ &= P(\mathbf{x})/P(\mathbf{x}_i) \\ &= P(x_i|x_1, \ldots, x_{i-1}, x_{i+1}, \ldots, x_n)/P(0|x_1, \ldots, x_{i-1}, x_{i+1}, \ldots, x_n), \quad (3.2)\end{aligned}$$

the solution to this problem immediately gives the most general form which may be taken by the conditional probability distribution at each site.

In dealing with the rather general situation described above, the Hammersley–Clifford theorem superseded the comparatively pedestrian results which had been obtained for "nearest-neighbour" systems on the k-dimensional finite cubic lattice (Spitzer, 1971; Besag, 1972a). However, the original method of proof is circuitous and requires the development of an operational calculus (the "blackening algebra"). A simple alternative statement and proof of the theorem rest upon the observation that for any probability distribution $P(\mathbf{x})$, subject to the above conditions, there exists an expansion of $Q(\mathbf{x})$, unique on Ω and of the form

$$\begin{aligned}Q(\mathbf{x}) = \sum_{1\le i\le n} x_i G_i(x_i) + \sum\sum_{1\le i<j\le n} x_i x_j G_{i,j}(x_i, x_j) \\ + \sum\sum\sum_{1\le i<j<k\le n} x_i x_j x_k G_{i,j,k}(x_i, x_j, x_k) + \cdots \\ + x_1 x_2 \cdots x_n G_{1,2,\ldots,n}(x_1, x_2, \ldots, x_n). \quad (3.3)\end{aligned}$$

For example, we have

$$x_i G_i(x_i) \equiv Q(0, \ldots, 0, x_i, 0, \ldots, 0) - Q(\mathbf{0}),$$

with analogous difference formulae for the higher order G-functions. With the above notation, Hammersley and Clifford's result may be stated in the following manner: *for any $1 \le i < j < \cdots < s \le n$, the function $G_{i,j,\ldots,s}$ in* (3.3) *may be non-null if and only if the sites $i, j, \ldots, s$ form a clique. Subject to this restriction, the G-functions may be chosen arbitrarily*. Thus, given the neighbours of each site, we can immediately write down the most general form for $Q(\mathbf{x})$ and hence for the conditional distributions. We shall see examples of this later on.

Proof of Theorem. It follows from equation (3.2) that, for any $\mathbf{x} \in \Omega$, $Q(\mathbf{x}) - Q(\mathbf{x}_i)$ can only depend upon x_i itself and the values at sites which are neighbours of site i. Without loss of generality, we shall only consider site 1 in detail. We then have, from equation (3.3),

$$Q(\mathbf{x}) - Q(\mathbf{x}_1) = x_1 \Big\{ G_1(x_1) + \sum_{2 \le j \le n} x_j G_{1,j}(x_1, x_j) + \sum\sum_{2 \le j < k \le n} x_j x_k G_{1,j,k}(x_1, x_j, x_k) + \cdots + x_2 x_3 \cdots x_n G_{1,2,\ldots,n}(x_1, x_2, \ldots, x_n) \Big\}.$$

Now suppose site $l(\neq 1)$ is *not* a neighbour of site 1. Then $Q(\mathbf{x}) - Q(\mathbf{x}_1)$ must be independent of x_l for all $\mathbf{x} \in \Omega$. Putting $x_i = 0$ for $i \neq 1$ or l, we immediately see that $G_{1,l}(x_1, x_l) = 0$ on Ω. Similarly, by other suitable choices of $\mathbf{x}$, it is easily seen successively that all 3-, 4-, $\ldots$, n-variable G-functions involving both x_1 and x_l must be null. The analogous result holds for any pair of sites which are not neighours of each other and hence, in general, $G_{i,j,\ldots,s}$ can only be non-null if the sites $i, j, \ldots, s$ form a clique.

On the other hand, any set of G-functions gives rise to a valid probability distribution $P(\mathbf{x})$ which satisfies the positivity condition. Also since $Q(\mathbf{x}) - Q(\mathbf{x}_i)$ depends only upon x_l if there is a non-null G-function involving both x_i and x_l, it follows that the same is true of $P(x_i|x_1, \ldots, x_{i-1}, x_{i+1}, \ldots, x_n)$. This completes the proof. □

We now consider some simple extensions of the theorem. Suppose firstly that the variates can take a denumerably infinite set of values. Then the theorem still holds if, in the second part, we impose the added restriction that the G-functions be chosen such that $\sum \exp Q(\mathbf{x})$ is finite, where the summation is over all $\mathbf{x} \in \Omega$. Similarly, if the variates each have absolutely continuous distributions and we interpret $P(\mathbf{x})$ and allied quantities as probability densities, the theorem holds provided we ensure that $\exp Q(\mathbf{x})$ is integrable over all $\mathbf{x}$. These additional requirements must not be taken lightly, as we shall see by examples in Section 4. Finally, we may consider the case of multivariate rather than univariate site variables. In particular, suppose that the random vector at site i has v_i components. Then we may replace that site by v_i notional sites, each of which is associated with a single component of the random vector. An appropriate system of neighbours may then be constructed and the univariate theorem be applied in the usual way. We shall not consider the multivariate situation any further in the present paper.

As a straightforward corollary to the theorem, it may easily be established that for any given Markov field

$$P(X_i = x_i, X_j = x_j, \ldots, X_s = x_s \,|\, \text{all other site values})$$

depends only upon $x_i, x_j, \ldots, x_s$ and the values at sites neighbouring sites

$i, j, \ldots, s$. In the Hammersley–Clifford terminology, the *local* and *global* Markovian properties are equivalent.

In practice, we shall usually find that the sites occur in a finite region of Euclidean space and that they often fall naturally into two sets: those which are internal to the system and those which form its boundary (or boundaries). In constructing a Markov field, it is quite likely that we are able to make reasonable assumptions concerning the conditional distribution associated with each of the internal sites but that problems arise at the boundary of the system. Such problems may usually be by-passed by considering the joint distribution of the internal site variables conditional upon fixed (observed) boundary values. We need then only specify the neighbours and associated conditional probability structure for each of the internal sites in order to define uniquely the above joint distribution. This is a particularly useful approach for lattice systems.

The positivity condition remains as yet unconquered and it would be of considerable theoretical interest to learn the effect of its relaxation. On the other hand, it is probably fair to say that the result would be of little practical significance in the analysis of spatial interaction with given site locations.

Finally, we note that, for discrete variables, a further proof of the Hammersley–Clifford theorem has been given by Grimmett (1973). This is apparently based upon the Möbius inversion theorem (Rota, 1964). Other references on the specification of Markov fields include Averintsev (1970), Preston (1973) and Sherman (1973).

4. Some Spatial Schemes Associated with the Exponential Family

In the next two sections, we become more specific in our discussion of spatial schemes. The present section deals with a particular subclass of Markov fields and with some of the models which are generated by it, whilst Sections 5.1, 5.2 and 5.3 are more concerned with practical aspects of conditional probability models. In Section 5.4, the simultaneous autoregressive approach (Mead, 1971; Ord, 1974) to finite spatial systems is discussed, again from the conditional probability viewpoint. Finally, in Section 5.5, stationary auto-normal models on the infinite regular lattice are defined and compared with the stationary simultaneous autoregressions of Whittle (1954).

In the remainder of this paper, we shall use the function $p_i(\cdot)$ to denote the conditional probability distribution (or density function) of X_i given all other site values. Thus $p_i(\cdot)$ is a function of x_i and of the values at sites neighbouring site i. Wherever possible, the arguments of $p_i(\cdot)$ will be omitted.

4.1. Auto-Models

Given n sites, labelled $1, \ldots, n$, and the set of neighbours for each, we have seen in Section 3 how the Hammersley–Clifford theorem generates the class of valid probability distributions associated with the site variables $X_1, \ldots, X_n$. Within this general framework, we shall in Section 4.2 consider particular schemes for which $Q(\mathbf{x})$ is well defined and has the representation

$$Q(\mathbf{x}) = \sum_{1 \le i \le n} x_i G_i(x_i) + \sum\sum_{1 \le i < j \le n} \beta_{i,j} x_i x_j, \tag{4.1}$$

where $\beta_{i,j} = 0$ unless sites i and j are neighbours of each other. Such schemes will be termed *auto-models*.

In order to motivate this definition, it is convenient to consider the wider formulation below. Suppose we make the following assumptions.

Assumption 1. The probability structure of the system is dependent only upon contributions from cliques containing no more than two sites. That is, when well defined, the expansion (3.3) becomes

$$Q(\mathbf{x}) = \sum_{1 \le i \le n} x_i G_i(x_i) + \sum\sum_{1 \le i < j \le n} x_i x_j G_{i,j}(x_i, x_j), \tag{4.2}$$

where $G_{i,j}(\cdot) \equiv 0$ unless sites i and j are neighbours.

Assumption 2. The conditional probability distribution associated with each of the sites belongs to the *exponential family* of distributions (Kendall and Stuart, 1961, p. 12). That is for each i,

$$\ln p_i(x_i; \ldots) = A_i(\cdot)B_i(x_i) + C_i(x_i) + D_i(\cdot), \tag{4.3}$$

where the functions B_i and C_i are of specified form and A_i and D_i are functions of the values at sites neighbouring site i. A valid choice of A_i determines the type of dependence upon neighbouring site values and D_i is then the appropriate normalizing function.

It is shown in Section 4.3 that as a direct consequence of Assumptions 1 and 2. A_i must satisfy

$$A_i(\cdot) \equiv \alpha_i + \sum_{j=1}^{n} \beta_{i,j} B_j(x_j), \tag{4.4}$$

where $\beta_{j,i} \equiv \beta_{i,j}$ and $\beta_{i,j} = 0$ unless sites i and j are neighbours of each other. Hence, it follows, when appropriate, that $G_{i,j}$ in equation (4.2) has the form

$$G_{i,j}(x_i, x_j) \equiv \beta_{i,j} H_i(x_i) H_j(x_j), \tag{4.5}$$

where $x_i H_i(x_i) = B_i(x_i) - B_i(0)$. Thus we generate the class of auto-models by making the additional requirement that, for each i, the function B_i is linear in x_i.

Superficially, auto-models might appear to form quite a useful subclass of Markov fields. Assumption 1 is not only satisfied for any rectangular lattice "nearest-neighbour" scheme but can also be taken as a fairly natural starting point in much wider lattice and non-lattice situations. Further, the linearity of B_i is satisfied by the most common members of the exponential family. However, the assumptions are, in fact, so restrictive, as seen through equation (4.4), that they often produce models which, in the end result, are devoid of any intuitive appeal at all. In Section 4.2, a range of auto-models has been included and hopefully illustrates both ends of the spectrum. Practical applications of two of the models will be discussed in later sections.

It is clear that, in terms of equation (4.1), auto-models have conditional probability structure satisfying

$$p_i(x_i;\ldots)/p_i(0;\ldots) = \exp\left[x_i\left\{G_i(x_i) + \sum_{j=1}^{n} \beta_{i,j}x_j\right\}\right], \tag{4.6}$$

where again $\beta_{j,i} \equiv \beta_{i,j}$ and $\beta_{i,j} = 0$ unless sites i and j are neighbours of each other. The models can further be classified according to the form which $p_i(\cdot)$ takes and this leads to the introduction of terms such as *auto-normal*, *auto-logistic* and *auto-binomial* to describe specific spatial schemes.

In the subsequent discussion, it will be assumed, unless otherwise stated, that any parameters $\beta_{i,j}$ are at least subject to the conditions following equation (4.6). Ranges of summation will be omitted wherever possible and these should then be apparent by comparison with equation (4.1) or (4.6).

4.2. Some Specific Auto-Models

4.2.1. *Binary Schemes*

For any finite system of binary (zero–one) variables, the only occasions upon which a given non-null G-function can contribute to $Q(\mathbf{x})$ in the expansion (3.3) are those upon which each of its arguments is unity. We may therefore replace all non-null G-functions by single arbitrary parameters, without any loss of generality, and this leads to the multivarite logistic models of Cox (1972). One would hope in practice that only a fairly limited number of non-zero parameters need to be included. In particular, if the only non-zero parameters are those associated with cliques consisting of single sites and of pairs of sites, we have an auto-logistic model for which we may write

$$Q(\mathbf{x}) = \sum \alpha_i x_i + \sum\sum \beta_{i,j}x_i x_j. \tag{4.7}$$

It follows that

$$p_i(\cdot) = \exp\left\{x_i\left(\alpha_i + \sum \beta_{i,j}x_j\right)\right\} \Big/ \left\{1 + \exp\left(\alpha_i + \sum \beta_{i,j}x_j\right)\right\}, \tag{4.8}$$

analogous to a classical logistic model (Cox, 1970, Chapter 1), except that here the explanatory variables are themselves observations on the process.

4.2.2. *Gaussian Schemes*

In many practical situations, especially those arising in plant ecology, it is reasonable to assume that the joint distribution of the site variables (plant yields), possibly after suitable transformation, is multivariate normal. It is evident that any such scheme is an *auto-normal* scheme. In particular, we shall consider schemes for which

$$p_i(\cdot) = (2\pi\sigma^2)^{-1/2} \exp\left[-\tfrac{1}{2}\sigma^{-2}\left\{x_i - \mu_i - \sum \beta_{i,j}(x_j - \mu_j)\right\}^2\right]. \tag{4.9}$$

Using the factorization (2.2) or otherwise, this leads to the joint density function,

$$P(\mathbf{x}) = (2\pi\sigma^2)^{-(1/2)n}|\mathbf{B}|^{1/2} \exp\{-\tfrac{1}{2}\sigma^{-2}(\mathbf{x}-\boldsymbol{\mu})^{\mathrm{T}}\mathbf{B}(\mathbf{x}-\boldsymbol{\mu})\}, \tag{4.10}$$

where $\boldsymbol{\mu}$ is the $n \times 1$ vector of arbitrary finite means, μ_i, and $\mathbf{B}$ is the $n \times n$ matrix whose diagonal elements are unity and whose off-diagonal (i, j) element is $-\beta_{i,j}$. Clearly $\mathbf{B}$ is symmetric but of course we also require $\mathbf{B}$ to be positive definite in order for the formulation to be valid.

At this point, it is perhaps worth indicating the distinction between the process (4.9) defined above, for which

$$E(X_i\,|\,\text{all other site values}) = \mu_i + \sum \beta_{i,j}(x_j - \mu_j), \tag{4.11}$$

and the process defined by the set of n *simultaneous* autoregressive equations, typically

$$X_i = \mu_i + \sum \beta_{i,j}(X_j - \mu_j) + \varepsilon_i, \tag{4.12}$$

where $\varepsilon_1, \ldots, \varepsilon_n$ are independent Gaussian variates, each with zero mean and variance σ^2. In contrast to equation (4.10), the latter process has joint probability density function,

$$P(\mathbf{x}) = (2\pi\sigma^2)^{-(1/2)n}|\mathbf{B}| \exp\{-\tfrac{1}{2}\sigma^{-2}(\mathbf{x}-\boldsymbol{\mu})^{\mathrm{T}}\mathbf{B}^{\mathrm{T}}\mathbf{B}(\mathbf{x}-\boldsymbol{\mu})\}, \tag{4.13}$$

where $\mathbf{B}$ is defined as before. Also, it is no longer necessary that $\beta_{j,i} \equiv \beta_{i,j}$, only that $\mathbf{B}$ should be non-singular. Further aspects of simultaneous autoregressive schemes will be discussed in Sections 5.4 and 5.5.

4.2.3 *Auto-Binomial Schemes*

Suppose that X_i has a conditional binomial distribution with associated "sample size" m_i and "probability of success" θ_i which is dependent upon the neighbouring site values. Then $H_i(x_i) \equiv 1$ and, under Assumption 1, the odds of "success" to "failure" must satisfy

$$\ln\{\theta_i/(1-\theta_i)\} = \alpha_i + \sum \beta_{i,j}x_j.$$

When $m_i = 1$ for all i, we again have the auto-logistic model.

4.2.4. *Auto-Poisson Schemes*

Suppose that X_i has a conditional Poisson distribution with mean μ_i dependent upon the neighobouring site values. Again $H_i(x_i) \equiv 1$ and, under Assumption 1, μ_i is subject to the form

$$\mu_i = \exp\left(\alpha_i + \sum \beta_{i,j}x_j\right).$$

Further, since the range of X_i is infinite, we must ensure that exp $Q(\mathbf{x})$ is summable over $\mathbf{x}$. We show below that this requires the further restriction $\beta_{i,j} \leq 0$ for all i and j.

We have

$$Q(\mathbf{x}) = \sum\{\alpha_i x_i - \ln(x!)\} + \sum\sum \beta_{i,j}x_i x_j.$$

Clearly exp $Q(\mathbf{x})$ must be summable when each $\beta_{i,j} = 0$ so the same holds when each $\beta_{i,j} \leq 0$. To show the necessity of the condition, we consider the distribution of the pair of variates (X_1, X_2) given that all other site values are equal to zero. The odds of the realization (x_1, x_2) to the realization (0,0) are then

$$\exp Q(x_1, x_2, 0, \ldots, 0) = \exp(\alpha_1 x_1 + \alpha_2 x_2 + \beta_{1,2}x_1 x_2)/(x_1!x_2!),$$

for non-negative integers x_1 and x_2. We certainly require that the sum of this quantity over all x_1 and x_2 converges and this is only true when $\beta_{1,2} \leq 0$. Similarly, we require $\beta_{i,j} \leq 0$ for all i and j. This restriction is severe and necessarily implies a "competitive" rather than "co-operative" interaction between auto-Poisson variates.

4.2.5. *Auto-Exponential Schemes*

Suppose that X_i has a conditional negative exponential distribution with mean μ_i dependent upon the values at sites neighbouring site i. Once more $H_i(x_i) \equiv 1$ and, under Assumption 1, μ_i must take the form $(\alpha_i + \sum \beta_{i,j}x_j)^{-1}$. The scheme is valid provided $\alpha_i > 0$ and $\beta_{i,j} \geq 0$ but the conditional probability structure appears to lack any form of intuitive appeal. Analogous statements hold for all gamma-type distributions.

4.3. Proof of Equation (4.4)

In order to establish the result (4.4) under Conditions 1 and 2, we begin by assuming that ln $p_i(0; \ldots)$ is well behaved, relaxing this condition later. For convenience, we shall write A_i and D_i of equation (4.3) as functions of

$$(x_i, \ldots, x_{i-1}, 0, x_{i+1}, \ldots, x_n)$$

although in reality they depend only upon the values at sites neighbouring site i. Since ln $p_i(0; \ldots)$ is well behaved, $Q(\mathbf{x})$ is well defined (under the positivity condition) and has the representation (4.2) according to Assumption 1. Equations (4.2) and (4.3) may now be related through equation (3.2). Putting $x_j = 0$ for all $j \neq i$, we obtain, for each i.

$$x_i G_i(x_i) = A_i(\mathbf{0})\{B_i(x_i) - B_i(0)\} + C_i(x_i) - C_i(0). \tag{4.14}$$

Now suppose sites 1 and 2 are neighbours of each other. Putting $x_j = 0$ for $j \geq 3$ and again using equation (3.2) to link (4.2) and (4.3), we obtain, for $i = 1$,

$$x_1 G_1(x_1) + x_1 x_2 G_{1,2}(x_1, x_2) = A_1(0, x_2, 0, \ldots, 0)\{B_1(x_1) - B_1(0)\} \\ + C_1(x_1) - C_1(0)$$

and, for $i = 2$.

$$x_2 G_2(x_2) + x_1 x_2 G_{1,2}(x_1, x_2) = A_2(x_1, 0, \ldots, 0)\{B_2(x_2) - B_2(0)\} \\ + C_2(x_2) - C_2(0).$$

Combining these two equations with (4.14), we deduce that

$$x_1 x_2 G_{1,2}(x_1, x_2) = \beta_{1,2}\{B_1(x_1) - B_1(0)\}\{B_2(x_2) - B_2(0)\},$$

where $\beta_{1,2}$ is a constant. More generally, if sites i and j are neighbours and $i < j$,

$$x_i x_j G_{i,j}(x_i, x_j) = \beta_{i,j}\{B_i(x_i) - B_i(0)\}\{B_j(x_j) - B_j(0)\}. \tag{4.15}$$

The result (4.4) is easily deduced from (4.14) and (4.15).

The condition that $\ln p_i(0; \ldots)$ is well behaved is not satisfied by all members of the exponential family. However, in cases where $\ln p_i(0; \ldots)$ degenerates as, for example, with most gamma distributions, we may use a simple transformation on the X_i's to affirm that (4.4) still holds. Suppose, without loss of generality, that $0 < p_i(1; \ldots) < \infty$ and in that case let $Y_i = \ln X_i$ at each site. Then the conditional probability structure of the process $\{Y_i\}$ also lies within the exponential family of distributions but there is no degeneracy associated with the value $Y_i = 0$. The previous arguments may then be applied to show that A_i still satisfies equation (4.4).

5. Some Two-Dimensional Spatial Schemes and Their Applications

5.1. Finite Lattice Schemes

In practice, the construction of conditional probability models on a finite regular lattice is simplified by the existence of a fairly natural hierarchy in the choice of neighbours for each site. For simplicity, and because it occurs most frequently in practice, we shall primarily discuss the rectangular lattice with sites defined by integer pairs (i, j) over some finite region. Where the notation becomes a little unwieldy, the reader may find it helpful to sketch and label the sites appropriately. The simplest model which allows for local stochastic interaction between the variates $X_{i,j}$ is then the *first-order* Markov scheme (or "nearest-neighbour" model) in which each interior site (i, j) is deemed to have four neighbours, namely $(i-1, j), (i+1, j), (i, j-1)$ and $(i, j+1)$. If, as suggested in Section 3, we now interpret $Q(\mathbf{x})$ as being concerned with the distribution of the internal site variables conditional upon given boundary values, the representation (3.3) in the Hammersley–Clifford theorem can be written

$$Q(\mathbf{x}) = \sum x_{i,j}\phi_{i,j}(x_{i,j}) + \sum x_{i,j}x_{i+1,j}\psi_{1,i,j}(x_{i,j}, x_{i+1,j})$$

$$+ \sum x_{i,j}x_{i,j+1}\psi_{2,i,j}(x_{i,j}, x_{i,j+1}), \tag{5.1}$$

where $\{\phi_{i,j}\}$, $\{\psi_{1,i,j}\}$ and $\{\psi_{2,i,j}\}$ are arbitrary sets of functions, subject to the summability of $Q(\mathbf{x})$, and the ranges of summation in (5.1) are such that each clique, involving at least one site internal to the system, contributes a single term to the representation. Writing (x, t, t', u, u') for the partial realization

$$(x_{i,j}, x_{i-1,j}, x_{i+1,j}, x_{i,j-1}, x_{i,j+1}),$$

the conditional probability structure at the site (i, j) is given by

$$p_{i,j}(x; t, t', u, u) = \exp\{f_{i,j}(x; t, t', u, u')\} \Big/ \sum \exp\{f_{i,j}(z; t, t', u, u')\}, \tag{5.2}$$

where

$$f_{i,j}(\cdot) = x\{\phi_{i,j}(x) + t\psi_{1,i-1,j}(t, x) + t'\psi_{1,i,j}(x, t')$$

$$+ u\psi_{2,i,j-1}(u, x) + u'\psi_{2,i,j}(x, u')\}$$

and the summation, or integration in the case of continuous variates, extends over all values z, possible at (i, j). In any given practical situation, the ϕ-, ψ_1- and ψ_2-functions can then be chosen to give an appropriate distributional form for $p_{i,j}(\cdot)$. For the scheme to be spatially homogeneous, these functions must be independent of position (i, j) on the lattice. We then have the special case discussed by Besag (1972a). If, further, $\psi_1 = \psi_2$, the scheme is said to be isotropic.

The idea of a first-order scheme may easily be extended to produce higher-order schemes. Thus a *second-order* scheme allows (i, j) to have the additional neighbours $(i-1, j-1)$, $(i+1, j+1)$, $(i-1, j+1)$ and $(i+1, j-1)$, whilst a *third-order* scheme further includes the sites $(i-2, j)$, $(i+2, j)$, $(i, j-2)$ and $(i, j+2)$. To obtain $Q(\mathbf{x})$, we merely add a contributory term for each clique which involves at least one site internal to the system. For example, a homogeneous second-order scheme has

$$\begin{aligned} Q(\mathbf{x}) = &\sum x_{i,j}\phi(\cdot) + \sum x_{i,j}x_{i+1,j}\psi_1(\cdot) + \sum x_{i,j}x_{i,j+1}\psi_2(\cdot) \\ &+ \sum x_{i,j}x_{i+1,j+1}\psi_3(\cdot) + \sum x_{i,j}x_{i+1,j-1}\psi_4(\cdot) \\ &+ \sum x_{i,j}x_{i+1,j}x_{i,j+1}\xi_1(\cdot) + \sum x_{i,j}x_{i+1,j}x_{i+1,j+1}\xi_2(\cdot) \\ &+ \sum x_{i,j}x_{i+1,j}x_{i+1,j-1}\xi_3(\cdot) + \sum x_{i,j}x_{i,j+1}x_{i+1,j+1}\xi_4(\cdot) \\ &+ \sum x_{i,j}x_{i+1,j}x_{i,j+1}x_{i+1,j+1}\delta(\cdot), \end{aligned}$$

where the arguments of each function are its individual multipliers; thus

$$\delta(\cdot) \equiv \delta(x_{i,j}, x_{i+1,j}, x_{i,j+1}, x_{i+1,j+1})$$

and so on. In specific examples, the apparent complexity of the expressions may be very much reduced. However, it is felt that, unless the variables are Gaussian, third- and higher-order schemes will almost always be too unwieldy to be of much practical use.

First- and second-order schemes may easily be constructed for other lattice systems in two or more dimensions. Amongst these, the plane triangle lattice is of particular interest, firstly because it frequently occurs in practice and secondly because a first-order scheme on a triangular lattice, for which each internal site has six neighbours, is likely to be more realistic than the corresponding scheme on a rectangular lattice.

5.2. Specific Finite Lattice Schemes

5.2.1. *Binary Data*

It is clear from equation (5.1) that the homogeneous first-order scheme for zero-one variables on a rectangular lattice is given by

$$Q(\mathbf{x}) = \alpha \sum x_{i,j} + \beta_1 \sum x_{i,j}x_{i+1,j} + \beta_2 \sum x_{i,j}x_{i,j+1},$$

where α, β_1 and β_2 are arbitrary parameters. This leads to the conditional probability structure

$$p_{i,j}(x; t, t', u, u') = \frac{\exp[x\{\alpha + \beta_1(t+t') + \beta_2(u+u')\}]}{1 + \exp\{\alpha + \beta_1(t+t') + \beta_2(u+u')\}}, \tag{5.3}$$

in the notation of Section 5.1. The scheme is necessarily auto-logistic.

For the second-order scheme, there are cliques of sizes three and four and there is no longer any need for the scheme to be auto-logistic. Thus, if we additionally write (v, v', w, w') for the partial realization $(x_{i-1,j-1}, x_{i+1,j-1}, x_{i-1,j+1}, x_{i+1,j+1})$, we find that $p_{i,j}(\cdot)$ is now given by an expression similar to (5.3) but with the terms in curly brackets { } replaced by

$$\alpha + \beta_1(t + t') + \beta_2(u + u') + \gamma_1(v + v') + \gamma_2(w + w') + \xi_1(tu + u'w + w't') + \xi_2(tv + v'u' + ut') + \xi_3(tw + w'u + u't') + \xi_4(tu' + uv + v't') + \eta(tuv + t'u'v' + tu'w + t'uw'). \quad (5.4)$$

The scheme is only auto-logistic if the ξ- and η-parameters are all zero.

Incidentally, this is a convenient point at which to mention the first-order binary scheme on a triangular lattice, for this can be thought of as a scheme on a rectangular lattice in which (i, j) has the *six* neighbours $(i - 1, j)$, $(i + 1, j)$, $(i, j - 1)$, $(i, j + 1)$, $(i - 1, j - 1)$ and $(i + 1, j + 1)$. The homogeneous first-order scheme is thus obtained from (5.4) by putting $\gamma_2 = \xi_1 = \xi_3 = \eta = 0$. The scheme is auto-logistic only if, in addition, $\xi_2 = \xi_4 = 0$.

Regarding applications of the rectangular lattice models, we shall, in Section 7.1, analyse Gleaves's *Plantago lanceolata* data using the first- and second-order isotropic auto-logistic schemes. However, none of the sets of data, cited in Section 1, appears to provide a convincing demonstration of low-order auto-logistic behaviour. It is hoped that more "appropriate" sets of data will become available in the future. A number of remarks are made in this context. Firstly, in order to carry out a detailed statistical analysis of spatial interaction, rather than merely test for independence or estimate the parameters of a model, it is usually the case that fairly extensive data are required. For example, spatial models have been fitted to Greig-Smith's data by Bartlett (1971b), using the spectral approximation technique of Bartlett and Besag (1969), and by Besag (1972c), using the coding technique of Section 6. The respective models are similar, though not equivalent, and each appears to give a fairly satisfactory fit. However, the last statement should be viewed with some scepticism since the goodness-of-fit tests available for such a small system (24 × 24) are very weak. This will be illustrated by the more detailed analysis of Gleaves's data.

Secondly, it is stressed that the lower-order homogeneous schemes, under discussion here, have been specifically designed with *local* stochastic interaction in mind; in particular, it is unreasonable to apply them in situations where there is evidence of gross heterogeneity over the lattice. For example, the hop plant data of Freeman (1953) display a fairly clear dichotomy between the upper and lower halves of the lattice, the former being relatively disease free (Bartlett, 1974). Thirdly, the use of lattice schemes on Greig-Smith's and Gleaves's data is, of course, an artifice: as remarked in Section 1, these examples are really concerned with spatial pattern rather than spatial interaction. Furthermore, as is well known, the size of quadrat used when collecting such data can profoundly influence the results of the subsequent statistical analysis. Incidentally, from a numerical viewpoint, it is most efficient to arrange the quadrat size so that 0's and 1's occur with approximately equal frequency. An alternative procedure might be to adopt some sort of nested analysis (Greig-Smith, 1964).

The criticisms above are not intended to paint a particularly gloomy picture but merely to point out some limitations of the models. It is maintained that auto-logistic analyses can be useful in practice; the models, having once been established, are easy to interpret and, even when rejected, can aid an understanding of the data and of the underlying spatial situation.

5.2.2. *Gaussian Variables*

It has already been stated in Section 2.2 that auto-normal schemes are of relevance to many ecological situations. For a finite rectangular lattice system, two homogeneous schemes are of particular practical interest. They are the first-order scheme for which $X_{i,j}$, given all other site values, is normally distributed with mean

$$\alpha + \beta_1(x_{i-1,j} + x_{i+1,j}) + \beta_2(x_{i,j-1} + x_{i,j+1}) \tag{5.5}$$

and constant variance σ^2 and the second-order scheme for which $X_{i,j}$, given all other site values, is normally distributed with mean

$$\begin{aligned} &\alpha + \beta_1(x_{i-1,j} + x_{i+1,j}) + \beta_2(x_{i,j-1} + x_{i,j+1}) \\ &\quad + \gamma_1(x_{i-1,j-1} + x_{i+1,j+1}) + \gamma_2(x_{i-1,j+1} + x_{i+1,j-1}) \end{aligned} \tag{5.6}$$

and constant variance, σ^2. Such schemes can, for example, be used for the analysis of crop yields in uniformity trials when, perhaps through local fluctuations in soil fertility or the influence of competition, it is no longer reasonable to assume statistical independence. This is illustrated in Section 7, using the classical wheat plots data of Mercer and Hall (1911).

In more general experimental situations, it is possible to set up inhomogeneous auto-normal schemes to account for stochastic interaction between the variables. For example, one can replace α in the expressions (5.5) and (5.6) by $\alpha_{i,j}$, allowing this to depend deterministically upon the treatment combination at (i, j), in the usual way. Such schemes can still be analysed by the coding methods which will be discussed in Section 6. It is suggested that there is a need for further research here, particularly into the use of specially constructed experimental designs which take advantage of both the model and the coding analysis.

At this point, it is perhaps worth while anticipating the results of Section 5.4 in order to re-emphasize the distinction between the present approach and that based upon *simultaneous* autoregressive schemes. Removing means for simplicity, suppose we consider the scheme defined by the equations,

$$X_{i,j} = \beta_1 X_{i-1,j} + \beta_1' X_{i+1,j} + \beta_2 X_{i,j-1} + \beta_2' X_{i,j+1} + \varepsilon_{i,j} \tag{5.7}$$

over some finite region, with appropriate adjustments at the boundary of the system, where the $\varepsilon_{i,j}$'s are independent Gaussian error variates with common variance. The analogous scheme on a finite triangular lattice has been examined by Mead (1967). It might well be assumed that, at least when

$\beta_1' = \beta_1$ and $\beta_2' = \beta_2$, the conditional expectation structure of the process (5.7) would tally with the expression (5.5), putting $\alpha = 0$. However, this is not at all the case: in fact, the process (5.7) has conditional expectation structure defined by

$$\begin{aligned}
&(1 + \beta_1^2 + \beta_1'^2 + \beta_2^2 + \beta_2'^2)E(X_{i,j}\,|\,\text{all other site values}) \\
&\quad = (\beta_1 + \beta_1')(x_{i-1,j} + x_{i+1,j}) + (\beta_2 + \beta_2')(x_{i,j-1} + x_{i,j+1}) \\
&\qquad - (\beta_1\beta_2' + \beta_1'\beta_2)(x_{i-1,j-1} + x_{i+1,j-1}) - (\beta_1\beta_2 + \beta_1'\beta_2')(x_{i-1,j+1} + x_{i-1,j-1}) \\
&\qquad - \beta_1\beta_1'(x_{i-2,j} + x_{i+2,j}) - \beta_2\beta_2'(x_{i,j-2} + x_{i,j+2}),
\end{aligned} \tag{5.8}$$

consistent with a special case (since there are only four independent β-parameters rather than six) in the class of *third-order* auto-normal schemes. The peculiar conditional expectation structure arises because of the bilateral nature of the autoregression: that is, in contrast with the unilateral time series situation, $\varepsilon_{i,j}$ is *not* independent of the remaining right-hand-side variables in (5.7). Some previous comments concerning the conditional probability structure of simultaneously defined schemes have been made by Bartlett (1971b), Besag (1972a) and Moran (1973a, b).

5.3. Non-Lattice Systems

We now turn to the construction of models for which there are a finite number of irregularly distributed, but co-planar, sites. As stated in Section 1, we shall only be concerned here with the distribution of the site variables X_i $(i = 1, \ldots, n)$, given the knowledge of their respective locations, and not with an investigation of the spatial pattern associated with the sites themselves. The first problem is in the choice of neighbours for each site. If the sites comprise a finite system of closed irregular regions in the form of a mosaic, such as counties or states in a country, it will usually be natural to include as neighbours of a given site i, those sites to which it is adjacent. In addition, it may be felt necessary to include more remote sites whose influence is, nevertheless, felt to be of direct consequence to the site i variable.

Alternatively, if the sites constitute a finite set of irregularly distributed points in the plane, a rather more arbitrary criterion of neighbourhood must be adopted. However, the situation can be reduced to the preceding one if we can find an intuitively plausible method of defining appropriate territories for each site. One possibility is to construct the Voronyi polygons (or Dirichlet cells) for the system. The polygon of site i is defined by the union of those points in the plane which lie nearer to site i than to any other site. This formulation clearly produces a unique set of non-overlapping convex territories, often capable of a crude physical or biological interpretation. It appears to have been first used in practice by Brown (1965) in a study of local stand density in pine forests. Brown interpreted the polygon of any particular tree as defining the "area potentially available" to it. If, in general, two sites are deemed to be neighbours only when their polygons are adjacent, it is evident that each internal site must have at least three neighbours and that cliques of more than four sites cannot occur. With this definition, Brown's pine trees each have approximately six neighbours, as might commonly be expected in situations where competitive influences tend to produce a naturally or artificially imposed regularity on the pattern of sites. A slight, but artificial, reduction in complexity occurs if we further stipulate that in order for two sites to be neighbours, the line joining them must pass through their common polygon side. Cliques can then contain no more than three members.

Mead (1971) and Ord (1974) have each used the Voronyi polygons of a system to set up and examine simultaneous autoregressive schemes such as (4.12).

Whatever the eventual choice of the neighbourhood criterion, we may derive the most general form for the available conditional probability structure in any particular situation by applying the Hammersley–Clifford theorem. Some specific schemes have been given in Section 4.2. In particular, we discuss the use of the auto-normal scheme (4.9). The first task is to reduce the dimensionality of the parameter space by relating the μ_i's and $\beta_{i,j}$'s in some intuitively reasonable way. In the case of point sites, suppose the Voronyi polygons are constructed and that $d_{i,j}$ represents the distance between neighbouring sites i and j whilst $l_{i,j}$ represents the length of their common polygon side. It is then often feasible to relate each μ_i and non-zero $\beta_{i,j}$ to the corresponding $d_{i,j}$ and $l_{i,j}$. The symmetry property of the $\beta_{i,j}$'s arises naturally. Specific suggestions for use in the scheme (4.12) have been made by Mead (1971) in the context of plant competition models. Analogous suggestions are made by Ord (1974) in a geographical context. These suggestions could equally be implemented in the case of conditional probability models.

5.4. Simultaneous Autoregressive Schemes

At various stages, reference has been made to the simultaneous autoregressive schemes (4.12) and (5.7). We now determine their associated conditional probability structure since this is a facet of the models which has occasionally been misunderstood in the past. In fact, it is convenient to widen the formulation somewhat by considering schemes of the form

$$\sum b_{i,j}X_j = Z_i \tag{5.9}$$

for $i = 1, \ldots, n$, or, in matrix notation, $\mathbf{BX} = \mathbf{Z}$. where $\mathbf{B}$ is an $n \times n$ non-singular matrix and $\mathbf{Z}$ is a vector of independent continuous random variables. In practice, the matrix $\mathbf{B}$ will often be fairly sparse. We neither demand that the Z_i's are identically distributed nor that they are Gaussian. Let $f_i(\cdot)$

denote the density function of Z_i. Then $X_1, \ldots, X_n$ have joint density,

$$P(\mathbf{x}) = \|\mathbf{B}\| f_1(\mathbf{b}_1^{\mathrm{T}}\mathbf{x}) f_2(\mathbf{b}_2^{\mathrm{T}}\mathbf{x}) \cdots f_n(\mathbf{b}_n^{\mathrm{T}}\mathbf{x}),$$

where $\mathbf{b}_i^{\mathrm{T}}$ denotes the ith row of $\mathbf{B}$. The conditional probability structure at site i is then immediately obtainable from equation (3.2) or an analogue thereof. In particular, the result (5.8) is easily deduced.

More generally, suppose we say that site $j \neq k$ is *acquainted* with site k if and only if, for some i, $b_{i,j} \neq 0$ and $b_{i,k} \neq 0$; that is, if and only if at least one of the equations (5.9) depends upon both X_j and X_k. Then it is easily seen that the conditional distribution of X_k can at most depend upon the values at sites acquainted with site k. That is, the neighbours of any site are included in its acquaintances.

In a given practical situation, the sets of acquaintances and neighbours of a site may well be identical but this is not necessarily so. Suppose, for example, we consider the process

$$X_{i,j} = \beta_1 X_{i-1,j} + \beta_2 X_{i,j-1} + \beta_3 X_{i-1,j-1} + \varepsilon_{i,j}, \tag{5.10}$$

defined, for convenience, over a $p \times q$ finite rectangular torus lattice, where the $\varepsilon_{i,j}$'s are independent Gaussian variables with zero means and common variances. Then (i, j) has acquaintances $(i-1, j)$, $(i+1, j)$, $(i, j-1)$, $(i, j+1)$, $(i-1, j+1)$, $(i+1, j-1)$, $(i-1, j+2)$ and $(i+1, j-2)$ provided β_1, β_2 and β_3 are non-zero. In general, these sites will also constitute the set of neighbours of (i, j). However, suppose $\beta_3 = \beta_1\beta_2$; then the sites $(i-1, j+1)$ and $(i+1, j-1)$ are no longer neighbours of (i, j). In fact, we shall find in Section 6 that this result provides a useful approach to problems of statistical inference, for it enables unilateral approximations to first-order auto-normal schemes to be constructed.

5.5. Stationary Auto-Normal Processes on an Infinite Lattice

We define a stationary Gaussian process $\{X_{i,j}: i, j = 0, \pm 1, \ldots\}$ to be a finite-order auto-normal process on the infinite rectangular lattice if it has autocovariance generating function (a.c.g.f.) equal to

$$K\Big(1 - \sum\sum b_{k,l} z_1^k z_2^l\Big)^{-1}, \tag{5.11}$$

where (i) only a finite number of the real coefficients $b_{k,l}$ are non-zero, (ii) $b_{0,0} = 0$, (iii) $b_{-k,-l} \equiv b_{k,l}$ and (iv) $\sum\sum b_{k,l} z_1^k z_2^l < 1$ whenever $|z_1| = |z_2| = 1$. K is a constant and is related to the variance of $X_{i,j}$. The ranges of summation are to be taken as $-\infty$ to $+\infty$ unless otherwise stated.

The existence of such processes was demonstrated by Rosanov (1967) and in certain special cases by Moran (1973a, b). Moran (1973b) included a simplified account of some of Rosanov's paper and we shall use this below to discuss the structure of the schemes. Firstly, however, we reintroduce the concept of neighbourhood. That is, for a stationary Gaussian process with a.c.g.f. of the form (5.11), we define the site $(i-k,\ j-l)$ to be a neighbour of (i,j) if and only if $b_{k,l} \neq 0$. We now show that this accords with our finite system definition.

It follows from equation (5.11) that, provided $|r|+|s|>0$,

$$\rho_{r,s} = \sum\sum b_{k,l}\rho_{r-k,s-l}, \tag{5.12}$$

where $\rho_{r,s}$ denotes the autocorrelation of lags r and s in i and j, respectively. Now let $\{\varepsilon_{i,j}: i,\ j=0, \pm 1, \ldots\}$ be a doubly infinite set of variates defined by

$$X_{i,j} = \alpha + \sum\sum b_{k,l}X_{i-k,j-l} + \varepsilon_{i,j}, \tag{5.13}$$

where $\alpha = (1-\sum\sum b_{k,l})\mu$ and $\mu = E(X_{i,j})$. Then the $\varepsilon_{i,j}$'s are stationary Gaussian variables with zero means and common variances σ^2, say. Also the equations (5.12) imply that $\varepsilon_{i,j}$ and $X_{i',j'}$ are uncorrelated provided $|i-i'|+|j-j'|>0$. This result together with (5.13), implies the following: given the values at any finite set of sites which includes the neighbours of (i,j), $X_{i,j}$ has conditional mean $\alpha+\sum\sum b_{k,l}x_{i-k,j-l}$ and conditional variance σ^2 independent of the actual surrounding values.

Thus, we have confirmed that the present criterion of neighbourhood is consistent with that for finite systems and that the properties of stationary, infinite lattice, auto-normal schemes are in accordance with those of the homogeneous, finite lattice schemes. In particular, we may define first-, second- and higher-order schemes analogous to those appearing in Section 5.2.2.

Finally, we make some remarks concerning the infinite lattice schemes proposed by Whittle (1954). Removing means for simplicity, Whittle considered simultaneously defined stationary processes in the class

$$\sum\sum a_{k,l}X_{i-k,j-l} = Z_{i,j}, \tag{5.14}$$

where $\{Z_{i,j}: i,\ j=0, \pm 1, \ldots\}$ is a doubly infinite set of independent Gaussian variates, each with zero mean and variance v. The scheme (5.14) has a.c.g.f.

$$v\left(\sum\sum a_{k,l}z_1^k z_2^l\right)^{-1}\left(\sum\sum a_{k,l}z_1^{-k}z_2^{-l}\right)^{-1}. \tag{5.15}$$

If the number of non-zero coefficients $a_{k,l}$ is finite, we shall refer to (5.14) as being a "finite-order" Whittle scheme. It is clear from (5.15) that any such scheme has a finite-order auto-normal representation. The converse is in general untrue: for example, even the first-order auto-normal scheme does not have a finite-order simuitaneous autoregressive representation unless β_1 or $\beta_2 = 0$. One is therefore led to pose the following question: when using finite-order schemes in the statistical analysis of spatial data, are there *a priori* reasons for restricting attention to the particular finite-order schemes generated by (5.14) or should the wider range of auto-normal models be considered? Note that when the number of sites is finite, there is, for Gaussian variates, a complete, but somewhat artificial, correspondence between the classes of simultaneous and conditional probability models.

A further point which is relevant, whether the number of sites is finite or infinite, is illustrated by the following example. The most general bilateral scheme used by Whittle in examining the wheat plot data of Mercer and Hall (1911) was the infinite lattice analogue of (5.7), namely

$$X_{i,j} = \beta_1 X_{i-1,j} + \beta_1' X_{i+1,j} + \beta_2 X_{i,j-1} + \beta_2' X_{i,j+1} + Z_{i,j}. \tag{5.16}$$

Firstly, this process again has the rather peculiar conditional expectation structure (5.8), but secondly, as noted by Whittle, there is an ambiguity in the identity of parameters. That is, if we interchange β_1 and β_1' and also β_2 and β_2', we obtain a process with the identical probability structure. For the scheme (4.12), the same holds true if we interchange $\mathbf{B}$ and $\mathbf{B}^{\mathrm{T}}$. This seems rather unsatisfactory. In the time series situation, the problem does not arise if one invokes the usual assumption that past influences future, not vice versa. With spatial schemes, the problem can be overcome if we are content to examine merely the conditional probability structure of the process, given by equation (5.8) in the present context. It is suggested that such considerations again support the use of the conditional probability approach to spatial systems. A further comment appears in Section 7 of the paper.

We note that first- and second-order stationary auto-normal schemes on the infinite lattice were first proposed by Lévy (1948) but that, as remarked by Moran (1973a), existence was assumed without formal justification. Moran himself concentrates almost exclusively on the first-order scheme.

6. Statistical Analysis of Lattice Systems

In this section, we propose some methods of parameter estimation and some goodness-of-fit tests applicable to spatial Markov schemes defined over a rectangular lattice. The methods may be extended to other regular lattice systems, notably the triangular lattice, and, in part, to some non-lattice situations. In practice, it would appear that, amongst lattice schemes, it is the ones of first and second order which are of most interest and it is these upon which we shall concentrate. It has already been established in Section 2 that, generally speaking, a direct approach to statistical inference through maximum likelihood is intractable because of the extremely awkward nature of the normalizing function. We therefore seek alternative techniques. The exceptional case occurs when the variates have an auto-normal structure, for which the normalizing function may often be evaluated numerically without too much effort, even in some non-lattice situations. Each of the methods will be illustrated in Section 7 of the paper.

6.1. Coding Methods on the Rectangular Lattice

We assume, in the notation of Section 4, that the conditional distributions, $p_{i,j}(\cdot)$, are of a given functional form but collectively contain a number of unknown parameters whose values are to be estimated on the basis of a single realization, **x**, of the system. Coding methods of parameter estimation were introduced by Besag (1972c), in the context of binary data, but they are equally available in more general situations.

In order to fit a first-order scheme, we begin by labelling the interior sites of the lattice, alternately $\times$ and $\cdot$, as shown in Figure 1. It is then immediately clear that, according to the first-order Markov assumption, the variables associated with the $\times$ sites, given the observed values at all other sites, are mutually independent. This results in the simple conditional likelihood,

$$\prod p_{i,j}(x_{i,j}; x_{i-1,j}, x_{i+1,j}, x_{i,j-1}, x_{i,j+1}),$$

for the $\times$ site values, the product being taken over all $\times$ sites. Conditional maximum-likelihood estimates of the unknown parameters can then be obtained in the usual way. Alternative estimates may be obtained by maximizing the likelihood function for the $\cdot$ site values conditional upon the remainder (or, that is, using a unit shift in the coding pattern). The two procedures are likely to be highly dependent but, nevertheless, it is reasonable, in practice, to carry out both and then combine the results appropriately.

In order to estimate the parameters of a second-order scheme, we may code the internal sites as shown in Figure 2. Again considering the joint distribution of the $\times$ site variables given the $\cdot$ site values, we may obtain conditional maximum-likelihood estimates of the parameters. By performing shifts of the entire coding framework over the lattice, four sets of estimates are available and these may then be combined appropriately.

Using the coding methods, we may easily construct likelihood-ratio tests to examine the goodness of fit of particular schemes. Here, we stress three points. Firstly, it is highly desirable that the wider class of schemes against which we test is one which has intuitive spatial appeal, otherwise the test is likely to be weak. This is, of course, an obvious comment but one which, in the limited statistical work on spatial analysis, has sometimes been neglected. Secondly, the two maximized likelihoods we obtain must be strictly comparable. For example, if the fit of a scheme of first order is being examined against one of second order, the resulting likelihood-ratio test will only be valid if both the schemes have been fitted to the same set of data—

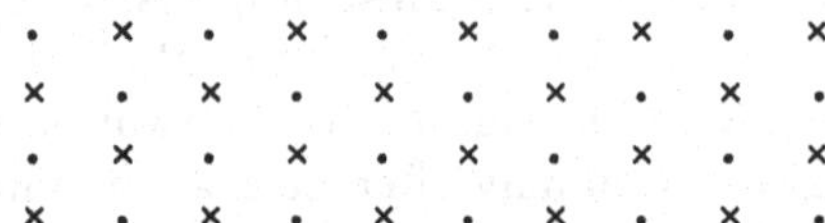

Figure 1. Coding pattern for a first-order scheme.

```
.  ×  .  ×  .  ×  .  ×  .  ×
.  .  .  .  .  .  .  .  .  .
.  ×  .  ×  .  ×  .  ×  .  ×
.  .  .  .  .  .  .  .  .  .
```

Figure 2. Coding pattern for a second-order scheme.

that is, using the Figure 2 coding in each case. Thirdly, there will be more than one test available (under shifts in coding) and these should be considered collectively. Whilst precise combination of the results may not be possible, they can usually be amalgamated in some conservative way. These points will be illustrated in Section 7.

The efficiency of coding techniques can to a limited extent be investigated following the methods of Ord (1974). Also of relevance are the papers by Ogawara (1951), Williams (1952) and Hannan (1955a, b) and some comments by Plackett (1960, p. 121), all on coding methods for Markov chains. The coding techniques will not, in general, be fully efficient but their great advantage lies in their simplicity and flexibility. Some results will be reported elsewhere but further investigation of the techniques is still required.

6.2. Unilateral Approximations on the Rectangular Lattice

An alternative estimation procedure for homogeneous first-order spatial schemes involves the construction of a simpler process which has approximately the required probability structure but which is much easier to handle. The approach is similar (equivalent for stationary auto-normal schemes) to that of Bartlett and Besag (1969). We begin by defining the set of *predecessors* of any site (i, j) in the positive quadrant to consist of those sites (k,l) on the lattice which satisfy either (i) $l < j$ or (ii) $l = j$ and $k < i$. We may then generate a *unilateral* stochastic process $\{X_{i,j}: i > 0, j > 0\}$ in the positive quadrant by specifying the distribution of each variable $X_{i,j}$ conditional upon the values at sites which are predecessors of (i, j). In practice, we shall allow the distribution of $X_{i,j}$ to depend only on a limited number of predecessor values. Such a process is a natural extension of a classical one-dimensional finite auto-regressive time series into two dimensions and is well defined if sufficient initial values are given. Special cases of such schemes have been discussed by Bartlett and Besag (1969), Bartlett (1971b) and Besag (1972b). By a judicious choice of the unilateral scheme, we may obtain a reasonable approximation to a given first-order spatial scheme. The more predecessor values we allow $X_{i,j}$ to depend upon, the better the approximation can be made. The great advantage of a unilateral scheme is that its likelihood function is easily written down and parameter estimation may be effected by straightforward maximum likelihood.

As the simplest general illustration, we consider unilateral processes of the form

$$P(x_{i,j}\,|\,\text{all predecessors}) = q(x_{i,j}; x_{i-1,j}, x_{i,j-1}).$$

The joint probability distribution of the variables $X_{i,j}$ $(1 \leq i \leq m, 1 \leq j \leq n)$ is given by

$$\prod_{i=1}^{m}\prod_{j=1}^{n} q(x_{i,j}; x_{i-1,j}, x_{i,j-1})$$

and, hence, for any interior site (i, j) we have, in the notation of Section 4, the bilateral structure

$$\frac{p_{i,j}(x; \ldots)}{p_{i,j}(x^*; \ldots)} = \frac{q(x; t, u)q(t'; x, w')q(u'; w, x)}{q(x^*; t, u)q(t'; x^*, w')q(u'; w, x^*)}.$$

That is, the conditional distribution, $p(x_{i,j}\,|\,\text{all other site values})$, depends not only upon $x_{i-1,j}$, $x_{i+1,j}$, $x_{i,j-1}$ and $x_{i,j+1}$ but also upon $x_{i-1,j+1}$ and $x_{i+1,j-1}$. Nevertheless, the primary dependence is upon the former set of values and, by a suitable choice of $q(\cdot)$, we may use the unilateral process as an approximation to a given homogeneous first-order spatial scheme. For a better approximation, we may consider unilateral processes of the form,

$$P(x_{i,j}\,|\,\text{all predecessors}) = q(x_{i,j}; x_{i-1,j}, x_{i,j-1}, x_{i-1,j-1})$$

and so on. The method will be illustrated for an auto-normal scheme in Section 7*.

6.3. Maximum-Likelihood Estimation for Auto-Normal Schemes

We begin by considering the estimation of the parameters in an auto-normal scheme of the form (4.9) but subject to the restriction $\boldsymbol{\mu} = \mathbf{0}$. We assume that the dimensionality of the parameter space is reduced through $\mathbf{B}$ having a particular structure and that σ^2 is both unknown and independent of the $\beta_{i,j}$'s. For a given realization $\mathbf{x}$, the corresponding likelihood function is then equal to

$$(2\pi\sigma^2)^{-1/2n}|\mathbf{B}|^{1/2} \exp(-\tfrac{1}{2}\sigma^{-2}\mathbf{x}^{\mathrm{T}}\mathbf{B}\mathbf{x}). \tag{6.1}$$

It follows that the maximum-likelihood estimate of σ^2 will be given by

$$\hat{\sigma}^2 = n^{-1}\mathbf{x}\hat{\mathbf{B}}x. \tag{6.2}$$

once $\hat{\mathbf{B}}$, the maximum-likelihood estimate of $\mathbf{B}$, has been found. Substituting (6.2) into (6.1), we find that $\hat{\mathbf{B}}$ may be obtained by minimizing

$$-n^{-1}\ln|\mathbf{B}| + \ln(\mathbf{x}^{\mathrm{T}}\mathbf{B}\mathbf{x}). \tag{6.3}$$

* Not reprinted here.

The problem of implementing maximum-likelihood estimation therefore rests upon the evaluation of the determinant, $|\mathbf{B}|$. We now examine how this relates to existing research into simultaneous autoregressions.

Suppose then that we temporarily abandon the auto-normal model above and decide instead to fit a simultaneous scheme of the form (4.12), again subject to $\boldsymbol{\mu} = \mathbf{0}$ and with $\mathbf{B}$ having the same structure as in (6.1). Provided (6.1) is valid so is the present, but different, scheme. The likelihood function now becomes

$$(2\pi\sigma^2)^{-1/2n}|\mathbf{B}|\exp(-\tfrac{1}{2}\sigma^{-2}\mathbf{x}^{\mathrm{T}}\mathbf{B}^{\mathrm{T}}\mathbf{B}\mathbf{x}) \tag{6.4}$$

and the new estimate of $\mathbf{B}$ must be found by minimizing

$$-2n^{-1}\ln|\mathbf{B}| + \ln(\mathbf{x}^{\mathrm{T}}\mathbf{B}^{\mathrm{T}}\mathbf{B}\mathbf{x}). \tag{6.5}$$

Again the only real difficulty centres upon the evaluation of the determinant $|\mathbf{B}|$, a point which we may, in a sense, now turn to advantage. Suppose that we wish to fit the auto-normal scheme associated with (6.1) to a given set of data. Then it follows that we may use existing approaches to fitting simultaneous autoregressive schemes provided that these can cope with the likelihood function (6.4). Indeed with minor modifications, we may use any existing computer programs. It is probably fair to say that thus far the simultaneous and conditional probability schools have tended to suggest the same structure for $\mathbf{B}$ in a given problem. This, together with the previous remarks, implies that it would be relatively straightforward to conduct a useful comparative investigation of the two approaches for some given sets of data.

As regards minimizing (6.5), computational progress has been made by Mead (1967) on small (triangular) lattices and by Ord (1974) in non-lattice situations where the number of sites is fairly limited (about 40 or less) and there are only one or two unknown parameters determining $\mathbf{B}$. The reader is referred to their papers for further details. As regards large lattices for which we may sometimes view the data as being a partial realization of a stationary Gaussian infinite lattice process, we may use the semi-analytical result of Whittle (1954) which is summarized below.

Whittle showed, for the simultaneous autoregression (5.14), that, given a partial realization of the process over n sites, the term $n^{-1}\ln|\mathbf{B}|$ in (6.5) can be approximated by the coefficient of $z_1^0 z_2^0$ in the power series expansion of

$$\ln\left(\sum\sum a_{k,l} z_1^k z_2^l\right).$$

There is no complication if the variates have equal, but non-zero, means. Thus, in order to fit a particular auto-normal scheme of the form (5.11) from a partial realization of the process over n sites, we need to minimize (6.3), where $n^{-1}\ln|\mathbf{B}|$ is the absolute term in the power series expansion of

$$\ln\left(1 - \sum\sum b_{k,l} z_1^k z_2^l\right)$$

and where, neglecting boundary effects,

$$\mathbf{x}^{\mathrm{T}}\mathbf{B}\mathbf{x} = C_{0,0} - \sum\sum b_{k,l}C_{k,l}$$

and $C_{k,l}$ denotes the empirical autocovariance of lags k and l in i and j, respectively (cf. Whittle, 1954).

For example, with the first-order scheme, analogous to (5.5), we minimize

$$-\Lambda(\boldsymbol{\beta}) + \ln(C_{0,0} - 2\beta_1 C_{1,0} - 2\beta_2 C_{0,1}),$$

where $\Lambda(\boldsymbol{\beta})$ is the absolute term in the power series expansion of

$$\ln\{1 - \beta_1(z_1 + z_1^{-1}) - \beta_2(z_2 + z_2^{-1})\}.$$

With the second-order scheme, analogous to (5.6), we minimize

$$-\Lambda(\boldsymbol{\beta}, \boldsymbol{\gamma}) + \ln(C_{0,0} - 2\beta_1 C_{1,0} - 2\beta_2 C_{0,1} - 2\gamma_1 C_{1,1} - 2\gamma_2 C_{1,-1}),$$

where $\Lambda(\boldsymbol{\beta}, \boldsymbol{\gamma})$ is the absolute term in the power series expansion of

$$\ln\{1 - \beta_1(z_1 + z_1^{-1}) - \beta_2(z_2 + z_2^{-1}) - \gamma_1(z_1 z_2 + z_1^{-1}z_2^{-1}) - \gamma_2(z_1 z_2^{-1} + z_1^{-1}z_2)\}.$$

The absolute terms can easily be evaluated for given parameter values by appropriate numerical Fourier inversion. The expression (6.3) may be minimized by, for example, the Newton–Raphson technique. Convergence, in the limited work thus far, has been extremely rapid. A numerical example is included in Section 7.

Finally, we note an analogy between the fitting of stationary auto-normal schemes in the analysis of spatial data and the fitting of autoregressive schemes in classical time-series analysis. That is, considering a particular scheme in the class (5.11), suppose that the corresponding autocorrelations are denoted by $\rho_{k,i}$. Then the effect of large-sample maximum-likelihood estimation is to ensure perfect agreement between $\rho_{k,l}$ and the corresponding sample autocorrelation $r_{k,l}$ whenever $b_{k,l} \neq 0$ in the original formulation. Thus, for a first-order scheme, the fit ensures that $\rho_{1,0} = r_{1,0}$ and $\rho_{0,1} = r_{0,1}$. For the second-order scheme, we additionally fit $\rho_{1,1} = r_{1,1}$ and $\rho_{1,-1} = r_{1,-1}$. In general, there is no such interpretation for simultaneously defined autoregressions. This may suggest that the auto-normal schemes are, in fact, a more natural extension of classical temporal autoregressions to spatial situations.

8. Concluding Remarks

In the preceding sections, an attempt has been made to establish that a conditional probability approach to spatial processes is not only feasible but is also desirable. It has been suggested, firstly, that the conditional probability approach has greater intuitive appeal to the practising statistician than the alternative joint probability approach: secondly, that the existence of the

Hammersley–Clifford theorem has almost entirely removed any consistency problems and, further, can easily be used as a tool for the construction of conditional probability models in many situations: thirdly, that the basic lattice models under the conditional probability approach yield naturally to a very simple parameter estimation procedure (the coding technique) and, at least for binary and Gaussian variates, to straightforward goodness-of-fit tests. For Gaussian variates, maximum likelihood appears equally available for both simultaneous and conditional probability models of similar complexity. As regards the joint probability approach, it is not clear to the present author how, outside the Gaussian situation, the models are to be used in practice. How, for example, would Gleaves's binary data be analysed?

On the other hand, the two examples discussed in Section 7 of the paper are far from convincing in demonstrating that simple conditional probability schemes provide satisfactory models for spatial processes. It is felt to be pertinent that, in each case, the data were derived from regions of the plane rather than point sites. There is clearly a need for more practical analyses to be undertaken. Some alternative suggestions on the specification of lattice models for aggregated data would also be of great interest.

Acknowledgements

I am indebted to those authors who have allowed me to quote their work prior to publication and also to the participants in the Liverpool "Spatial Pattern and Interaction" meeting in November 1973 for valuable discussions, especially Tim Gleaves who collected the *Plantago lanceolata* data and carried out the associated preliminary computations. Finally, I should like to express my thanks to Professor M.S. Bartlett for his continued interest and encouragement.

References

Averintsev, M.B. (1970). On a method of describing complete parameter fields. *Problemy Peredachi Informatsii*, **6**, 100–109.

Bartlett, M.S. (1955). *An Introduction to Stochastic Processes*. Cambridge University Press, Cambridge.

Bartlett, M.S. (1967). Inference and stochastic processes. *J. Roy. Statist. Soc. Ser. A*. **130**, 457–477.

Bartlett, M.S. (1968). A further note on nearest neighbour models. *J. Roy. Statist. Soc. Ser. A*, **131**. 579–580.

Bartlett, M.S. (1971a). Physical nearest-neighbour models and non-linear time series. *J. Appl. Probab.*, **8**, 222–232.

Bartlett, M.S. (1971b). Two-dimensional nearest-neighbour systems and their ecological applications. *Statist. Ecology Ser.*, vol. 1, pp. 179–194. Pennsylvania State University Press.

Bartlett, M.S. (1974). The statistical analysis of spatial pattern. Presented at the Third Conference on Stochastic Processes and their Applications, Sheffield, August 1973.

Bartlett, M.S. and Besag, J.E. (1969). Correlation properties of some nearest-neighbour models. *Bull. Internat. Statist. Inst.*, **43**, Book 2, 191–193.

Batchelor, L.D. and Reed, H.S. (1918). Relation of the variability of yields of fruit trees to the accuracy of field trials. *J. Agrtc. Res.*, **12**, 245–283.

Besag, J.E. (1972a). Nearest-neighbour systems and the auto-logistic model for binary data. *J. Roy. Statist. Soc. Ser. B*, **34**, 75–83.

Besag, J.E. (1972b). On the correlation structure of some two-dimensional stationary processes. *Biometrika*, **59**, 43–48.

Besag, J.E. (1972c). On the statistical analysis of nearest-neighbour systems. Proceedings of the European Meeting of Statisticians. Budapest, August 1972 (to appear).

Brook, D. (1964). On the distinction between the conditional probability and the joint probability approaches in the specification of nearest-neighbour systems. *Biometrika*, **51**, 481–483.

Brown, G.S. (1965). Point density in stems per acre. *New Zealand. For. Serv. Res.*, Note **38**, 1–11.

Clarke, R.D. (1946). An application of the Poisson distribution. *J. Inst. Actuar.*, **72**, 481.

Cochran, W.G. (1936). The statistical analysis of the distribution of field counts of diseased plants. *J. Roy. Statist. Soc. Suppl.*, **3**, 49–67.

Cox, D.R. (1970). *Analysis of Binary Data.* Methuen, London.

Cox, D.R. (1972). The analysis of multivariate binary data. *Appl. Statist.*, **21**, 113–120.

Feller, W. (1957). *An Introduction to Probability Theory and its Applications*, Vol. 1. Wiley, New York.

Freeman, G.H. (1953). Spread of disease in a rectangular plantation with vacancies. *Biometrika*, **40**, 287–305.

Gleaves, J.T. (1973). Unpublished Ph.D. Thesis, University of Liverpool.

Greig-Smith, P. (1964). *Quantitative Plant Ecology*, 2nd ed. Butterworth, London.

Grimmett, G.R. (1973). A theorem about random fields. *Bull. London Math. Soc.*, **5**, 81–84.

Hammersley, J.M. and Clifford, P. (1971). Markov fields on finite graphs and lattices (unpublished).

Hannan, E.J. (1955a). Exact tests for serial correlation. *Biometrika*, **42**, 133–142.

Hannan, E.J. (1955b). An exact test for correlation between time series. *Biometrika*, **42**, 316–326.

Kendall, M.G. and Stuart, A. (1961). *The Advanced Theory of Statistics*, Vol. 2. Griffin, London.

Levy, P. (1948). Chaines doubles de Markoff et fonctions aléatoires de deux variables. *C.R. Acad. Sci. Paris*, **226**, 53–55.

Matui, I. (1968). Statistical study of the distribution of scattered villages in two regions of the Tonami plain, Toyami prefecture. In *Spatial Analysis* (Berry, B.J. and Marble, D.F., eds.). Prentice-Hall, Engelwood Cliffs, N.J.

Mead, R. (1966). A relationship between individual plant spacing and yield. *Ann. Bot.*, **30**, 301–309.

Mead, R. (1967). A mathematical model for the estimation of interplant competition. *Biometrics*, **23**, 189–205.

Mead, R. (1968). Measurement of competition between individual plants in a population. *J. Ecol.*, **56**, 35–45.

Mead, R. (1971). Models for interplant competition in irregularly distributed populations. In *Statistical Ecology*, Vol. 2, pp. 13–32. Pennsylvania State University Press.

Mercer, W.B. and Hall, A.D. (1911). The experimental error of field trials. *J. Agric. Sci.*, **4**, 107–132.

Moran, P.A.P. (1973a). A Gaussian Markovian process on a square lattice. *J. Appl. Probab.*, **10**, 54–62.

Moran, P.A.P. (1973b). Necessary conditions for Markovian processes on a lattice. *J. Appl. Probab.*, **10**, 605–612.

Ogawara, M. (1951). A note on the test of serial correlation coefficients. *Ann. Math. Statist.*, **22**, 115–118.

Ord, J.K. (1974). Estimation methods for models of spatial interaction. *J. Amer. Statist. Assoc.* (to appear).

O'Sullivan, P.M. (1969). *Transportation Networks and the Irish Economy*. L.S.E. Geographical Papers No. 4. Weidenfeld and Nicholson, London.

Patankar, V.N. (1954). The goodness of fit of frequency distributions obtained from stochastic processes. *Biometrika*, **41**, 450–462.

Plackett, R.L. (1960). *Principles of Regression Analysis*. Clarendon Press, Oxford.

Preston, C.J. (1973). Generalised Gibbs states and Markov random fields. *Adv. Appl. Probab.*, **5**, 242–261.

Rosanov, Yu.A. (1967). On the Gaussian homogeneous fields with given conditional distributions. *Theor. Probab. Appl.*, **12**, 381–391.

Rota, G.C. (1964). On the foundations of combinatorial theory. *Z. Wahrsch. Verw. Gebcete*, **2**, 340–368.

Sherman, S. (1973). Markov random fields and Gibbs random fields. *Israel J. Math.*, **14**, 92–103.

Spitzer, F. (1971). Markov random fields and Gibbs ensembles. *Amer. Math. Monthly*, **78**, 142–154.

Whittle, P. (1954). On stationary processes in the plane. *Biometrika*, **41**, 434–449.

Whittle, P. (1963). Stochastic processes in several dimensions. *Bull. Internat. Statist. Inst.*, **40**, 974–994.

Williams, R.M. (1952). Experimental designs for serially correlated observations. *Biometrika*, **39**, 151–167.

Introduction to Levit (1974) On the Optimality of Some Statistical Estimates

J.F. Pfanzagl

The Prehistory of the Subject

The intuitive appeal of maximum likelihood estimators (equivalently: Bayes estimators with uniform prior) made it seem easy to establish their optimality as a mathematical theorem, at least asymptotically. A first result in this direction is due to Edgeworth (1908), who proves the maximum likelihood estimator of a location parameter to be asymptotically optimal among all estimators which are solutions to estimating equations. Various attempts by Fisher did not achieve much more than generalizing Edgeworth's result from location parameters to arbitrary one-dimensional parameters. It took surprisingly long to realize that the idea of asymptotic efficiency as originally conceived was too naive. The reason for this is, perhaps, that the admirable intuition of the early statisticians was not kept in control by the requirement of mathematical rigor.

It became clear only later through the celebrated superefficiency example of Hodges (see LeCam (1953, p. 280)) that asymptotic optimality in the naive sense is not feasible: Any estimator sequence $\vartheta^{(n)}$ can be modified such that the asymptotic variance of $n^{1/2}(\vartheta^{(n)} - \vartheta_0)$ becomes 0 at a given parameter value ϑ_0 without changing its asymptotic performance for $\vartheta \neq \vartheta_0$.

Several approaches have been invented to rescue the idea of asymptotic optimality:

(i) To show that superefficiency can occur on a negligible part of the parameter set only.
(ii) To consider only estimators for which the distribution of $n^{1/2}(\vartheta^{(n)} - \vartheta)$ converges to a limit distribution locally uniformly in ϑ.

(iii) To prove a minimax theorem, say

$$\liminf_{n\to\infty} \sup_{\vartheta\in \mathbb{N}} E_\vartheta \ell(n^{1/2}(\vartheta^{(n)} - \vartheta)) \geq \int \ell(u)\, dN_{(0,I(\vartheta_0)^{-1})}(u)$$

for certain loss function ℓ and a neighborhood V of ϑ_0. $I(\vartheta_0)$ is the informative matrix.

Only remedies (ii) and (iii) can be easily extended to nonparametric models. Levit chooses approach (iii). A first version of the minimax theorem is due to LeCam (1953, p. 237, Theorem 14). Two decades later, this result was available (Hájek, 1972, p. 186, Theorem 4.1) under weak regularity conditions on the family $F_\vartheta, \vartheta \in \Theta \subset \mathbb{R}^k$, which guarantee LAN. For $k = 1$, this minimax theorem reads as follows

$$\lim_{a\to\infty} \liminf_{n\to\infty} \sup_{|\vartheta-\vartheta_0|<n^{-1/2}a} E_\vartheta \ell(n^{1/2}(\vartheta^{(n)} - \vartheta)) \geq \int \ell(u)\, dN_{(0,I(\vartheta_0)^{-1})}(u),$$

with the addendum that equality holds for some nonconstant loss functions ℓ if $n^{1/2}(\vartheta^{(n)} - \vartheta)$ is approximable in measure by $I(\vartheta)^{-1} n^{-1/2} \sum_{i=1}^n p^\bullet(x_i, \vartheta)/p(x_i, \vartheta)$.

By these results, an asymptotic bound for the concentration of estimators in parametric models was firmly established. This bound, expressed by intrinsic properties of the family $\{F_\vartheta: \vartheta \in \Theta\}$, does not refer to any particular estimator. Yet it gives a (partial) answer to the question originally posed: In highly regular models, maximum likelihood estimators do attain this bound and are, therefore, asymptotically optimal.

Nonparametric Bounds

Now let $\mathscr{F}$ be an arbitrary family of distributions, and let $\varphi: \mathscr{F} \to \mathbb{R}$ be some functional. For a long time, it was a matter of course that even in this general framework certain functionals (like quantiles or moments) admit $\sqrt{n}$-consistent estimators (usually the functional of the empirical distribution). There was, however, no asymptotic bound, determined by intrinsic properties of the family of distributions and the functional, and hence no means for judging the quality of such estimators.

Stein (1956, p. 188) suggested how such bounds could be obtained: To consider for each distribution $F_0 \in \mathscr{F}$ one-parametric subfamilies passing through F_0, and to determine the asymptotic bound in the nonparametric model from the largest bound among the one-parametric submodels. Of course, there is no guarantee that a nonparametric bound obtained in this way is large enough. Hence this procedure leads to a conclusive result only if estimators attaining this bound can be found.

Stein applies his idea to three examples, all of which are of the same type: They are—in modern terminology—semiparametric and adaptive (i.e., the bound obtained with the nonparametric component unknown is the same as the bound with the nonparametric component known). This particular situation makes, strictly speaking, the determination of nonparametric bounds superfluous. What would be needed are adaptive estimators (attaining—in the nonparametric model—the bound for the parametric model). For one of these examples, the estimation of the median of an unknown symmetric distribution, Stein suggests how such an estimator could be obtained. (This problem was first solved by van Eeden (1970). The best results now available are Stone (1975) and Beran (1978).)

Stein's heuristic idea escaped notice for almost two decades. It was Levit's paper (1974) in which this idea was converted into a workable technique to obtain bounds for the asymptotic performance of estimators of functionals defined on nonparametric families.

Let $\mathscr{F}$ denote the family of distributions, and let $\varphi\colon \mathscr{F} \to \mathbb{R}$ be the functional to be estimated. Given a distribution $F \in \mathscr{F}$, Levit considers paths through F, i.e., one-parametric families $F_h \in \mathscr{F}$, $|h| \leq h_0$, with $F_0 = F$. These paths are assumed to be Hellinger differentiable at F, i.e., there exists a Hellinger derivative $\dot{\xi}$, defined by

$$\lim_{h\to 0} E_h\left[h^{-1}\left(\sqrt{\frac{dF_h}{dF}}-1\right)-\dot{\xi}\right]^2 = 0. \tag{1}$$

Recall that $E_h(\dot{\xi}) = 0$ and $E_h(\dot{\xi}^2) < \infty$. ($\dot{\xi}$ depends on the path, of course, even though this does not find its expression in the notation.) The parametrization of the family F_h is standardized such that

$$\varphi(F_h) = \varphi(F) + h, \tag{2}$$

which leads to $(d/dh)\varphi(F_h) = 1$.

The infimum of $4E_F(\dot{\xi}^2)$ over all differentiable paths is denoted by $I(F)$. (The unusual factor 4 results from Levit's definition of the Hellinger derivative, with $\dot{\xi}$ instead of $\frac{1}{2}\dot{\xi}$ in (1).) If T_n is an estimator for $\varphi(F)$, Levit considers $T_n - \varphi(F)$ as an estimator for the parameter h in the one-parametric subfamily. Applying Hájek's minimax theorem to the estimation problem for the "most difficult" path [the one for which $E_F(\dot{\xi}^2)$ is maximal], Levit obtains (see pp. 220/1, Theorem 1.1) that

$$\liminf_{n\to\infty} \sup_{F\in V} E_F(\ell(n^{1/2}(T_n - \varphi(F)))) \geq \sup_{F\in V} \int \ell(u)\, dN_{(0,I(F)^{-1})}(u), \tag{3}$$

where V is an arbitrary open set. It is clear from the proof that it suffices to consider a sequence V_n of neighborhoods, shrinking to a given $F_0 \in \mathscr{F}$. In this case, it is natural to replace the right-hand side of (3) by $\int \ell(u) dN_{(0,I(F_0)^{-1})}(u)$.

If equality holds in (3) for some nonconstant loss function, then (see pp. 223, 224) $n^{1/2}(T_n - \varphi(F))$ is approximable in measure by $(2/I(F))n^{-1/2}$ $\sum_1^n \dot{\xi}(X_i)$.

This general result is illustrated by application to the functional

$$\varphi(F) = \int \cdots \int \phi(x_1, \ldots, x_m)\, dF(x_1) \ldots dF(x_m).$$

Levit shows (see inequalities (2.3) and (2.6)) that

$$I(F) = 1/m^2\sigma^2(F),$$

where

$$\sigma^2(F) = E_F(\varphi_1 - \varphi(F))^2,$$

and

$$\varphi_1(x) = \int \phi(x, x_2, \ldots, x_m)\, dF(x_2) \ldots dF(x_m).$$

Using earlier results of Hoeffding, Levit proves (p. 231) that this bound is attained by the relevant U-statistic.

Levit (1975) sums up the theoretical results from Levit (1974) and establishes the asymptotic optimality of minimum contrast estimators for minimum contrast functionals $\varphi(F)$, defined by $\int \psi(x, \varphi(F))\, dF(x) = 0$.

The final step in the elaboration of a nonparametric asymptotic theory follows in Koshevnik and Levit (1976). This paper introduces the concept of a "tangent cone" as the minimal closed subspace containing the derivatives of all differentiable paths passing through F. Moreover, the constrictive condition (2) is replaced by

$$\lim_{h \to \infty} h^{-1}[\varphi(F_h) - \varphi(F)] = E_F(\dot{\varphi}\dot{\xi}). \tag{4}$$

In this representation, $\dot{\xi}$ is the derivative of the path F_h; $\dot{\varphi}$ is the canonical gradient, i.e., the unique element of the tangent cone rendering a representation (4). (In fact, the functional now attains its values in $\mathbb{R}^k$. It is denoted by ϕ, and φ is used for the canonical gradient.)

It seems to be of interest that derivatives of functionals occur in statistical literature as early as 1955, in an attempt of Kallianpur and Rao (1955) to prove the asymptotic efficiency of maximum likelihood estimators for one-dimensional parameters. Koshevnik and Levit are aware of this paper, but they are considerate enough not to mention that it is technically insufficient and fails to establish the intended result.

The generalization to $\mathbb{R}^k$-valued functionals poses no problem. It provides the basis for the extension to infinite-dimensional functionals in later papers.

The papers mentioned above restrict themselves to a limited number of examples. In most of these examples, the family of distributions is "large," so that the tangent cone at F consists of all F-square-integrable functions with F-expectation zero. It was left to other authors to fully exploit the fruitfulness of this approach: To study families with more interesting tangent cones, and to determine bounds for a great many functionals.

The present state of affairs is reflected in the thorough monograph of Bickel et al. (1993).

About the Author

B.Ya. Levit was born in Russia, 1946. He obtained his Ph.D. degree in 1972 under the supervision of Hasminskii. The paper was published in the *Proceedings of the Prague Conference on Asymptotic Statistics*, organized by J. Hájek. Since Levit was unable to attend this conference, his paper was read by Chibisov. Since 1991, Levit has been Associate Professor at the University of Utrecht. His current interest is, still, in asymptotic statistics.

References

Beran, R. (1978). An efficient and robust adaptive estimator of location. *Ann. Statists.*, **6**, 292–313.

Bickel, P.J., Klaassen, Ch.A.J., Ritov, Y., and Wellner, J.A. (1993). *Efficient and Adaptive Estimation for Semiparametric Models*. John Hopkins University Press, Baltimore.

Edgeworth, F.Y. (1908, 1909). On the probable errors of frequency constants. *J. Roy. Statist. Soc.*, **71**, 381–397, 499–512, 651–678; Addendum, **72**, 81–90.

Hájek, J. (1972). Local asymptotic minimax and admissibility in estimation. *Proc. Sixth Berkeley Symp. Math. Statist. Probab.*, **1**, 175–194.

Kallianpur, G. and Rao, C.R. (1955). On Fisher's lower bound to asymptotic variance of a consistent estimate. *Sankyā*, **15**, 331–342.

LeCam, L. (1953). On some asymptotic properties of maximum likelihood estimates and related Bayes estimates. *Univ. California Publ. Statist.*, **1**, 277–320.

Levit, B.Ya. (1974). On optimality of some statistical estimates. *Proceedings of the Prague Symposium on Asymptotic Statistics*, Prague, Czechoslovakia Vol. 2 (J. Hájek, ed.), pp. 215–238. Charles University Press.

Levit, B.Ya. (1975). On the efficiency of a class of non-parametric estimates. *Theory Probab. Appl.*, **20**, 723–740.

Koshevnik, Yu.A., and Levit, B.Ya. (1976). On a non-parametric analogue of the information matrix. *Theory Probab. Appl.*, **21**, 738–753.

Stein, C. (1956). Efficient nonparametric testing and estimation. *Proc. Third Berkeley Symp. Math. Statist. Probab.*, Vol. 1, pp. 187–195.

Stone, Ch.J. (1975). Adaptive maximum likelihood estimators of a location parameter. *Ann. Statist.*, **3**, 267–284.

van Eeden, C. (1970). Efficiency-robust estimation of location. *Ann. Math. Statist.*, **41**, 172–181.

On Optimality of Some Statistical Estimates

B.Ya. Levit
Academy of Sciences, Moscow

Introduction

The progress recently achieved in parametric estimation is closely related to success in obtaining the information-type lower bounds for risks. In this regard, we can refer to paper [1], where some quite general and strong results have been obtained.

This comment is intended to illustrate the difficulties arising in nonparametric estimation. The lack of bounds analogous to the information inequalities gives no possibility of verifying the optimality of any nonparametric procedure and that deprives the whole theory of its basis.

Meantime the possibility of an application of some parametric methods to nonparametric problems was pointed out by Stein [2], with the aim of obtaining the information-type inequalities. It seems that Stein's idea, though stated heuristically, is rather fruitful. The following can be considered in essence as the development and some applications of this idea.

The results derived below are applicable to a class of nonparametric problems that can naturally be defined as *regular* ones. In Section 1, for this type of problem, a functional is defined playing the role of the classical Fisher information and coinciding with it in the special case of parametric estimation. In this way we arrive at an extension of the results of [1] to nonparametric problems and, following [1], define the notion of local asymptotic minimax (LAM) estimates. Moreover, a recent result due to Ibragimov and Has'minskii concerning sequential estimation [3] can also be extended showing that, in general, no asymptotical gain can be obtained by the use of sequential methods, for the regular estimation problems.

In this paper, we consider in detail only one application of the general results mentioned above, namely, to proving the LAM property for a well-known class of statistics, called U-statistics. Some remarks are needed here on the conception of optimality usually used in connection with these estimates. It is well known (see [4]) that U-statistics have the property of minimum risk, for any nonnegative convex loss function, in the class of all unbiased estimates of so-called "regular parameters" (see (2.1) below). This property sometimes is used as the definition of the efficiency of estimators. However, there is a number of reasons for which such definition, based on consideration of the unbiased estimators only, is unsatisfactory.

First of all, even in the estimation of "regular parameters," the class of unbiased estimators is rather small; any such estimator coincides with U-statistics after symmetrization [4]. Moreover, in a variety of well-known nonparametric problems, the unbiased estimators do no exist at all, and that gives additional reason for studying "biased" estimation.

On the other hand, we can recognize situations (e.g., see Theorem 3.2 below) when even in estimation of regular parameters there exist biased estimates, which are in a sense better than any unbiased estimator. Finally, in estimation, not only convex loss function are of interest (e.g., the loss function $1(x) = \text{sign}(|x| - a)$) which undermines again the advantage of unbiasedness.

In Section 2–3 we shall treat the optimality of U-statistics without the hypothesis of unbiasedness. The presented results establish their LAM property in the case when only minimum information is available about the unknown distribution (such as the existence of moments, smoothness of density function, and so on).

A remark at the end of Section 1 shows that the general approach allows us to consider both types of estimation problems—parametric and nonparametric—from a naturally unified point of view. Some other applications concerning estimates such as sample median, least squares estimates, and others will be considered in another paper.

1. Lower Bounds in Nonparametric Estimation

Let $X_1, X_2, \ldots$ be independent identically distributed observations in a measurable space $(\mathscr{X}, \mathscr{B})$ with an unknown distribution F. Suppose that F belongs to a class $\mathscr{F}$ of distributions on $\mathscr{B}$ and consider the problem of estimation of a real functional $\Phi(F)$, defined of $\mathscr{F}$.

Relating to the content of the problem at hand, and also to the appearance of $\mathscr{F}$ and $\Phi(F)$, there is often defined a measure of closeness on $\mathscr{F}$. But for the time being, we shall only assume that there is defined a topology $\mathscr{R}$ on $\mathscr{F}$.

The following condition is quite natural in statistical inference.

Condition 0. The functional $\Phi(F)$ is nonconstant on any open set $V \in \mathscr{R}$.

However, this condition is unsufficient for our purposes, and we shall impose a somewhat more restrictive

Condition 1. For any $F \in \mathscr{F}$ except, possibly, $F \in W$, where $W \subset \mathscr{F}$ is a closed set with int $W = \phi$, there exists a family $f = \{F_h\}$ of distributions in $\mathscr{F}(|h| < h_0 = h_0(f))$ such that

$$F_0 = F, \tag{1.1a}$$

$$\text{the family } f \text{ is continuous at } h = 0, \tag{1.1b}$$

$$\Phi(F_h) = \Phi(F) + h. \tag{1.1c}$$

It is clear that Condition 1 is related to Condition 0, being more restrictive. In fact, the following is easily established.

Proposition 1.1. *Let $(\mathscr{F}, \mathscr{R})$ be a subspace of linear topological space $(\mathscr{F}', \mathscr{R}')$, where $\mathscr{F}'$ is the linear hull of $\mathscr{F}$. Suppose that $\Phi(F)$ is defined on $\mathscr{F}'$ and admits a Gâteaux derivative*

$$\varphi(F, F') = \left.\frac{d\Phi(F + tF')}{dt}\right|_{t=0} \qquad (F' \in \mathscr{F}).$$

which is continuous in F.

Then Conditions 0 and 1 are equivalent.

Depending on how we interpret the estimation problem, the class $\mathscr{F}$ and the function $\Phi(F)$, there is usually a meaningful notion of closeness between elements of $\mathscr{F}$. For the time being, we just assume that a topology $\mathscr{R}$ is defined on $\mathscr{F}$.

Let f be a family, satisfying (1.1a, b), or a curve in $\mathscr{F}$, containing F. Denote by F_h^c the absolutely continuous part of F_h with respect to F and define

$$\xi_h = \sqrt{\frac{dF_h^c}{dF}(X)}, \tag{1.2}$$

where X is distributed according to F. Suppose that the random process ξ_h, defined for $|h| < h_0(f)$ possesses the mean square derivative at $h = 0$,

$$\lim_{h \to 0} E_F\{h^{-1}(\xi_h - 1) - \dot{\xi}\}^2 = 0. \tag{1.3}$$

Here and in the sequel the index F means that the expectation (or probability) corresponds to the distribution F.

Definition 1.1. Let (1.3) be fulfilled and let

$$\overline{\lim_{h \to 0}}\, h^{-2} \int (\sqrt{dF_h} - \sqrt{dF})^2 \le E_F \dot{\xi}^2. \tag{1.4}$$

Then call

$$I_f(F) = 4E_F\dot{\xi}^2$$

the information corresponding to the family f. Otherwise, let $I_f(F) = \infty$.

Definition 1.2. We shall call

$$I(F) = \inf I_f(F),$$

where inf is taken over all families, satisfying (1.1a)–(1.1c), the Fisher information in estimating $\Phi(F)$ or, briefly, information.

It is clear that $I(F)$ thus defined depends both on the distribution set $\mathscr{F}$ and on the topology $\mathscr{R}$. Note also that $I^{-1/2}(F)$ is in a sense the norm of the gradient of the functional $\Phi(F)$. Note also the following.

Remark 1.1. The condition (1.4) is closely related to (1.3). Actually, it follows from (1.3), if F_h is absolutely continuous with respect to F. Besides, both these conditions hold if the family F_h is dominated by a measure and the corresponding density P_h is absolutely continuous in h and admits continuous in h classical Fisher information $I(h)$ (see [1]).

Remark 1.2. Theorems 1.1–1.3 presented below still hold if the condition (1.4) is substituted by

$$\overline{\lim_{h\to 0}}\, h^{-2}\int(\sqrt{dF_h} - \sqrt{dF})^2 < \infty. \tag{1.5}$$

Under the last condition we obtain another quantity $I_1(F)$ in Definition 1.2,

$$I_1(F) \leq I(F).$$

But for our purposes it will be sufficient to deal with $I(F)$, which sometimes is more readily computable.

Now following [1], let us define the class L of loss functions $\ell(\cdot)$, which satisfy the following conditions:

$$\ell(y) = \ell(|y|),$$
$$\ell(y) \leq \ell(z), \qquad |y| \leq |z|,$$
$$\int_{-\infty}^{\infty} \ell(y)\exp\{-\tfrac{1}{2}\lambda y^2\}\,dy < \infty, \qquad \lambda > 0,$$
$$\ell(0) = 0.$$

The next result is heavily based on Theorem 4.1 of [1].

Theorem 1.1. *Let $T_n = T_n(X_1, \ldots, X_n)$ be any estimate of $\Phi(F)$ (a measur-*

able function of $X_1, \ldots, X_n$). Then for any open set $V \in \mathscr{R}$ such that

$$I_V = \inf_{F \in V} I(F) > 0, \tag{1.6}$$

$$\sup_{F \in V} E_F \ell(\sqrt{n}(T_n - \Phi(F))) \geq \sqrt{\frac{I_V}{2\pi}} \int \ell(x) \exp\left\{-\frac{I_V x^2}{2}\right\} dx + \sigma(1), \tag{1.7}$$

where $\sigma(1) \to 0$ as $n \to \infty$ uniformly in T_n.

Proof. Let $I_V < \infty$—otherwise the statement is trivial. For $F \in V$, let f be a family of distributions satisfying (1.1a)–(1.1c) for which

$$I_f(F) < I(F) + \delta, \qquad \delta > 0. \tag{1.8}$$

Then for a sufficiently small a we have

$$\begin{aligned} \sup_{F \in V} E_F \ell(\sqrt{n}(T_n - \Phi(F))) &\geq \sup_{|h|<a} E_{F_h} \ell(\sqrt{n}(T_n - \Phi(F_h))) \\ &\geq \sup_{|h|<a} E_{F_h} \ell(\sqrt{n}(T_n - \Phi(F) - h)) \\ &= \sup_{|h|<a} E_{F_h} \ell(\sqrt{n}(T'_n - h)). \end{aligned} \tag{1.9}$$

Here $T'_n = T_n - \Phi(F)$ can be treated as an estimator (a measurable function!) in estimating parameter h for the family F_h. Therefore we can use Theorem 4.1 of [1] to estimate the right-hand side of (1.9). It is necessary only to check the condition of this theorem—the so-called local asymptotic normality of the family F_h at $h = 0$. But this follows from [5], with the only exception that the Radon–Nikodym derivative $(dP_n, \Theta_n)/(dP_n, \Theta)$ in [5] must be replaced by the density of the absolutely continuous part of P_n, Θ_n with respect to $P_{n,\Theta}$. Note that the condition imposed in [5] that F_h be absolutely continuous with respect to F was used only to establish that

$$E_F \dot{\xi} = 0. \tag{1.10}$$

But in our case this follows from (1.3) and (1.4) (or (1.5)) (see [6]).

Therefore, according to [1],

$$\sup_{|h|<a} E_{F_h} \ell(\sqrt{n}(T'_n - h)) \geq \sqrt{\frac{I_f(F)}{2\pi}} \int \ell(x) \exp\left\{-\frac{I_f(F) x^2}{2}\right\} dx + e(1) \qquad (n \to \infty)$$

uniformly in T'_n or from (1.9)

$$\varliminf_n \inf_{T_n} \sup_{F \in V} E_F \ell(\sqrt{n}(T_n - \Phi(F))) \geq \sqrt{\frac{I_f(F)}{2\pi}} \int \ell(x) \exp\left\{-\frac{I_f(F) x^2}{2}\right\} dx.$$

Since $F \in V$ is arbitrary, $I_f(F)$ differs from $I(F)$ by an arbitrary δ (see (1.8)), and the last integral is continuous in $I_f(F)$

$$\varliminf_{n} \inf_{T_n} \sup_{F \in V} E_f \ell(\sqrt{n}(T_n - \Phi(F))) \geq \sqrt{\frac{I_V}{2\pi}} \int \ell(x) \exp\left\{-\frac{I_V x^2}{2}\right\} dx. \quad \square$$

The result of [1] mentioned in Theorem 1.1 has recently been extended by Ibragimov and Hasminskii in [3] to the case of sequential estimation, for the loss functions of the form $\ell(x) = |x|^\alpha, \alpha > 0$. Relying upon this result we obtain, much in the same way as in Theorem 1.1, the following statement.

Theorem 1.2. *Let $T_n = T(X_1, \ldots, X_\tau)$ be any sequential estimator for which the stopping time τ satisfies the relation*

$$\sup_{F \in V} E_F \tau \leq n. \tag{1.11}$$

Assume $\ell(x)$ be of the form $|x|^\alpha, \alpha > 0$, and (1.6) fulfilled. Then the relation (1.7) holds as $n \to \infty$ uniformly in T_n, satisfying (1.11).

Remark 1.3. It follows from the proof that if there exists sequence $F_k \in V$ such that $I(F_k) > 0$ while $I(F_k) \to 0$,

$$\varliminf_{n} \inf_{T_n} \sup_{F \in V} E_F \ell(\sqrt{n}(T_n - \Phi(F))) = \sup_{x} \ell(x)$$

for the estimates T_n, considered in Theorems 1.1–1.2.

A necessary condition has been found in [1] for an estimate to achieve equality in (1.7). From this it follows that an estimate with asymptotically minimal risk has to be asymptotically normal. Let us demonstrate that an analogous result holds also in nonparametric set-up.

We will say that a relation holds for a sufficiently small vicinity of F if there exists an open set V_0 containing F such that this relation is true with respect to the set $V_0 \cap V$, for any other vicinity V of F.

Let for a sequence of estimates $T_n = T_n(X_1, \ldots, X_n)$

$$\lim_{n} \sup_{G \in V} E_G \ell(\sqrt{n}(T_n - \Phi(G))) = \sqrt{\frac{I_V}{2\pi}} \int \ell(x) \exp\left\{-\frac{I_V x^2}{2}\right\} dx \tag{1.12}$$

for any sufficiently small vicinity of F.

Suppose there exists a sequence of curves $f_k = \{F_h^k\}, |h| < h_k, k = 1, 2, \ldots,$ satisfying conditions (1.1), (1.3), (1.4) with $\dot{\xi} = \dot{\xi}_k$, and a function $\dot{\xi}_F = \dot{\xi}_F(x), x \in X$, such that

$$\lim_{k} E_F \{\dot{\xi}_k - \dot{\xi}_F\}^2 = 0 \tag{1.13}$$

and

$$4E_F\{\dot{\xi}_F\}^2 = I(F).$$

Define

$$\Delta_{n,F} = \frac{2}{\sqrt{n}I(F)} \sum_{i=1}^{n} \dot{\xi}_F(X_i). \tag{1.14}$$

Theorem 1.3. *Let $I(F)$ be continuous (in the sense of $\mathscr{R}$), and $I(F) > 0$. Then for any T_n satisfying* (1.12)

$$\sqrt{n}(T_n - \Phi(F)) - \Delta_{n,F} \to 0, \qquad n \to \infty, \tag{1.15}$$

in P_F-probability, providing $\ell(x)$ is nonconstant.

Proof. Suppose (1.15) does not hold, i.e., for some $\delta > 0$ and a subsequence $m \in \infty$

$$P_F\{|\sqrt{m}(T_m - \Phi(F)) - \Delta_{n,F}| \geq 2\varepsilon\} \geq 2\varepsilon. \tag{1.16}$$

Since from (1.10), (1.13) we have

$$\lim_{k\to\infty} \lim_{n\to\infty} E_F\left\{\frac{1}{\sqrt{n}} \sum_{i=1}^{n} (\dot{\xi}_k(x_i) - \dot{\xi}_F(x_i))\right\}^2 = 0$$

and $I_{f_k}(F) \to I(F)$, it follows from (1.16) that for k and m sufficiently large

$$P_F\left\{\left|\sqrt{m}(T_m - \Phi(F)) - \frac{2}{\sqrt{m}I_{f_k}(F)} \sum_{i=1}^{m} \dot{\xi}_k(x_i)\right| > \varepsilon\right\} > \varepsilon. \tag{1.17}$$

Next consider the family F_h^k; Theorem 4.2 in [1] implies that if (1.17) holds, then there exists $\alpha > 0$ continuously depending only on ε and $I_{f_k}(F)$ such that

$$\lim_{\delta\to 0} \underline{\lim}_{m} \sup_{|h|<\sigma} E_{F_h} k\ell(\sqrt{m}(T_m - \Phi(F_h^k)))$$

$$\leq \sqrt{\frac{I_{f_k}(F)}{2\pi}} \int \ell(x) \exp\left\{-\frac{I_{f_k}(F)x^2}{2}\right\} dx + \alpha.$$

On the other hand, from (1.12) and the continuity of $I(F)$, it follows that for any $\gamma > 0$ and for k sufficiently large

$$\lim_{\delta\to 0} \overline{\lim_{n}} \sup_{|h|<\delta} E_{F_h} k\ell(\sqrt{n}(T_n - \Phi(F_n^k)))$$

$$\leq \sqrt{\frac{I_{f_k}(F)}{2}} \int \ell(x) \exp\left\{-\frac{I_{f_K}(F)x^2}{2}\right\} dx + \gamma.$$

The contradiction proves the theorem.

Assume that $0 < I_V < \infty$ for a sufficiently small neighbourhood of any $F \in \mathscr{F}$;[1] we arrive then at a definition analogous to that in the parametric case.

Definition 1.3. The sequence of the estimators $T_n = T_n(X_1, \ldots, X_n)$ is called $(\mathscr{F}, \mathscr{R}, \ell)$—locally asymptotically minimax (LAM) if for a sufficiently small neighbourbood V of any $F \in \mathscr{F}$

$$\underline{\lim}_n \left\{ \inf_{T_n'} \sup_{G \in V} E_G \ell(\sqrt{n}(T_n' - \Phi(G))) - \sup_{G \in V} E_G \ell(n(T_n - \Phi(G))) \right\} \geq 0,$$

where inf is taken w.r.t. to any estimate T_n' based on $X_1, \ldots, X_n$.

Replacing T_n and T_n' in this definition by sequential estimators satisfying (1.11), we obtain the analogous definition of the LAM sequential estimator.

From Theorems 1.1–1.3 it follows:

Consequence 1.1. Let (1.12) be fulfilled for any sufficiently small neighbourhood of $F \in \mathscr{F}$. Then T_n is $(\mathscr{F}, \mathscr{R}, \ell)$ locally asymptotically minimax. It is also $(\mathscr{F}, \mathscr{R}, \ell)$ LAM estimate in the class of all sequential procedures, satisfying (1.11), if $\ell(x)$ is of the form $|x|^\alpha$, $\alpha > 0$. Moreover, T_n is asymptotically normal $(\Phi(F), 1/nI(F))$ provided the conditions of Theorem 1.3 are fulfilled.

Concluding this section we will give a simple remark which will be useful later on in evaluating $I(F)$.

Remark 1.4. Let $f = \{F_h\}$ be a family of distributions satisfying (1.1a, b), (1.4), let $\Phi(F_h)$ be continuously differentiable in h,

$$\left. \frac{d\Phi(F_h)}{dh} \right|_{h=0} = b > 0$$

and

$$I_f(F) = a.$$

Then

$$I(F) \leq a/b^2.$$

The proof is easy and can therefore be omitted.

It remains to note that the parametric estimation problems can be treated as a special case of the general scheme. Specifically, the conditions in [1] are the ones under which the $I(F)$ defined above, in the case of a single real parameter, coincides with the classical Fisher information. For the case of a multidimensional parameter the derivation of an informational inequality by the use of method essentially similar to ours is presented in [2].

[1] Note that both cases when it fails may occur; it is natural to consider condition $0 < I_V < \infty$ as describing the *regular* class of estimation problems.

In the next sections we will apply the above results to verification of the asymptotic optimality of some common non-parametric procedures. The question of principle, but in general not of evidence, is that in these examples, the bounds obtained are exact, i.e. asymptotically unimprovable.

2. Estimation of Linear Functionals

We shall consider now the estimation of the functionals of the following type:

$$\Phi(F) = \int_{X \ldots X} \cdots \int \varphi(x_1, \ldots, x_m)\, dF(x_1) \ldots dF(x_m), \tag{2.1}$$

where φ is some real measurable function, which is supposed to be known to the observer. Without loss of generality we may assume this function to be symmetric in its arguments.

In this section we will derive the lower bounds for risk of any estimator of $\Phi(F)$. For this let us try to evaluate the quantity $I(F)$ defined in the previous section.

Denote

$$\begin{gathered}\varphi_1(x) = \varphi_{1,F}(x) = \int \cdots \int \varphi(x, x_2, \ldots, x_m)\, dF(x_2) \ldots dF(x_m), \\ \sigma^2(F) = E_F\{\varphi_1(x_1) - \Phi(F)\}^2.\end{gathered} \tag{2.2}$$

The conditions below will imply the finiteness of $\sigma^2(F)$.

Usually in evaluating the informational quantity $I(F)$ it occurs to be easier to give a lower bound for $I(F)$.

At this point the following statement is of interest, which we will present here without proof.

Theorem 2.1. *Let $E_G\varphi^2$ be bounded in G in some vicinity of F and $\sigma^2(F) > 0$. Then*

$$I(F) \geq \frac{1}{m^2\sigma^2(F)}. \tag{2.3}$$

Note that both the sign of equality and strict inequality may occur in (2.3). It depends on the information at hand about the distribution set $\mathcal{F}$ or, in other words, on how vast the set $\mathcal{F}$ is. However, we will be interested here in conditions for there to be equality in (2.3).

The meaning of the corresponding restrictions is easier to understand if it is assume that $m = 1$ and φ is bounded. For this case setting

$$\frac{dF_h}{dF} = 1 + h(\Psi(x) - E_F\Psi) \qquad (|h| < h_0 = h_0(\Psi)), \tag{2.4}$$

where

$$\Psi(x) = \sigma^{-2}(F)\varphi(x)$$

we have

$$\Phi(F_h) = \Phi(F) + h,$$

$$I_f(F) = E_F\{\Psi - E_F\Psi\}^2 = \sigma^{-2}(F).$$

Thus the equality follows in (2.3) (cf. Definition 1.2).

Of course if φ is unbounded, and the relation (2.4) does not define a family of distributions. Nevertheless, we can use an appropriate approximation of φ in this case.

Theorem 2.2. *Suppose that, together with a distribution F, the space* $(\mathscr{F}, \mathscr{R})$ *contains a continuous, at* $h = 0$, *family of distributions* f_n, $n = 1, 2, \ldots,$ *of the form* (2.4) *where* $\Psi = \Psi_n$ *is a sequence of bounded functions such that*

$$\lim_n E_F\{\Psi_n - \varphi_1\}^2 = 0. \tag{2.5}$$

Then

$$I(F) \leq \frac{1}{m^2\sigma^2(F)}. \tag{2.6}$$

Proof. It is clear that the family (2.4) satisfies (1.1a, b), (1.3), (1.4), and

$$\begin{aligned} I_{f_n}(F) &= E_F\{\Psi_n - E_F\Psi_n\}^2 = E_F\{\varphi_1 - E_F\varphi_1\}^2 + \sigma(1) \\ &= \sigma^2(F) + \sigma(1), \qquad n \to \infty. \end{aligned} \tag{2.7}$$

We have also

$$\begin{aligned} \Phi(F_h) - \Phi(F) &= \int \cdots \int (\varphi - \Phi(F)) \left[\prod_{i=1}^m dF_h(x_i) - \prod_{i=1}^m dF(x_i) \right] \\ &= \int \cdots \int (\varphi - \Phi(F)) \left[\prod_{i=1}^m (1 + h(\Psi_n(x_i) - E_F\Psi_n)) - 1 \right] \prod_{i=1}^m dF(x_i) \\ &= mh \int (\varphi_1 - \Phi(F))(\Psi_n - E_F\Psi_n)\, dF(x) + c_2h^2 + \cdots + c_m h^m, \end{aligned}$$

where

$$\max_{2 \leq i \leq m} |c_i| = c_n < \infty.$$

Recalling Remark 1.4 and setting for this

$$b = b_n = m \int (\varphi_1 - \Phi(F))(\Psi_n - E_F\Psi_n)\, dF(x) = \sigma^2(F)m + \sigma(1), \qquad n \to \infty,$$

we have in virtue of (2.7)

$$I(F) \leq \frac{I_{f_n}(F)}{b_n^2} = \frac{1}{m^2\sigma^2(F)} + \sigma(1), \qquad n \to \infty.$$

Since n is arbitrary here, the required inequality (2.6) follows.

From the last theorem we obtain, e.g., the following

Corollary 2.1. *Let $\mathscr{F}$ be the set of distributions with respect to which:* (a) *function $|\varphi_1|^s (s \geq 2)$ is uniformly integrable, or;* (b) *for a given $s > 2$ the moments $E_F|\varphi_1|^s$ are bounded, and the topology in $\mathscr{F}$ is defined by the distance in variation. Then the inequality* (2.6) *holds.*

Now it is interesting to see to what extent we can further restrict the space $\mathscr{F}$ and strengthen the topology $\mathscr{R}$ so that the statement of Theorem 2.2 still holds. Though it seems there is no final answer to this question let us consider the following example. Suppose there exist density function $p(\cdot)$ with respect to Lebesgue measure for the family $\mathscr{F}$ of distributions in R^1 which possesses, say, continuous bounded derivatives of order k. Also assume a distance is defined in $\mathscr{F}$, in the sense of closeness in uniform metric of density functions themselves and their k derivatives.

Corollary 2.2. *Let $(\mathscr{F}, \mathscr{R})$ be the space defined above, $\sigma^2(F) < \infty$ for $F \in \mathscr{F}$ and the function φ_1 has a continuous derivative of order k. Then the inequality* (2.6) *holds.*

This follows, clearly, from the possibility of approximating the function φ_1 in (2.5) by sufficiently smooth bounded functions Ψ_n.

Remark 2.1. Evidently the last statement also holds if, for example, it is additionally required that for $p \in \mathscr{F}$ there exists a finite Fisher information

$$\int \frac{(p'(x))^2}{p(x)}\, dx$$

and the convergence $p_\alpha \to p$ in the sense of $\mathscr{R}$ implies that

$$\int \left(\frac{p'_\alpha}{\sqrt{p_\alpha}} - \frac{p}{\sqrt{p}} \right)^2 dx \to 0.$$

Theorems 1.1, 1.2, and 2.2 imply the following.

Corollary 2.3. *Suppose the assumptions of Theorem 2.2 are fulfilled. Then for any open set $V \in \mathscr{R}$, loss function $\ell(\cdot) \in L$, and an arbitrary sequence of estimates T_n*

$$\varliminf_{n} \inf_{T_n} \sup_{F \in V} E_F \ell(\sqrt{n}(T_n - \Phi(F))) \geq \frac{1}{\sqrt{2\pi} m \sigma_V} \int \ell(x) \exp\left\{-\frac{x^2}{2m^2\sigma_V^2}\right\} dx,$$

where

$$\sigma_V^2 = \sup_{F \in V} \sigma^2(F).$$

The same inequality holds in the case of sequential estimates T_n, satisfying (1.11), for the loss function $\ell(x) = |x|^\alpha, \alpha > 0$.

3. LAM Property of U-Statistics

A well known class of statistics is generated by the estimators of the functional (2.1) called U-statistics. These are of the form

$$U_n = U_n(X_1, \ldots, X_n) = \frac{1}{\binom{n}{m}} \sum_{S_n} \varphi(x_{\alpha_1}, \cdots, x_{\alpha_m}), \tag{3.1}$$

where

$$S_n = \{(\alpha_1, \ldots, \alpha_m) \colon 1 \leq \alpha_1 < \alpha_2 < \cdots < \alpha_m \leq n\}.$$

In the pioneering paper [7] and in a number of works following it, different asymptotic properties of these estimates were proved. Our aim is to prove that U-statistics are locally asymptotically minimax for a wide class of loss functions. The main task will be in verifying (1.12). Note that simultaneously the unimprovability of bounds obtained in Section 2 will be established.

We will need an additional restriction on the set $\mathscr{F}$; namely we will assume below that the function φ^2 is uniformly integrable in some vicinity of each $F \in \mathscr{F}$.

Denote $\tilde{L}$ the subclass of function $\ell(x) \in L$ which are continuous at $x = 0$.

Theorem 3.1. *Let*

(1) $|\varphi|^2$ *is uniformly integrable in some vicinity of any $F \in \mathscr{F}$;*
(2) *the space $(\mathscr{F}, \mathscr{R})$ satisfies the assumption of Theorem 2.2; and*
(3) *for some $K > 0, \Delta > 0$, the function $\ell(\cdot) \in \tilde{L}$ satisfies inequality*

$$\ell(x) \leq k(1 + |x|^{2-\Delta}). \tag{3.2}$$

Then the estimator U_n defined by (3.1) *is $(\mathscr{F}, \mathscr{R}, \ell)$ locally asymptotically minimax.*

PROOF. Clearly it is sufficient to show that uniformly in a small vicinity of F

$$E_G\ell(\sqrt{n}(U_n - \Phi(G))) \to \frac{1}{\sqrt{2\pi}m\sigma(G)} \int \ell(x) \exp\left\{-\frac{x^2}{2m^2\sigma^2(G)}\right\} dx. \quad (3.3)$$

For this the following statement is appropriate, which is a specification of a result of [7]. Define $Z_n = \sqrt{n}(U_n - \Phi(G))$ and let $N(g)$ be a standard normal c.d.f.

Lemma 3.1. *For each* $\varepsilon > 0$

$$\lim_n \sup_{|g|>\varepsilon} \left| P_G\{Z_n < g\} - N\left(\frac{g}{m\sigma(G)}\right)\right| = 0 \quad (3.4)$$

uniformly in functions φ *and distributions* G *such that*

$$\lim_{M\to\infty} \sup_{\varphi,G} \int_{|\varphi_1|>M} \varphi_1^2(x)G(dx) = 0. \quad (3.5)$$

PROOF. Denote

$$Y_n = \frac{m}{\sqrt{n}} \sum_{\alpha=1}^{n} (\varphi_1(X_\alpha) - \Phi(G)).$$

According to [7]

$$E_G(Z_n - Y_n)^2 \le \frac{C}{n^2}, \quad (3.6)$$

where C depends only on $E_G\varphi^2(X_1, \ldots, X_m)$ and so far can be chosen independently in G for a small vicinity of F. Since for any $\varepsilon > 0$ and $|h| < \delta(\varepsilon)$

$$\sup_{\sigma|g|>\varepsilon} \left| N\left(\frac{g+h}{\sigma}\right) - N\left(\left(\frac{g}{\sigma}\right)\right)\right| < \varepsilon$$

it is sufficient to show, by the virtue of (3.6), that (3.4) holds with Z_n replaced by Y_n.

Let $\gamma > 0$. If $\sigma(G)$ does not exceed sufficiently small $\delta(\gamma, \varepsilon) > 0$, then for each n

$$\sup_{|g|>\varepsilon} \left| P_G\{Y_n < g\} - N\left(\frac{g}{m\sigma(G)}\right)\right| < \gamma.$$

Therefore it remains to consider those G for which $\sigma(G) \ge \delta(\gamma, \varepsilon)$.

Use the expansion of characteristic function in the form

$$Ee^{itX} = E\left\{1 + itx - \frac{t^2X^2}{2} + \theta(X)t^2X^2g(|tX|)\right\},$$

where $|\theta| < 1$, $0 \le g \le 1$, and $g(x)$ is a continuous, monotone for $x > 0$ function. From this it easily follows that

$$\sup_{\varphi,G} E_G|e^{itY_n} - e^{-(m^2\sigma^2(G)t^2/2)}| < t^2\nu_n(t),$$

where $v_n(t)$ is a bounded function tending to zero as $n \to \infty$ and sup corresponds to those φ, F for which (3.5) is satisfied. Now the validity of (3.4) for Y_n follows from the well-known result about the relation between closeness of distribution functions and their characteristic functions (see, for example, [8, p. 299]). Thus the statement of Lemma 3.1 is proved.

Now we have to show that φ_1^2 is also uniformly integrable in a vicinity of any F. Indeed,

$$\int_{|\varphi_1| \geq M} \varphi_1^2(X_1)\, dP_G = \int_{|\varphi_1(X_1)| \geq M} E_G^2\{\varphi(X_1, \ldots, X_m)/X_1\}\, dP_G$$

$$\leq \int_{|\varphi_1(X_1)| \geq M} E_G\{\varphi^2/X_1\}\, dP_G = \int_{|\varphi_1(X_\lambda)| \geq M} \varphi^2\, dP_G$$

$$\leq \int_{|\varphi| \geq \sqrt{M}} \varphi^2\, dP_G + \int_{|\varphi| < \sqrt{M}, |\varphi_1| \geq M} \varphi^2\, dP_G$$

$$\leq \sigma(1) + MP_G\{|\varphi_1| \geq M\} \leq \sigma(1) + M^{-1}E_G|\varphi_1|^2$$

$$\leq \sigma(1) + M^{-1}E_G|\varphi|^2 = o(1),$$

as $M \to \infty$ uniformly in φ, G, satisfying (3.5) with φ_1 replaced by φ.

The uniform integrability of $\ell(Z_n)$ (cf. (3.2) and (3.6)), continuity of $\ell(\cdot)$ a.e. including zero and (3.4) implies (3.3). The theorem is thus proved. □

We note in turn that for the loss functions which do not satisfy (3.2) the risk corresponding to U_n can be, in general, infinite. Of course, for each loss function satisfying the inequality.

$$\ell(x) \leq K(1 + |x|^k), \tag{3.7}$$

the optimality under assumption of boundness of $E_F|\varphi|^{k+\Delta}$ can be formulated (for this (3.6) can be replaced by an analogous relation of higher order, of the type obtained in [9]).

However, we will consider slightly modified estimates U_n' which though being in general biased as estimates of $\Phi(F)$ are, as we will see, LAM estimators simultaneously for all loss functions which increase not faster than some power of x.

Denote

$$\varphi^{(n)}(x_1, \ldots, x_m) = \begin{cases} \varphi(x_1, \ldots, x_m) & \text{if } |\varphi| < \sqrt{n}, \\ 0 & \text{otherwise,} \end{cases} \tag{3.8}$$

and let

$$U_n' = \frac{1}{\binom{n}{m}} \sum_{S_n} \varphi^{(n)}(X_{\alpha_1}, \ldots, X_{\alpha_m}). \tag{3.9}$$

Theorem 3.2. *Suppose conditions* (1), (2) *are fulfilled and for some* κ, $k > 0$, *the loss function* $\ell \in \tilde{L}$ *satisfies* (3.7). *Then the estimate* (3.9), *is* $(\mathcal{F}, \mathcal{R}, \ell)$ *locally asymptotically minimax.*

Proof. Below all the limit relations and constants are uniform in some vicinity of F, a circumstance which we will not always explicitly indicate. Denote

$$\varphi_1^{(n)} = E_G\{\varphi^{(n)}(X_1, \ldots, X_m)/X_1 = x\},$$

$$\Phi_n(G) = E_G\varphi_1^{(n)}(X_1) = E_G\,\varphi^{(n)}(X_1, \ldots, X_m),$$

$$Z_n' = \sqrt{n}(U_n' - \Phi(G)),$$

$$Z_n'' = \sqrt{n}(U_n' - \Phi_n(G)).$$

Note that uniformly in a vicinity of F

$$|\Phi(G) - \Phi_n(G) = \left| \int_{|\varphi| \geq \sqrt{n}} \varphi\, dG(x_1) \ldots dG(x_m) \right| \le \frac{1}{\sqrt{n}} \int_{|\varphi| \geq \sqrt{n}} \varphi^2 dG(x_1) \ldots dG(x_m) = o(1/\sqrt{n}). \tag{3.10}$$

Therefore

$$Z_n' - Z_n'' \to 0 \tag{3.11}$$

also uniformly. As will be shown for any $k > 0$ and n sufficiently large

$$E_G\,|Z_n''|^k < c(k). \tag{3.12}$$

Thus from (3.10), it also follows that

$$E_G\,|Z_n'|^k \le E_G\,|Z_n''|^k + o(1) \qquad (n \to \infty). \tag{3.13}$$

According to Lemma 3.1, Z_n'' is uniformly asymptotically normal $(0, m^2\sigma^2(G))$. Thus by the virtue of (3.11), (3.13) it is sufficient to show that (3.12) is valid since then we can proceed just in the same way as at the very end of reasoning in the proof of Theorem 3.1.

For this let us define another estimate

$$T_n = \frac{1}{\tilde{n}} \sum_{i=1}^{\tilde{n}} \xi_i^{(n)},$$

where $\tilde{n} = [n/m]$,

$$\xi_i^{(n)} = \varphi^{(n)}(X_{(i-1)m+1}, \ldots, X_{im}).$$

Note that T_n is an unbiased estimate of $\Phi_n(G)$, for which U_n' is the U-statistic. Therefore, using the property of U-statistics mentioned in the Introduction to this paper, we have, for any $k > 0$,

$$E_G\,|Z_n''|^k = n^{k/2} E\,G|U_n' - \Phi_n(G)|^k \le n^{k/2} E_G\,|T_n - \Phi_n(G)|^k.$$

(see, for example, [4]). Now it remains to show that the moments of order k of the variable

$$\sqrt{n}E_G(T_n - \Phi_n(G)) = \sqrt{\frac{n}{\tilde{n}}}\left(\frac{1}{\sqrt{\tilde{n}}}\sum_{i=1}^{\tilde{n}}(\xi_i^{(n)} - \Phi_n(G))\right)$$

are bounded. Evidently we can restrict ourselves to the cases when $k = 1, 2, \ldots$.

The conditions under which the moments of a properly normed sum of independent variables tend to the corresponding moments of normal distribution were studied, e.g., in [10]. To use the result of this work we need the following estimates. From (3.8), for any $\eta > 0$, we have

$$\tilde{n} \int\limits_{|\xi_1^{(n)}/\sqrt{\tilde{n}}|>\eta} \left|\frac{\xi_i^{(n)}}{\sqrt{\tilde{n}}}\right|^k dP_G \leq \frac{\tilde{n}\cdot n^{(k-2)/2}}{\tilde{n}^{k/2}} \int\limits_{|\xi_i^{(n)}|>\eta\sqrt{\tilde{n}}} |\xi_i^{(n)}|^2\, dP_G = o(1) \quad (n \to \infty), \tag{3.14}$$

uniformly in a vicinity of F and

$$\int\limits_{|\xi_i^{(n)}|<\eta\sqrt{\tilde{n}}} |\xi_i^{(n)}|^2\, dP_G - \left(\int\limits_{|\xi_i^{(n)}|<\eta\sqrt{\tilde{n}}} \xi_i^{(n)} dP_G\right)^2$$

$$= \int \varphi^2 dP_G - \left(\int \varphi\, dP_G\right)^2 + \sigma(1) = \sigma^2(G) + o(1) \qquad (n \to \infty) \tag{3.15}$$

also uniformly. It follows from (3.14), (3.15), according to [10], that for any $k = 1, 2, \ldots$ the corresponding moments

$$E_G\left(\frac{1}{\sqrt{\tilde{n}}}\sum_{i=1}^{\tilde{n}}(\xi_i^{(n)} - \Phi_n(G))\right)^k$$

converge to the moments of the same order of the normal distribution with the $(0, \sigma^2(G))$.

Finally, the convergence is uniform in corresponding neighbourhood of F as it follows from the reasoning in [10]. Thus the inequality (3.12) follows, which in turn implies Theorem 3.2 as we have shown.

Acknowledgement

The author would like to express his gratitude to Dr. R.Z. Hasminskii for his encouraging discussions and interest in this work.

References

[1] Hájek J. (1972). Local asymptotic minimax and admissibility in estimation. *Proc. 6th Berkeley Sympos. Math. Statist. and Probability*, **1**, pp. 175–194.

[2] Stein, C. (1956). Efficient nonparametric testing and estimation. *Proc. 3rd Berkeley Sympos. Math. Statist. and Probability*, **1**.

[3] Ibragimov, I.A. and Khas'minskii, R.Z. (1974). On sequental estimation. *Theory Probab. Appl.*, **19**, 233–244.

[4] Fraser, D.A.S. (1957). *Nonparametric Methods in Estimation*. Wiley, New York.

[5] Roussas G.G. (1965). Asymptotic inference in Markov processes. *Ann. Math. Statist.*, **36**, 978–992.

[6] LeCam, L. (1970). On the assumptions used to prove asymptotic normality of maximum likelihood estimates. *Ann. Math. Statist.*, **41**, 802–828.

[7] Hoeffding W. (1948). A class of statistics with asymptotically normal distribution. *Ann. Math. Statist.*, **19**, 293–325.

[8] Loève M. (1961). *Probability Theory*, 2nd ed. van Nostrand, New York.

[9] Grams W.F. and Serfling R.J. (1973). Convergence rates for U-statistics and related statistics. *Ann. Statist.*, **1**, 153–160.

[10] Zaremba, S.K. (1958). Note on the central limit theorem. *Math. Z.*, **69**, 295–298.

Introduction to Aalen (1978) Nonparametric Inference for a Family of Counting Processes

Ian W. McKeague
Florida State University

Odd Olaf Aalen was born in 1947 in Oslo, Norway, and he grew up there. He completed a Masters degree thesis at the University of Oslo in 1972 under the guidance of Jan Hoem. This thesis developed some methods for estimating the efficacy and risks involved in the use of intrauterine contraceptive devices (IUDs). Hoem had suggested that time-continuous Markov chains might be useful for modeling the progression of events experienced by a woman following the insertion of an IUD, as this was a model that Hoem had found useful in other contexts. The need to fit such models nonparametrically started Aalen thinking about the approach that he was to develop fully in his ground-breaking 1978 paper that is reprinted here.

The approach pioneered by Aalen has had a lasting impact on event history analysis and on survival analysis especially. By focusing on "dynamic" aspects and how the past affects the future, Aalen found that counting processes provide a natural and flexible way of setting up statistical models for a progression of point events through time. Such models are specified in terms of the intensity $\lambda(t)$ of a basic counting process $N(t)$ that is defined to be the number of events observed to occur by time t. The intensity is the instantaneous rate of observable point events at time t, given the history up to time t, and it can be readily designed to depend on covariate processes or on the past of the counting process in ways that are suited to a particular application.

Prior to Aalen's work, the seminal paper of Cox (1972) (reprinted in Volume II) had shown the importance of the instantaneous hazard rate as a natural way of specifying regression models for right-censored survival data. Aalen's fundamental contribution was to recognize how the hazard function enters into the intensity of the basic counting process $N(t)$ for right censored data, leading to greatly simplified derivations of the asymptotic properties of various estimators under quite general patterns of censorship, and to a full appreciation of the conceptual simplicity of "dynamic modeling." Among other things, it then became much easier to study the product-limit survival function estimator due to Kaplan and Meier (1958) (reprinted in Volume II), the cumulative hazard function estimator due to Nelson (1969), and Cox's maximum partial likelihood estimator (Andersen and Gill, 1982). The statistical theory of event history analysis is now well developed and thorough accounts of it can be found in the monographs by Fleming and Harrington (1991) and Andersen et al. (1993).

In his masters thesis, Aalen had proposed an estimator for the cumulative hazard function in the context of the classical competing risks model of survival analysis. Consider a survival time T with corresponding survival function S and failure rate function $\alpha(t) = f(t)/S(t)$, where T is assumed to have a density f. Heuristically, $\alpha(t)\,dt = P(T \in [t, t+dt)|T \geq t)$. In the classical competing risks model we have data on the possibly right-censored survival time $X = \min(T, C)$ and the "censoring indicator" $\delta = I(T \leq C)$, where C is a censoring time that is independent of T. Aalen's estimator of the cumulative hazard function $A(t) = \int_0^t \alpha(s)\,ds$, based on a random sample $\{(X_i, \delta_i): i = 1, \dots, n\}$ from the distribution of (X, δ), is given by

$$\hat{A}(t) = \int_0^t \frac{dN(s)}{Y(s)},$$

where $N(t) = \#\{i: X_i \leq t, \delta_i = 1\}$ is the counting process that registers uncensored failures through time, and $Y(t) = \#\{i: X_i \geq t\}$ is the number of individuals at risk at time t. Here $1/Y(s)$ is interpreted to be 0 whenever $Y(s) = 0$. This estimator was to become known as the Nelson–Aalen estimator; it had previously been proposed by Nelson (1969) in the equivalent form

$$\hat{A}(t) = \sum_{i: X_i \leq t} \frac{\delta_i}{Y(X_i)}.$$

Soon after completing his masters thesis (Aalen, 1972), Aalen realized that this estimator must work in more general settings beyond the competing risks model. He first considered non-time-homogeneous Markov chains on a finite state space. The role of $\alpha(t)$ was then played by the intensity that an individual makes a transition from a given state i to a given state j. To estimate the corresponding cumulative transition intensity $A(t)$ based on a sample of independent realizations of the chain, $N(t)$ was taken to be the total number of observed $i \to j$ transitions until time t, and $Y(t)$ as the observed number of individuals in state i at time t. The Nelson–Aalen estimator was then defined exactly as above. Aalen derived some results for this estimator [see Aalen (1976)], but without the use of martingale theory the proofs were cumbersome.

The most unsatisfying aspect of this first extension, however, was that it didn't go far enough: the Markovian framework appeared to be unnecessarily restrictive. The only *essential* assumption seemed to be that the force of transition $\alpha(t)$ should act only on the individuals observed to be in state i at time t, irrespective of how they got there or of the kind of censoring mechanism in effect. Unfortunately, the theory of point processes available in 1972 was mainly restricted to the stationary case and did not allow for the intensity to depend on the past history of the process. But that was about to change.

The Berkeley Ph.D. dissertation of Pierre Brémaud (1972) introduced martingale methods into the theory of point processes for the first time, allowing the intensity to depend almost arbitrarily on the past history of the counting process (the only restrictions being that it is "predictable" and nonnegative). Coincidentally, Aalen came to Berkeley as a Ph.D. student in 1973, but it was to be another year before he was to hear of Brémaud's work. By 1974 he was being supervised by Lucian Le Cam, who suggested that the paper by McLeish (1974) on discrete-time martingale central limit theorems might be relevant. Some time later, Aalen mentioned his search for a suitable theory of point processes to David Brillinger, who then recalled some papers he had recently received from the Electronics Research Laboratory at Berkeley—these were Brémaud's dissertation and some technical reports by R. Boel, P. Varaiya, and E. Wong [later published as Boel et al. (1975a, b)]. Along with a technical report of Dolivo (1974), these papers gave Aalen the general counting process framework that he had been seeking.

In the context of the competing risks model, Aalen was able to show that the basic counting process $N(t)$ has intensity $\lambda(t) = \alpha(t)Y(t)$, which is to say that

$$M(t) = N(t) - \int_0^t \lambda(s)\,ds$$

is a martingale with respect to the history of the data available at time t (a history is a nested family of σ-fields indexed by t, often called a filtration). This was viewed as a special case of the multiplicative intensity model $\lambda(t) = \alpha(t)Y(t)$ for the intensity of a general counting process. Here $\alpha(t)$ is an unknown nonrandom function, and the process $Y(t)$ is predictable with respect to a given history $\mathcal{F}_t$ that includes all the data available at time t. The precise definition of "predictable" is technically complicated, being a part of the general theory of stochastic processes, but it essentially means that $Y(t)$ is determined by the collection $\mathcal{F}_{t-}$ of events occurring *before* time t.

It soon became clear that the counting process/martingale framework is extremely powerful and that it can often greatly simplify notation and calculations that get out of hand with more elementary methods. The martingale property of $M(t)$ can be expressed heuristically as $E(dM(t)|\mathcal{F}_{t-}) = 0$; this property is a consequence of the interpretation of $\lambda(t)$ as the instantaneous rate of occurrence of unit jumps in the counting process:

$$E(dN(t)|\mathcal{F}_{t-}) = P\{N(t) - N(t-) = 1|\mathcal{F}_{t-}\} = \lambda(t)\,dt.$$

The key to understanding the behavior of the Nelson–Aalen estimator was that $\hat{A}(t) - A(t)$ can be written as the integral $m(t) = \int_0^t Y(s)^{-1} dM(s)$, where we assume for simplicity that $Y(s)$ does not vanish. This integral is itself a martingale, since the predictability of $Y(t)$ and a basic property of conditional expectation give that

$$E(dm(t)|\mathscr{F}_{t-}) = Y(t)^{-1}E(dM(t)|\mathscr{F}_{t-}) = 0.$$

We can then read-off results about $\hat{A}(t)$ from standard results in martingale theory. The mean and variance are easily calculated; $\hat{A}(t)$ is found to be (approximately) unbiased.

Using Le Cam's suggestion of the relevance of McLeish's (1974) paper, Aalen next developed a central limit theorem for this type of martingale integral (see Aalen (1977)). In the context of the competing risks model, this led to the result that $\sqrt{n}(\hat{A}(t) - A(t))$ converges in distribution to a zero-mean Gaussian process. Aalen also realized that two-sample tests for censored survival data could be treated in a similar fashion, leading to a unified approach to the extensive literature on that subject, and later expanded to include k-sample tests by Andersen et al. (1982).

Following the completion of his Ph.D. dissertation (Aalen, 1975), Aalen visited the University of Copenhagen for some months. This helped spread the word on the counting process approach, and Copenhagen has been a major center for research in counting process methods in statistics ever since. Neils Keiding, Per Kragh Andersen, and Søren Johansen in Copenhagen, and Richard Gill in the Netherlands, were especially inspired by the approach in its early days, and they have been among the most active contributors to its subsequent development.

From 1977 to 1981, Aalen was an associate professor at the University of Tromsø in northern Norway. Since then he has been a professor of medical statistics at the University of Oslo, where he has primarily been involved with applications. He still occasionally contributes to the literature on counting processes in survival analysis. In a series of papers (Aalen, 1980, 1989, 1993) he developed an extension of the multiplicative intensity model to allow for the process $Y(t)$ to be multidimensional so that information on covariates can be included. Aalen calls this model the linear hazard model [for a review, see McKeague (1996)] and has strongly advocated it as an alternative to the popular Cox (1972) proportional hazards model for the regression analysis of censored survival data. He has also contributed to the growing literature on frailty models; see, e.g., Aalen (1994). In these models, the basic counting processes are assumed to have intensity of the form $\lambda(t) = Z\alpha(t)Y(t)$, as in the multiplicative intensity model, except that Z is a random variable that explains shared frailty within a group of individuals and whose distribution is assumed to belong to some parametric family [see Chapter IX of Andersen et al. (1993)].

Recently, Aalen (1996) wrote an article for a festschrift in honor of Le Cam's seventieth birthday. This article contains a nontechnical discussion of the merits of dynamic modeling and provides many historical details concerning his 1978 paper, some of which are recounted here. Although some important types of censoring cannot be dealt with effectively using counting process methods (e.g., interval censoring or current status data, in which the basic counting process $N(t)$ is not observable), the influence of Aalen's break-through continues to be felt. The April 1996 issue of the *Annals of Statistics*, for instance, contains six papers on censored data, two of which use counting process methods in some way. The reader should consult the excellent monograph of Andersen et al. (1993) for an in-depth survey of the extensive literature spawned by Aalen's breakthrough.

References

Aalen, O.O. (1972). Estimering av risikorater for prevensjonsmidlet "spiralen" (in Norwegian). Masters thesis, Institute of Mathematics, University of Oslo.

Aalen, O.O. (1975). Statistical inference for a family of counting processes. Ph.D. dissertation, Department of Statistics, University of California, Berkeley.

Aalen, O.O. (1976). Nonparametric inference in connection with multiple decrement models. *Scand. J. Statist.*, **3**, 15–27.

Aalen, O.O. (1977). Weak convergence of stochastic integrals related to counting processes. *Z. Wahrsch. Verw. Gebiete*, **38**, 261–277. Correction: (1979), **48**, 347.

Aalen, O.O. (1980). A model for nonparametric regression analysis of counting processes. *Lecture Notes in Statistics*, vol **2**, pp 1–25. Mathematical Statistics and Probability Theory (W. Klonecki, A. Kozek, and J. Rosiński, eds.). Springer-Verlag, New York.

Aalen, O.O. (1989). A linear regression model for the analysis of life times. *Statist. Medicine*, **8**, 907–925.

Aalen, O.O. (1993). Further results on the nonparametric linear regression model in survival analysis. *Statist. Medicine*, **12**, 1569–1588.

Aalen, O.O. (1994). Effects of frailty in survival analysis. *Statist. Methods Medical Res.*, **3**, 227–243.

Aalen, O.O. (1996). Counting processes and dynamic modelling. To appear in a festschrift in honor of Lucien Le Cam's 70th birthday.

Andersen, P.K., Borgan, Ø., Gill, R.D., and Keiding, N. (1982). Linear nonparametric tests for comparison of counting processes, with applications to censored survival data (with discussion). *Internat Statist. Rev.*, **50**, 219–258. Correction: **52**, p. 225.

Andersen, P.K., Borgan, Ø, Gill, R.D., and Keiding, N. (1993). *Statistical Models Based on Counting Processes*. Springer-Verlag, New York.

Andersen, P.K. and Gill, R.D. (1982). Cox's regression model for counting processes: a large sample study. *Ann. Statist.*, **10**, 1100–1120.

Boel, R., Varaiya, P., and Wong, E. (1975a). Martingales on jump processes, I: Representation results. *SIAM J. Control*, **13**, 999–1021.

Boel, R., Varaiya, P., and Wong, E. (1975b). Martingales on jump processes, II: Applications. *SIAM J. Control*, **13**, 1022–1061.

Brémaud, P. (1972). A martingale approach to point processes. Ph.D. dissertation, Electronics Research Laboratory, University of California, Berkeley.

Cox, D.R. (1972). Regression models and life tables (with discussion). *J. Roy. Statist. Soc. Ser. B*, **34**, 187–220.

Dolivo, F. (1974). Counting processes and integrated conditional rates: A martingale approach with application to detection. Ph.D. dissertation, University of Michigan.

Fleming, T.R. and Harrington, D.P. (1991). *Counting Processes and Survival Analysis*. Wiley, New York.

Kaplan, E.L. and Meier, P. (1958). Nonparametric estimation from incomplete observations. *J. Amer. Statist. Assoc.*, **53**, 457–481.

McKeague, I.W. (1996). Aalen's additive risk model. To appear in *Encyclopedia of Statistical Sciences* (S. Kotz and C.B. Read, eds.). Wiley, New York.

McLeish, D.L. (1974). Dependent central limit theorems and invariance principles. *Ann. Probab.*, **2**, 620–628.

Nelson, W. (1969). Hazard plotting for incomplete failure data. *J. Qual. Technol.*, **1**, 27–52.

Nonparametric Estimation of Partial Transition Probabilities in Multiple Decrement Models*

Odd Aalen
University of Oslo and University of California, Berkeley

Abstract

Nonparametric estimators are proposed for transition probabilities in partial Markov chains relative to multiple decrement models. The estimators are generalizations of the product limit estimator. We study the bias of the estimators, prove a strong consistency result and derive asymptotic normality of the estimators considered as stochastic processes. We also compute their efficiency relative to the maximum likelihood estimators in the case of constant forces of transition.

1. Introduction

The multiple decrement, or competing risks, model is an old tool in actuarial science, demography and medical statistics. Nelson (1969), Altshuler (1970), Hoel (1972), Peterson (1975) and Aalen (1976) have studied empirical, or nonparametric, statistical analyses for such models. The methods applied are related to the product-limit estimator of Kaplan and Meier (1958). The present paper is a continuation of Aalen (1976). We give nonparametric estimators of *general partial transition probabilities*. For theoretical study of these quantities, see, e.g., Hoem (1969). Peterson (1975) independently of us suggests the same kind of estimator. Our theoretical results are, however, not contained in his paper.

* This paper was prepared with the partial support of National Science Foundation Grant GP 38485.

Note that the assumptions made in our competing risks model correspond to what is often termed "independence" of risks.

Formally a multiple decrement model may be described as a time-continuous Markov chain with one transient state labeled 0 and m absorbing states numbered from 1 to m. We define $P_i(t)$; $i = 0, 1, \ldots, m$; to be the probability that the process is in state i at time t given that it started in state 0 at time 0. The force of transition (see, e.g., Hoem (1969)) or infinitesimal transition probability from state 0 to state i at time t is given by

$$\alpha_i(t) = P_i'(t)/P_0(t), \qquad i = 1, \ldots, m,$$

provided the derivative exists. We make the following assumption (cf. Feller (1957, Section XVII.9)):

Assumption. $\alpha_i(t)$ exists and is continuous everywhere for $i = 1, \ldots, m$.

The word "nonparametric" will in this paper mean that no further assumption is made about the $\alpha_i(t)$.

By a partial chain we mean the model that occurs if we put $\alpha_j(t) \equiv 0$ for all $j \notin A$ where A is some subset of $\{1, \ldots, m\}$. We want to estimate the transition probabilities in the partial chain. The above mentioned papers study such estimation in the case that A contains only one state.

We introduce some notation. The total forces of transition to the set of states $\{1, \ldots, m\}$ and to subset A of states are given by

$$\delta(t) = \sum_{i=1}^{m} \alpha_i(t) \qquad \text{and} \qquad \delta_A(t) = \sum_{i \in A} \alpha_i(t)$$

respectively. The cumulative forces of transition are given by

$$\beta_A(t) = \int_0^t \delta_A(s)\, ds, \qquad \beta_i(t) = \int_0^t \alpha_i(s)\, ds.$$

Let $q_A(t)$ be the probability, relative to the partial model, of not leaving state 0 in the time interval $[0, t]$. We have $q_A(t) = \exp(-\beta_A(t))$. Let $p_A(t) = 1 - q_A(t)$ and $p_i(t) = p_{\{i\}}(t)$. Finally, define $p(t) = P_0(t)$ and $r(t) = p(t)^{-1}$.

The probability of transition $0 \to i$ in the time interval $[0, t]$ in the partial chain corresponding to A is given by

$$P_i(t, A) = \int_0^t \alpha_i(s) \exp\left(-\int_0^s \delta_A(u)\, du\right) ds.$$

(Of course we must have $i \in A$.)

We will make the following assumption about the experiment and the observation: we observe continuously, over the time interval $[0, 1]$, n independent processes of the kind described above, each with the same set of forces of transition. Every process is assumed to be in state 0 at time 0.

Use the following notation: $N_i(t)$ is the number of processes in state i at time t. We define the $N_i(t)$ to be right-continuous for $i > 0$. Let $N_A(t) =$

$\sum_{i \in A} N_i(t)$, and let $M(t) = N_0(t)$ and define this to be a left-continuous process. Define:

$$R(t) = M(t)^{-1} \quad \text{if} \quad M(t) > 0,$$
$$= 0 \qquad \text{if} \quad M(t) = 0.$$

As usual "a.s." denotes "almost surely," while $X_n \to_p X$ denotes convergence in probability. Let $L(X)$ denote the distribution of X, and let $I(B)$ denote the indicator function of the set B.

If we want to stress the dependence on n we will write $M_n(t), N_{i,n}(t)$ and similarly for the other quantities.

2. Estimation

Write:

$$P_i(t, A) = \int_0^t q_A(s)\, d\beta_i(s).$$

We will estimate this quantity by substituting estimators for the functions $q_A(s)$ and $\beta_i(s)$.

If we think of the set A as one single state, then we can use Kaplan and Meier's (1958) product limit estimator for estimating $q_A(t)$. It can be written in the following form:

$$\hat{q}_A(t) = \exp \int_0^t \log(1 - R(s))\, dN_A(s). \tag{2.1}$$

This integral, and the similar integrals below, are to be taken as Lebesgue-Stieltjes integrals over the time interval $[0, t]$.

For $\beta_i(t)$ we can use the closely related estimator studies by Nelson (1969) and Aalen (1976):

$$\hat{\beta}_i(t) = \int_0^t R(s)\, dN_i(s). \tag{2.2}$$

We suggest the following estimator for $P_i(t, A)$:

$$\hat{P}_i(t, A) = \int_0^t \hat{q}_A(s - 0)\, d\hat{\beta}_i(s). \tag{2.3}$$

Alternatively we can write the estimator in the form:

$$\hat{P}_i(t, A) = \int_0^t \hat{q}_A(s - 0) R(s))\, dN_i(s). \tag{2.4}$$

Clearly, one might suggest other versions of this estimator. Instead of substituting $\hat{q}_A(t)$ for $q_A(t)$ one could use $\exp[-\int_0^t R(u) dN_A(u)]$ while $\beta_i(t)$

might be estimated by $-\int_0^t \log[1 - R(u)]dN_i(u)$. (This was suggested by a referee.) When we prefer the estimator (2.3) this has the following reason: firstly, it may be shown that $\hat{P}_i(t, \{i\})$ coincides with Kaplan and Meier's estimator of $p_i(t) = P_i(t, \{i\})$. Clearly, this ought to be the case since our intention is to generalize that estimator. Secondly, consider the case $A = \{1, \ldots, m\}$. The $P_i(t, A) = P_i(t)$, and hence it is reasonable to require that the estimator $\hat{P}_i(t, A)$ specializes to $(1/n)N_i(t)$. That this is indeed the case may be shown with some computation.

By results of Kaplan and Meier (1958), Breslow and Crowley (1974), Meier (1975) and Aalen (1976) $\hat{q}_A(t)$ and $\hat{\beta}_i(t)$ are known to have nice properties, and it is reasonable to assume that these carry over to $\hat{P}_i(t, A)$. In this paper we will mainly concentrate on large sample properties, but first we will give the following results:

Proposition 1. *$\hat{P}_i(t, A)$ is based on minimal sufficient and complete statistics.*

This proposition is an immediate consequence of Theorem 3.1 of Aalen (1976).

Proposition 2. *Let $\hat{p}_i(t) = \hat{p}_i(t, \{i\})$. The following holds:*

(i) $0 \leqq p_i(t) - E\hat{p}_i(t) \leqq (1 - p(t))^n \beta_i(t)$;
(ii) $|E\hat{p}_i(t, A) - P_i(t, A)| \leqq (1 - p(t))^n \beta_i(t)(1 + \beta_A(t))$.

Note that according to this proposition the bias of $\hat{p}_i(t)$ and $\hat{P}_i(t, A)$ converges exponentially to 0 when $n \to \infty$. Moreover, Proposition 1 implies that these estimators are uniformly minimum variance estimators for their expectations. These facts coupled together indicate that the estimators should be reasonable candidates.

In the case $A = \{1, \ldots, m\}$ we have $\hat{P}_i(t, A) = (1/n)N_i(t)$ and hence it is of course unbiased. By Proposition 1 we have in this case a uniformly minimum variance unbiased estimator.

Proposition 2 is easily proved by the technique used in the proof of Theorem 4.1 in Aalen (1976). For completeness the proof will be given in the Appendix. The same method of proof may be used to derive approximate expressions of variances and covariances. However, in this paper we will only give the variances and covariances of the limiting distribution.

3. Consistency

We will prove the following consistency result. (All limits below are taken with respect to n.)

Theorem 1. *When $n \to \infty$ the following holds:*

$$\sup_{0 \leqq qt \leqq ql} \frac{n^{1/2}}{\log n} |\hat{P}_{i,n}(t, A) - P_i(t, A)| \to 0 \quad a.s.$$

For the proof we need some intermediate results. The first one is a part of Lemma 2.2 of Barlow and van Zwet (1970).

Lemma 1. *Let $X_1, X_2, \ldots, X_n$ be independent and identically distributed with continuous distribution function $F(x)$. Let $F_n(x)$ be the empirical distribution function. Then*

$$\frac{n^{1/2}}{\log n} \sup_x |F_n(x) - F(x)| \to 0 \quad a.s.$$

We next state Lemma 1 of Breslow and Crowley (1974).

Lemma 2. *Let $\hat{\beta}_A(t) = \sum_{i \in A} \hat{\beta}_i(t)$. If $M(1) > 0$, then*

$$0 < -\log \hat{q}_A(t) - \hat{\beta}_A(t) < \frac{n - M(1)}{nM(1)}.$$

The first part of the next proposition is a strengthening of Theorem 6.1 in Aalen (1976). The second part gives a strong consistency result for the product limit estimator.

Proposition 3. *The following limits hold when $n \to \infty$:*

(i) $\sup_{0 \leqq t \leqq 1} (n^{1/2}/\log n) |\hat{\beta}_{i,n}(t) - \beta_i(t)| \to 0$ *a.s.*,
(ii) $\sup_{0 \leqq t \leqq 1} (n^{1/2}/\log n) |\hat{q}_{A,n}(t) - q_A(t)| \to 0$ *a.s.*

PROOF. Put $c_n = n^{1/2}/\log n$. The supremas below are all taken over the set $0 \leqq qt \leqq q1$. We have

$$\hat{\beta}_{i,n}(t) - \beta_i(t) = \int_0^t (nR_n(s) - r(s))\, d\left(\frac{1}{n} N_{i,n}(s)\right)$$
$$+ \int_p^t r(s)\, d\left(\frac{1}{n} N_{i,n}(s) - P_i(s)\right).$$

Hence:

$$\sup c_n |\hat{\beta}_{i,n}(t) - \beta_i(t)| \leqq qX_n + Y_n,$$

where

$$X_n = \sup c_n |nR_n(t) - r(t)|,$$

$$Y_n = \sup \left| \int_0^t r(s)\, d\left[c_n \left(\frac{1}{n} N_{i,n}(s) - P_i(s)\right)\right]\right|.$$

It is enough to show that $X_n \to 0$ and $Y_n \to 0$ a.s. We have:

$$X_n \leqq \frac{\sup c_n \left|\frac{1}{n} M_n(t) - p(t)\right|}{\frac{1}{n} M_n(1) p(1)} I(M_n(1) > 0) + 2n^{3/2} r(1) I(M_n(1) = 0).$$

The first part goes to 0 a.s. by Lemma 1. The second part goes to 0 a.s. by the Borel–Cantelli lemma and the fact that $M_n(1)$ is binomial $(n, p(1))$. Partial integration gives us for Y_n:

$$Y_n = \sup\left| r(t)c_n\left(\frac{1}{n} N_{i,n}(t) - P_i(t)\right) - \int_0^t c_n\left(\frac{1}{n} N_{i,n}(s) - P_i(s)\right) dr(s)\right|$$

$$\leq 2r(1)\sup c_n\left|\frac{1}{n} N_{i,n}(t) - P_i(t)\right|.$$

Hence, by Lemma 1, also $Y_n \to 0$ a.s. This proves (i). For the proof of (ii) we first use (i) and Lemma 2 to establish the following:

$$c_n I(M_n(1) > 0)\sup_t|\log \hat{q}_{A,n}(t) - \log q_A(t)| \to 0 \quad \text{a.s.}$$

By applying the mean value theorem we get:

$$c_n I(M_n(1) > 0)\sup_t|\hat{q}_{A,n}(t) - q_A(t)| \to 0 \quad \text{a.s.}$$

Using the fact that $M_n(1)$ is binomially distributed and applying the Borel-Cantelli lemma we get:

$$c_n I(M_n(1) > 0)\sup_t|\hat{q}_{A,n}(t) - q_A(t)| \to 0 \quad \text{a.s.}$$ □

Proof of Theorem 1. We can write:

$$\hat{P}_{i,n}(t, A) - P_i(t, A) = \int_0^t \hat{q}_{A,n}(s-0)\, d\hat{\beta}_{i,n}(s) - \int_0^t q_A(s)\, d\beta_i(s)$$

$$= \int_0^t (\hat{q}_{A,n}(s-0) - q_A(s))\, d\hat{\beta}_{i,n}(s)$$

$$+ \int_0^t q_A(s)\, d(\hat{\beta}_{i,n}(s) - \beta_i(s)).$$

By treating separately the two integrals of the last expression Theorem 1 may be proved in the same way as part (i) of Proposition 3. □

4. Weak Convergence

In this section we will study convergence in distribution of the stochastic processes $\hat{P}_i(t, A)$. Throughout the section t will be limited to the interval [0, 1]. Let D be the function space considered in Section 14 of Billingsley (1968) and let ρ be the metric d_0 defined there. In this section the term "weak convergence" will be used with respect to the product metric ρ_k on the product space D^k for appropriate values of k. Let C be the subset of D consisting of all continuous functions on [0, 1], and let λ be the usual uniform metric on this space. Let λ_k be the product metric on the product space

C^k. It is well known that λ_k and ρ_k coincide on C^k. It is also known that if $x_n \in D^k$ and $x \in C^k$, then $x_n \to x$ in the λ_k-metric if and only if $x_n \to x$ in the ρ_k-metric. For both these facts see e.g., Billingsley (1968, page 112). They will be repeatedly used below without further mentioning. A consequence of the last mentioned fact is the if Z_n and Z are random elements of D and C respectively, then $\rho_k(Z_n, Z) \to_p 0$ if and only if $\lambda_k(Z_n, Z) \to_p 0$.

Let $X_1, \dots, X_m$ be independent Gaussian processes on the time interval $[0, 1]$, each with independent increments and expectation 0. Let the variances be given by $\operatorname{Var} X_i(t) = \int_0^t \alpha_i(s) r(s)\, ds$. We choose versions of $X_1, \dots, X_m$ with continuous sample paths.

Theorem 2. *The vector consisting of all processes of the form* $Y_{i,n}(\bullet, A) = n^{1/2}(\hat{P}_{i,n}(\bullet, A) - P_i(\bullet, A))$ *for* $i \in A$ *and* $A \subset \{1, \dots, M\}$ *converges weakly to the vector consisting of the Gaussian processes* $Y_i(\bullet, A)$ *defined by:*

$$-Y_i(t, A) = \int_0^t \left[\int_s^t q_A(u)\alpha_i(u)\, du - q_A(s) \right] dX_i(s) + \sum_{j \in A - \{i\}} \int_0^t \int_s^t q_A(u)\alpha_i(u)\, du\, dX_j(s),$$

where the integrals are stochastic integrals in quadratic mean.

Remark. Stochastic integrals in quadratic mean are defined, e.g., in Cramér and Leadbetter (1967, Section 5.3) where their properties are discussed. One should note that the representation of the Y-processes as stochastic integrals over the X-processes makes it very simple to compute moments. For instance we have:

$$-EY_i(t, A) = \int_0^t \left[\int_s^t q_A(u)\alpha_i(u)\, du - q_A(s) \right] dEX_i(s) + \sum_{j \in A - \{i\}} \int_0^t \int_s^t q_A(u)\alpha_i(u)\, du\, dEX_j(s) = 0.$$

Since the X_i are independent processes and have independent increments, we have:

$$\begin{aligned}
\operatorname{Var} Y_i(t, A) &= \int_0^t \left[\int_s^t q_A(u)\alpha_i(u)\, du - q_A(s) \right]^2 d(\operatorname{Var} X_i(s)) \\
&\quad + \sum_{j \in A - \{i\}} \int_0^t \left[\int_s^t q_A(u)\alpha_i(u)\, du \right]^2 d(\operatorname{Var} X_i(s)) \\
&= \int_0^t \left[\int_s^t q_A(u)\alpha_i(u)\, du - q_A(s) \right]^2 \alpha_i(s) r(s)\, ds \\
&\quad + \sum_{j \in A - \{i\}} \int_0^t \left[\int_s^t q_A(u)\alpha_i(u)\, du \right]^2 \alpha_j(s) r(s)\, ds.
\end{aligned}$$

It is clear how covariances between pairs of $Y_i(t, A)$ for different i, t and A can be computed in the same easy way. The expressions will not be given here, we will just note that if $i \in A$ and $j \in B$, then:

$$A \cap B = \varnothing \Rightarrow Y_i(\bullet, A) \quad \text{is independent of} \quad Y_j(\bullet, B).$$

PROOF. The proof will follow the Pyke and Shorack (1968) approach. Define $X_{i,n}(t) = n^{1/2}(\hat{\beta}_{i,n}(t) - \beta_i(t))$ and $Z_{A,n}(t) = n^{1/2}(\hat{q}_{A,n}(t) - q_A(t))$.

We can write:

$$Y_{i,n}(t, A) = \int_0^t Z_{A,n}(s)\, d\beta_i(s) + \int_0^t q_A(s)\, dX_{i,n}(s) + n^{-1/2} \int_0^t Z_{A,n}(s)\, dX_{i,n}(s). \tag{4.1}$$

(These integrals and the integrals below are still defined as Lebesgue-Stieltjes or Lebesgue integrals until otherwise stated.)

For proving the theorem we will use the following proposition which is proved as Theorem 8.2 in Aalen (1976). Put $\mathbf{X} = [X_1, \ldots, X_m]$ and $\mathbf{X}(n) = [X_{1,n}, \ldots, X_{m,n}]$. ∎

Proposition 4. $\mathbf{X}(n)$ *converges weakly to* $\mathbf{X}$.

To exploit this result we must express the $Y_{i,n}(\bullet, A)$ in terms of the $X_{j,n}$. We will write $U_n = o_p(1)$ for random elements U_n of D^k when $\rho_k(U_n, 0)$ (or equivalently $\lambda_k(U_n, 0)$) converges in probability to 0.

By the arguments used in the proof of Proposition 3, part (ii) we have:

$$\begin{aligned} Z_{A,n} &= -q_A n^1(\hat{\beta}_{A,n} - \beta_A) + o_p(1) \\ &= -q_A \sum_{j \in A} X_{j,n} + o_p(1). \end{aligned}$$

Using this (4.1) can be rewritten in the following way:

$$\begin{aligned} Y_{i,n}(t, A) = &- \int_0^t q_A(s) \sum_{j \in A} X_{j,n}(s)\alpha_i(s)\, ds \\ &+ \int_0^t q_A(s)\, dX_{i,n}(s) - S_{i,n}(t, A) + o_p(1) \end{aligned} \tag{4.2}$$

where

$$S_{i,n}(t, A) = n^{-1/2} \sum_{j \in A} \int_0^t q_A(s) X_{j,n}(s)\, dX_{i,n}(s).$$

We now want to prove that $S_{i,n}(\bullet, A) = o_p(1)$. Here we follow Pyke and Shorack (1968) in using item 3.1.1 in Skorohod (1956). D is complete and separable with metric ρ and the same holds therefore for D^k with metric ρ_k. Thus the mentioned result of Skorohod ensures us that there exists a probability space representation of the processes

$$\mathbf{X} = [X_1, \dots, X_m] \quad \text{and} \quad \mathbf{X}(n) = [X_{1,n}, \dots, X_{m,n}]$$

such that $\rho_k(\mathbf{K}(n), \mathbf{X}) \to 0$ a.s., and we use this representation in the following.

All supremas below are taken over $t \in [0,1]$. We have:

$$\begin{aligned}\sup|S_{i,n}(t,A)| &\leqq \sum_{j\in A} n^{-1/2} \sup\left|\int_0^t q_A(s)X_{j,n}(s)\,dX_{i,n}(s)\right| \\ &\leqq \sum_{j\in A} n^{-1/2} \sup\left|\int_0^t q_A(s)(X_{j,n}(s) - X_j(s))\,dX_{i,n}(s)\right| \\ &\quad + \sum_{j\in A} n^{-1/2} \sup\left|\int_0^t q_A(s)X_j(s)\,dX_{i,n}(s)\right| \\ &\leqq \sum_{j\in A} \sup|X_{j,n}(t) - X_j(t)|(\hat{\beta}_{i,n}(1) + \beta_i(1)) \\ &\quad + \sum_{j\in A} \sup\left|\int_0^t q_A(s)X_j(s)\,d(\hat{\beta}_{i,n}(s) - \beta_i(s))\right|.\end{aligned}$$

The first term in the last expression converges in probability to 0. For the second term we use the method described in the proof of Theorem 4 in Breslow and Crowley (1974). Consider a subset Ω_0 of the underlying probability space such that $P(\Omega_0) = 1$, and such that for $\omega \in \Omega_0$ and X_j are uniformly continuous on [0, 1], and $\rho(\hat{\beta}_{i,n}, \beta_i)$ converges to 0. Choose a partition (depending on ω) of [0, 1] into K intervals $I_k = (\xi_{k-1}, \xi_k]$ such that

$$\sup_j \sup_{t\in i_k} |X_j(t)q_A(t) - X_j(\xi_k)q_A(\xi_k)| < \varepsilon$$

for $k = 1, \dots, K$. Then the second term above is bounded by

$$\varepsilon(\hat{\beta}_{i,n}(1) + \beta_i(1)) + 2K\lambda(X_j q_A, 0)\lambda(\hat{\beta}_{i,n}, \beta_i),$$

which tends to $2\beta_i(1)\varepsilon$ when $n \to \infty$. Since ε is arbitrary, this shows that the second term also converges in probability to 0. Hence by (4.2) we can write:

$$Y_{i,n}(t,A) = -\sum_{j\in A}\int_0^t q_A(s)X_{j,n}(s)\alpha_i(s)\,ds + \int_0^t q_A(s)\,dX_{i,n}(s) + o_p(1).$$

By partial integration rewrite the second term above and get:

$$\begin{aligned}Y_{i,n}(t,A) = &-\sum_{j\in A}\int_0^t q_A(s)X_{j,n}(s)\alpha_i(s)\,ds + q_A(t)X_{i,n}(t) \\ &- \int_0^t X_{i,n}(s)\,d[q_A(s)] + o_p(1).\end{aligned}$$

We are now ready to use Proposition 4. Let $\mathbf{Y}(n)$ be the vector consisting of all $Y_{i,n}(\bullet, A)$ in some order. It is easily figured out that the vector has $l = m2^{m-1}$ components. If we regard $\mathbf{Y}(n)$ as a function of $\mathbf{X}(n)$, then we have a mapping from D^m to D^l. If we use the metrics λ_m and λ_l on these spaces then the mapping is obviously continuous. This is also the case for the metrics ρ_n and ρ_k if the function is restricted to C^m. Since now $\mathbf{X}$ is a.s. an element of C^m, it follows from Theorem 5.1 of Billingsley (1968) and from Proposition 4 above that $\mathbf{Y}(n)$ converges weakly to the vector consisting of the components:

$$U_i(t, A) = -\sum_{j \in A} \int_0^t q_A(s) X_j(s) \alpha_i(s)\, ds + q_A(t) X_i(t) - \int_0^t X_i(s)\, d[q_A(s)].$$

So far all our integrals have been Lebesgue–Stieltjes or Lebesgue integrals. However, as shown on page 90 in Cramér and Leadbetter (1967), the integrals in the last expression can alternatively be taken as stochastic integrals in quadratic mean without changing their value. By using the partial integration formula 5.3.7 in Cramér and Leadbetter (1967) we conclude that $U_i(t, A) = Y_i(t, A)$. □

5. Estimation of the Asymptotic Variance

We will suggest an estimator of Var $Y_i(t, A)$, which is given in the remark after Theorem 2. We first rewrite the expression for the variance in the following form:

$$\operatorname{Var} Y_i(t, A) = \int_0^t [P_i(t, A) - P_i(s, A) - q_A(s)]^2 r(s)\, d\beta_i(s)$$

$$+ \sum_{j \in A - \{i\}} \int_0^t [P_i(t, A) - P_i(s, A)]^2 r(s) d\beta_j(s).$$

Estimators of $q_A(t), \beta_j(t)$ and $P_i(t, A)$ are given by (2.1), (2.2) and (2.3). Since $M(t)$ is binomial $(n, r(t)^{-1})$, it is reasonable to estimate $r(t)$ by $nR(t)$. Hence, by the same principle as was used for estimating $P_i(t, A)$, we assert that the following is a reasonable estimator of Var $Y_i(t, A)$.

$$n \int_0^t [\hat{P}_i(t, A) - \hat{P}_i(s, A) - \hat{q}_A(s)]^2 R(s)\, d\hat{\beta}_i(s)$$

$$+ \sum_{j \in A - \{i\}} n \int_0^t [\hat{P}_i(t, A) - \hat{P}_i(s, A)]^2 R(s)\, d\hat{\beta}_j(s).$$

Alternatively we can write:

$$n\int_0^t [\hat{P}_i(t,A) - \hat{P}_i(s,A) - \hat{q}_A(s)]^2 R^2(s)\, dN_i(s)$$

$$+ \sum_{j\in A-\{i\}} n\int_0^t [\hat{P}_i(t,A) - \hat{P}_i(s,A)]^2 R^2(s)\, dN_j(s).$$

By the same kind of arguments as in Section 3 it is easily shown that this estimator converges almost surely to Var $Y_i(t,A)$, uniformly in t.

Of course, relevant covariance functions of different kinds can be estimated in a similar way.

6. Asymptotic Relative Efficiency

In this section we will assume that all forces of transition are constant on the time interval $[0,t]$, i.e.,

$$\alpha_i(s) = \alpha_i \quad \forall s \in [0,t], \qquad i = 1,\ldots,m, \quad 0 \leqq t \leqq 1,$$

where the α_i are positive numbers. Let $\alpha_{i,t}^*$ be the maximum likelihood estimator of α_i based on complete observation over the time interval $[0,t]$. The $\alpha_{i,t}^*$ are the so-called "occurrence/exposure" rates for the α_i. They are given by:

$$\alpha_{i,t}^* = N_i(t) \Big/ \int_0^t M(s)\, ds,$$

(see, e.g., Hoem (1971)).

Let $\delta = \sum_{j=1}^m \alpha_j$, $\delta = \sum_{j\in A} \alpha_j$, $\delta_t^* = \sum_{j=1}^m \alpha_{j,t}^*$, $\delta_{A,t}^* = \sum_{j\in A} \alpha_{j,t}^*$. The maximum likelihood estimators of the $P_i(t,A)$ are given by:

$$P_i^*(t,A) = \frac{\alpha_{i,t}^*}{\delta_{A,t}^*}(1 - e^{-\delta_{A,t}^*}).$$

In this section we will study the asymptotic efficiency of $\hat{P}_i(t,A)$ relative to $P_i^*(t,A)$ in the sense that we will compare the variances of the asymptotic distributions. We will denote these by asVar $\hat{P}_i(t,A)$ and asVar $P_i^*(t,A)$ respectively. We regard i and A as fixed and introduce parameters a and b by $\delta_A = a\delta$ and $\alpha_i = b\delta_A$. One notes that $0 \leqq a \leqq 1$ and $0 \leqq b \leqq 1$.

By applying a Taylor series development and using results in Hoem (1971) we find:

$$\text{asVar}\, P_i^*(t,A) = (1 - e^{-\delta t})^{-1} a^{-1} b[a^2 b\delta^2 t^2 e^{-2a\delta t} + (1-b)(1 - e^{-a\delta t})^2].$$

From the variance formula in the remark to Theorem 2 we can compute:

$$\text{asVar}\, \hat{P}_i(t,A) = (1-2a)^{-1} ab[(1-2ab)e^{\delta t(1-2a)} - b(1-2a)e^{-2a\delta t} - 1 + b].$$

The efficiency is the quotient between asVar $P_i^*(t, A)$ and asVar $\hat{P}_i(t, A)$:

$$e(a, b, \delta, t) = \frac{(1-2a)[a^2 b\delta^2 t^2 e^{-2a\delta t} + (1-b)(1-e^{-a\delta t})^2]}{a^2(1-e^{-\delta t})[(1-2ab)e^{\delta t(1-2a)} - b(1-2a)e^{-2a\delta t} - 1 + b]}.$$

Table 1 gives values of this function for some values of a, b and δt.

It is seen from the expression above that

$$f(a, b, \delta) = \lim_{t\to\infty} e(a, b, \delta, t) = 0, \qquad a < \tfrac{1}{2} \quad \text{or} \quad b = 1,$$

$$= \frac{2a-1}{a^2}, \qquad a \geqq \tfrac{1}{2} \quad \text{and} \quad b < 1.$$

One should note the discontinuity of f at $b = 1$. In particular $f(1, b, \delta) = 1$ for $b < 1$ while $f(1, 1, \delta) = 0$. This complements results given by Sverdrup (1965, page 195) for a single decrement model, where of course $b = 1$.

Table 1 together with the asymptotic results in the last paragraph seem to indicate that the relative efficiency of the nonparametric estimators is good when either δt is small or when $b < 1$ and a is relatively close to 1.

Table 1. Asymptotic Efficiency.

			$\delta t = 0.5$		
	$a = 0.20$	$a = 0.40$	$a = 0.60$	$a = 0.80$	$a = 1.00$
$b = 0.25$	0.985	0.990	0.993	0.996	0.997
$b = 0.50$	0.983	0.987	0.989	0.991	0.992
$b = 0.75$	0.981	0.983	0.985	0.986	0.987
$b = 1.00$	0.979	0.979	0.979	0.979	0.979
			$\delta t = 1.0$		
	$a = 0.20$	$a = 0.40$	$a = 0.60$	$a = 0.80$	$a = 1.00$
$b = 0.25$	0.942	0.961	0.976	0.987	0.991
$b = 0.50$	0.936	0.950	0.963	0.973	0.979
$b = 0.75$	0.929	0.937	0.945	0.953	0.958
$b = 1.00$	0.921	0.921	0.921	0.921	0.921
			$\delta t = 2.0$		
	$a = 0.20$	$a = 0.40$	$a = 0.60$	$a = 0.80$	$a = 1.00$
$b = 0.25$	0.795	0.866	0.928	0.970	0.988
$b = 0.50$	0.775	0.834	0.892	0.940	0.967
$b = 0.75$	0.752	0.789	0.835	0.882	0.920
$b = 1.00$	0.724	0.724	0.724	0.724	0.724
			$\delta t = 3.0$		
	$a = 0.20$	$a = 0.40$	$a = 0.60$	$a = 0.80$	$a = 1.00$
$b = 0.25$	0.610	0.746	0.874	0.960	0.992
$b = 0.50$	0.580	0.697	0.827	0.928	0.976
$b = 0.75$	0.542	0.622	0.736	0.856	0.935
$b = 1.00$	0.496	0.496	0.496	0.496	0.496

Acknowledgment

This work grew out of a part of my master's thesis (Aalen, 1972) which I wrote under the supervision of Professor Jan M. Hoem. Professor L. Le Cam has been very helpful in answering questions about the weak convergence theory. Also, Professor P. Bickel has made some useful comments. Finally I am very grateful to the referees for correcting some mistakes and pointing out several obscurities.

Appendix

PROOF OF PROPOSITION 2. We will first show:

$$0 \leqq E\hat{q}_A(t) - q_A(t) \leqq \beta_A(t)(1 - p(t))^n. \tag{A.1}$$

In Section 1 $M(t)$, and hence $R(t)$, was defined to be left-continuous. By using this fact together with (2.1) we can write:

$$\hat{q}_A(t+h) = \hat{q}_A(t)(1 - R(t)I(t,h))(1 - U(t,h))$$

where $I(t,h)$ is 1 if there is at least one transition to A in the time interval $(t, t+h)$ and 0 otherwise, and

$$\Pr(u(t,h) \neq 0) = o(h), \qquad \Pr(0 \leqq U(t,h) \leqq 1) = 1.$$

Hence, if we put $f(t) = E\hat{q}_A(t)$, we get

$$f(t+h) = f(t) - E(\hat{q}_A(t)R(t)I(t,h)) + o(h).$$

Now $\Pr[I(t,h) = 1|M(t)] = M(t)\delta_A(t)h + o(h)$, hence:

$$f(t+h) = f(t) - h\delta_A(t)E(\hat{q}_A(t)R(t)M(t)) + o(h);$$

$$f(t) = f(t-h) - h\delta_A(t-h)E[\hat{q}_A(t-h)R(t-h)M(t-h)] + o(h).$$

Hence: $f'(t) = -\delta_A(t)E[\hat{q}_A(t)R(t)M(t)]$.

Define K by $K(x) = 1$ if $x = 0$ and $K(x) = 0$ otherwise. We then get

$$f'(t) = -\delta_A(t)f(t) + \delta_A(t)E[\hat{q}_A(t)K(M(t))].$$

Solving for $f(t)$ and exploiting the condition $f(0) = 1$ gives us:

$$f(t) = q_A(t) + \int_0^t E[\hat{q}_A(s)K(M(s))] \exp\left(-\int_s^t \delta_A(u)\,du\right)\delta_A(s)\,ds.$$

Hence:

$$0 \leqq f(t) - q_A(t) \leqq EK(M(t))\beta_A(t)$$

which is equivalent to (A.1). By now using a method similar to the one above we can easily show:

$$E\hat{P}_i(t,A) - P_i(t,A) = \int_0^t (E\hat{q}_A(s) - q_A(s))\alpha_i(s)\,ds - \int_0^t E[\hat{q}_A(s)K(M(s))]\alpha_i(s)\,ds.$$

Hence by (A.1):

$$\begin{aligned} |E\hat{P}_i(t,A) - P_i(t,A)| &\leqq \int_0^t (1-p(s))^n \beta_A(s)\alpha_i(s)\,ds \\ &\quad + \int_0^t E[\hat{q}_A(s)K(M(s))]\alpha_i(s)\,ds \\ &\leqq (1-p(t))^n\beta_A(t)\beta_i(t) + E(K(M(t)))\beta_i(t) \\ &= (1-p(t))^n\beta_i(t)(1+\beta_A(t)). \end{aligned}$$

This proves Proposition 2. □

References

[1] Aalen, O. (1972). Estimering av risikorater for prevensjons-midlet "spiralen." (In Norwegian.) Graduate thesis in statistics, University of Oslo.

[2] Aalen, O. (1976). Nonparametric inference in connection with multiple decrement models. *Scand. J. Statist.*, **3**, 15–27.

[3] Altshuler, B. (1970). Theory for the measurement for competing risks in animal experiments. *Math. Biosci.*, **6**, 1–11.

[4] Barlow, R.B. and van Zwet, W.R. (1970). Asymptotic properties of isotonic estimators for the generalized failure rate function. In *Nonparametric Techniques in Statistical Inference* (M.L. Puri, ed.). Cambridge, Cambridge University Press.

[5] Billingsley, P. (1968). *Convergence of Probability Measures*. Wiley, New York.

[6] Breslow, N. and Crowley, J. (1974). A large sample study of the life table and product limit estimates under random censorship. *Ann. Statist.*, **2**, 437–453.

[7] Chiang, C.L. (1968). *Introduction to Stochastic Processes in Biostatistics*. Wiley, New York.

[8] Cramèr, H. and Leadbetter, M.R. (1967). *Stationary and Related Stochastic Processes*. Wiley, New York.

[9] Dvoretzky, A., Kiefer, J., and Wolfowitz, J. (1956). Asymptotic minimax character of the sample distribution function and of the classical multinomial estimator. *Ann. Math. Statist.*, **27**, 642–669.

[10] Feller, W. (1957). *An Introduction to Probability Theory and its Applications*, vol. **I**, 2nd ed. Wiley, New York.

[11] Hoel, D.G. (1972). A representation of mortality data by competing risks. *Biometrics*, **28**, 475–488.

[12] Hoem, J.M. (1969). Purged and partial Markov chains. *Skand. Aktuarietidskr.*, **52**, 147–155.
[13] Hoem, J.M. (1971). Point estimation of forces of transition in demographic models. *J. Roy. Statist. Soc. Ser. B*, **33**, 275–289.
[14] Kaplan, E.L. and Meier, P. (1958). Nonparametric estimation from incomplete observations. *J. Amer. Statist. Assoc.*, **53**, 457–481.
[15] Meier, P. (1975). Estimation of a distribution function from incomplete observations. In *Perspectives in Probability* (J. Gani, ed.), pp. 67–87. Applied Probability Trust.
[16] Nelson, W. (1969). Hazard plotting for incomplete failure data. *J. Qual. Technol.*, **1**, 27–52.
[17] Peterson, A.V. (1975). Nonparametric estimation in the competing risks problem. Technical Report No. 73, Stanford University.
[18] Pyke, R. and Shorack, G. (1968). Weak convergence of a two sample empirical process and a new approach to the Chernoff–Savage theorems. *Ann. Math. Statist.*, **39**, 755–771.
[19] Simonsen, W. (1966). *Forsikringsmatematik*. Hefte I. Københavns Universitets Fond til Tilvejebringelse af Laeremidler.
[20] Skorohod, A.V. (1956). Limit theorems for stochastic processes. *Theory Probab. Appl.*, **1**, 261–290.
[21] Sverdrup, E. (1965). Estimates and test procedures in connection with stochastic models for deaths, recoveries and transfers between different states of health. *Skand. Aktuarietidskr.*, **48**, 184–211.
[22] Zwinggi, E. (1945). *Versicherungsmathematik*. Verlag Birkhäuser, Basel.

Introduction to Morris (1982) Natural Exponential Families with Quadratic Variance Functions

G.G. Letac
Universite Paul Sabatier

Morris' paper has distinctly two parts: the first one sets the stage for natural exponential families, or NEF (Sections 1, 2, 7, 9, and 10); the other one concentrates on NEF with quadratic variance functions, called QVF by Morris. Let us call their set "Morris class," he deserves it.

To define a NEF on $\mathbb{R}$, Morris starts from an arbitrary Stieltjes measure dF on $\mathbb{R}$ (implicitely such that the interval

$$\Theta = \{\theta \in \mathbb{R}; \exp \psi(\theta) = \int \exp(\theta x)\, dF(x) < +\infty\}$$

has a nonempty interior). The NEF generated by dF is the set of probabilities on $\mathbb{R}$

$$dF_\theta(x) = \exp(\theta x - \psi(\theta))\, dF(x),$$

when θ varies in Θ. Since ψ is real analytic, and since

$$\psi'(\theta) = \int x\, dF_\theta(x)$$

in his formula (2.5), then $\Omega = \psi'(\Theta)$ can be called the domain of the means of the NEF. Since ψ is a strictly convex function (well, the trivial case where dF is a Dirac mass has to be excluded), the $\theta \mapsto \mu = \psi'(\theta)$ is one-to-one from Θ to Ω. We introduce now the main character with $V(\mu)$, the variance of $dF_{(\psi')^{-1}(\mu)}$. This function V on Ω is called the variance function. Since $\psi'' = V(\psi')$, up to an unimportant constant, the Laplace transform $\exp \psi$ can be retrieved by this differential equation and V characterizes the NEF.

* This paper was prepared with the partial support of National Science Foundation Grant GP 38485.

Of course, there is some sloppiness here. Since Θ may not be an open interval (consider the case of the stable law

$$dF(x) = (\sqrt{2\pi})^{-1} p x^{-3/2} \exp\left(-\frac{p^2}{2x}\right) \prod_{(0,+\infty)}(x)\, dx,$$

where $\Theta = (-\infty, 0])$, then (2.5) can be infinite at the ends, Ω may not be open (as claimed 11 lines after (2.9)) and can contain points at infinity. The cure is to replace Θ by its interior, as done by Mora (1986), to have correct statements. Another sloppiness is to declare—again 11 lines after (2.9)—that Ω is equal to the interior of the closed convex support C of dF. This is clearly false; look at the examples:

$$dF(x) = (1 + x^4)^{-1} \exp(|x|)\, dx,$$

or

$$dF(x) = \sum_{n=1}^{\infty} \frac{1}{n^3}\, \delta_n(dx).$$

Quoting Efron (1978) is even surprising, since this paper by Efron shows (in $\mathbb{R}^2$) that the domains of the means are not necessarily convex. In other words, the phenomena of steepness is ignored by Morris. Barndorff-Nielsen (1978) tells the true story. On $\mathbb{R}$ it reads: if

$$\operatorname{int} \Theta = (\alpha, \beta), \qquad \Omega = (a, b), \qquad C = (a', b'),$$

then $b < b'$ if and only if β belongs to Θ and $\psi'(\beta)$ is infinite. But honestly, a simple proof of the theorem of steepness would be desirable. The one given by Barndorff-Nielsen is scattered in two or three different chapters, and although it is dealing with analytic functions, it uses the heavy theory of nonsmooth convex functions of Rockafellar.

Anyway, these slips are minor. Morris isolates and unifies important concepts (NEF, Θ, Ω, V, large deviations in one dimension, orthogonal polynomials as related to NEF) in a much clearer way that anybody before. Actually many definitions of exponential families given in the literature are quite messy, with two main defects: (i) the domain of the parameters is not clearly specified; and (ii) we have to face artificial generalities, like

$$P_\alpha(d\omega) = a(\alpha) b(\omega) \exp(\varphi_1(\alpha) t_1(\omega) + \cdots + \varphi_n(\alpha) t_n(\omega)) m(d\omega),$$

where α is in $\mathbb{R}^n$. Instead of playing with the above monster, Morris says (at least for $n = 1$): first consider the image dF in $\mathbb{R}^n$ of the measure $b(\omega)m(d\omega)$ by $\omega \mapsto x = T(\omega) = (t_1(\omega), \ldots, t_n(\omega))$, then use the natural parameters $(\theta_1, \ldots, \theta_n) = (\varphi_1(\alpha), \ldots, \varphi_n(\alpha))$ and write $\exp -\psi(\theta)$ instead of $a(\alpha)$: then you will get a tool well adapted to theoretical work. Note also that Morris ignores the incorrect expressions "The model belongs to *the* exponential family" (instead of "the model is *an* exponential family") and "This distribution is a power series distribution" (instead of "this model is a NEF concentrated on $\mathbb{N}$): a distribution is not a model, i.e., a set of distributions; and any distribution on $\mathbb{N}$ belongs to a NEF, thus would be a "power series distribution."

The second part (Sections 3, 4, 5, 6, 7*, and 8) deals with the Morris class of NEF, i.e., with NEF such that the variance function in the restriction to Ω of the polynomial with degree ≤ 2. He proves that there exists six types of such NEF, after setting on page 68 that two NEF will be said to be equivalent (i.e., of the same type) if we pass from one to the other by "linear transformation" and "convolution and division." (More accurately "linear" should be "affine," and "convolution and division" is merely replacing ψ by $\lambda\psi$, where λ can be real, not only rational.)

It has been pointed out (e.g., Shanbhag (1994)) that classification is not that new. Consider actually the following statements about the NEF $\{dF_\theta; \theta \in \Theta\}$

(A) The NEF is either normal, Poisson, Gamma, Negative-Binomial, Binomial, or GHS.
(B) V is quadratic.
(C) If X and Y are independent with distributions F_{θ_0} and $F_{\theta_0}^{-\lambda}$, then $\mathbb{E}(X|X+Y)$ is at most quadratic in $X+Y$.
(D) If $m_0 = \int x\,dF_{\theta_0}$ and if $(P_n)_{n=0}^{\infty}$ are the associated orthogonal monic polynomials, then there exist $R > 0$ and two analytic functions a and b on $(-R, R)$ such that for all x in $\mathbb{R}$

$$\sum_{n=0}^{\infty} \frac{z^n}{n!} P_n(x) = \exp(a(z)(x - m_0) + b(z)).$$

(E) Write $f(x,m) = \exp x(\psi'^{-1}(m) - \psi((\psi')^{-1}(m)))$ for m in Ω and consider the polynomial $J_n(x) = (\partial/\partial m)^n f(x,m)_{m=m_0}$. Then the $(J_n)_{n=0}^{\infty}$ are orthonormal with respect to dF_{θ_0}.
(F) The Shanbhag condition, about the associated Bhattacharya's matrices.

These six conditions are equivalent.

Implications from (A) toward the others is always proved by inspection. But the oldest result of this nature is due to Meixner (1934), who proves (D) $\Rightarrow$ (A). For this reason, the Morris class could be called the Meixner class. (C) $\Rightarrow$ (A) is due to Laha and Lukacs (1959), (E) $\Rightarrow$ (A) is due to Feinsilver (1986)—but see Letac (1992) for a clearer proof—(F) $\Rightarrow$ (A) is due to Shanbhag (1972/1979). With no exception, these results are proved via a form of (B) which is "ψ'' is a quadratic polynomial in ψ'" where the important concept of variance function does not appear clearly. Morris considers the proof of (B) $\Rightarrow$ (A) as obvious. As it is, I have just the minor complaint that the fact that Ω is what it should be is not quite clear. Actually, it should rely on a "principle of maximal analyticity" for the Laplace transforms

* Section 7 contains general formulae connecting moments and cumulants and moments, and is relevant to both parts of the paper.

which can be found for instance in Kawata (1972) or in Theorem 3.1 of Letac and Mora (1990). Note that (E) $\Rightarrow$ (A) is mentioned by Morris (four lines before Corollary 1 on page 77) but E is taken for all m_0, which makes a stronger hypothesis than that of Feinsilver.

Morris devotes Sections 4 and 5 to the GHS distribution, which has also been studied by Lai and Vere-Jones (1975). Fifteen years after, this GHS is still a little mystery. How can a distribution, which occurs so naturally in a computation, be lacking of probabilistic interpretation and thus of statistical usefulness? Let me recall that its only known appearance is by taking a standard plane Brownian motion $Z(t) = X(t) + iY(t)$, by considering $T = \inf\{t; |Y(t)| = 1\}$ and—with the help of the martingale $\exp iZ(t)$—by showing that $\mathbb{E}(\exp \theta T) = (\cos\theta)^{-1}$ for $|\theta| < \pi/2$. But extension of a similar interpretation to $\cos\theta_0(\cos(\theta_0 + \theta))^{-1}$ is even lacking.

In the recent history of statistics, the Morris paper has played two roles: first, it has changed our way of seeing exponential families, by providing clearer concepts, by throwing away useless objects, and keeping the kernel: NEF, domain of the means, parametrization by the mean. Second, it has opened the new topic of variance functions. Although their study has been considered as highly questionable by some statisticians (Mc Cullagh, 1991; Barndorff-Nielsen, 1991), the variance functions appear to be:

a compact and powerful way to characterize a given NEF; for instance, the magnificent and largely ignored result by Olkin and Rubin (1962) has now a much simpler proof, by considering the variance functions of Wishart families (Casalis and Letac, 1996).

a systematic tool, to give some order in the zoo of distribution functions: see Bar-Lev and Enis (1987) and Jorgensen (1987) for the stable laws, Casalis (1991) and (1997) for multivariate quadratic NEF, Mora (1986) and Letac and Mora (1990) for the cubic variance functions, which are associated in the literature to Schrödinger, Abel, Borel-Tanner, Consul, D. Kendall, and Takaćs. (These cubic distributions are the hitting times of basic processes like Brownian, Poisson, or gamma processes). We can even dream of an ideal Johnson and Kotz treatise, where distributions would be logically ordered, by considering first the distributions with the simplest variance functions, and second the distributions which are the beta distributions of the preceeding ones, as sketched by Morris (1983) (for instance, the hypergeometric distribution on $\{0, \ldots, n\}$ is the beta of the binomial).

Nobody who studies exponential families can now afford to avoid reading the masterpiece written by C. Morris.

References

Bar-Lev, S. and Enis, P. (1987). Reproducibility and natural exponential families with power variance functions. *Ann. Statist.*, **15**, 1507–1522.

Barndorff-Nielsen, O. (1978). *Information and Exponential Families in Statistical Theory*. Wiley, New York.
Barndorff-Nielsen, O. (1991). *Proceedings of the 48th Session of the International Statistical Institute*, **54**, Book 4.
Casalis, M. (1991). Les familles exponentielles à variance quadratique homogène sont des lois de Wishart sur un cône symétrique. *C. R. Acad. Sci. Paris. Sér. I*, **312**, 537–540
Casalis, M. (1997). The $2d+4$ simple quadratic natural exponential families on $\mathbb{R}^d$. *Ann. Statist.*, **25**.
Casalis, M. and Letac, G. (1996). The Lukacs–Olkin–Rubin characterization of Wishart distributions on symmetric cones. *Ann. Statist.*, **24**, 763–786.
Efron, B. (1978). The geometry of exponential families. *Ann. Statist.*, **6**, 362–376.
Feinsilver, P. (1996). Some classes of orthogonal polynomials associated with martingales. *Proc. Amer. Math. Soc.*, **98**, 298–302.
Jorgensen, B. (1987). Exponential dispersion models. *J. Roy. Statist. Soc. Ser. B*, **49**, 127–162.
Kawata (1972). *Fourier Analysis and Probability*. Wiley, New York.
Laha, R.G. and Lukacs, E. (1960). On a problem connected with quadratic regression. *Biometrika*, **47**, 335–343.
Lai, C.D. and Vere-Jones, D. (1975). Odd man out. The Meixner hypergeometric distribution. *Austral. J. Statist.*, **21**, 256–265.
Letac, G. (1992). *Lectures on Natural Exponential Families and Their Variance Functions*. Monografias de Matemática, vol. **50**. Instituto de Matemática Pura et Aplicada, Rio de Janeiro.
Letac, G. and Mora, M. (1990). Natural real exponential families with cubic variance functions. *Ann. Statist.*, **18**, 1–37.
McCullagh, P. (1991). *Proceedings of the 48th Session of the International Statistical Institute*, vol. **54**, Book 4.
Meixner, J. (1934). Orthogonale Polynomsysteme mit einer besonderen Gestalt der erzeugenden function. *J. London Math. Soc.*, **9**, 6–13.
Mora, M. (1986). Classification des fonctions variance cubiques des familles exponentielles sur $\mathbb{R}$. *C.R. Acad. Sci. Paris Sér. I*, **302**, 587–590.
Morris, C.N. (1983). Natural exponential families with quadratic variance functions: Statistical theory. *Ann. Statist.*, **10**, 517–529.
Olkin, I. and Rubin, H. (1962). A characterization of the Wishart distribution. *Ann. Math. Statist.*, **33**, 1272–1280.
Shanbhag, D. (1972). Some characterizations based on the Bhattacharyya matrix. *J. Appl. Probal.*, **9**, 580–587.
Shanbhag, D. (1979). Diagonality of the Bhattacharyya matrix as a characterization. *Teor. Veroyatnos.i Primenen.*, **24**, 424–427.
Shanbhag, D. (1994). Review of Letac (1992). *Math. Rev.*, **94f**: 60020.

Natural Exponential Families with Quadratic Variance Functions*

Carl N. Morris
University of Texas, Austin

Abstract

The normal, Possion, gamma, binomial, and negative binomial distributions are univariate natural exponential families with quadratic variance functions (the variance is at most a quadratic function of the mean). Only one other such family exists. Much theory is unified for these six natural exponential families by appeal to their quadratic variance property, including infinite divisibility, cumulants, orthogonal polynomials, large deviations, and limits in distribution.

1. Introduction

The normal, Poisson, gamma, binomial, and negative binomial distributions enjoy wide application and many useful mathematical properties. What makes them so special? This paper says two things: (i) they are *natural exponential families* (NEFs); and (ii) they have *quadratic variance functions* (QVF), i.e., the variance $V(\mu)$ is, at most, a quadratic function of the mean μ for each of these distributions.

Section 2 provides background on general exponential families, making two points. First, because of some confusion about the definition of exponential families, the terms "natural exponential families" and "natural observations" are introduced here to specify those exponential families and

* Support for this work provided by NSF Grant No. MCS-8003416 at the University of Texas, and by a grant from the Department of Health and Human Services to The Rand Corporation, where the author is currently a consultant, is gratefully acknowledged.

random variables whose convolutions comprise one exponential family. Second, the "variance function" $V(\mu)$ is introduced as a quantity that characterizes the NEF.

Only six univariate, one-parameter families (and linear functions of them) are natural exponential families having a QVF. The five famous ones are listed in the initial paragraph. The sixth is derived in Section 3 as the NEF generated by the hyperbolic secant distribution. Section 4 shows this sixth family contains infinitely divisible, generally skewed, continuous distributions, with support $(-\infty, \infty)$.

In Sections 6 through 10, natural exponential families with quadratic variance functions (NEF–QVF) are examined in a unified way with respect to infinite divisibility, cumulants, orthogonal polynomials, large deviations, and limits in distribution. Other insights are obtained concerning the possible limit laws (Section 10), and the self-generating nature of infinite divisibility in NEF–QVF distributions.

This paper concentrates on general NEF–QVF development, emphasizing the importance of the variance function $V(\mu)$, the new distributions, and the five unified results. Additional theory for NEF–QVF distributions, e.g., concerning classical estimation theory, Bayesian estimation theory, and regression structure, will be treated in a sequel to this paper. Authors who have established certain statistical results for NEF–QVF distributions include Shanbag (1972) and Blight and Rao (1974) for Bhattacharya bounds; Laha and Lukacs (1960) and Bolger and Harkness (1965) for quadratic regression; and Duan (1979) for conjugate priors.

2. Natural Exponential Families and Variance Functions

A parametric family of distributions with natural parameter set $\Theta \subset R$ (the real line) is a univariate exponential family if random variables Y governed by these distributions satisfy

$$P_\theta(Y \in A) = \int_A \exp\{\theta T(y) - \psi(\theta)\}\xi(dy) \tag{2.1}$$

for some measure ξ not depending on $\theta \in \Theta$, $A \subset R$ a measurable set, and T a real valued measurable function (Lehmann, 1959, Barndorff-Nielsen, 1978). Any factor of the density of y not depending on θ is absorbed into ξ. In (2.1), θ is called the *natural parameter* and Θ, which is the largest set (an interval, possibly infinite) for which (2.1) is finite when $A = R$, is called the *natural parameter space*. Often θ is a nonlinear function of some more familiar parameter. The function $\psi(\theta)$ is determined by ξ so that (2.1) has unit probability if $A = R$.

The natural observation in (2.1) is $X = T(Y)$. Its distribution belongs to a *natural exponential family* (NEF).

$$P_\theta(X \in A) = \int_A \exp\{\theta x - \psi(\theta)\}\, dF(x) \tag{2.2}$$

with F a Stieltjes measure on R. If $0 \in \Theta$ and $\psi(0) = 0$ then F is a cumulative distribution function (CDF). Otherwise, choose any $\theta_0 \in \Theta$ and let $dF_0(x) \equiv \exp\{\theta_0 x - \psi(\theta_0)\}\, dF(x)$. Then F_0 is a CDF, and via (2.2) generates the same exponential family as F. Hence F in (2.2) is taken to be a CDF without loss of generality.

The modifier "natural" is needed to distinguish NEFs from exponential families not satisfying (2.2). For example, $Y \sim \text{Beta}(m\mu, m(1-\mu))$ and $Y \sim \text{Lognormal}(\alpha, \sigma^2)$ satisfy (2.1) with $T(y) = m\log(y/(1-y)), \theta = \mu$ and $T(y) = \log(y), \theta = \alpha/\sigma^2$, respectively, but they are not NEFs. Convolutions of NEFs, $\sum X_t = \sum T(Y_t)$, are exponential families with natural parameter θ (and NEFs), but $\sum Y_t$, is not, unless T is linear. Cumulants of $X = T(Y)$ are derivatives of $\psi(\theta)$, but no simple expression yields cumulants of Y. Thus the results being developed here pertain only to NEFs.

Every univariate CDF F_0 possessing a moment generating function (MGF) in a neighborhood of zero *generates* a NEF, as follows. Define the cumulant generating function (CGF) $\psi(\theta)$ on Θ, the largest interval for which (2.3) exists, by

$$\psi(\theta) \equiv \log \int \exp(\theta x)\, dF_0(x), \qquad \theta \in \Theta. \tag{2.3}$$

Then the CDFs $F_\theta, \theta \in \Theta$, defined by

$$dF_\theta(x) \equiv \exp(\theta x - \psi(\theta))\, dF_0(x) \tag{2.4}$$

form a NEF, with F_θ a CDF. The NEF so generated was called a *conjugate family*, (Khinchin, 1949) predating its use by Bayesians for quite a different purpose, and plays an important role in expansion theory (Barndorff-Nielsen and Cox, 1979). Given any $\theta^* \in \Theta$. F_θ^* generates the same NEF (2.3) and (2.4). Thus the NEF is closed in that it can be generated by any of its members. The mean, variance, and CGF are

$$\mu = E_\theta X = \int x\, dF_\theta(x) = \psi'(\theta), \tag{2.5}$$

$$V(\mu) = \text{Var}_\theta(X) = \int (x-\mu)^2 dF_\theta(x) = \psi''(\theta), \tag{2.6}$$

$$\psi_\theta(t) = \log \int \exp(tx)\, dF_\theta(x) = \psi(t+\theta) - \psi(\theta). \tag{2.7}$$

In (2.3), $\psi(0) = 0$ since F_0 is a distribution function, so $F_\theta = F_0$ when $\theta = 0$ in (2.4). The results in (2.5) and (2.6) follow for $r = 1, 2$ from (2.7) and the property of the CGF that the rth cumulant, C_r of F_θ is given by

$$C_r \equiv \left. \frac{d^r \psi_\theta(t)}{dt^r} \right|_{t=0} = \psi^{(r)}(\theta). \tag{2.8}$$

The range of $\psi'(\theta)$ over $\theta \in \Theta$ will be denoted Ω, i.e., $\Omega = \psi'(\Theta)$. Ω is an interval (possibly infinite) of R. In applications, the mean $\mu \in \Omega$ of the distribution is a more standard parameterization that $\theta \in \Theta$ for most NEFs.

The variance $\psi''(\theta)$ depends on $\mu = \psi'(\theta)$ as does every function of θ because μ is a $1-1$ function of θ since $\psi''(\theta) > 0$. Formula (2.6) expresses this fact and we name $V(\mu)$, together with its domain Ω, the *variance function* (VF) of the NEF.

The variance function (Ω, V) characterizes the NEF, but no particular member of the NEF, because it determines the CGF, and hence the characteristic function, as follows. Given $V(\bullet)$ and μ, with $0 < V(\mu_0) < \infty$, define Ω as the largest interval containing μ_0 such that $0 < V(m) < \infty$ for all $m \in \Omega$. (Any other $\mu_0 \in \Omega$ regenerates Ω in the same way.) Now define $\psi(\bullet)$ by

$$\psi\left(\int_{\mu_0}^{\mu} \frac{dm}{V(m)}\right) = \int_{\mu_0}^{\mu} \frac{m\,dm}{V(m)} \tag{2.9}$$

for all $\mu \in \Omega$. Note that $\theta = \int_{\mu_0}^{\mu} dm/V(m)$ and the range of θ as μ varies over Ω is Θ, the natural parameter space. Validity of (2.9) follows from differentiation with respect to μ. In (2.9), $\psi(0) = 0, \psi'(0) = \mu_0$, and $\psi''(0) = V(\mu_0)$.

Observe that V without Ω may not characterize the NEF. For example, $V(\mu) = \mu^2$ characterizes two different NEFs, one with $\Omega_1 = (-\infty, 0)$ and the other with $\Omega_2 = (0, \infty)$. These correspond to the usual exponential distribution (Ω_2) and the negative of the usual exponential distribution (Ω_1), the sets being separated by points (in this case one point) with $V(\mu) \le 0$. If F_0 is not the CDF of a degenerate distribution in (2.3) and (2.4), $V(\mu)$ is strictly positive throughout Ω.

The mean-space Ω of a NEF is the interior of the smallest interval containing the support $\mathscr{X}$ of F in (2.2). Thus $\Omega = \mathscr{X}$ if $\mathscr{X}$ is an open interval, and the closure of Ω always contains $\mathscr{X}$ (Efron, 1978, Section 7).

The variance function characterizes the particular NEF within all NEFs, but not within a wider family of distributions. For example, the beta and lognormal distributions mentioned after (2.2) have QVF with

$$V_1(\mu) = \mu(1-\mu)/(m+1)$$

and

$$V_2(\mu) = \mu^2\{\exp(\sigma^2) - 1\}, \qquad \mu \equiv \exp(\alpha + \sigma^2/2),$$

respectively. These VFs match those of the binomial and gamma, respectively, and so fail to characterize the family within all exponential families.

Formula (2.8) is easily modified to generate cumulants of any NEF in terms of the mean $\mu = \psi'(\theta)$. If $C_r(\mu)$ is the rth cumulant expressed in terms of μ, then $C_1(\mu) = \mu$, $C_2(\mu) = V(\mu)$, and

$$C_{r+1}(\mu) = V(\mu)C_r'(\mu), \qquad r \geq 2, \tag{2.10}$$

primes denoting derivatives wrt μ. We have

$$C_{r+1}(\mu) = \psi^{(r+1)}(\theta) = d\psi^{(r)}(\theta)/d\theta$$
$$= \left\{\frac{d\psi^{(r)}(\theta)}{d\mu}\right\}\left(\frac{d\mu}{d\theta}\right) = C_r(\mu)C_2(\mu),$$

which is (2.10). Thus, suppressing dependence on μ,

$$C_3(\mu) = V'V, \qquad C_4(\mu) = V(V')^2 + V^2V'', \tag{2.11}$$

are expressed in terms of derivatives of the variance function, as can be higher cumulants, using higher derivatives of $V(\mu)$.

3. NEFs with Quadratic Variance Functions

The development in Section 2 shows that many NEFs exist. A few have quadratic variance function (QVF)

$$V(\mu) = v_0 + v_1\mu + v_2\mu^2. \tag{3.1}$$

Those with QVF include the normal $N(\mu, \sigma^2)$ with $V(\mu) = \sigma^2$ (constant variance function): the Poisson, Poiss(λ) with $\mu = \lambda$, $V(\mu) = \mu$ (linear); the gamma Gam$(r, \lambda), \mu = r\lambda$, $V(\mu) = r\lambda^2 = \mu^2/r$; the binomial, Bin$(r,p), \mu = rp$, $V(\mu) = rpq = -\mu^2/r + \mu(q = 1 - p)$; and the negative binomial NB(r,p), the number of successes before the rth failure, p =probability of success, $\mu = rp/q$, $V(\mu) = rp/q^2 = \mu^2/r + \mu(q \equiv 1 - p)$. Table 1 lists properties of these distributions.

The four operations: (i) linear transformation $X \to (X - b)/c$; (ii) convolution; (iii) division, producing $\mathscr{P}(X_1)$ from $\mathscr{P}(X)$ with $X = X_1 + \cdots + X_n$, X_t iid, as discussed below, $\mathscr{P}(Z)$ signifying the law of Z; and (iv) generation of the NEF as in (2.3), (2.4), all produce NEFs, usually different from that of X, carrying each member of the original NEF to the same new NEF. Thus NEFs are equivalance classes under these operations. These operations also preserve the QVF property, as shown next, and so preserve NEF–QVF structure.

If X has a NEF distribution (2.4) and $V(\mu)$ is quadratic, as in (3.1), we shall say X has a NEF–QVF distribution. Let $c \neq 0$ and b be constants, and let $X^* = (X - b)/c$ be a linear transformation of X with mean, $\mu^* = (\mu - b)/c$. Then $V^*(\mu^*) \equiv \text{Var}(X^*) = V(\mu)/c^2 = V(c\mu^* + b)/c^2$, so

$$V^*(\mu^*) = V(b)/c^2 + V'(b)\mu^*/c + v_2(\mu^*)^2. \tag{3.2}$$

Thus $v_2 \to v_2, v_1 \to V'(b)/c$, and $v_0 \to V(b)/c^2$ if $X \to (X - b)/c$.

Table 1. Univariate Natural Exponential Families with Quadratic Variance Functions.

	Normal $N(\lambda, \sigma^2)$	Poisson Poiss(λ)	Gamma Gam(r, λ)
Density	$\frac{1}{\sigma\sqrt{2\pi}} e^{-(x-\lambda)^2/2\sigma^2}$	$\frac{\lambda^x e^{-\lambda}}{x!}$	$\left(\frac{x}{\lambda}\right)^{r-1} \frac{e^{-x/\lambda}}{\lambda\Gamma(r)}$
	$-\infty < x < \infty$		$0 < x < \infty$
	$-\infty < \lambda < \infty$	$x = 0, 1, 2, \ldots.$	$0 < \lambda < \infty$
	$\sigma^2 > 0$	$0 < \lambda < \infty$	$r > 0$
Inf. Divis. (Sec. 6)	Yes	Yes	Yes
Elem. Distn.	$N(\lambda, 1)$	Poiss(λ)	Exponential(λ)
θ	λ/σ^2	$\log \lambda$	$-1/\lambda$
Θ	$(-\infty, \infty)$	$(-\infty, \infty)$	$(-\infty, 0)$
$\psi(\theta)$	$\frac{\sigma^2\theta^2}{2} = \frac{\lambda^2}{2\sigma^2}$	$e^\theta = \lambda$	$-r \log(-\theta)$ $= r \log(\lambda)$
mean $= \mu = \psi'(\theta)$	$\lambda = \theta\sigma^2$	$\lambda = e^\theta$	$r\lambda = -r/\theta$
Ω	$(-\infty, \infty)$	$(0, \infty)$	$(0, \infty)$
$V(\mu) = \psi''(\theta)$		$\lambda = e^\theta$	$r\lambda^2 = r/\theta^2$
$= v_2\mu^2 + v_1\mu + v_0$	$v_0 = \sigma^2$	$= \mu$	$= \mu^2/r$
$d = v_1^2 - 4v_0v_2$	0	1	0
$\log E_\theta e^{t,X}$ $= \psi(t+\theta) - \psi(\theta)$	$t\lambda + t^2\sigma^2/2$	$\lambda(e^t - 1)$	$-r \log(1 - t\lambda)$
Orthog. Polynomials (Sec. 8)	Hermite	Poisson–Charlier	Generalized Laguerre
Exponential Large Deviations (Sec. 9), $\varepsilon > 0$	No Conditions	$\lambda \geq \varepsilon$	$r \geq \varepsilon$
Limit Laws (Sec. 10)	Normal	Normal	Normal

(*continued*)

Define $d \equiv v_1^2 - 4v_0v_2$ to be the *discriminant of* $V(\mu)$. Since

$$\{V'(\mu)\}^2 = (2v_2\mu + v_1)^2 = 4v_2V(\mu) + d, \tag{3.3}$$

the discriminant d^* of (3.2) is $[V'(b)/c]^2 - 4v_2V(b)/c^2 = d/c^2$. Thus d is unaltered by translations, and $d \to d/c^2$ if $X \to (X - b)/c$.

Now let $X_i, i = 1, \ldots, n$, be independent identically distributed (iid) as a NEF–QVF, with $V(\mu)$ as in (3.1). Define $X^* = (X_1 + \cdots + X_n - nb)/c$. Then X^* has a NEF–QVF distribution, $\mu^* \equiv EX^* = n(\mu - b)/c, \Omega^* = (\Omega - b)n/c$, and

$$\begin{aligned} V^*(\mu^*) &= nV(b + c\mu^*/n)/c^2 \\ &= nV(b)/c^2 + V'(b)\mu^*/c + v_2(\mu^*)^2/n. \end{aligned} \tag{3.4}$$

Thus, if $X \to (X_1 + \cdots + X_n - nb)/c$, then

$$v_0 \to nV(b)/c^2, \qquad v_1 \to V'(b)/c, \qquad v_2 \to v_2/n, \tag{3.5}$$

Table 1 (*continued*)

	Binomial Bin(r,p)	Negative Binomial NB(r,p)	NEF–GHS NEF–GHS(r,λ)
Density	$\binom{r}{x}p^x q^{r-x}$	$\dfrac{\Gamma(x+r)}{\Gamma(r)x!}p^x q^r$	Sec. 5
	$x=0,1,\ldots,r$	$x=0,1,2,\ldots.$	$-\infty<x<\infty$
	$0<p<1$	$0<p<1$	$-\infty<\lambda<\infty$
	$r=1,2,\ldots.$	$r>0$	$r>0$
Inf. Divis. (Sec. 6)	No	Yes	Yes
Elem. Distn.	Bernoulli (p)	Geometric (p)	NEF–HS(λ)
θ	$\log(p/q)$	$\log(p)$	$\tan^{-1}(\lambda)$
Θ	$(-\infty,\infty)$	$(-\infty,0)$	$\left(-\frac{\pi}{2},\frac{\pi}{2}\right)$
$\psi(\theta)$	$-r\log(q)$	$-r\log(q)$	$-r\log(\cos(\theta))$
	$=r\log(1+e^\theta)$	$=-r\log(1-e^\theta)$	$=(r/2)\log(1+\lambda^2)$
mean $=\mu=\psi'(\theta)$	$rp=r/(1+e^{-\theta})$	$\dfrac{rp}{q}=\dfrac{r}{e^{-\theta}-1}$	$r\lambda=r\tan(\theta)$
Ω	$(0,r)$	$(0,\infty)$	$(-\infty,\infty)$
$V(\mu)=\psi''(\theta)$	$rpq=\dfrac{re^\theta}{(1+e^\theta)^2}$	rp/q^2	$r(1+\lambda^2)$
$=v_2\mu^2+v_1\mu+v_0$	$=\mu^2/r+\mu$	$=\mu^2/r+\mu$	$=\mu^2/r+r$
$d=v_1^2-4v_0v_2$	1	1	-4
$\log E_\theta e^{t,X}$ $=\psi(t+\theta)-\psi(\theta)$	$r\log(pe^t+q)$	$r\log\left(\dfrac{q}{1-pe^t}\right)$	$-r\log(\cos(t)$ $-\lambda\sin(t))$
Orthog. Polynomials (Sec. 8)	Krawtchouk	Meixner	Pollaczek
Exponential Large Deviations (Sec. 9), $\varepsilon>0$	$\varepsilon\le p\le 1-\varepsilon$	$r,p\ge\varepsilon$	$r\ge\varepsilon$
Limit Laws (Sec. 10)	Normal Poisson	Normal Poisson Gamma	Normal Gamma

and the discriminant

$$d\to d/c^2. \tag{3.6}$$

Formulas (3.4) and (3.5) show that the QVF property is preserved and how the VF changes under convolution and linear transformation.

The convolution operation sometimes can be reversed. If X follows a NEF–QVF distribution, suppose n can be given so that $X=X_1^*+\cdots+X_n^*$ for iid variables $\{X_1^*\}$. This is possible for any n with infinitely divisible distributions, but otherwise only in certain cases. Distributions permitting this for some $n\ge 2$ are termed *divisible* here, paralleling the infinitely divisible terminology. The operation producing $\mathscr{P}(X_1^*)$ from $\mathscr{P}(X)$ is termed *division*, and X_1^* or $\mathscr{P}(X_1^*)$ is called a *divisor* of X or of $\mathscr{P}(x)$. The VF of X_1^* is obtained from that of X by reversing (3.5), taking $b = 0, c = 1$. Then $v_0 \to v_0/n, v_1 \to v_1$, and $v_2 \to nv_2$ under division.

If more generally for some $n, X = c(X_1^* + \cdots + X_n^*) + b$, i.e., X_1^* is the n-divisor of $(X - b)/c$, then $\mu^* = EX_1^* = (\mu - b)/nc$ and $V^*(\mu^*) = \text{Var}(X_1^*) = V(b + cn\mu^*)/nc^2$. This is (3.4), replacing n by $1/n$. Thus (3.5) unifies n-convolutions and n-divisions of X, using $1/n$ to represent division. By combining these two operations, n can take fractional values, and for infinitely divisible distributions n can be any positive real number. Neither the convolution nor the division operation affects the discriminant d, as shown by (3.6).

Finally, in NEF–QVF distributions, $\psi(\theta)$ is related to θ and $V(\mu)$ by

$$\log(V(\mu)/V(\mu_0)) = 2v_2[\psi(\theta) - \psi(\theta_0)] + v_1[\theta - \theta_0], \tag{3.7}$$

with $\mu = \psi'(\theta), \mu_0 = \psi'(\theta_0)$. Formula (3.7) is non-trivial except in the normal case ($v_1 = v_2 = 0$), and is proved by differentiating wrt θ.

4. Finding All NEFs with QVF: The Sixth Family

Because the normal distribution $N(\mu, \sigma^2)$ has $V(\mu) = \sigma^2$, and the VF characterizes the NEF, it follows that the normal distribution is the unique NEF with constant variance function ($v_2 = v_1 = 0$). Similarly, the Poisson distributions, including linear transformations of the usual Poisson, are the only NEFs with strictly linear variance function, $V(\mu) = v_1\mu + v_0, v_1 \neq 0$.

Let us characterize all *strictly* quadratic NEF–QVF distributions. Suppose X has a NEF–QVF distribution with VF (3.1), and $v_2 \neq 0$. Define $X^* = aV'(X)$, taking $a = 1$ if $d = 0, a = |dv_2|^{1/2}$ otherwise. Then X^* is a linear function of X with variance function given by (3.2),

$$V^*(\mu^*) = s + v_2(\mu^*)^2, \qquad s = -\text{sgn}(dv_2). \tag{4.1}$$

We may regard (4.1) as the *canonical* member under linear transformations of the NEF–QVF with $v_2 \neq 0$ specified ($v_1 = 0, s = 0, \pm 1$). All other quadratic VFs, i.e., all other v_0, v_1 can be obtained from the canonical VF (4.1) by the linear transformation $X = (X^*/a - v_1)/2v_2$, the inverse of the transformation $X^* = aV'(X)$.

Six cases with $v_2 \neq 0$ in (4.1) correspond to combinations of $v_2 < 0$, $v_2 > 0$ and $s = -1, 0, 1$. The two having $v_2 < 0$ with $s = -1$ and $s = 0$ make $V(\mu) < 0$, so are impossible. Three others correspond to (linear transformations of) the gamma ($v_2 > 0, s = 0$), the binomial ($v_2 < 0, s = 1$), and the negative binomial ($v_2 > 0, s = -1$), cf. Table 1.

The case $v_2 > 0, s = 1$ remains. For $v_2 = 1$, the missing distribution is the natural observation of a beta exponential family. Let $y \sim \text{Beta}(0.5 +$

$\theta/\pi, 0.5-\theta/\pi), |\theta| < \pi/2$. The natural observation for this exponential family is $x \equiv \log\{y/(1-y)\}/\pi$, having density

$$f_{1,\theta}(x) \equiv \frac{\exp[\theta x + \log\{\cos(\theta)\}]}{2\cosh(\pi x/2)} \tag{4.2}$$

with respect to Lebesgue measure and support $-\infty < x < \infty$. The proof that X has this density follows almost immediately from the reflection formula (Abramowitz and Stegun, 1965, page 256), which implies that $\beta(0.5+t, 0.5-t) = \pi/\cos(\pi t)$.

The mean and variance of (4.2) are derived by differentiating the CGF $\psi(\theta) = -\log(\cos(\theta))$ to get $EX = \mu = \psi'(\theta) = \tan(\theta)$, Var $X = V(\mu) = \psi''(\theta) = \csc^2(\theta) = 1 + \mu^2$, so X has a NEF–QVF distribution. Convolutions (r times) and infinite divisibility of (4.2), to be discussed in the next section, yield all distributions with $V(\mu) = r + \mu^2/r$, per the discussion surrounding (3.5), for any $r > 0$.

Linear transformations yield other distributions for a specified $v_2 = 1/r > 0$, producing arbitrary v_0, v_1, subject to $d = v_1^2 - 4v_0v_2 < 0$; from (3.3), linear transformations must preserve the sign of the discriminant, and $d = -4 < 0$ from $V(\mu)$ above. Then if X^* has $V^*(\mu^*) = r + (\mu^*)^2/r, X = X^*(v_0/r - v_1^2/4)^{1/2} - rv_1/2$ has VF

$$V(\mu) = v_0 + v_1\mu + \mu^2/r, \qquad r > 0, \qquad v_1^2 < 4v_0/r. \tag{4.3}$$

When $\theta = 0$, (4.2) is called the "hyperbolic secant" (HS) distribution (Johnson and Kotz, 1970). Convolution and infinite division when $\theta = 0$ produces the "generalized hyperbolic secant" (GHS) distributions (Baten, 1934; Harkness and Harkness, 1968), all symmetric distributions.

The NEFs generated by the HS distribution (and generated by their linear transformations) appear to be new. They have VFs given by (4.3), and are skewed when $\theta \neq 0$. The natural exponential family generated by the generalized hyperbolic secant distributions will be referred to as the *NEF–GHS* distributions, reserving *NEF–HS* for (4.2). Properties of these new distributions are developed in the following sections.

We have just found that all univariate natural exponential families with quadratic variance functions are the normal (constant). Poisson (linear), gamma, binomial, negative binomial, and NEF–GHS distributions and linear transformations of these. These six distributions are summarized in Table 1.

Each of these distributions contains a subfamily, which we term the family of *elementary distributions*, with leading coefficient in $V(\mu)$ having unit magnitude ± 1. These are the normal distribution with unit variance, ($v_0 = 1$), the usual Poisson ($v_1 = 1$), the exponential ($v_2 = 1$ in the gamma), the Bernoulli ($v_2 = -1, r = 1$ in the binomial), the geometric ($v_2 = 1, r = 1$ in the negative binomial), and the NEF–HS (4.2).

Remarkably, just six simple distributions, one elementary distribution from each class, including the $N(0, 1)$, the Poisson ($\lambda = 1$), the exponential (mean = 1), the Bernoulli ($p = 1/2$), the geometric ($p = 1/2$), and the hyperbolic secant ($\theta = 0$), generate six of the main families of distributions in statistics via the (commutative) processes of:

1. linear transformation: $X \to (X - b)/c$;
2. generation of the NEF, per (2.3)–(2.4);
3. convolution; and
4. infinite division (division in the binomial case).

A fifth process, not commuting with the above, produces many other named univariate exponential families including the lognormal, the beta, the extreme value, the Pareto, and the Weibull distributions, which are not NEFs. These are derived as:

5. nonlinear transformations of NEF–QVF distributions, i.e., $X \to Y = T^{-1}(X)$, T as in (2.1).

5. The NEF–GHS Distributions

The hyperbolic secant density $f_{1,0}(x)$ in (4.2) with $\theta = 0$ is symmetric about 0 with unit variance and CDF

$$F_{1,0}(x) = \frac{1}{\pi} \operatorname{arc\,tan}\left(\sin\mathrm{h}\, \frac{\pi x}{2}\right) + \frac{1}{2} \tag{5.1}$$

on $-\infty < x < \infty$ (Johnson and Kotz, 1970). Convolutions and divisions of this distribution, which is infinitely divisible (Feller, 1971, also proved in Section 6) are said to have the *generalized hyperbolic secant distribution* (GHS) (Johnson and Kotz, 1970). These densities, all symmetric about 0, take fairly simple form for integral convolutions $f_{r,0}(x) = f_{1,0}(x) * \cdots * f_{1,0}(x)$ r times. Then

$$f_{2,0}(x) = \frac{x}{2 \sin\mathrm{h}(\pi x/2)}, \tag{5.2}$$

and for $r \geq 1$, integers only,

$$f_{r+2,0}(x) = \frac{x^2 + r^2}{r(r+1)} f_{r,0}(x). \tag{5.3}$$

Thus $f_{r,0}(x)$ is a polynomial of degree r divided by $\cosh(\pi x/2)$ if r is an odd integer, and divided by $\sinh(\pi x/2)$ if r is even. The $f_{r,0}(x)$ densities when r is not an integer are expressible as infinite products

$$f_{r,0}(x) = \frac{2^{r-2}}{\Gamma(r)} \prod_{j=0}^{\infty} \{1 + x^2/(r + 2j)^2\}^{-1}$$

(Johnson and Kotz, 1970).

The MGF of $f_{r,0}(x)$ is $1/\cos^r(t)$ and the characteristic function $1/\cosh^r(t)$. For $r = 1$, then, the characteristic function is $1/\cosh(t)$, the same as the density, making the hyperbolic secant density $f_{1,0}(x)$ "self-conjugate," like the normal density.

We note that $f_{1,0}(x)$ is the distribution of $(2/\pi)\log|C_1|$ and $f_{r,0}(x)$ is of $(2/\pi)\log|C_1 C_2 \ldots C_r|$ with $C_1, C_2, \ldots, C_r$, independent Cauchy random variables. This follows easily for $r = 1$ from the fact that C_1 is the ratio of two independent normals, so $Y = C_1^2/(1 + C_1^2) \sim \text{Beta}(0.5, 0.5)$, and then the argument preceding (4.2) shows $X = \log\{Y/(1 - Y)\}/\pi = 2\log|C_1|/\pi$ has density $f_{1,0}(x)$. Convolutions give the result for general r. Thus, the product of independent Cauchy distributed variables is simply related to (5.3).

Now let us generate the NEF $f_{r,\theta}(x)$ from $f_{r,0}(x)$ by using (2.3) and (2.4) to get

$$f_{r,\theta}(x) \equiv \exp\{\theta x + r\log\cos(\theta)\}f_{r,0}(x), \qquad |\theta| < \pi/2, \tag{5.4}$$

with the CGF $\psi(\theta) = -r\log\{\cos(\theta)\}$ obtained from the MGF of $f_{r,0}(x)$. It has MGF $\{\cos(\theta)/\cos(\theta + t)\}^r$, mean $\mu = \psi'(\theta) = r\tan(\theta)$ and variance $V(\mu) = \psi''(\theta) = \mu^2/r + r$ with discriminant $d = -4$. It follows from (5.3) and (5.4) that

$$f_{r+2,\theta}(x) = \cos^r(\theta)\frac{x^2 + r^2}{r(r+1)} f_{r,\theta}(x). \tag{5.5}$$

Let $\lambda \equiv \tan(\theta)$ be the mean of the elementary density $f_{1,0}(x)$ (cf. Section 4). Then for any $r > 0$,

$$f^*_{r,\lambda}(x) = (1 + \lambda^2)\exp\{x\tan(\lambda)\}f_{r,0}(x) \tag{5.6}$$

may be a more useful parameterization, $-\infty < \lambda < \infty$. The GHS distributions ($\lambda = 0$) are bell-shaped with exponentially decaying tails. For $\lambda \neq 0, f^*_{r,\lambda}(x)$ has skewness coefficient $2\lambda/(r + r\lambda^2)^{1/2}$, bounded above and below (as $\lambda \to \infty$) by $\pm 2/r^{1/2}$, the skewness of $\text{Gam}(r, \alpha)$. In fact, as $\lambda \to \pm\infty, f^*_{r,\lambda}(x)$ is approximated by the density of $\pm G$ with $G \sim \text{Gam}(r, |\lambda|)$, a fact proved in Section 10. Thus λ, or θ, is a shape parameter measuring the nonsymmetry of $f^*_{r,\lambda}(x)$, and compromising between a bell-shaped distribution when $\lambda = 0$ and a gamma distribution as $|\lambda| \to \infty$. Of course, as $r \to \infty$, this bell-shaped distribution approaches the normal. With $r = 1$, the compromise is between the HS distribution and an exponential (or minus an exponential), so as $\lambda \to \infty$, the probability that $X < 0$ vanishes rapidly while the modal value is positive but very close to zero. For any fixed r and $\theta \geq 0$, the right tail of $f_{r,\theta}(x)$ behaves like the tail of a $\text{Gam}(r, 1/(\pi/2 - \theta))$ distribution.

Other properties of the NEF–GHS distributions are developed in the next sections as part of the general NEF–QVF theory.

6. Infinite Divisibility of NEF–QVF Distributions

The binomial distribution is not infinitely divisible, for no bounded distribution can be, but the other five NEF-QVF distributions are (Feller, 1971). A new proof is given here for all five distributions simultaneously, revealing these five to be "self-generating," a term defined below.

Khinchin's characterization (Gnedenko and Kolmogorov, 1954) that a distribution possessing a MGF be infinitely divisible amounts to requiring its CGF $\psi(t)$ satisfy

$$\psi''(t)/\psi''(0) = E\exp(tY) \tag{6.1}$$

for some random variable Y. We show for NEF–QVF distributions that Y may be taken as $Y = V'(\bar{X}), \bar{X} = (X_1 + X_2)/2$ being the average of two iid elementary distributions (cf. Section 4) in the family. In this case, $V'(\bar{X})$ is linear (or constant), V being quadratic; and $X_1 + X_2$, being a convolution, belongs to a NEF.

A family infinitely divisible distributions will be called *self-generating* if the random variable Y in (6.1) is a convolution of members of the family.

Theorem 1. NEF–QVF *distributions are infinitely divisible, provided* $v_2 \geq 0$. *The elementary* NEF–QVF *distributions are self-generating.*

PROOF. First consider the elementary distributions (cf. Section 4), so $v_2 = 0$ or $v_2 = 1$. Let F_0 be the CDF in (2.4) of the distribution with $\theta = 0$, and $\psi(t)$ its CGF. With $\mu = \psi'(\theta)$ we have for F_θ in (2.4).

$$\begin{aligned} d\{\log \psi''(\theta)\}/d\theta &= V(\mu)d\{\log V(\mu)\}/d\mu = V'(\mu) \\ &= 2v_2\mu + v_1 = 2v_2\psi'(\theta) + v_1. \end{aligned}$$

Thus, integrating wrt θ, using $\psi(0) = 0$ to determine the constant of integration, and exponentiating:

$$\psi''(\theta) = \psi''(0)\exp\{2v_2\psi(\theta) + v_1\theta\}. \tag{6.2}$$

Let X_1 and X_2 be iid as F_0. Then

$$Y = V'(\bar{X}) = v_2(X_1 + X_2) + v_1 \quad \text{has MGF} \quad E\exp(tY) = \exp\{2\psi(v_2 t) + v_1 t\}.$$

This is (6.2) at $\theta = t$, because $v_2\psi(\theta) = \psi(v_2\theta)$ when $v_2 = 0$ or $v_2 = 1$. Hence F_0, an elementary CDF, is infinitely divisible. Then nonelementary distributions, with $v_2 \neq 1$, are also infinitely divisible, being convolutions of divisors of the elementary distribution. This completes the proof. □

7. Cumulants and Moments

The following theorem provides a simple relation between moments and cumulants that replaces complicated formulae relating moments to cumulants (e.g., expressions given in Kendall and Stuart, 1963). It holds in general, not only for exponential families.

Theorem 2. *Suppose X has m moments, $EX^r \equiv M'_r$, and cumulants C_r. Then for $1 \leq r \leq m$,*

$$M'_r = \sum_{0}^{r-1} \binom{r-1}{i} M'_1 C_{r-1}. \tag{7.1}$$

Central moments $M_r = E(X - M'_1)^r$ satisfy

$$M_1 = 0, \qquad M_2 = C_2, \qquad M_3 = C_3,$$

and for $r \geq 4$,

$$M_r = C_r + \sum_{2}^{r-2} \binom{r-1}{i} M_1 C_{r-1}. \tag{7.2}$$

Proof. First show, by induction, that the rth derivative is

$$\phi^{(r)}(t) = \sum_{0}^{r-1} \binom{r-1}{j} \psi^{(j)}(t) \psi^{(r-j)}(t)$$

if $\phi(t)$ is the chf of X and $\phi(t) = \log \phi(t)$. Evaluating this at $t = 0$ yields (7.1). Replacing X by $X - \mu$ before computing ϕ and ψ, in which case $\phi'(0) = \psi'(0) = 0$ and substituting $t = 0$ yields (7.2). □

The cumulants $C_r = C_r(\mu)$ of NEFs can be computed from $V = V(\mu)$ and its derivatives, as observed in (2.11). If the NEF has QVF then $(V')^2 = 4v_2 V + d$ by (3.3) and $V'' = 2v_2$ is constant. It follows from (2.11) and (7.2) that, with QVF, the fourth cumulant and moment satisfies

$$C_4 = 6v_2 V^2 + dv \qquad \text{and} \qquad M_4 = (3 + 6v_2) V^2 + dV. \tag{7.3}$$

Of course, $V^{(r)}(\mu) = 0$ for QVF for all $r \geq 3$, leading to polynomial dependence of all cumulants (and moments) on $V(\mu)$, as Theorem 3 will show.

Definition 1. The NEF–QVF cumulant coefficients $c_{m,i}$ are defined for integers $m = 0, 1, 2, \ldots$ and $0 \leq i \leq m$ by $c_{m,0} = c_{m,m} = 1$ and for any $1 \leq i \leq m - 1$,

$$c_{m,i} = c_{m-1,i-1} + (i+1)^2 c_{m-1,i}. \tag{7.4}$$

Table 2. Values of c_{mt} in (7.4).

m \ i	0	1	2	3	4	5	6	7
0	1	0	0	0	0	0	0	0
1	1	1	0	0	0	0	0	0
2	1	5	1	0	0	0	0	0
3	1	21	14	1	0	0	0	0
4	1	85	147	30	1	0	0	0
5	1	341	1408	627	55	1	0	0
6	1	1365	13013	11440	2002	91	1	0
7	1	5461	118482	196053	61490	5278	140	1

Note that Pascal's triangle for the binomial coefficients is generated by similar recursive formulae, suppressing $(i+1)^2$ in (7.4). Table 2 displays enough $c_{m,i}$ values to generate the first 17 cumulants, using Theorem 3, and hence the first 17 moments, using Theorem 2.

Theorem 3. *For* NEF–QVF *distributions, with* $m = 1, 2, \ldots$

$$\phi^{(2m)}(\theta) = V \sum_{0}^{m-1} c_{m-1,i}(2i+1)! v_2^i d^{m-1-i} V^i, \tag{7.5}$$

$$\phi^{(2m+1)}(\theta) = \frac{1}{2} V V' \sum_{0}^{m-1} c_{m-1,i}(2i+2)! v_2^i d^{m-1-i} V^i. \tag{7.6}$$

PROOF. For any function $T = T(V)$ having derivative T' wrt V, we have

$$\frac{d}{d\theta} T(V) = \frac{d\mu}{d\theta} \frac{dV}{d\mu} \frac{dT(V)}{dV} = VV'T'(V), \tag{7.7}$$

$$\begin{aligned} \frac{d}{d\theta}\{V'T(V)\} &= V\{V''T(V) + (V')^2 T'(V)\} \\ &= 2v_2 VT(V) + V(4v_2 V + d)T'(V). \end{aligned} \tag{7.8}$$

Now (7.5) clearly holds for $m = 1, \psi^{(2)}(\theta) = V$. Assume it holds for m and let $T(V)$ be the r.h.s. of (7.5), a polynomial of degree m in V. Then application of (7.7) to (7.5) term-by-term gives (7.6) immediately. Next, differentiate (7.6) term-by-term using (7.8) with $T(V) = V^{i+1}$ to get

$$\frac{1}{2} \sum_{0}^{m-1} c_{m-1,i}(2i+2)! v_2^i d^{m-1-i} \{2v_2 V^{i+2} + (4v_2 V + d)(i+1) V^{i+1}\}.$$

Rearranging terms to get the coefficients of V^{i+1}, and using (7.4), yields (7.5). The proof follows by induction on m. □

Theorem 3 shows that for NEF–QVF distributions, the rth cumulant, and also the rth moment (using Theorems 2 and 3), is a polynomial of degree at most r in μ. It also shows that when $r = 2m$, the cumulants and moments are polynomials of degree m in $V = V(\mu)$. For r odd, an extra (linear) factor $V'(\mu)$ is required. Moreover, C_r^2 is a polynomial of degree r in V even when r is odd, because the square C_r^2 defined by (7.6) involves the product of $(V')^2 = 4v_2 V + d$ and a polynomial of degree $r - 1$.

The first eight NEF–QVF cumulants written out from Theorem 3 are

$$\begin{aligned} &C_2 = V, \qquad C_3 = V'V, \qquad C_4 = 6v_2 V^2 + dV, \\ &C_5 = V'V(12v_2 V + d), \qquad C_6 = 120v_2^2 V^3 + 30v_2 dV^2 + d^2 V, \\ &C_7 = V'V(360v_2^2 V^2 + 60v_2 dV + d^2), \\ &C_8 = 5040v_2^3 V^4 + 1680v_2^2 dV^3 + 126v_2 d^2 V^2 + d^3 V. \end{aligned} \tag{7.9}$$

Using (7.1) and (7.2) with this information yields the central moments

$$\begin{aligned} &M_2 = V, \qquad M_3 = V'V, \qquad M_4 = (3 + 6v_2)V^2 + dV, \\ &M_5 = C_5 + 10C_2C_3 = V'V((10 + 12v_2)V + d), \\ &M_6 = (15 + 130v_2 + 120v_2^2)V^3 + (25 + 30v_2)dV^2 + d^2 V. \end{aligned} \tag{7.10}$$

The coefficients of skewness $\gamma_3 \equiv C_3/V^{1.5}$ and kurtosis $\gamma_4 \equiv C_4/V^2$ are

$$\gamma_3 = V'V^{-1/2}, \qquad \gamma_3^2 = 4v_2 + d/V, \qquad \gamma_4 = 6v_2 + d/V = 2v_2 + \gamma_3^2, \tag{7.11}$$

for NEF–QVF distributions.

The normal is the only symmetric NEF, with or without QVF, because the skewness, $V'(\mu)V^{-1/2}(\mu)$, vanishes for all μ only if $V(\mu)$ is constant. Thus, no symmetric distribution other than the normal (including the GHS, logistic, etc.) can generate a symmetric NEF–QVF family. This follows easily from (2.11).

In cases with $v_2 d = 0$, i.e., the normal ($v_2 = d = 0$), the Poisson ($v_2 = 0$), and the gamma ($d = 0$), (7.5) and (7.6) are well-known cumulant expressions. However, these expressions do unify results for the six distributions, and in the binomial ($v_2 = -1/r, d = 1$), negative binomial ($v_2 = 1/r, d = 1$), and NEF–GHS ($v_2 = 1/r, d = -4$) cases. Theorem 3 provides useful new expressions for the cumulants.

8. Orthogonal Polynomials for NEF–QVF Distributions

Let $f(x, \theta)$ be a NEF–QVF density proportional to $\exp(x\theta - \psi(\theta))$ relative to some measure as in (2.2). Define

$$P_m(x, \mu) = V^m(\mu)\left\{\frac{d^m}{d\mu^m} f(x, \theta)\right\}/f(x, \theta) \tag{8.1}$$

for $m = 1, 2 \ldots$. Derivatives in (8.1) are taken with respect to the mean μ, not θ. We have

$$P_0(x,\mu) = 1, \quad P_1(x,\mu) = x - \mu, \quad P_2(x,\mu) = (x-\mu)^2 - V'(\mu)(x-\mu) - V(\mu),$$

and will show that $P_m(x, \mu)$ is a polynomial of degree m in both x and μ with leading term x^m, and that $\{P_m\}$ is a family of orthogonal polynomials.

From (8.1) it follows immediately that (with arguments suppressed).

$$P_{m+1} = V^{m+1} f^{-1} d(P_m f V^m)/du = (P_1 - mV')P_m + VP'_m, \qquad m \geq 1, \quad (8.2)$$

with $P'_m = \partial P_m(x,\mu)/\partial\mu$, $P^{(r)} \equiv \partial^r P_m(x,\mu)/\partial\mu^r$.

Theorem 4. *The set* $\{P_m(x,\mu) : m = 0, 1, \ldots\}$ *is, for* NEF–QVF *families, an orthogonal system of polynomials with respect to* $f(x,\theta) = \exp\{x\theta - \psi(\theta)\}$. $P_m(x,\mu)$ *has exact degree* m *in both* μ *and* x *with leading term* x^m. *It is generated by*

$$P_{m+1} = (P_1 - mV')P_m - m\{1 + (m-1)v_2\}VP_{m-1} \qquad (8.3)$$

for $m \geq 1$ *with* $P_0 = 1, P_1 = x - \mu$. *Define* $a_0 = 1$ *and for* $m \geq 1$,

$$a_m \equiv m! \prod_{t-m}^{m-1} (1 + iv_2). \qquad (8.4)$$

Then for $m \leq 1, r = 0, 1, \ldots, m$, *the derivatives wrt* μ *are*

$$P_m^{(r)} = (-1)^r (a_m/a_{m-r}) P_{m-r}. \qquad (8.5)$$

Finally, $E_\theta P_m = 0$ *for* $m \geq 1$ *and*

$$E_\theta P_m P_n = \delta_{mn} a_m V^m, \qquad m, n \geq 0. \qquad (8.6)$$

Proof. Define $b_m \equiv (m+1)(1 + mv_2)$. We start by proving (8.5) for $r = 1$, $a_m/a_{m-1} = b_{m-1}$. Now $P'_m = -b_{m-1}P_{m-1}$ holds for $m = 1$, so assume it holds for $m \geq 1$ and use it in (8.2) to write

$$P_{m+1} = (P_1 - mV')P_m - b_{m-1}VP_{m-1}.$$

Differentiating this wrt μ,

$$\begin{aligned}
P'_{m+1} &= (-1 - mV'')P_m + (P_1 - mV')P'_m - b_{m-1}V'P_{m-1} - b_{m-1}VP'_{m-1} \\
&= -(1 + 2mv_2)P_m - b_{m-1}(P_1 - mV')P_{m-1} - b_{m-1}V'P_{m-1} - b_{m-1}VP'_{m-1} \\
&= -(1 + 2mv_2)P_m - b_{m-1}[(P_1 - (m-1)V')P_{m-1} + VP'_{m-1}] \\
&= -(1 + 2mv_2 + b_{m-1})P_m = -b_m P_m \quad \text{(from (8.2))}.
\end{aligned}$$

Induction on m now proves $P_m^{(1)} - b_{m-1}P_{m-1}$. By iterating this $r - 1$ more times,

$$P_m^{(r)} = (-1)^r (b_{m-1} \ldots b_{m-r}) P_{m-r} = (-1)^r (a_m/a_{m-r}) P_{m-r},$$

proving (8.5).

Equation (8.3) follows from (8.2) and (8.5) with $r = 1$. That $P_m(x, \mu)$ is a polynomial of exact degree m in both x and μ with leading term x^m follows inductively from (8.3). Now for any $n < m$, using (8.1)

$$E_\theta X^n P_m = V^m \int x^n f^{(m)}(x, \theta)\, dF(x) = V^m \frac{\partial^m}{\partial \mu^m} E_\theta X^n = 0,$$

because Section 7 revealed $E_\theta X^n$ to be a polynomial of degree at most n in $\mu = \psi'(\theta)$. It follows that P_m is orthogonal to every polynomial of lower degree, and so (8.6) holds for $m \neq n$.

To prove (8.6) for $m = n \geq 1$, multiply (8.3) by P_{m-1} and take expectations to show $EP_1 P_m P_{m-1} = b_{m-1} VEP_{m-1}$. Repeat this procedure with P_{m-1} to get $EP_{m+1}^2 = EP_1 P_m P_{m+1}$. Putting these together, $EP_m^2 = b_{m-1} VEP_{m-1}^2$. Iterating,

$$EP_m^2 = b_{m-1} \ldots b_0 V^m EP_0 = a_m V^m. \qquad \square$$

The polynomials of this section are known individually as the Hermite (normal distribution). Poisson–Charlier (Poisson distribution), Generalized Laguerre (gamma), Krawtchouk (binomial), Meixner (negative binomial), and Pollaczek (GHS) polynomials (symmetric GHS subfamilies only, not the NEF–GHS) (Szego, 1975). The main new results here lie in unifying these six polynomial systems and in the new polynomials and results provided by (8.3) through (8.6) for nonsymmetric members of the NEF–GHS family.

The system (8.1) forms a set of polynomials of the indicated degree for any natural exponential family. However $EP_2 P_3 = V'''(\mu) V^3(\mu)$, for example, which is not zero unless $V(\mu)$ is quadratic. Thus, these polynomials form an orthogonal system only if QVF holds.

Finally, two useful facts follow from Theorem 4. The first (8.7) relates the orthogonal polynomials of different members in the exponential family. The second (8.8) provides the expectation of a polynomial defined with the "wrong" parameter.

Corollary 1. *Let $\mu, \mu_0 \in \Omega$ be given for two distributions in the same* NEF–QVF. *Then for any $m = 0, 1, \ldots,$*

$$P_m(x, \mu_0) = a_m \sum_0^m \frac{(\mu - \mu_0)^{m-r}}{(m-r)!} \frac{1}{a_r} P_r(x, \mu) \tag{8.7}$$

with a_m defined by (8.4). *Hence*

$$E_\mu P_m(X, \mu_0) = \frac{a_m}{m!} (\mu - \mu_0)^m. \tag{8.8}$$

Proof. Expand $P_m(x, \mu_0)$ in a Taylor series (of order m only) around μ and use (8.5) to get

$$P_m(x, \mu_0) = \sum_0^m \frac{(\mu_0 - \mu)^r}{r!} P_m^{(r)}(x, \mu) = \sum_0^m \frac{(\mu - \mu_0)^r}{r!} \frac{a_m}{a_{m-r}} P_{m-r}(x, \mu).$$

Interchanging r and $m-r$ yields (8.7). Then (8.8) follows because $E_\mu P_r(X,\mu) = \delta_{0r}$. □

9. A Large Deviation Theorem

This section proves the sharpest possible large deviation theorem for the entire NEF–QVF class. First we have a lemma applying to any NEF, not just those with QVF.

Lemma. *Let X have a NEF distribution. Then for all $t \geq 0$, writing $\sigma^2 = V(\mu)$,*

$$P\left(\frac{X-\mu}{\sigma} \geq t\right) \leq \exp(-B(t)), \qquad B(t) \equiv \sigma^2 \int_0^t \frac{(t-w)\,dw}{V(\mu+w\sigma)}. \tag{9.1}$$

Proof. For simplicity, assume $\theta = 0$ so $\psi(0) = 0, \psi'(0) = \mu, \psi''(0) = \sigma^2$. For any $w \geq 0$, $P(X-\mu-t\sigma \geq 0) \leq E\exp\{w[X-\mu-t\sigma]\} = \exp\{\psi(w) - w(\mu+t\sigma)\}$. This is minimized at $w = w(t)$ satisfying $\psi'(w(t)) = \mu + t\sigma$. Thus $w(0) = 0$, since $\psi'(0) = \mu$. Define $B(t) = w(t)(\mu+t\sigma) - \psi(w(t))$ as the negative of the minimum value. Note $B(0) = 0$. Now $B'(t) = w(t)\sigma$, after simplification, so $B'(0) = 0$, and $B''(t) = w'(t)\sigma$. Differentiating $\psi'(w(t)) = \mu + t\sigma$ with respect to t, $\psi''(w(t))w'(t) = \sigma$. But $\psi''(w(t)) = V(\psi'(w(t))) = V(\mu+t\sigma)$. Since $\sigma^2 = V(\mu), B''(t) = \sigma^2/V(\mu+t\sigma)$. Using Taylor's theorem with integral remainder for expansion about $t = 0$, and $B(0) = B'(0) = 0$, gives $B(t) = \int_0^t (t-w)B''(w)\,dw$. This is (9.1). □

With QVF distributions we can go further, for then in (9.1),

$$V(\mu+t\sigma)/\sigma^2 = \{V(\mu) + t\sigma V'(\mu) + v_2 t^2\sigma^2\}/\sigma^2 = 1 + \gamma t + v_2 t^2,$$

with $\gamma = V'(\mu)/\sigma$ the skewness coefficient of X. Suppose $\gamma, v_2 \leq C < \infty$, a constant, taking $C \geq 1$ for convenience. Then for all $t \geq 0, C \geq 1, 0 < 1 + \gamma t + v_2 t^2 \leq 1.5C(1+t^2)$ (consider cases $t \leq 1$ and $t \geq 1$ separately). Formula (9.1) for $B(t)$ gives

$$B(t) = \int_0^t \frac{(t-w)\,dw}{1+\gamma w + v_2 w^2} \geq \frac{1}{1.5C}\int_0^t \frac{(t-w)\,dw}{1+w^2}$$
$$= [t \cdot \text{arc}\tan(t) - 0.5\log(1+t^2)]/1.5C \leq \pi t/3C. \tag{9.2}$$

This behaves like $\pi t/3C$ for t large.

Theorem 5. *Suppose γ, the skewness, and v_2 are bounded above by absolute constants as $t \to \infty$. If X has a* NEF–QVF *distribution, there exists an absolute constant $b > 0$ such that*

$$P\left(\frac{X-\mu}{\sigma} \geq t\right) \leq \exp(-bt) \tag{9.3}$$

for all $t \geq 1$. If $|\gamma|$ and v_2 are both bounded above, then there exists $b_0 > 0$ such that

$$P\left(\left|\frac{X-\mu}{\sigma}\right| \geq t\right) \leq \exp(-b_0 t) \tag{9.4}$$

for all $|t| \geq 1$.

Proof. (9.3) follows from the discussion preceding the theorem, taking $b = \pi/3C$ *if* γ, $v_2 \leq C$. Then (9.4) follows from (9.3) by $X \to -X$ which sends $\gamma \to -\gamma$ and leaves v_2 unchanged. □

The skewness $\gamma^2 = 4v_2 + d/V(\mu)$ from (7.11), and is invariant under linear transformations $X \to (X-b)/c$. Thus Theorem 5 requires v_2 bounded above, and if $d > 0$, $V(\mu)$ bounded away from zero. This happens, for example if there exists $\varepsilon > 0$ fixed such that (in the notation of Table 1): $\lambda \geq \varepsilon$ (Poisson); $r \geq \varepsilon$ (gamma and NEF–GHS); $p(1-p) \geq \varepsilon$ (binomial); $p \geq \varepsilon$ and $r \geq \varepsilon$ (negative binomial); and no conditions (normal). These conditions are noted in one of the last lines of Table 1.

If $X_1, X_2, \cdots$ all have NEF–QVF distributions, $EX_t = \mu_t$, Var $X_t = \sigma_t^2$, then

$$\lim_{k\to\infty} P(\max_{1\leq t\leq k}|X_t - \mu_t|/\sigma_t \geq t_k) = 0 \tag{9.5}$$

as $k \to \infty$ if $t_k \to \infty$ faster than log k, and if v_{2t} and $|\gamma_t|$ are uniformly bounded. This follows from (9.4) because the l.h.s. of (9.5) is bounded by $\sum_{t=1}^{k} P(|X_t - \mu_t| \geq t_k\sigma_t)$. This fact is used in (Morris, 1967) to prove that certain tests in NEF–QVF distributions are ε-Bayes for large k.

The exponential bounds (9.3), (9.4) are the best possible for NEF–QVF distributions because the three elementary distributions with $v_2 = 1$ (the exponential, geometric, and hyperbolic secant) whose CDF's can be evaluated, have exponential tails exactly. Improvements are possible in the normal and Poisson cases, however. For example, in the normal case $B(t) = (1/2)t^2$ from (9.2) and insertion of this in (9.1) gives a sharper than exponential bound for large t.

10. Limits in Distribution

If $V(\mu)$ takes an approximate form, $V^*(\mu)$, say, it is reasonable that the NEF corresponding to $V(\mu)$ should have approximately the distribution of the NEF corresponding to $V^*(\mu)$. That is, because $V(\mu)$ characterizes the NEF, nearness of V to V^* forces nearness of the characteristic functions. This also implies convergence of all moments and cumulants because they are functions of $V(\mu)$, via (7.5), (7.6) and (7.2), or through (2.9), the MGF.

For example, suppose $X \sim \mathrm{Bin}(r,p), \mu = rp, V(\mu) = -\mu^2/r + \mu$. Then $V(\mu) \to V^*(\mu) \equiv \mu$ as $r \to \infty$ if $p \to 0$ and μ is fixed. Similarly $X \sim \mathrm{NB}(r,p), \mu = rp/(1-p)$, and $V(\mu) = \mu^2/r + \mu \to V^*(\mu) = \mu$ as $r \to \infty$ and $p \to 0$, holding μ constant. Because $V^*(\mu) = \mu$ is the Poisson variance function, we thereby obtain the familiar Poisson limits for the binomial and negative binomial distributions as $r \to \infty$ with rp fixed.

Other distributional approximations hold. For $\mathrm{NB}(r,p)$ let $\mu = rp/(1-p)$ be large and r be fixed (i.e., $p \to 1$). Then $V(\mu) = u^2/r + \mu = (\mu^2/r)(1 + r/u) \sim V^*(\mu) \equiv \mu^2/r$ in ratio. Since V^* is the VF of $\mathrm{Gam}(r, \mu/r)$, we have $\mathrm{NB}(r,p) = \mathrm{Gam}(r, 1/(1-p))$ for p near 1. For $r = 1$ this means Geometric $(p) \doteq$ Exponential $(1/(1-p))$ as $p \to 1$.

For similar reasons with r fixed, the NEF–GHS family with $\mu = r\lambda$, $V(\mu) = r + \mu^2/r$ has approximately a $\mathrm{Gam}(r, \mu/r) = \mathrm{Gam}(r, \lambda)$ distribution for μ (and λ) large (i.e. if $r/\mu = 1/\lambda \to 0$), for then $V(\mu) \sim V^*(\mu) = \mu^2/r$, the VF of $\mathrm{Gam}(r, \mu/r)$. This assertion appears at the end of Section 5.

The central limit theorem (CLT) and weak law of large numbers (WLLN) also follow for NEFs by showing the VF of the appropriately scaled $\sum_1^n X_t$ is approximately constant (CLT) or vanishes (WLLN) as $n \to \infty$. In other words, the expectation μ^* of $X^* = \sum(X_t - \alpha)/n^{1/2}$ is bounded if α satisfies $|\alpha - \mu| < C/n^{1/2}$, C a constant. Then $V^*(\mu^*) = \mathrm{Var}(X^*) = V(\alpha + \mu^*/n^{1/2}) \doteq V(\alpha)$ is nearly a constant function of μ^* for n large. Thus, X^* is asymptotically normal. Alternatively, if $X^* = \sum X_t/n$, then $\mu^* = EX^* = \mu$, and $V^*(\mu^*) = V(\mu)/n \to 0$ as $n \to \infty$. Because this limit is the variance of a constant, the WLLN is proved.

Convolutions and scalings of X map $X \to \sum_1^n X_t/c$ and $V \to V^*(\mu) = nV(c\mu/n)/c^2$ and so $v_t^* = v_t c^{t-2}/n^{1-t}$; cf. (3.4) and 3.5 with $b = 0$. Limit theorems for such transformations can never increase the complexity of V (because $v_t = 0 \Rightarrow v_t^* = 0$), or the order (degree) of V, if V is a polynomial. Thus the normal and Poisson distributions can be limits of the *strictly* quadratic NEF–QVF distributions, but not conversely. Interesting limit theorems make V^* *less* complex that V; i.e., some of the v_t^* vanish in the limit. The three limit distributions just considered (the Poisson, gamma, and normal) do this with $V^*(\mu)$ a monomial. Because such limits never increase the order of a polynomial V, the normal, with constant variance function, is the most widely reached nontrivial limit law. Of course the WLLN, with $V(\mu)$ vanishing, is even more widely achieved.

Within NEF–QVF families, the discriminant d changes to $d^* = d/c^2$ under convolution and linear transformations $X \to \sum(X_i - b)/c$. If $c \to \infty$, limits of these distributions have $d^* = 0$, i.e., must be normal or gamma limits. Otherwise limit distributions must preserve the sign of d. Hence the Poisson, binomial, and negative binomial distributions, all having $d > 0$, can be limits of one another. The NEF–GHS distributions, being the only NEF–QVF distributions with $d < 0$, cannot be limits of any other NEF–QVF distributions.

References

Abramowitz, M. and Stegun, I.A. (eds.) (1965). *Handbook of Mathematical Functions.* Dover, New York.

Barndorff-Nielsen, O. (1978). *Information and Exponential Families in Statistical Theory.* Wiley, New York.

Barndorff-Nielsen, O. and Cox, D.R. (1979). Edgeworth and saddle-point approximations with statistical applications. *J. Roy. Statist. Soc. Ser. B*, **41**, No. 3, 279–312.

Baten, W.D. (1934). The probability law for the sum of n independent variables, each subject to the law $(1/2h)\text{sech}(\pi x/2h)$. *Bull. Amer. Math. Soc.*, **40**, 284–290.

Blight, B.J.N., and Rao, P.V. (1974). The convergence of Bhattacharya bounds. *Biometrika*, **61**, 137–142.

Bolger, E.M. and Harkness, W.L. (1965). Characterizations of some distributions by conditional moments. *Ann. Math. Statist.*, **36**, No. 2, 703–705.

Duan, N. (1979). Significance test for prior distributions: The modified efficient score test and its asymptotic theory. Technical Report No. 135. Department of Statistics, Stanford University.

Efron, B. (1978). The geometry of exponential families. *Ann. Statist.*, **6**, 362–376.

Feller, W. (1971). *An Introduction to Probability Theory and its Applications*, Vol. II., 2nd ed. Wiley, New York.

Gnedenko, B.V., and Kolmogorov, A.N. (1954). *Limit Distributions for Sums of Independent Random Variables.* Addison-Wesley, Cambridge, MA.

Harkness, W.L. and Harkness, M.L. (1968). Generalized hyperbolic secant distributions. *J. Amer. Statist. Assoc.*, **63**, 329–337.

Johnson, N.L. and Kotz, S. (1970). *Continuous Univariate Distributions-2.* Houghton-Mifflin, Boston.

Kendall, M. and Stuart, A. (1963). *The Advanced Theory of Statistics*, Vol. I. *Distribution Theory*, 2nd ed. Griffin, London.

Khinchin, A.I. (1949). *Mathematical Foundations of Statistical Mechanics.* Dover, New York.

Laha, R.G. and Lukacs, E. (1960). On a problem connected with quadratic regression. *Biometrika*, **47**, 335–343.

Lehmann, E. (1959). *Testing Statistical Hypotheses.* Wiley, New York.

Morris, C. (1967). A class of epsilon Bayes tests of a simple null hypothesis on many parameters in exponential families. Technical Report No. 57. Department of Statistics, Stanford University.

Shanbag, D.N. (1972). Some characterizations based on the Bhattacharya matrix. *J. Appl. Probab.*, **9**, 580–587.

Szego, G. (1975). *Orthogonal Polynomials*, Vol. 23, 4th ed., American Mathematical Society, Colloquium Publications, Providence, RI.

Introduction to Barndorff-Nielsen (1983) On a Formula for the Distribution of the Maximum Likelihood Estimator

T. DiCiccio

This searching and far-reaching paper has helped shape the area of inference for parametric models during the past decade. By turns historical and forward in perspective, scholarly and innovative in content, and precise and speculative in presentation, this influential work has proven to be a rich source of insights, results, and open issues for numerous researchers.

The initial contribution of the paper is to introduce and elucidate an approximation to the conditional density of the maximum likelihood estimator. Consider an observed random vector $x = (x_1, \ldots, x_n)$ whose distribution depends on an unknown d-dimensional parameter $\omega = (\omega^1, \ldots, \omega^d)$. Suppose that $(\hat{\omega}, a)$ is a sufficient statistic, where $\hat{\omega}$ is the maximum likelihood estimator and a is an ancillary statistic. The remarkably simple and accurate p^*-formula, which approximates the conditional density of $\hat{\omega}$ given a, is

$$p^*(\hat{\omega}; \omega|a) = c(a)|-l_{\omega\omega}(\hat{\omega}; \hat{\omega}, a)|^{1/2} \exp\{l(\omega; \hat{\omega}, a) - l(\hat{\omega}; \hat{\omega}, a)\},$$

where $c(a)$ is a normalizing constant, $l(\omega; \hat{\omega}, a)$ is the log likelihood for ω expressed in terms of $(\hat{\omega}, a)$ and $l_{\omega\omega}(\omega; \hat{\omega}, a) = \partial^2 l(\omega; \hat{\omega}, a)/\partial\omega^T \partial\omega$.

The normalizing constant $c(a)$ is parametrization invariant. Replacing $c(a)$ by the approximation $(2\pi)^{-d/2}$ produces what is now known as the $p^\dagger$-formula (Barndorff-Nielsen and Cox, 1994). The log likelihood function for ω obtained from the $p^\dagger$-formula is exact, whereas the p^*-formula does not necessarily have this property. Under reparametrization, both the p^*- and $p^\dagger$-formulas transform according to the usual change-of-variables rule for densities; in this sense, they too are parametrization invariant.

When the maximum likelihood estimator $\hat{\omega}$ is sufficient, there is no need to specify an ancillary statistic a, and conditioning does not arise. This situation occurs when ω is the mean parameter in a full exponential family; the

$p^{\dagger}$-formula reproduces the familiar saddlepoint approximation (Daniels, 1954, 1980; Barndorff-Nielsen and Cox, 1979; Reid, 1988; Field and Ronchetti, 1990), and the p^*-formula is its normalized version. It is well known that the saddlepoint approximation has relative error of order $O(n^{-1})$ and that normalizing reduces this error to order $O(n^{-3/2})$.

It is carefully shown in the paper that for a very general class of group transformation models, which includes location, location-scale, and linear regression models, the maximal invariant is a suitable ancillary statistic and that the p^*-formula is exact. Exactness cases are important in part because they help distinguish the p^*-formula from other asymptotic approximations, such as Edgeworth expansions.

Beyond these important special cases of full exponential families and transformation models, exactly ancillary statistics are typically necessary and unavailable. The p^*-formula can be applied, however, by constructing a statistic a such that, to a high enough order, a is approximately ancillary and $(\hat{\omega}, a)$ is approximately sufficient. Illuminating expositions of approximate sufficiency and ancillarity are given by Barndorff-Nielsen and Cox (1994) and McCullagh (1984, 1987).

It is now widely accepted that, in broad generality, the p^*- and $p^{\dagger}$-formulas have relative error of order $O(n^{-3/2})$ and $O(n^{-1})$, respectively, provided the ancillary a is chosen appropriately. This accuracy is asserted in the paper with supporting examples, and previous work by Barndorff-Nielsen (1980) and Durbin (1980) is cited for justification. More comprehensive demonstrations are given by McCullagh (1984, 1987). A very general treatment of the p^*-formula is offered by Skovgaard (1990), and Barndorff-Nielsen and Cox (1994) derive the p^*- and $p^{\dagger}$-formulas in the context of curved exponential families.

Approximate ancillaries are not unique. For example, in curved exponential families, there are the Efron–Hinkley ancillary (Efron and Hinkley, 1978; Skovgaard, 1985), the score ancillary (Barndorff-Nielsen, 1980), and the directed likelihood ancillary (Barndorff-Nielsen, 1984). An intriguing aspect of the p^*- and $p^{\dagger}$-formulas is that they are valid with constant order of asymptotic error for broad classes of approximate ancillaries.

The second major contribution of the paper is the development and investigation of modified profile likelihood. Suppose that the parameter ω is partitioned in the form $\omega = (\psi, \kappa)$, where ψ is a nuisance parameter and κ is a κ-dimensional parameter of interest. The profile log likelihood function is $l^p(\kappa) = l(\hat{\psi}_\kappa, \kappa)$, where $l(\psi, \kappa)$ is the log likelihood function for (ψ, κ) based on x, and $\hat{\psi}_\kappa$ is the constrained maximum likelihood estimator of ψ for fixed κ. Note that $l^p(\kappa)$ is maximized at the maximum likelihood estimator $\hat{\kappa}$.

Likelihood-based methods for approximate inference arising from first-order asymptotic considerations in the absence of nuisance parameters are usually extended to the nuisance parameter case by treating the profile log likelihood as a genuine log likelihood for κ. In particular, the hypothesis

$\kappa = \kappa_0$ can be tested using the profile log likelihood ratio statistic

$$w(\kappa_0) = 2\{l^p(\hat{\kappa}) - l^p(\kappa_0)\} = 2\{l(\hat{\psi}, \hat{\kappa}) - l(\hat{\psi}_0, \kappa_0)\},$$

where $\hat{\psi}_0 = \hat{\psi}_{\kappa_0}$; typically, to error of order $O(n^{-1})$, the distribution of $w(\kappa_0)$ is chi-squared with κ degrees of freedom. Confidence regions having coverage error of order $O(n^{-1})$ can be constructed by using the chi-squared approximation for $w(\kappa_0)$.

When κ is scalar, methods based on $w(\kappa_0)$ produce approximate two-sided tests and confidence intervals. Approximate one-sided tests and confidence limits can be obtained using the directed likelihood

$$r(\kappa_0) = \text{sgn}(\hat{\kappa} - \kappa_0)\sqrt{w(\kappa_0)},$$

which typically has the standard normal distribution to error of order $O(n^{-1/2})$.

The profile log likelihood is not a genuine log likelihood function. Indeed, the profile score $l^p_\kappa(\kappa) = \partial l^p(\kappa)/\partial\kappa$, is not unbiased; in general, $E\{l^p_\kappa(\kappa); \psi, \kappa\} = O(1)$. Methods of inference derived by regarding $l^p(\kappa)$ as a true log likelihood can be seriously deficient, particularly when the information available about the nuisance parameter ψ is slight. This situation is likely to arise when the dimension of ψ is large or the sample size is small. In such cases, with high probability, the profile log likelihood can be centered around a completely erroneous point, resulting in inferences that are unacceptable. An extreme example of this phenomenon is the Neyman–Scott problem, discussed in the paper, that involves n pairs of normally distributed variables: all of the variables are independent and have a common variance, but each pair has a different mean. This example is somewhat unusual because the dimension of the nuisance parameter increases with the sample size. Here, the maximum likelihood estimator of the variance is inconsistent; it converges in probability to one-half of the true parameter value. Thus, the profile log likelihood will not be situated appropriately for reasonable inferences.

Examples of this type point to the need for improving the profile log likelihood so that it behaves more like a genuine log likelihood. An accepted solution to the Neyman–Scott problem can be obtained as a marginal likelihood, as described in the paper. Any adjusted profile log likelihood should mimic exact solutions in those problems where estimation of nuisance parameters can be taken into account by conditioning or marginalization.

Modified profile likelihood is developed in the paper through a marginalization argument, although the possibility of using a conditioning argument is also mentioned. For the marginalization argument, it is assumed that, for an ancillary statistic a, $(\hat{\psi}, \hat{\kappa}, a)$ is sufficient and that the joint conditional density of $(\hat{\psi}, \hat{\kappa})$ given a can be factorized as

$$p(\hat{\psi}, \hat{\kappa}; \psi, \kappa|a) = p(\hat{\psi}; \psi, \kappa|\hat{\kappa}, a)p(\hat{\kappa}; \kappa|a) = p(\hat{\psi}_\kappa; \psi, \kappa|\hat{\kappa}, a)\left|\frac{\partial\hat{\psi}_\kappa}{\partial\hat{\psi}}\right| p(\hat{\kappa}; \kappa|a),$$

where, to calculate the partial derivative matrix $\partial\hat{\psi}_\kappa/\partial\hat{\psi}$, $\hat{\psi}_\kappa$ is viewed as a function of $(\hat{\psi}, \hat{\kappa}, a)$ and κ. Given a, the marginal density of $\hat{\kappa}$ can be used for inference about κ. An approximation to $p(\hat{\kappa}; \kappa|a)$ is obtained by applying the p^*-formula to both $p(\hat{\psi}, \hat{\kappa}; \psi, \kappa|a)$ and $p(\hat{\psi}_\kappa; \psi, \kappa|\hat{\kappa}, a)$; the modified profile log likelihood that emerges from this approximation is

$$l^{mp}(\kappa) = l^p(\kappa) - \tfrac{1}{2}\log|-l_{\psi\psi}(\hat{\psi}_\kappa, \kappa)| - \log\left|\frac{\partial\hat{\psi}_\kappa}{\partial\hat{\psi}}\right|$$

$$= l^p(\kappa) + \tfrac{1}{2}\log|-l_{\psi\psi}(\hat{\psi}_\kappa, \kappa)| - \log|l_{\psi;\hat{\psi}}(\hat{\psi}_\kappa, \kappa)|,$$

where $l_{\psi\psi}(\psi, \kappa) = \partial^2 l(\psi, \kappa)/\partial\psi^T\partial\psi$ and $l_{\psi;\hat{\psi}}(\psi, \kappa) = \partial^2 l(\psi, \kappa; \hat{\psi}, \hat{\kappa}, a)/\partial\psi^T\partial\hat{\psi}$. The latter matrix of mixed second-order partial derivatives requires that the log likelihood function $l(\psi, \kappa)$ be expressed in terms of the sufficient statistic $(\hat{\psi}, \hat{\kappa}, a)$.

In the paper, modified profile likelihood is shown to be parametrization invariant. A variety of illustrative examples are provided, where modified profile likelihood is shown to reproduce appropriate marginal and conditional likelihoods either exactly or very closely. When κ consists of canonical parameters in a full exponential family, conditioning is a widely accepted approach to eliminate nuisance parameters (Reid, 1995); Barndorff-Nielsen and Cox (1979) approximate the relevant conditional density by a double application of the saddlepoint method and thereby deduce a simple yet accurate approximation to the conditional likelihood. Modified profile likelihood is shown to coincide with this double saddlepoint approximation. Even in the extreme Neyman–Scott problem, modified profile likelihood is shown to agree with the appropriate marginal likelihood.

Barndorff-Nielsen and Cox (1994) describe the derivation of modified profile likelihood through two conditioning arguments. For the first argument, it is assumed that the joint conditional density of $(\hat{\psi}, \hat{\kappa})$ given a can be factorized as

$$p(\hat{\psi}, \hat{\kappa}; \psi, \kappa|a) = p(\hat{\psi}; \psi, \kappa|a)p(\hat{\kappa}; \kappa|\hat{\psi}, a),$$

so that inference about κ can be based on the conditional density $p(\hat{\kappa}; \kappa|\hat{\psi}, a)$. The second argument, a generalization of the first, assumes that the joint distribution of $\hat{\kappa}$ and $\hat{\psi}_\kappa$ has the factorization

$$p(\hat{\psi}_\kappa, \hat{\kappa}; \psi, \kappa|a) = p(\hat{\psi}_\kappa; \psi, \kappa|a)p(\hat{\kappa}; \kappa|\hat{\psi}_\kappa, a);$$

the appropriate conditional density for inference is $p(\hat{\kappa}; \kappa|\hat{\psi}_\kappa, a)$. In both cases, to develop modified profile likelihood, the relevant inferential density is approximated by applying the p^*-formula to the other two densities.

The second conditioning argument is very similar to one used by Cox and Reid (1987) to develop their adjusted profile likelihood, often called conditional profile likelihood. They assume that ψ and κ are orthogonal, that is, $i_{\psi\kappa} = E\{-\partial^2 l(\psi, \kappa)/\partial\psi^T\partial\kappa; \psi\kappa\} = 0$. In principle, orthogonality can always be achieved when κ is scalar. The Cox and Reid adjusted profile log likelihood is

$$l^{cp}(\kappa) = l^p(\kappa) - \tfrac{1}{2}\log| - l_{\psi\psi}(\hat{\psi}_\kappa, \kappa)|,$$

which agrees with the modified profile log likelihood, except for the log-Jacobian term. The this extra term in $l^{mp}(\kappa)$ is negligible under orthogonality. Typically, $\hat{\psi}_\kappa = \hat{\psi} + O_p(n^{-1/2})$ for values of κ that differ from the true value by order $O(n^{-1/2})$; however, under orthogonality, $\hat{\psi}_\kappa = \hat{\psi} + O_p(n^{-1})$. Thus, the Jacobian is $1 + O_p(n^{-1})$, and the omitted log-Jacobian is $O_p(n^{-1})$. Although there is some arbitrariness in the definition of the adjusted profile likelihood because orthogonal parameterizations are not unique, distinct versions of $l^{cp}(\kappa)$ necessarily agree to error of order $O_p(n^{-1})$.

When κ consists of canonical parameters in an exponential family, the complementary expectation parameter is orthogonal to κ. For this choice of nuisance parameter, the Cox and Reid adjusted profile likelihood agrees with modified profile likelihood. The important feature here is that $\hat{\psi}_\kappa = \hat{\psi}$; in such cases, the log-Jacobian term in the modified profile log likelihood is identically zero. Modified profile likelihood and adjusted profile likelihood are compared in detail by Barndorff-Nielsen and McCullagh (1993).

An important feature of the modified profile likelihood is that its score is more nearly unbiased than is the usual profile score. Indeed, $E\{\partial l^{mp}(\kappa)/\partial\kappa; \psi, \kappa\} = O(n^{-1})$, and the same is true for the $l^{cp}(\kappa)$. Another method for adjusting profile likelihood based directly on reducing the bias of the profile score is developed by McCullagh and Tibshirani (1990), who also consider scale adjustments. An important line of recent research has been the development of adjustments to profile likelihood that do not explicitly require special constructs such as orthogonal parameters or ancillary statistics (Barndorff-Nielsen, 1994).

Barndorff-Nielsen and Cox (1984) use the p^*-formula to give an elegant and penetrating proof of the efficacy of Bartlett correction. The expectation of the profile log likelihood ratio statistic for testing $\kappa = \kappa_0$ typically can be expanded as

$$E\{w(\kappa_0); \psi_0, \kappa_0\} = k\{1 + m(\psi_0, \kappa_0)/n\} + O(n^{-2}),$$

where (ψ_0, κ_0) is the true value of the parameter. Asymptotically, the distributions of both $w(\kappa_0)$ and the adjusted statistic $w'(\kappa_0) = \{1 + m(\psi_0, \kappa_0)/n\}^{-1} w(\kappa_0)$ are chi-squared with k degrees of freedom; however, the expectation of $w'(\kappa_0)$ is closer to the mean of χ^2_k. In continuous cases, the surprising outcome of this simple adjustment is that it brings the entire distribution, not just the mean, closer to the limiting one. While the error in the chi-squared approximation to the distribution of $w(\kappa_0)$ is of order $O(n^{-1})$, the error for $w'(\kappa_0)$ is of order $O(n^{-2})$. In practice, this accuracy persists if the unknown nuisance parameter is replaced by a $\sqrt{n}$-consistent estimator.

Multiplicative terms that achieve this improvement in the chi-squared approximation are called Bartlett adjustment factors. Similar correction cannot be achieved for other asymptotically chi-squared test statistics based on the maximum likelihood estimator and the profile score.

In their proof, Barndorff-Nielsen and Cox (1984) show that Bartlett correction is valid conditionally, as well as unconditionally, and they give a very useful formula for the Bartlett adjustment factor. Their version of the adjusted statistic is

$$w'(\kappa_0) = (\bar{c}/\bar{c}_0)^{2/k} w'(\kappa_0),$$

where $\bar{c} = (2\pi)^{d/2} c$, $\bar{c}_0 = (2\pi)^{(d-k)/2} c_0$, c is the normalizing constant in the p^*-formula for $\hat{\omega}$ in the full model, and c_0 is the normalizing constant in the p^*-formula for $\hat{\psi}_0$ in the submodel specified by the null hypothesis. In a number of examples, the adjustment factor based on the p^*-formula performs better than do more traditional adjustments involving direct expansions of $E\{w(\kappa_0); \psi_0, \kappa_0\}$. The derivation based on the p^*-formula shows that $w'(\kappa_0)$ is conditionally distributed as χ^2_k to error of order $O(n^{-3/2})$; Barndorff-Nielsen and Hall (1988) show that the error is actually $O(n^{-2})$. Jensen (1993) gives a survey of Bartlett correction.

The p^*-formula has also been fruitfully applied to improve upon the directed likelihood $r(\kappa_0)$. In a companion paper to the present one, Barndorff-Nielsen (1986) derives a statistic $u(\kappa_0)$, independent of the nuisance parameter, such that the conditional distribution of

$$r^*(\kappa_0) = r(\kappa_0) + r(\kappa_0)^{-1} \log\{u(\kappa_0)/r(\kappa_0)\}$$

is standard normal to error of order $O(n^{-3/2})$. The statistic $u(\kappa_0)$ satisfies $u(\kappa_0) = r(\kappa_0) + O_p(n^{-1/2})$, and hence, $r^*(\kappa_0) = r(\kappa_0) + O_p(n^{-1/2})$.

Using $r^*(\kappa_0)$ in practice to test hypotheses or to construct confidence limits usually requires calculation of $\Phi\{r^*(\kappa_0)\}$, where Φ is the standard normal distribution function. To error of order $O(n^{-3/2})$, this quantity can be approximated by

$$\Phi\{r_0^*\} \approx \Phi\{r_0\} + \{r_0^{-1} - u_0^{-1}\}\phi\{r_0\},$$

where $r_0^* = r^*(\kappa_0)$, $r_0 = r(\kappa_0)$, $u_0 = u(\kappa_0)$, and ϕ is the standard normal density function. This approximation is closely related to the Lugannani and Rice (1980) tail probability formula. The version $\Phi\{r^*(\kappa_0)\}$ has the advantage, however, of never being less than 0 or exceeding 1.

When nuisance parameters are absent, so that $\omega = \kappa$, the formula for $u(\kappa_0)$ becomes (Barndorff-Nielsen, 1988; Fraser, 1990)

$$u_0 = \{-l_{\kappa\kappa}(\hat{\kappa}; \hat{\kappa}, a)\}^{-1/2}\{l_{;\hat{\kappa}}(\hat{\kappa}; \hat{\kappa}, a) - l_{;\hat{\kappa}}(\kappa_0; \hat{\kappa}, a)\},$$

where $l_{\kappa\kappa}(\kappa; \hat{\kappa}, a) = \partial^2 l(\kappa; \hat{\kappa}, a)/\partial\kappa^2$ and $l_{;\hat{\kappa}}(\kappa; \hat{\kappa}, a) = \partial l(\kappa; \hat{\kappa}, a)/\partial\hat{\kappa}$. In similar notation, the formula for the general case $\omega = (\psi, \kappa)$ is (Barndorff-Nielsen, 1991)

$$u_0 = \frac{|l_{;\hat{\omega}}(\hat{\omega}; \hat{\omega}, a) - l_{;\hat{\omega}}(\hat{\psi}_0, \kappa_0; \hat{\omega}, a) l_{\kappa;\hat{\omega}}(\hat{\psi}_0, \kappa_0; \hat{\omega}, a)|}{|-l_{\psi\psi}(\hat{\psi}_0, \kappa_0; \hat{\omega}, a)|^{1/2} |-l_{\omega\omega}(\hat{\omega}; \hat{\omega}, a)|^{1/2}}.$$

In a wide range of examples, the standard normal approximation to the distribution of r_0^* if found to be extraordinarily accurate. Jensen (1992) has shown that this approximation performs well in large-deviation regions. Pierce and Peters (1992) examine the correction $r_0^{-1}\log(u_0/r_0)$ in the definition of r_0^*, when κ is a component of the canonical parameter in a full exponential family, by decomposing it into the sum of two terms, one of which has substantial effect while the other is usually insignificant; Davison (1988) and Skovgaard (1987) also consider r_0^* in this context.

Implementation of r_0^* is impeded in many circumstances because an ancillary a must be specified explicitly to calculate u_0. Consequently, alternatives to u_0 have been sought that do not require ancillary statistics. Such quantities are developed by Barndorff-Nielsen and Chamberlin (1994) and by DiCiccio and Martin (1993), who take a Bayesian approach based on noninformative priors. The price to be paid is accuracy; typically, the corresponding versions of r_0^* have the standard normal distribution to error of order $O(n^{-1})$ only.

For any adjusted profile likelihood $l^a(\kappa)$, it is natural to consider the analogues of $w(\kappa_0)$ and $r(\kappa_0)$ given by $w_a(\kappa_0) = 2\{l^a(\hat{\kappa}^a) - l^a(\kappa_0)\}$ and $r_a(\kappa_0) = \text{sgn}(\hat{\kappa}^a - \kappa_0)\sqrt{w_a(\kappa_0)}$, where $\hat{\kappa}^a$ maximizes $l^a(\kappa)$. In many instances, the numerical accuracy of chi-squared and standard normal approximations for $w_a(\kappa_0)$ and $r_a(\kappa_0)$ is greatly superior than for $w(\kappa_0)$ and $r(\kappa_0)$. In wide generality, Bartlett correction is available for $w_a(\kappa_0)$ (Ghosh and Mukerjee, 1990; DiCiccio and Stern, 1994). In the context of inference about a component of the canonical parameter in a full exponential family, Fraser, Reid, and Wong (1991) develop $r_{mp}^*(\kappa_0)$ to improve upon $r_{mp}(\kappa_0)$; see also Fraser (1991). For general models, Barndorff-Nielsen and Chamberlin (1994) and DiCiccio and Martin (1993) consider versions of $r_a^*(\kappa_0)$ that do not explicitly require ancillary statistics, but for these versions, the standard normal approximation has error of order $O(n^{-1})$ only.

Bickel and Ghosh (1990) give a proof of the Bartlett correction using a Bayesian approach, and Sweeting (1995) considers r_0^* from a Bayesian perspective. Bayesian and conditional frequentist inference are closely related; see Efron (1993) for a recent contribution. The p^*-formula and related concepts are extremely useful for exploring these connections. For example, they will become increasingly important for establishing and investigating noninformative priors (Casella, DiCiccio, and Wells, 1995).

The strong impact of Barndorff-Nielsen's monumental paper is evident, and its broad influence will endure.

References

Barndorff-Nielsen, O.E. (1980). Conditionality resolutions. *Biometrika*, **67**, 293–310.

Barndorff-Nielsen, O.E. (1984). On conditionality resolution and the likelihood ratio for curved exponential models. *Scand. J. Statist.*, **11**, 157–170. Corrigendum: *Scand. J. Statist.*, **12** (1985), 191.

Barndorff-Nielsen, O.E. (1986). Inference on full or partial parameters based on the standardized signed log likelihood ratio. *Biometrika*, **73**, 307–322.

Barndorff-Nielsen, O.E. (1988). Discussion of paper by N. Reid. *Statist. Sci.*, **3**, 228–229.

Barndorff-Nielsen, O.E. (1991). Modified signed log likelihood ratio. *Biometrika*, **78**, 557–563.

Barndorff-Nielsen, O.E. (1994). Adjusted versions of profile likelihood and directed likelihood, and extended likelihood. *J. Roy. Statist. Soc. Ser. B*, **56**, 125–140.

Barndorff-Nielsen, O.E. and Chamberlin, S.R. (1994). Stable and invariant adjusted directed likelihoods. *Biometrika*, **81**, 485–500.

Barndorff-Nielsen, O.E. and Cox, D.R. (1979). Edgeworth and saddlepoint approximations with statistical applications (with discussion). *J. Roy. Statist. Soc. Ser. B*, **41**, 279–312.

Barndorff-Nielsen, O.E. and Cox, D.R. (1984). Bartlett adjustments to the likelihood ratio statistic and the distribution of the maximum likelihood estimator. *J. Roy. Statist. Soc. Ser. B*, **46**, 483–495.

Barndorff-Nielsen, O.E. and Cox, D.R. (1994). *Inference and Asymptotics*. Chapman & Hall, London.

Barndorff-Nielsen, O.E. and Hall, P. (1988). On the level-error after Bartlett adjustment of the likelihood ratio statistic. *Biometrika*, **75**, 374–378.

Barndorff-Nielsen, O.E. and McCullagh, P. (1993). A note on the relation between modified profile likelihood and the Cox–Reid adjusted profile likelihood. *Biometrika*, **80**, 321–328.

Bickel, P.J. and Ghosh, J.K. (1990). A decomposition for the likelihood ratio statistic and the Bartlett correction—A Bayesian argument. *Ann. Statist.*, **18**, 1070–1090.

Casella, G., DiCiccio, T.J., and Wells, M.T. (1995). Discussion of paper by N. Reid. *Statist. Sci.*, **10**, 179–185.

Cox, D.R. and Reid, N. (1987). Parameter orthogonality and approximate conditional inference (with discussion). *J. Roy. Statist. Soc. Ser. B*, **49**, 1–18 (Discussion 18–39).

Daniels, H.E. (1954). Saddlepoint approximations in statistics. *Ann. Math. Statist.*, **25**, 631–650.

Daniels, H.E. (1980). Exact saddlepoint approximations. *Biometrika*, **67**, 53–58.

Davison, A.C. (1988). Approximate conditional inference in generalized linear models. *J. Roy. Statist. Soc. Ser. B*, **50**, 445–461.

DiCiccio, T.J. and Martin, M.A. (1993). Simple modifications for signed roots of likelihood ratio statistics. *J. Roy. Statist. Soc. Ser. B*, **55**, 305–316.

DiCiccio, T.J. and Stern, S.E. (1994). Frequentist and Bayesian Bartlett correction of test statistics based on adjusted profile likelihoods. *J. Roy. Statist. Soc. Ser. B*, **56**, 397–408.

Durbin, J. (1980). Approximations for densities of sufficient estimators. *Biometrika*, **67**, 311–333.

Efron, B. (1993). Bayes and likelihood calculations from confidence intervals. *Biometrika*, **80**, 3–26.

Efron, B. and Hinkley, D.V. (1978). Assessing the accuracy of the maximum likelihood estimator: Observed versus expected Fisher information (with discussion). *Biometrika*, **65**, 457–487.

Field, C. and Ronchetti, E. (1990). *Small Sample Asymptotics*. Lecture Notes—Monograph Series, vol. 13. Institute of Mathematical Statistics, Hayward, CA.

Fraser, D.A.S. (1990). Tail probabilities from observed likelihoods. *Biometrika*, **77**, 65–76.

Fraser, D.A.S. (1991). Statistical inference: Likelihood to significance. *J. Amer. Statist. Assoc.*, **86**, 258–265.

Fraser, D.A.S., Reid, N., and Wong, A. (1991). Exponential linear models: A two-pass procedure for saddlepoint approximation. *J. Roy. Statist. Soc. Ser. B*, **53**, 483–492.

Ghosh, J.K. and Mukerjee, R. (1992). Bayesian and Frequentist Bartlett corrections for likelihood ratio and conditional likelihood ratio tests. *J. Roy. Statist. Soc. Ser. B*, **54**, 867–875.

Jensen, J.L. (1992). The modified signed likelihood ratio statistic and saddlepoint approximations. *Biometrika*, **79**, 693–703.

Jensen, J.L. (1993). A historical sketch and some new results on the improved log likelihood ratio statistic. *Scand. J. Statist.*, **20**, 1–15.

Lugannani, R. and Rice, S. (1980). Saddlepoint approximation for the distribution of the sum of independent random variables. *Adv. Appl. Probab.*, **12**, 475–490.

McCullagh, P. (1984). Local sufficiency. *Biometrika*, **71**, 233–244.

McCullagh, P. (1987). *Tensor Methods in Statistics*. Chapman & Hall, London.

McCullagh, P. and Tibshirani, R. (1990). A simple method for the adjustment of profile likelihoods. *J. Roy. Statist. Soc. Ser. B*, **52**, 325–344.

Pierce, D.A. and Peters, D. (1992). Practical use of higher order asymptotics for multiparameter exponential families (with discussion). *J. Roy. Statist. Soc. Ser. B*, **54**, 701–737.

Reid, N. (1988). Saddlepoint methods and statistical inference (with discussion). *Statist. Sci.*, **3**, 213–238.

Reid, N. (1995). The roles of conditioning in inference (with discussion). *Statist. Sci.*, **10**, 138–196.

Skovgaard, I.M. (1985). A second-order investigation of asymptotic ancillarity. *Ann. Statist.*, **13**, 534–551.

Skovgaard, I.M. (1987). Saddlepoint expansions for conditional distributions. *J. Appl. Probab.*, **24**, 875–887.

Skovgaard, I.M. (1990). On the density of minimum contrast estimators. *Ann. Statist.*, **18**, 779–789.

Sweeting, T.J. (1995). A framework for Bayesian and likelihood approximations in statistics. *Biometrika*, **82**, 1–24.

On a Formula for the Distribution of the Maximum Likelihood Estimator

O.E. Barndorff-Nielsen
Department of Theoretical Statistics, Aarhus University, Aarhus, Denmark

Summary

A simple formula for the conditional distribution of the maximum likelihood estimator given a maximal ancillary statistic is discussed and exemplified. The formula is generally accurate to order $O(n^{-1})$ or even $O(n^{-3/2})$, and for many important models it is, in fact, exact. After some preliminary discussion of the formula and of certain relevant aspects of likelihood, the formula is used to motivate the definition of a modified profile likelihood whose inferential properties are illustrated. The question of when the distribution formula is exact is considered, and in this connexion several new examples of exactness, including a bivariate generalization of the inverse Gaussian distribution, are adduced. The formula is shown also to be exact for arbitrary transformation models. To prove this it has been necessary to extend the basic theory of transformation models to cover the cases where the group action is not free. This extension, which appears of interest in itself, also allows of a generalization of a useful formula for the marginal likelihood for the index parameter of a composite transformation model.

1. Introduction

In an earlier paper (Barndorff-Nielsen, 1980) the formula

$$p(\hat{\omega};\omega|a) \doteq c|\hat{j}|^{1/2}\bar{L}, \qquad (1.1)$$

which provides an exact or approximate expression for the conditional distribution of the maximum likelihood estimator, was discussed to some extent. The formula is a synthesis and extension of various results by Fisher (1934), Fraser (1968), Daniels (1954), Barndorff-Nielsen and Cox (1979), Cox (1980), Hinkley (1980) and Durbin (1980a). The expression on the left-hand side of (1.1) is the probability function of the maximum likelihood estimator $\hat{\omega}$ of the parameter ω, given a maximal ancillary statistic a. On the right-hand side, $\bar{L}$ denotes the normed likelihood function,

$$\bar{L} = L/\hat{L} = L(\omega)/L(\hat{\omega}),$$

and $|\hat{j}|$ is the determinant of the observed information evaluated at the maximum likelihood point, that is $\hat{j} = j(\hat{\omega})$, where $j(\omega)$ is the observed information, which equals minus the matrix l of second-order derivatives of the log likelihood function $l(\omega) = \log L(\omega)$. Finally, c is a norming constant determined so as to make the total probability mass of $c|\hat{j}|^{1/2}\bar{L}$, given a, equal to one. In general c is a function of both a and ω, that is $c = c(a;\omega)$, though in important cases c does not, in fact, depend on ω or is even an absolute constant. It is convenient to have the notation

$$p^*(\hat{\omega};\omega|a) = c|\hat{j}|^{1/2}\bar{L} \tag{1.2}$$

for the right-hand side of (1.1).

Two important properties of relation (1.1) are that it is parameterization invariant and that it does not depend on which version of the likelihood function one considers. To be more specific, the first property means that reparameterization and calculation of p^* from the transformed likelihood function yields the same result as first calculating p^* and then transforming p^* by the ordinary rule for transformation of probability density functions. This follows by means of formula (2.4) below, and that formula, incidentally, also implies that the norming constant c in (1.2) is, by itself, parameterization invariant. The import of the second property, referred to above, is that before calculating p^* one can freely drop from an original likelihood any factor which depends on the observations only. Note however that, in (1.1) and (1.2), L and $\bar{L}$ vary with $(\hat{\omega}, a)$ as well as with ω.

To illustrate formula (1.1) and to clarify the rather condensed notation of the expression on the right-hand sides of (1.1) and (1.2) we consider three quite different but prototypical examples.

Example 1.1 (Location-Scale Model). Let $x_1,\ldots,x_n$ be a sample of size $n > 1$ from a location-scale distribution on the real line, i.e. the probability function for $x = (x_1,\ldots,x_n)$ is

$$p(x;\mu,\sigma) = \sigma^{-n}\prod_{i=1}^{n} f\{(x_i-\mu)/\sigma\},$$

with f a known probability density function on R. We assume that the maximum likelihood estimate $(\hat{\mu}, \hat{\sigma})$ of (μ, σ) exists uniquely with probability one, and we let

$$a = (a_1, \ldots, a_n) = \left(\frac{x_1 - \hat{\mu}}{\hat{\sigma}}, \ldots, \frac{x_n - \hat{\mu}}{\hat{\sigma}}\right).$$

The statistic a, called the configuration of the sample by R.A. Fisher, is distribution constant and, as shown by Fisher (1934), the probability density function for $(\hat{\mu}, \hat{\sigma})$ given the ancillary a may be written

$$p(\hat{\mu}, \hat{\sigma}; \mu, \sigma | a) = c_0(a)\hat{\sigma}^{n-2}p(x; \mu, \sigma),$$

where on the right-hand side x is to be expressed as $x = (\hat{\mu} + \hat{\sigma}a_1, \ldots, \hat{\mu} + \hat{\sigma}a_n)$ and where $c_0(a)$ is a certain factor depending on a only.

Now, if we take $L = L(\mu, \sigma) = L(\mu, \sigma; x) = p(x; \mu, \sigma)$, then $\hat{L} = \hat{\sigma}^{-n}p(a; 0, 1)$ and

$$l = \log L = -n \log \sigma + \sum_{i=1}^{n} \log f\{(x_i - \mu)/\sigma\},$$

and hence $|j| = D(a)\hat{\sigma}^{-4}$, where

$$D(a) = \left\{\sum g''(a_i)\right\}\left\{n + \sum a_i^2 g''(a_i)\right\} - \left\{\sum a_i g''(a_i)\right\}^2$$

and $g(x) = -\log f(x)$. The main term on the right-hand side of (1.1) is thus of the form

$$|\hat{j}|^{1/2}\bar{L} = c_1(a)\hat{\sigma}^{n-2}p(x; \mu, \sigma),$$

with $c_1(a) = D(a)^{1/2}/p(a; 0, 1)$. It follows that the normalizing factor c in (1.1) is given by $c = c(a) = c_0(a)/c_1(a)$ and we have exactly

$$p(\hat{\mu}, \hat{\sigma}; \mu, \sigma | a) = c|\hat{j}|^{1/2}\bar{L}.$$

EXAMPLE 1.2 (von Mises–Fisher Model). Consider the $(\kappa - 1)$-dimensional von Mises–Fisher distribution

$$p(x; \xi, \kappa) = a_k(\kappa) \exp(\kappa\xi \cdot x), \tag{1.3}$$

where x and ξ are unit vectors of dimension k and where the norming constant $a_\kappa(\kappa)$ is given by

$$a_k(\kappa) = \kappa^{(1/2)k-1}/\{(2\pi)^{(1/2)k}I_{(1/2)k-1}(\kappa)\},$$

$I_{(1/2)k-1}$ being a Bessel function. Let $x_1, \ldots, x_n$ be a sample from this distribution. The sum $x. = x_1 + \cdots + x_n$ is sufficient and we have $\hat{\xi} = x./r$, where $r = |x.|$ is the resultant length. As is well known, and obvious from the definition of r, the distribution of the resultant length depends on κ only. We may therefore use formula (1.1) to obtain an expression for the distribution of $\hat{\xi}$ given r, by presuming κ to be known, in which case r is ancillary. Since (1.1) is independent of the version of the likelihood function we simply take

$$L(\xi) = \exp(\kappa\xi \cdot x.) = \exp(r\kappa\xi \cdot \hat{\xi})$$

and find $\hat{L} = e^r$ and $|\hat{j}| = D(r\kappa)^{k-1}$, where D is an absolute constant. Hence $|\hat{j}|^{1/2}\bar{L}$ is of the form $c_1(r, \kappa)\exp(r\kappa\xi \cdot \hat{\xi})$ and comparing this expression with (1.3) one sees immediately that the norming constant c in (1.1) must be given by $c = a_k(r\kappa)/c_1(r, \kappa)$, that is

$$c|\hat{j}|^{1/2}\bar{L} = a_k(r\kappa)\exp(r\kappa\xi \cdot \hat{\xi}),$$

which is the model function of the von Mises–Fisher distribution with direction ξ and precision $r\kappa$. As is well known this is, in fact, the conditional distribution for $\hat{\xi}$ given r. Thus (1.1) is exact in the present instance too.

The fact that (1.1) is exact in the above two examples is no coincidence. As will be discussed in §4 equality holds in (1.1) quite generally for transformation models.

We might also have applied (1.1) to derive an expression, this time approximate, for the joint distribution of $\hat{\xi}$ and $\hat{\kappa}$ in Example 1.2. Since $(\hat{\xi}, \hat{\kappa})$ is minimal sufficient no conditioning is involved. The resulting formula is a special case of the renormalized saddlepoint approximation for exponential families, discussed *inter alia* by Daniels (1954, 1980) and Barndorff-Nielsen and Cox (1979). In fact, as we now indicate, for exponential families (1.1) is equivalent to this approximation. Consider a full exponential model

$$p(x; \theta) = \exp\{\theta \cdot s(x) - K(\theta) - \varphi(x)\}, \tag{1.4}$$

where θ and s are vectors of dimension k; let $x_1, \ldots, x_n$ be a sample from (1.4) and set $t = s(x_1) + \cdots + s(x_n)$. The saddlepoint approximation to the distribution of t is then (Barndorff-Nielsen and Cox, 1979, p. 299)

$$p(t; \theta) \doteq (2\pi)^{-(1/2)k}|i(\hat{\theta})|^{-1/2}\exp[(\theta - \hat{\theta}) \cdot t + n\{K(\hat{\theta}) - K(\theta)\}] \tag{1.5}$$

and the ratio of the two expressions on either side of the approximation sign is of the form $1 + O(n^{-1})$. In (1.5), $i(\theta)$ denotes the expected, or Fisher, information on θ. Since $L(\theta) = \exp\{\theta \cdot t - nK(\theta)\}$, we may rewrite (1.5) as

$$p(t; \theta) \doteq (2\pi)^{-(1/2)k}|i(\hat{\theta})|^{-1/2}\bar{L}, \tag{1.6}$$

and, as (1.4) is full, we have $i(\hat{\theta}) = j(\hat{\theta})$, that is we may substitute j for i in (1.6). The renormalized saddlepoint approximation for t is therefore

$$p(t; \theta) \doteq = c|j(\hat{\theta})|^{-1/2}\bar{L}. \tag{1.7}$$

The renormalization usually reduces the relative error from $O(n^{-1})$ to $O(n^{-3/2})$. Finally, we reparametrize to the mean parameter $\tau = E_\theta(t)$, that satisfies $\hat{\tau} = t$, and we note that $j(\hat{\tau}) = \{j(\hat{\theta})\}^{-1}$, where $j(\hat{\tau})$ and $j(\hat{\theta})$ are the observed information matrices relative to the two different parametrizations; a more specific notation would be $j_\tau(\hat{\tau}) = \{j_\theta(\hat{\theta})\}^{-1}$. Using this we find that (1.7) takes the form

$$p(\hat{\tau};\tau) \doteq c|j(\hat{\tau})|^{1/2}\bar{L} \tag{1.8}$$

with the same c as in (1.7). This relation coincides with (1.1).

It follows from results of Durbin (1980a, b) that even if the model is not exponential, formula (1.1) will generally be accurate to order $O(n^{-1})$, and often in fact to order $O(n^{-3/2})$, provided $\hat{\omega}$ is minimal sufficient. Here the basic observations $x_1, \ldots, x_n$ need not be independent and identically distributed but may, for instance, be realizations from a stationary time series. It also follows from Durbin (1980a) that, in these circumstances where no conditioning is involved, if $p(\hat{\omega};\omega)$ and $p^*(\hat{\omega};\omega)$ are integrated over the same subset of the set of possible $\hat{\omega}$ values then, under minor regularity conditions, the difference of the integrals is of order $O(n^{-1})$.

Our last introductory example is concerned with the inverse Gaussian distribution and falls in the saddlepoint category discussed above.

Example 1.3 (Inverse Gaussian Model). The model function of the family of inverse Gaussian distributions is

$$p(x;\chi,\psi) = \frac{\sqrt{\chi}}{\sqrt{2\pi}} \exp\{\sqrt{\chi\psi}\}x^{-3/2} \exp\{-\tfrac{1}{2}(\chi x^{-1} + \psi x)\}, \tag{1.9}$$

where $x > 0, \chi > 0$ and $\psi \geq 0$. We denote this model by $N^-(\chi,\psi)$. The joint distribution of a sample $x_1, \ldots, x_n$ from $N^-(\chi,\psi)$ is

$$p(x_1,\ldots,x_n;\chi,\psi) = \frac{\chi^{1/2n}}{2\pi^{1/2n}} \exp\{n\sqrt{\chi\psi}\} \prod_{i=1}^{n} x_i^{-3/2} \exp\{-\tfrac{1}{2}n(\chi\tilde{x} + \psi\bar{x})\}, \tag{1.10}$$

where $\tilde{x} = n^{-1}\sum x_i^{-1}$ and $\bar{x} = n^{-1}\sum x_i$.

Suppose first that χ is known. We then have an exponential model of order 1 and with no ancillary statistics. Let $\mu = \sqrt{\chi/\psi} = E(\bar{x})$ and let us calculate the right-hand side of (1.1) using μ as the parameter. Taking the simplest possible version of $L(\mu)$ from (1.10), we have $L(\mu) = \exp\{n\chi(\mu^{-1} - \frac{1}{2}\bar{x}\mu^{-2})\}$, and since $\hat{\mu} = \bar{x}$ we find $\hat{j} = n\chi\bar{x}^{-3}$ and

$$|j(\hat{\mu})|^{1/2}\bar{L}(\hat{\mu}) = \bar{x}^{-3/2}\exp\{n\sqrt{\chi\psi}\}\exp\{-\tfrac{1}{2}(n\chi\bar{x}^{-1} + n\psi\bar{x})\}.$$

Comparing with (1.9) one sees that the norming constant c of (1.1) must equal $(2\pi)^{-1/2}$. The distribution for $\hat{\mu} = \bar{x}$ determined by the right-hand side of (1.1) is therefore $N^-(n\chi, n\psi)$ and, as is well known, this is the actual distribution of $\bar{x}$. We have here a further instance of exactness of (1.1), and it is to be noted that the present model is not of the transformation type.

Because of the parameterization invariance of (1.1) we might alternatively have proceeded by taking ψ as the parameter, obtaining an expression for the distribution of $\hat{\psi}$ from (1.1) and finally, if desired, transforming this to a distribution for $\hat{\mu} = \sqrt{\hat{\chi}/\hat{\psi}}$.

Next, let both χ and ψ be considered as unknown. We have $\omega = (\chi, \psi)$ as parameter, there are ancillary statistics and the model is not of transformation type. It turns out that, again, relation (1.1) is exact, as may be shown by using the well known facts that $\bar{x} \sim N^{-}(n\chi, n\psi)$ and $n(\tilde{x} - \bar{x}^{-1}) \sim \Gamma(\frac{1}{2}n - \frac{1}{2}, \frac{1}{2}\chi)$. Here we use $\Gamma(\lambda, \alpha)$ to denote the gamma distribution

$$\{\alpha^{\lambda}/\Gamma(\lambda)\}x^{\lambda-1}\exp(\alpha x).$$

Further instances of exactness of (1.1) will be presented in §4, including a bivariate generalization of the inverse Gaussian distribution. As will be discussed, these examples have various important properties of independent interest. The discussion of exactness as a whole indicates that, for continuous variates, if there exists an explicit formula for the distribution of the maximum likelihood estimator then (1.1) will yield this expression exactly.

The results of Durbin (1980a) and Barndorff-Nielsen (1980) show that the accuracy of (1.1) under repeated observations will be $O(n^{-3/2})$ under very broad assumptions. Even when no exactly ancillary statistic a exists, (1.1) will often give an accurate approximation provided a is approximately ancillary. In particular, it was shown by Barndorff-Nielsen (1980) that, for curved exponential families and with a as the so-called affine ancillary, relation (1.1) is accurate to order $O(n^{-1})$ in relative error. It should be noted that, generally, when an exact ancillary does exist for a curved exponential model the affine ancillary will itself be exact; see Barndorff-Nielsen (1980), Barndorff-Nielsen et al. (1982).

In §2 we discuss some aspects of likelihood, requisite for the rest of the paper. A certain derivation based on (1.1) leads, in §3, to the definition of a modified profile likelihood with, generally, better inferential properties than the ordinary profile likelihood function. The question of when (1.1) is exact forms the subject of §4. A simple and useful formula for the marginal likelihood function of the index parameter from an arbitrary composite transformation model is briefly discussed in §5.

In general we use p to denote probability density with respect to a dominating measure μ, and, in (1.1), μ is understood to be geometric measure on the range space of $\hat{\omega}$, that is μ is there typically either Lebesgue measure or counting measure. Our vectors will be column vectors and we denote the transpose of a matrix M by M^{T}.

We now adduce a number of further comments on formula (1.1) and on $p^{*}(\hat{\omega}; \omega|a)$.

Most of the results discussed above and in the rest of this paper hold equally if expected information $\hat{i} = i(\hat{\omega})$ is employed in (1.1) and (1.2) instead of observed information $\hat{j}$. Here $i(\omega) = E_{\omega}\{j(\omega)\}$. To what extent it is advantageous to use one of $\hat{j}$ and $\hat{i}$ rather than the other depends on the circumstances.

It is assumed in (1.1) that $(\hat{\omega}, a)$ is in one-to-one correspondence with the minimal sufficient statistic t of the underlying model $\mathcal{M}$. The sample space of the model $\mathcal{M}$ is denoted by $\mathcal{X}$, the basic data is $x \in \mathcal{X}$, and Ω stands for the set of possible values of ω. We generally denote the dimensions of t and ω by k and d, respectively, and then the dimension of a is $k - d$. Note that when $k = d$ no conditioning is involved in (1.1). As mentioned previously, the ancillary statistic a is allowed to be only approximately distribution constant.

Calculation of p^* requires knowledge of the normed likelihood function only. As already pointed out, it therefore makes no difference which version L of the likelihood function is taken as the starting point for deriving p^*. As a consequence of this we have that p^*, in addition to being parameterization invariant, is also invariant under one-to-one transformations of the data x. Note also that, since $\bar{L}$ is minimal sufficient and $(\hat{\omega}, a)$ is assumed to be in one-to-one correspondence with t, the right-hand side of (1.1) in fact depends on x through $(\hat{\omega}, a)$ only.

The formula (1.2) for p^* is based on the unconditional likelihood function L. An alternative way of obtaining an, approximate or exact, expression for $p(\hat{\omega}; \omega|a)$ would be to insert the conditional likelihood function in (1.2). The two expressions for $p(\hat{\omega}; \omega|a)$ coincide when a is exactly ancillary, but not in general otherwise.

The norming constant c in p^* can often be well approximated by a constant c_0 independent of both a and ω. In fact, due to the asymptotic normality of $\hat{\omega}$ and the asymptotic independence of $\hat{\omega}$ and a, which hold under broad assumptions, c_0 can frequently be taken to be $(2\pi)^{-(1/2)d}$. The resulting modification of (1.1), that is $p(\hat{\omega}; \omega|a) \doteq p^0(\hat{\omega}; \omega|a)$, say, where $p^0(\hat{\omega}; \omega|a) = c_0|\hat{j}|^{1/2}\bar{L}$, is exact only in exceptional cases but has the advantage over (1.1) of yielding always the exact likelihood function for ω.

We conclude this section by establishing some basic terminology for transformation models.

A transformation model $\mathcal{M}$ consists of a family $\mathcal{P}$ of probability measures on a sample space $\mathcal{X}$ for which there exists a class $\mathcal{F}$ of transformations of $\mathcal{X}$ and a member P of $\mathcal{P}$ such that $\mathcal{P} = \{fP: f \in \mathcal{F}\}$, where fP is the lifting of P by the transformation f, that is $(fP)(A) = P(f^{-1}(A))$; by a transformation of $\mathcal{X}$ we mean a one-to-one mapping of $\mathcal{X}$ onto itself. More generally, we speak of a model $\mathcal{M}$ as a composite transformation model if $\mathcal{M}$ can be viewed as a collection $\mathcal{M} = \{\mathcal{M}_\kappa: \kappa \in \mathcal{K}\}$ of models $\mathcal{M}_\kappa$ that are all transformational with respect to one and the same $\mathcal{F}$. Usually, and we shall assume that this is the case, the class $\mathcal{F}$ is determined as follows. A group G is said to act on a space $\mathcal{X}$ if there is a mapping γ of G into the set of all transformations of $\mathcal{X}$ with the property that $\gamma(g'g) = \gamma(g) \circ \gamma(\acute{g})$ for all g and $\acute{g}$ in G, where $\circ$ denotes composition of mappings. The class $\mathcal{F}$ is then supposed to be of the form $\mathcal{F} = \gamma(G)$ for some group G acting on the sample space $\mathscr{X}$ by an action γ. For simplicity, with $x \in \mathscr{X}$ and μ a measure on $\mathscr{X}$ we write gx for $\gamma(g)(x)$ and $g\mu$ for $\gamma(g)\mu$, provided there is no risk of ambiguity. We shall furthermore assume that there exists a measure μ on $\mathscr{X}$ which dominates $\mathscr{P}$ and which is invariant under the action of G, that is $g\mu = \mu$ for every $g \in G$. Then, writing $p(x; g)$ for $dgP/d\mu$ and $p(x)$ for $p(x; e)$, where e is the identity element of G, we have

$$p(x; g) = p(g^{-1}x) \quad \langle \mu \rangle, \tag{1.11}$$

where the symbol $\langle \cdot \rangle$ is used to indicate the dominating measure considered. For any $x \in \mathscr{X}$ the set $Gx = \{gx: g \in G\}$ of points traversed by x under the action of G is termed the orbit of x. The sample space $\mathscr{X}$ is thus partitioned into disjoint orbits, and if on each orbit we select a point u, to be called the orbit representative, then any point x in $\mathscr{X}$ can be determined by specifying the representative u of Gx and an element $z \in G$ such that $x = zu$. In this way x has, as it were, been expressed in new coordinates (z, u) and we speak of (z, u) as an orbital decomposition of x.

The orbit representative, or any one-to-one transformation thereof, is a maximal invariant, and hence ancillary, statistic, and inference under the model proceeds by first conditioning on that statistic.

2. Likelihood and Quasilikelihood

A common device in drawing inference about a parameter $\psi = \psi(\omega)$ of interest is to specify some function $L^\dagger(\psi)$, say, which depends on ω through ψ only and which, to some extent at least, is used as if the inference base had ψ as the full parameter and had likelihood function $L^\dagger(\psi)$. We shall refer to such functions $L^\dagger$ as quasilikelihood functions, including proper likelihood functions within this concept. Examples of quasilikelihood functions are marginal and conditional likelihood functions, and the more general concept of a partial likelihood function (Cox, 1975), profile, or partially maximized, likelihood functions, and "quasilikelihood" functions in the sense of Wedderburn (1974). A further instance termed "modified profile likelihood" functions will be discussed in §3.

A quasilikelihood function yields a maximum quasilikelihood estimate, and an, approximate or exact, expression for the distribution of this estimate may, at least in certain cases, be obtained by means of the formula (1.2) for p^*, by using the quasilikelihood function on the right-hand side of (1.2), also for calculating $\hat{j}$. Note that if the quasilikelihood function can be recognized as belonging to a known family of likelihoods the expression for p^* may often be written down without any calculations; see Example 3.3.

In the rest of the present section we establish some notation and a few results for ordinary and marginal likelihoods, which will be used later.

The likelihood function $L(\omega)$ is determined only up to a multiplicative constant b which may depend on the observation x, and in writing equations for likelihoods we freely drop or introduce multiplicative constants, thus somewhat abusing the equality sign by allowing expressions like $L = bL$.

Suppose $\psi = \psi(\omega)$ is a one-to-one transformation of the parameter ω, to a new d-dimensional parameter ψ. Then

$$\dot{l}(\psi) = \frac{\partial \omega^{\mathrm{T}}}{\partial \psi} \dot{l}(\omega),$$

and hence

$$\ddot{l}(\psi) = \frac{\partial \omega^{\mathrm{T}}}{\partial \psi} \ddot{l}(\omega) \frac{\partial \omega}{\partial \psi^{\mathrm{T}}} + \frac{\partial^2 \omega}{\partial \psi \partial \psi^{\mathrm{T}}} \cdot \dot{l}(\omega), \tag{2.1}$$

with the dot product $\cdot$ as defined by Barndorff-Nielsen (1978, §1.2), i.e. with $\omega = (\omega_1, \ldots, \omega_d)$ we have

$$\frac{\partial^2 \omega}{\partial \psi \partial \psi^{\mathrm{T}}} \cdot \dot{l}(\omega) = \sum_{i=1}^{d} \frac{\partial l}{\partial \omega_i} \frac{\partial \omega_i}{\partial \psi \partial \psi^{\mathrm{T}}}.$$

From relation (2.1) we find

$$i(\psi) = \frac{\partial \omega^{\mathrm{T}}}{\partial \psi} i(\omega) \frac{\partial \omega}{\partial \psi^{\mathrm{T}}}, \qquad j(\hat{\psi}) = \frac{\partial \omega^{\mathrm{T}}}{\partial \psi} (\hat{\partial}) j(\hat{\omega}) \frac{\partial \omega}{\partial \psi^{\mathrm{T}}} (\hat{\psi}). \tag{2.2}$$

These two formulae have in turn the consequence that

$$\left| \frac{\partial \omega^{\mathrm{T}}}{\partial \psi} (\hat{\psi}) \right| = |i(\hat{\psi})^{1/2} / |i(\hat{\omega})|^{1/2} = |j(\hat{\psi})|^{1/2} / |j(\hat{\omega})|^{1/2}. \tag{2.3}$$

Relation (2.3) may be interpreted as saying that the measures on Ω defined by $|i(\hat{\omega})^{1/2} d\hat{\omega}$ and by $|j(\hat{\omega})|^{1/2} d\hat{\omega}$ are parameterization invariant. Incidentally, this invariance of $|i(\hat{\omega})|^{1/2} d\hat{\omega}$ is, essentially, the basis for Jeffreys's (1961, §3.10) definition of invariant prior distributions. For transformation models these measures have the crucial property of being invariant under the group of transformations of Ω induced via $\hat{\omega}$; see §4.

If u is a statistic we speak of the likelihood function which would have been obtained if only the value of u, and not the basic data x, had been observed as the marginal likelihood function based on u.

Quite generally, the probability function for u may be expressed as

$$p(u; \omega) = E_{\omega_0} \left\{ \frac{L(\omega)}{L(\omega_0)} \middle| u \right\} p(u; \omega_0), \tag{2.4}$$

where ω_0 denotes an arbitrarily fixed value of ω and it is assumed that $p(x; \omega) > 0$ for all x and ω. In a considerable range of cases the conditional mean value in (2.4) can be determined simply and the problem of finding $p(u; \omega)$ is thus reduced to that of deriving an expression for $p(u; \omega_0)$, where

ω_0 may be chosen at will to simplify the calculation. This technique will be instrumental in Example 4.2 below. Other applications are given, for instance, by Barndorff-Nielsen (1978a, §8.2(iii)) and Fraser (1979, §8.1.3). In the present context note that the marginal likelihood function is expressible as

$$L(\omega; u) = E_{\omega_0}\left\{\frac{L(\omega)}{L(\omega_0)}\bigg| u\right\}. \tag{2.5}$$

For composite transformation models and with u as the maximal invariant statistic, an alternative useful formula can be established that expresses the marginal likelihood for the index parameter in terms of an integral with respect to invariant measure on the transformation parameter space. It is possible to recast this latter formula as

$$L(\kappa; u) = \int L(\psi, \kappa; x)\Delta(\psi)^{-1}|i_{\kappa_0}(\psi)|^{1/2} d\psi, \tag{2.6}$$

where κ is the index parameter, ψ is the transformation parameter, Δ is the modular function of the group, and $i_{\kappa_0}(\psi)$ is the expected information on ψ for some arbitrary fixed value κ_0 of the index parameter. There is an analogous formula in terms of observed information. The derivation is indicated in §5.

3. Modified Profile Likelihood

Consider a model with parameter (ψ, κ) and minimal sufficient statistic t, and suppose there exists a statistic u such that both $(\hat{\psi}, u)$ and $(\hat{\psi}, \hat{\kappa})$ are one-to-one transformations of t and such that the distribution of u depends on κ only.

Denoting the maximum likelihood estimator of ψ for fixed κ by $\hat{\psi}_\kappa$, we have

$$\begin{aligned} p(u; \kappa) &= p(\hat{\psi}, u; \psi, \kappa)/p(\hat{\psi}; \psi, \kappa|u) \\ &= \left\{p(\hat{\psi}, \hat{\kappa}; \psi, \kappa)\left|\frac{\partial(\hat{\psi}, \hat{\kappa})}{\partial(\hat{\psi}, u)}\right|\right\} \Big/ \left\{p(\hat{\psi}_\kappa; \psi, \kappa|u)\left|\frac{\partial\hat{\psi}_\kappa}{\partial\hat{\psi}}\right|\right\}, \end{aligned}$$

where $\partial\hat{\psi}_\kappa/\partial\hat{\psi}$ is the matrix of partial derivatives of $\hat{\psi}_\kappa$ with respect to $\hat{\psi}$, the estimator $\hat{\psi}_\kappa$ being considered as a function of $\hat{\psi}$ and u. Using (1.1) to approximate $p(\hat{\psi}, \hat{\kappa}; \psi, \kappa)$ and $p(\hat{\psi}_\kappa; \psi, \kappa|u)$, we obtain an approximation to $p(u; \kappa)$ and hence also to the marginal likelihood $L(\kappa; u)$. Our concern here is with the latter. Making the further approximation of treating the norming constants as being independent of ψ and κ(§§1, 4) we find, omitting terms depending only on the observation,

$$L(\kappa; u) \simeq \left|\frac{\partial\hat{\psi}}{\partial\hat{\psi}_\kappa}\right| |\hat{j}_\kappa|^{-1/2}\tilde{L}(\kappa), \tag{3.1}$$

where j_κ is the observed information for ψ when κ is fixed and $\tilde{L}(\kappa)$ is the profile, or partially maximized, likelihood for κ, that is $\tilde{L}(\kappa) = L(\hat{\psi}_\kappa, \kappa)$.

The expression on the right-hand side of (3.1) is meaningful whether the distribution of u depends on κ only or not. The expression is also well-defined even if $(\hat{\psi}, \hat{\kappa})$ is not minimal sufficient so long as there exists an ancillary statistic a such that $(\hat{\psi}, \hat{\kappa}, a)$ is minimal sufficient. We shall term it the modified profile likelihood for κ and denote it by $L^0(\kappa)$, that is

$$L^0(\kappa) = L^0(\kappa; u) = \left|\frac{\partial\hat{\psi}}{\partial\hat{\psi}_\kappa}\right| |\hat{j}_\kappa|^{-1/2}\tilde{L}(\kappa), \tag{3.2}$$

The dependence on u should be noted. It will often be natural to take $u = (\hat{\kappa}, a)$, where a is ancillary for (ψ, κ). In view of (2.3) the modified profile likelihood is parameterization invariant. Of the two modifying factors $|\hat{j}_\kappa|^{-1/2}$ and $|\partial\hat{\psi}/\partial\hat{\psi}_\kappa|$ in (3.2), the former corresponds to a variance stabilizing transformation of the parameter ψ while the latter may be thought of as compensating for a possible mathematical "inelegance" of the parametrization. The value of κ, where $L^0(\kappa)$ takes its maximum value, will be denoted by $\hat{\kappa}^0$.

The question we address in the present section is to what extent the modified profile likelihood function L^0 is useful as a quasilikelihood function, in the sense in which quasilikelihood functions were defined in §2. We do this mainly by way of a variety of examples. As will become apparent the modified likelihood in many cases approximates closely to, or coincides with, an appropriate marginal and conditional likelihood, and it is often simpler to calculate. It would no doubt be possible to establish some general theorems to this effect; that will not be aimed at here, though some brief indications will be given.

As with formula (1.1) it may be advantageous to use expected information $\hat{i}_\kappa$ in (3.2) instead of observed information $\hat{j}_\kappa$. The resulting expression is again parametrization invariant. We shall denote it by L^e, that is

$$L^e(\kappa) = \left|\frac{\partial\hat{\psi}}{\partial\hat{\psi}_\kappa}\right| |\hat{i}_\kappa|^{-1/2}\tilde{L}(\kappa). \tag{3.3}$$

An alternative form of (3.2) is sometimes convenient. If we set $l_1 = \partial l/\partial\psi$ the likelihood equation for $\hat{\psi}_\kappa$ may be written $l_1(\hat{\psi}_\kappa, \kappa; \hat{\psi}, u) = 0$, and differentiation of this with respect to $\hat{\psi}_\kappa$ yields

$$\frac{\partial l_1}{\partial\hat{\psi}_\kappa} + \frac{\partial l_1}{\partial\hat{\psi}}\frac{\partial\hat{\psi}^{\mathrm{T}}}{\partial\hat{\psi}_\kappa} = 0$$

or, equivalently,

$$\frac{\partial\hat{\psi}}{\partial\hat{\psi}_\kappa} = \left\{\frac{\partial^2 l(\hat{\psi}_\kappa, \kappa; \hat{\psi}, u)}{\partial\hat{\psi}_\kappa \partial\hat{\psi}^{\mathrm{T}}}\right\}^{-1} \hat{j}_\kappa.$$

Hence

$$L^0(\kappa) = \left|\frac{\partial^2 l(\hat\psi_\kappa, \kappa; \hat\psi, u)}{\partial\hat\psi_\kappa \partial\hat\psi^{\mathrm{T}}}\right|^{-1} |\hat{j}_\kappa|^{1/2} \tilde{L}(\kappa). \tag{3.4}$$

Example 3.1 (Normal Distribution). Let $x_1, \ldots, x_n$ be a random sample from the normal distribution $N(\xi, \sigma^2)$. The modified profile likelihood for σ^2 is

$$L^0(\sigma^2) = \sigma^{-(n-1)} \exp\{-\tfrac{1}{2}\sigma^{-2} \textstyle\sum (x_i - \bar{x})^2\}. \tag{3.5}$$

The estimate of σ^2 obtained by maximizing $L^0(\sigma^2)$ is the usual estimate

$$s^2 = (n-1)^{-1} \sum (x_i - \bar{x})^2.$$

In fact, $L^0(\sigma^2)$ equals the marginal likelihood function $L(\sigma^2; s^2)$.

In calculating (3.5) from (3.2) we have $\kappa = \sigma^2$ and $u = s^2$, and it is natural to choose $\psi = \xi$ whence $\hat\psi = \hat\psi_\kappa = \bar{x}$. To illustrate the role of the adjusting factor $|\partial\hat\psi/\partial\hat\psi_\kappa|$, suppose instead that we had taken $\psi = (\xi - c)/\sigma$ for some known constant c; that is ψ determines the probability mass of the $N(\xi, \sigma^2)$ distribution above the "tolerance limit" c. Then $\hat\psi = (\bar{x} - c)/\hat\sigma$ and $\hat\psi_\kappa = (\bar{x} - c)/\sigma = (\sqrt{\hat\kappa}/\sqrt{\kappa})\hat\psi$ so that $|\partial\hat\psi/\partial\hat\psi_\kappa| = \sqrt{\kappa}/\sqrt{\hat\kappa}$. This factor is cancelled out by a similar factor in $\hat{j}_\kappa$, introduced by the change from $\psi = \xi$ to $\psi = (\xi - c)/\sigma$.

Example 3.2 (Inverse Gaussian Model). The modified profile likelihood function for the parameter χ, based on a sample of n from the $N^-(\chi, \psi)$ distribution, is found from (1.10) to be

$$L^0(\chi) = \chi^{(1/2)n - 1/2} \exp(-\tfrac{1}{2}\chi w), \tag{3.6}$$

where $w = \sum(x_i^{-1} - \bar{x}^{-1})$. We have $L^0(\chi) = L(\chi : w)$, as might be expected from Example 1.3. The profile likelihood $\tilde{L}(\chi)$ differs from (3.6) by a factor of $\chi^{1/2}$.

The distributions for s^2, of Example 3.1, and w, of Example 3.2, obtained by using the modified profile likelihood L^0 as input in (1.2) are in fact the correct distributions of s^2 and w.

Example 3.3 (von Mises–Fisher Model). For a sample $x_1, \ldots, x_n$ from the von Mises–Fisher distribution (1.3) the modified profile likelihood for the precision parameter κ is

$$L^0(\kappa) = L^0(\kappa; r) = \{a_k(\kappa)\}^n \kappa^{-(1/2)k + 1/2} \exp(r\kappa),$$

where $r = |x_.|$ is the resultant length.

This may be compared with the marginal likelihood of κ based on r:

$$L(\kappa; r) = \{a_\kappa(\kappa)\}^n / a_\kappa(r\kappa).$$

The two likelihood functions $L^0(\kappa)$ and $L(\kappa; r)$ differ only to the extent of the approximation

$$I_v(x) \doteq \frac{1}{\sqrt{(2\pi)}} x^{-1/2} e^x \tag{3.7}$$

which is asymptotically valid, to order $O(x^{-1})$, for $x \to \infty$, that is, as $r\kappa \to \infty$,

$$L(\kappa; r) = L^0(\kappa; r)[1 + O\{(r\kappa)^{-1}\}].$$

If the asymptotic formula (3.7) is used to approximate $a_k(\kappa)$ in the expression for $L^0(\kappa)$ one obtains

$$L^0(\kappa) \doteq \kappa^{(1/2)(n-1)(k-1)} \exp\{-(n-r)\kappa\}.$$

The right-hand side is recognized as the likelihood function of the $\Gamma\{\frac{1}{2}(n-1)(k-1), \kappa\}$-distribution, corresponding to an observation $n-r$. Substitution of this latter likelihood function into formula (1.2) will therefore yield $\Gamma\{\frac{1}{2}(n-1)(k-1), \kappa\}$ as a distribution for $n-r$. We have thus, by a new route, arrived at the usual χ^2 approximation to the distribution of $2(n-r)$, which is valid for $\kappa \to \infty$.

Example 3.4 (Normal Regression). Let $x_1, \ldots, x_n$ be independent and such that $x_i \sim N(\mu + \beta \cdot t_i, \sigma^2)$ where the known covariate vectors t_i are of dimension r and satisfy $t_1 + \cdots + t_n = 0$ and rank $[t_1, \ldots, t_n] = r$. The modified profile likelihood for the main effect μ, as determined by the specifications $\psi = (\beta, \sigma^2)$, $\kappa = \mu$ and $u = \hat{\kappa}$, may be written

$$L^0(\mu) = \{1 + t^2/(n-1-r)\}^{-(1/2)n+(1/2)r+1}, \tag{3.8}$$

where $t = \sqrt{n(\bar{x} - \mu)}/\sqrt{\sum(x_i - \hat{\mu} - \hat{\beta} \cdot t_i)^2/(n-1-r)}$ is the usual Student's test statistic for μ. Note that (3.8) differs from the probability density function for t only by a norming constant and in that the latter has $-\frac{1}{2}n + \frac{1}{2}r$ rather than $-\frac{1}{2}n + \frac{1}{2}r + 1$ in the exponent. The expression for $L^0(\mu)$ should also be compared to that of the profile likelihood for μ,

$$\tilde{L}(\mu) = \{1 + t^2/(n-1-r)\}^{-(1/2)n}.$$

In the not infrequent case where $\hat{\psi}_\kappa = \hat{\psi}$ for all κ the modified profile likelihood is

$$L^0(\kappa) = |\hat{j}_\kappa|^{-1/2} \tilde{L}(\kappa). \tag{3.9}$$

This is true, in particular, if the model function is of the form

$$p(x; \psi, \kappa) = b(x; \kappa) \exp\{\alpha(\kappa)\varphi(x; \psi)\}, \tag{3.10}$$

and then

$$L^0(\kappa) = |\alpha(\kappa)|^{-(1/2)d} b(x; \kappa) \exp\{\alpha(\kappa)\varphi(x; \hat{\psi})\}, \tag{3.11}$$

where $\alpha(\kappa)$ is a real-valued function of κ and d is the dimension of ψ. The generalized linear models of Nelder and Wedderburn (1972) have the structure (3.10). The same is true of a considerable class of composite exponential transformation models, and this includes the von Mises–Fisher model and the hyperboloid model. Yet a third type of examples are provided by the symmetric hyperbolic distributions

$$p(x;\xi,\kappa) = \{2\delta K_1(\delta\kappa)\}^{-1} \exp\left[-\kappa\sqrt{\delta^2 + (x-\xi)^2}\right] \tag{3.12}$$

provided that $\delta \geq 0$ is taken as fixed. For $\delta = 0$, equation (3.12) is the Laplace distribution.

Another important class of cases for which $\hat{\psi}_\kappa = \hat{\psi}$, so that $L^0(\kappa)$ is given by (3.9), is that of steep, and in particular regular, exponential models with (ψ, κ) as a mixed parameter. More specifically, suppose that the exponential model function is

$$a(\theta)b(x)\exp(\theta \cdot t); \tag{3.13}$$

let $t = (u, v)$ and $\theta = (\kappa, \lambda)$ be similar partitions of the canonical statistic and the canonical parameter, respectively, and let the corresponding partition of the mean τ of t be $t = (\varphi, \psi)$. Then $\hat{\psi}_\kappa = \hat{\psi} = v$ and, denoting the variance matrix of v by $\sum_{22}(\kappa, \psi)$, we find that $L^0(\kappa) = |\sum_{22}(\kappa, v)|^{1/2}\tilde{L}(\kappa)$ or

$$L^0(\kappa) = \left|\sum_{22}(\kappa, v)\right|^{1/2} a(\kappa, v)\exp\{\kappa \cdot u + \lambda(\kappa, v) \cdot v\}, \tag{3.14}$$

where we now conceive also of a and λ as functions of the mixed parameter (κ, ψ).

A comparison of (3.14) and formula (4.8) of Barndorff-Nielsen and Cox (1979) shows that in the present case, of exponential models, the modified profile likelihood function (3.14) is equal to the approximation to the conditional likelihood function of κ given v determined by the double saddlepoint approximation to the conditional probability function of u given v. In fact, while the definition of modified profile likelihood was motivated above by a marginalization argument, it could equally have been derived from (1.1) by a conditionality consideration. This is an important property. For an illustration of the accuracy of the double saddlepoint approximation to the conditional likelihood see §6.2 of Barndorff-Nielsen and Cox (1979).

Suppose now that we have m samples $x_{i1}, \ldots, x_{in_i}$ $(i = 1, \ldots, m)$ from one and the same model, exponential or not, the parameters for the ith sample being κ and ψ_i, that is ψ varies from sample to sample while κ is the same throughout. As is well known, the maximum likelihood estimate of κ, which of course equals the estimate obtained from the profile likelihood $\tilde{L}(\kappa)$, will in general be inefficient or even inconsistent. The two cases of normal variates x_{ij} with either a common mean ξ or a common variance σ^2 illustrate this (Neyman and Scott, 1948). It is therefore of some interest to investigate what inferences on κ might be provided in situations of this kind by the modified profile likelihood $L^0(\kappa)$ when this is treated as a quasilikelihood for κ. We do not undertake a systematic investigation of this point here, but will discuss some indicative examples.

EXAMPLE 3.5 (Precision of Duplicate Measurements). Duplicate measurements are taken of n objects using one and the same measurement device. This results in observations x_{ij}, where $i = 1, \ldots, n$ and $j = 1, 2$. It is assumed that x_{ij} follows a location-scale model, that is $x_{ij} \sim \sigma^{-1} f\{(x_{ij} - \mu_i)/\sigma\}$ with f a known probability density function on the real line. Letting a denote the configuration ancillary $\{(x_{11} - \hat{\mu}_1)/\hat{\sigma}, \ldots, (x_{n1} - \hat{\mu}_n)/\hat{\sigma}, (x_{n2} - \hat{\mu}_n)/\hat{\sigma}\}$ and taking $u = (\hat{\sigma}, a)$, we find the modified profile likelihood for σ to be

$$L^0(\sigma) = \sigma^{-n} \prod_i [g''\{(x_{i1} - \hat{\mu}_{i\sigma})/\sigma\} + g''\{(x_{i2} - \hat{\mu}_{i\sigma})/\sigma\}]^{-1/2} \prod_{i,j} f\{(x_{ij} - \hat{\mu}_{i\sigma})/\sigma\}. \tag{3.15}$$

In the present case the factor $|\partial\hat{\psi}/\hat{\psi}_\kappa|$ equals 1 because for fixed a, σ and $\hat{\sigma}$ the estimator $\hat{\mu}_{i\sigma}$ is a translate of $\hat{\mu}_i$.

In particular, for the normal error law $f(x) = (2\pi)^{-1/2} e^{-(1/2)x^2}$ we have

$$L^0(\sigma) = \sigma^{-n} \exp\left\{-\frac{1}{4}\sigma^{-2} \sum_i (x_{i1} - x_{i2})^2\right\}.$$

This equals the marginal likelihood for σ based on $s^2 = (2n)^{-1} \sum_i (x_{i1} - x_{i2})^2$, and it yields the usual estimate s^2 for σ^2, whereas the full maximum likelihood estimate $\hat{\sigma}^2 = \frac{1}{2}s^2$ is not even consistent (Neyman and Scott, 1948).

If expected information $\hat{i}_\kappa$ rather than observed information $\hat{j}_\kappa$ is employed, we obtain instead of (3.15) the simpler expression

$$L^e(\sigma) = \sigma^{-n} \prod_{i,j} f\{(x_{ij} - \hat{\mu}_{i\sigma})/\sigma\}.$$

The modified profile likelihoods $L^0(\sigma)$ and $L^e(\sigma)$ may be compared to the marginal likelihood for σ which, on account of (2.6), is given by

$$L(\sigma; a_0) = \sigma^{-n} \prod_{i=1}^n \int_{-\infty}^{\infty} f(x_{i1}/\sigma - \zeta_i) f(x_{i2}/\sigma - \zeta_i)\, d\zeta_i, \tag{3.16}$$

where $a_0 = (x_{11} - x_{12}, \ldots, x_{n1} - x_{n2})$. While these three quasilikelihoods do generally differ, in the standard normal case they coincide, and the same is true for the extreme value distribution $f(x) = e^{-x}\exp(-e^{-x})$. In fact, in the latter case we have

$$L^0(\sigma) = L^e(\sigma) = L(\sigma; a) = \sigma^{-n} \exp(-x_{\cdot\cdot}/\sigma) \prod_i \left\{\sum_j \exp(-x_{ij}/\sigma)\right\}^{-2}.$$

EXAMPLE 3.6 (Weighted Mean Problem). For $i = 1, \ldots, m$, let $x_{i1}, \ldots, x_{in_i}$ be a sample of size n_i from the normal distribution $N(\xi, \sigma_i^2)$. The problem is to draw inference on ξ from these m samples. As shown by Neyman and Scott (1948), under mild regularity conditions the maximum likelihood estimate of ξ is consistent but not asymptotically efficient for general n_i as m tends to infinity. In fact, it is not known whether an asymptotically efficient estimator exists in this case. However Neyman and Scott (1948) constructed an estimator which, subject to certain regularity assumptions, is asymptotically more efficient than the maximum likelihood estimator, except when all the n_i are equal, in which case the two estimators have the same asymptotic variance.

The profile likelihood for ξ is

$$\tilde{L}(\xi) = \prod_{i=1}^{m} \left\{ \sum_{j=1}^{n_i} (x_{ij} - \xi)^2 \right\}^{-(1/2)n_i} \tag{3.17}$$

and the corresponding likelihood equation, for the maximum likelihood estimate $\hat{\xi}$ of ξ, may be written

$$\sum_{i=1}^{m} n_i \frac{\bar{x}_{i\cdot} - \xi}{s_i'^2 + (\bar{x}_{i\cdot} - \xi)^2} = 0, \tag{3.18}$$

where $\bar{x}_{i\cdot} = (x_{i1} + \cdots + x_{in_i})/n_i$ and $s_i'^2 = \sum_i (x_{ij} - \bar{x}_{i\cdot})^2/n_i$. By contrast, the more efficient estimator established by Neyman and Scott (1948) is defined as the solution to the equation

$$\sum_{i=1}^{m} (n_i - 2) \frac{\bar{x}_{i\cdot} - \xi}{s_i'^2 + (\bar{x}_{i\cdot} - \xi)^2} = 0. \tag{3.19}$$

The same estimation equation for ξ had actually been proposed earlier by Bartlett (1936).

A third estimator, advocated by Kalbfleisch and Sprott (1970), is obtained by replacing the factor n_i in (3.18) or $n_i - 2$ in (3.19) by $n_i - 1$. While this estimator, in contrast to the Bartlett–Neyman–Scott estimator, does incorporate information from samples of size $n_i = 2$ it is generally less efficient asymptotically if the n_i's are greater than 2. This follows from the calculations of Neyman and Scott (1948), who considered an estimation equation of the form

$$\sum_{i=1}^{m} w_i \frac{\bar{x}_{i\cdot} - \xi}{s_i'^2 + (\bar{x}_{i\cdot} - \xi)^2} = 0, \tag{3.20}$$

with general weights w_i. The proposals by Bartlett (1936) and Kalbfleisch and Sprott (1970) were based on certain notions of sufficiency or ancillarity of quantities which depend on the parameters as well as on the observations.

A recent investigation by Chaubey and Gabor (1981) shows that $\tilde{L}(\xi)$ may well be multimodal, and in particular cases $\tilde{L}(\xi)$ takes its maximal value at several of the mode points. Chaubey and Gabor (1981) considered the two-sample situation $m = 2$ and studied what they called a marginal likelihood for ξ. That likelihood turns out to be equal to the profile likelihood (3.17). Consequently (3.18), (3.19) and (3.20) may have several solutions.

In discussing the derivation of a modified profile likelihood for ξ we first note that, with $\kappa = \xi$ and $\psi = (\sigma_1^2, \ldots, \sigma_m^2)$,

$$|\hat{j}_\xi| = \prod_{i=1}^{m} \left\{ \sum_{j=1}^{n_i} (x_{ij} - \xi)^2 \right\}^{-2}.$$

Furthermore, we have $\hat{\sigma}_i^2 = \sum_j (x_{ij} - \hat{\xi})^2 / n_i$ and $\hat{\sigma}_{i\xi}^2 = \sum_j (x_{ij} - \xi)^2 / n_i$. For $m = 1$, we have $\hat{\sigma}_{1\xi}^2 = \hat{\sigma}_1^2 + (\hat{\xi} - \xi)^2$ and hence $|\partial\hat{\psi}/\partial\hat{\psi}_\xi| = 1$. To calculate $|\partial\hat{\psi}/\partial\hat{\psi}_\xi|$ for an $m > 1$ we would have to specify an ancillary statistic of dimension $m - 1$ for $(\xi, \sigma_1^2, \ldots, \sigma_m^2)$. No such exactly ancillary statistic exists, as is known from the Behrens–Fisher problem. Furthermore, the forms of the various tests for the Behrens–Fisher problem proposed by Welch show that an approximately ancillary statistic for $(\xi, \sigma_1^2, \ldots, \sigma_m^2)$ must be so intractable analytically as to make calculation of the corresponding $|\partial\hat{\psi}/\partial\hat{\psi}_\xi|$ prohibitive.

However, it is perhaps noteworthy that if we were to ignore the factor $|\partial\hat{\psi}/\partial\hat{\psi}_\xi|$ and adopt $|\hat{j}_\xi|^{-1/2}\tilde{L}(\xi)$ as a quasilikelihood for ξ then, since

$$|\hat{j}_\xi|^{-1/2}\tilde{L}(\xi) = \prod_{i=1}^{m} \{s_i'^2 + (\bar{x}_{i\cdot} - \xi)^2\}^{-(1/2)n_i+1},$$

we would obtain the same estimation equation for ξ as suggested by Bartlett and Neyman and Scott, that is (3.19).

If the m samples are from the exponential model (3.13), with (κ, ψ_i) as the mixed parameter for the ith sample, we obtain by (3.14)

$$L^0(\kappa) = \prod_{i=1}^{m} \left[\left| \sum_{22} (\kappa, \hat{\psi}_i) \right|^{1/2} a^{n_i}(\kappa, \hat{\psi}_i) \exp\{\kappa \cdot u_{i\cdot} + \lambda(\kappa, \hat{\psi}_i) \cdot v_{i\cdot}\} \right], \quad (3.21)$$

where

$$u_{i\cdot} = u(x_{i1}) + \cdots + u(x_{in_i}), \qquad v_i = v(x_{i1}) + \cdots + v(x_{in_i}), \qquad \hat{\psi}_i = \bar{v}_{i\cdot} = n_i^{-1} v_{i\cdot}.$$

Example 3.7 (Inverse Gaussian Samples). With (3.13) as the inverse Gaussian model (1.9) and with $\kappa = \chi$ as the parameter of interest, we find from (3.21), assuming for simplicity that $n_1 = \cdots = n_m = n$,

$$L^0(\chi) = \chi^{(1/2)m(n-1)} \exp(-\tfrac{1}{2}\chi \textstyle\sum w_i), \quad (3.22)$$

where

$$w_i = \sum_{j=1}^{n} (x_{ij} - \bar{x}_{i\cdot}^{-1}).$$

The modified profile likelihood (3.22) coincides with the marginal likelihood for χ based on $(w_1, \ldots, w_m)$, which would ordinarily be used for inference on χ. One has

$$\hat{\chi}^0 = \left\{ \frac{1}{m(n-1)} \sum_{i=1}^{m} w_i \right\}^{-1},$$

which is consistent for m and/or n tending to infinity. The unconditional maximum likelihood estimate of χ is

$$\hat{\chi} = \left(\frac{1}{mn} \sum_{i=1}^{m} w_i \right)^{-1}$$

and this is not consistent for $m \to \infty$ and n fixed.

Examples 3.5 and 3.7 show that in some, highly regular, cases with increasing number of nuisance parameters the modified profile likelihood yields optimal estimates of the parameter of interest. In general, the estimate $\hat{\kappa}^0$ derived from $L^0(\kappa)$ will not be consistent for $m \to \infty$ and bounded n_i. However, it will generally be "more nearly consistent" than $\hat{\kappa}$. Otherwise expressed, in a situation where the n_i increase slowly to infinity with m both $\hat{\kappa}^0$ and $\hat{\kappa}$ are consistent but $\hat{\kappa}^0$ tends to ψ considerably faster than $\hat{\kappa}$. To illustrate this, let us consider the case of m matched pairs.

EXAMPLE 3.8 (Matched Pairs). Suppose that we have observed m two-by-two tables of the form

y_i	z_i	s_i
$1 - y_i$	$1 - z_i$	$2 - s_i$
1	1	2

Here $n_i = 1, x_i = (y_i, z_i)$, κ is the log odds ratio and $\psi_i = E(s_i)$ $(i = 1, \ldots, m)$. Suppose, furthermore, that all pairs are discordant, that is $s_1 = \cdots = s_m = 1$, and let c and d denote the number of pairs such that $(y_i, z_i) = (1, 0)$, $(y_i, z_i) = (0, 1)$ respectively. Then the modified profile likelihood for κ is

$$L^0(\kappa) = \exp\left\{\frac{1}{2}(c-d)\kappa\right\}\left\{\exp\left(\frac{1}{4}\kappa\right) + \exp\left(-\frac{1}{4}\kappa\right)\right\}^{-3(c+d)}.$$

Let $\rho = e^{\kappa}$, that is ρ is the odds ratio. Following Breslow (1981), we find that asymptotically, as $m \to \infty$, $\hat{\rho}^0$ and $\hat{\rho}$ satisfy $\hat{\rho}^0 = \{(5\rho + 1)/(\rho + 5)\}^2$ and $\hat{\rho} = \rho^2$. Table 1 gives some numerical values.

Even in this extreme case of matched pairs the estimate from the modified profile likelihood is a substantial improvement over the unconditional maximum likelihood estimate, particularly in the region of values of ρ of greatest practical interest. This, in conjunction with the data example of

Table 1. Matched Pairs. Odds Ratio, ρ, and Two Probability Limits.

ρ	1	1.5	2.	3	5	10	15	20
$\hat{\rho}^0$	1	1.71	2.47	4.00	6.76	11.56	14.44	16.32
$\hat{\rho}$	1	2.25	4	9	25	100	225	400

criminal twins (Barndorff-Nielsen and Cox, 1979, §6.2), rather strongly indicates that for many sparse contingency table problems modified profile likelihood estimates of the interaction parameters will be satisfactory approximations to the conditional maximum likelihood estimates.

Modified profile likelihoods may also be of interest in robustness investigations, in line with the parametric approach as discussed, for instance, by Fraser (1979, §2.5).

For instance, suppose that we are considering a location-scale model whose error law has probability density function f and that we wish to investigate robustness against deviations form f, the relevant type of deviations being embodied as a family of alternative error laws f_κ, where κ is a parameter indexing these laws. As part of the robustness investigation one may study the marginal likelihood for κ based on the configuration statistic u, and this is expressible as

$$L(\kappa; u) = \int\int \sigma^{-n} \prod_{i=1}^{n} f_\kappa\{(x_i - \mu)/\sigma\}\sigma^{-1}\, d\sigma\, d\mu; \tag{3.23}$$

see, for instance, formula (2.6). Alternatively, it is a possiblity to adopt the modified profile likelihood for κ as quasilikelihood. In view of Example 1.1 we find

$$L^0(\kappa) = \hat{\sigma}_\kappa D_\kappa(a_\kappa)^{-1/2}\tilde{L}(\kappa), \tag{3.24}$$

with $D_\kappa(a_\kappa)$ defined as $D(a)$, in Example 1.1, but in terms of f_κ rather than f. Here we have used

$$\left| \frac{\partial(\hat{\mu}, \hat{\sigma})}{\partial(\hat{\mu}_\kappa, \hat{\sigma}_\kappa)} \right| = \hat{\sigma}/\hat{\sigma}_\kappa.$$

From a computational viewpoint (3.24) has the advantage over (3.23) of not involving a multiple integral. It is straightforward to generalize (3.24) from the simple location-scale case to linear regression specifications.

More generally still, for an arbitrary composite transformation model with index parameter κ the modified profile likelihood fox κ may be written

$$L^0(\kappa) = \Delta(\hat{h}_\kappa)^{-1}|j_\kappa(e; u_\kappa)|^{-1/2}\tilde{L}(\kappa), \tag{3.25}$$

where $j_\kappa(e; u_\kappa)$ denotes observed information on the transformation parameter for given κ, evaluated at the identity transformation and with $u_\kappa = \hat{h}_\kappa^{-1}x$, which is a maximal invariant, playing the role of the observation. This follows from results to be briefly discussed at the end of §4 but we omit the derivation. Formula (3.25) should be compared with (2.6) for the marginal likelihood or, rather, to the alternative form obtained by substituting observed information for expected information in (2.6).

In important exponential family situations the integral in the latter form can be approximated by Laplace's method, the resulting expression being precisely the modified profile likelihood. We shall not develop this theme here, but it may be pointed out in this connexion that a number of the approximations to marginal likelihoods in multivariate normal analysis discussed by Muirhead (1978) coincide with the respective modifed profile likelihoods and that this exemplifies the indicated general result for exponential models.

4. Exactness Cases

Several examples of models for which (1.1) is exact were discussed in §1. Here was shall in a more systematic way address the question of when $p^*(\hat{\omega}; \omega|a)$ equals $p(\hat{\omega}; \omega|a)$. While some useful general results can be established, a complete characterization is not available. It would be of particular interest to be able to characterize all models such that (1.1) is exact for samples of arbitrary size from the model. The examples referred to above as well as the further examples to be discussed in the present section indicates that such models are bound to have properties that make them of special statistical importance. There seems also reason to believe that if for a given model a simple explicit expression for the distribution of the maximum likelihood estimator is available for arbitrary sample size and if the random variates are continuous then (1.1) will be exact.

It follows from a result of Daniels (1980) that the only linear exponential models of order 1 for which (1.1) is exact for all sample sizes are the normal distribution with known variance, the gamma distribution with known shape parameter, and the inverse Gaussian distribution with known χ. Any model obtained from one of these by a one-to-one transformation of the observation x does possess the exactness property as well. We may conclude that the models just mentioned are the only exponential models of order 1 having (1.1) exact for all sample sizes. No similar results are known for exponential models of order greater than 1.

The main general result, to be discussed briefly at the end of this section, is that (1.1) is exact for all transformation models. The location-scale model and the von Mises–Fisher model with a known precision parameter, considered in Examples 1.1 and 1.2, are transformation models, as are most or all tractable Gaussian models with structured mean vectors and/or variance matrices, such as those discussed by Andersson (1975). The hyperboloid model (Barndorff-Nielsen, 1978b; Jensen, 1981) also affords an interesting transformation model example. That model is analogous to the von Mises–Fisher model but pertains to observations on the unit hyperboloid H^{k-1} in R^k rather than the unit sphere S^{k-1}. A further reason for drawing the hyperboloid model into the picture is that in the special case $k = 3$ and with both the "precision" parameter and the "direction" parameter considered as unknown we have a model with $k = d = 3$ that is not a transformation model but nevertheless have (1.1) exact. A similar result does not hold for the von Mises–Fisher model.

EXAMPLE 4.1 (Hyperboloid Model). The hyperboloid model of exponential order 3 may be defined as the joint distribution of two random variables u and v with probability density function

$$p(u, v; \chi, \varphi, \kappa) = (2\pi)^{-1}\kappa e^{\kappa} \exp[-\kappa\{\cosh\chi \cosh u - \sinh\chi \sinh u \cos(v - \varphi)\}], \tag{4.1}$$

where $0 \leq u < \infty$, $0 \leq v < 2\pi$ and where the parameters satisfy $0 \leq \chi < \infty$, $0 \leq \varphi < 2\pi$ and $0 < \kappa < \infty$. For a sample $(u_1, v_1), \ldots, (u_n, v_n)$ of size $n > 1$, from (4.1) one finds that $p^*(\hat{\chi}, \hat{\varphi}, \hat{\kappa}; \chi, \varphi, \kappa)$, as calculated from (1.2), is exactly equal to the probability density function of $(\hat{\chi}, \hat{\varphi}, \hat{\kappa})$. The proof relies on the facts (Jensen, 1981) that, if one defines r by

$$r = \left\{\left(\sum \cosh u_i\right)^2 - \left(\sum \sinh u_i \cos v_i\right)^2 - \left(\sum \sinh u_i \sin v_i\right)^2\right\}^{1/2},$$

then $r^2 = n(\hat{\kappa}^{-1} + 1)$ and $r - n \sim \Gamma(n - 1, \kappa)$ and, furthermore, the conditional distribution of $(\hat{\chi}, \hat{\varphi})$ given r is of the hyperboloid type (4.1) but with κ replaced by $r\kappa$. The statistic r is the analogue of the resultant length for the von Mises–Fisher distribution.

Examples 1.3 and 4.1 have provided instances of exponential models with $k = d$ and of order k for which (1.1) is exact, even though the models are not transformational. The values of $k = d$ in question were 1, 2 and 3, respectively. Further instances, of higher order, will occur below.

First, however, we make some additional general observations. It is convenient to introduce the notation

$$M = |\hat{j}|^{-1/2}\hat{L}, \tag{4.2}$$

and when $p(\hat{\omega}, a; \omega)$, the joint probability function of $\hat{\omega}$ and a, is relative to Lebesgue measure we shall write $L_0, \hat{L}_0, M_0$, etc. in case the likelihood function employed is the particular version

$$L_0 = p(\hat{\omega}, a; \omega). \tag{4.3}$$

Suppose a is exactly ancillary, and restrict consideration to the case where the probability functions involved are relative to the relevant Lebesgue measures. Then (1.1) is exact if and only if the function

$M_0 = M_0(\hat{\omega}, a)$ depends on a only. Incidentally, in this case the norming constant c in (1.2) depends on a only and, furthermore,

$$p(a) = c^{-1}M_0. \tag{4.4}$$

To see this, observe that

$$p^*(\hat{\omega}; \omega|a) = cM_0^{-1}p(a)p(\hat{\omega}; \omega|a),$$

from which the results follow.

In the particular case where $\hat{\omega}$ itself is minimal sufficient, so that a is degenerate and no conditioning is involved, the necessary and sufficient condition for exactness of (1.1) is that M_0 be constant.

If we have a number of models $\mathscr{M}_1, \ldots, \mathscr{M}_r$ with (k, d) values $(k_1, d_1), \ldots, (k_r, d_r)$ and such that (1.1) is exact for each of these models then the model obtained by independent combination of $\mathscr{M}_1, \ldots, \mathscr{M}_r$ has (k, d) value $(k_1 + \cdots + k_r, d_1 + \cdots + d_r)$ and (1.1) exact.

Sometimes a more intriguing coupling of models is possible. Suppose $p(u; \gamma)$ is a model function for which (1.1) is exact and, for each fixed u, let $p(v; \delta|u)$ be a model function, also with (1.1) exact. The joined model

$$p(u, v; \gamma, \delta) = p(u; \gamma)p(v; \delta|u)$$

will then, under certain assumptions, have (1.1) exact. More specifically, assuming that the probability functions are relative to Lebesgue measures and using the general condition for exactness mentioned above, it may be shown that $p(u, v; \gamma, \delta)$ equals $p^*(u, v; \gamma, \delta)$ provided the following three conditions are satisfied:

(i) the (k, d) values for all the models $p(u; \gamma)$, $p(v; \delta|u)$ and $p(u, v; \gamma, \delta)$ have $d = k$;
(ii) u, v and (u, v) are minimal sufficient in the respective models;
(iii) the norming constant in the expression for $p^*(v; \delta|u)$ does not depend on u.

EXAMPLE 4.2 (Combination of Inverse Gaussian Distributions). Let $\chi, \psi, \kappa, \lambda$ denote parameters and define

$$p(u; \chi, \psi) = \frac{\sqrt{\chi}}{\sqrt{2\pi}} \exp\{\sqrt{\chi\psi}\}u^{-3/2} \exp\{-\tfrac{1}{2}(\chi u^{-1} + \psi u)\},$$

$$p(u; \kappa, \lambda|u) = \frac{\sqrt{\kappa}}{\sqrt{2\pi}} \exp\{u\sqrt{\kappa\lambda}\}uv^{-3/2} \exp\{-\tfrac{1}{2}(\kappa u^2 v^{-1} + \lambda v)\};$$

i.e. we have $u \sim N^-(\chi, \psi)$ and $v|u \sim N^-(\kappa u^2, \lambda)$. The joint distribution of u and v is

$$p(u, v; \chi, \psi, \kappa, \lambda) = (2\pi)^{-1}\sqrt{\chi\kappa} \exp\{\sqrt{\chi\psi}\}u^{-1/2}v^{-3/2} \times \exp(-\tfrac{1}{2}[\chi u^{-1} + \{\psi - 2\sqrt{\kappa\lambda}\}u + \kappa u^2 v^{-1} + \lambda v]). \tag{4.5}$$

Suppose for the moment that χ and κ are known. By Example 1.3 and the sufficient condition for exactness given above one sees that

$$p(u, v; \chi, \psi, \kappa, \lambda) = p^*(u, v; \chi, \psi, \kappa, \lambda),$$

where the right-hand side is given by

$$p^*(u, v; \chi, \psi, \kappa, \lambda) = p^*(\hat{\psi}, \hat{\lambda}; \psi, \lambda)\left|\frac{\partial(\hat{\psi}, \hat{\lambda})^{\mathrm{T}}}{\partial(u, v)}\right|; \tag{4.6}$$

we have suppressed the dependence on χ and κ, supposed known, on the right-hand side of (4.6). More generally, the same conclusion would hold had the conditional distribution of v been specified by $v|u \sim N^-\{\kappa d(u), \lambda\}$ for an arbitrary positive function $d(u)$.

The model (4.5), with all four parameters unknown, will be denoted by $[N^-, N^-](\chi, \psi, \kappa, \lambda)$. It has some interesting properties, analogous to those of $N^-(\chi, \psi)$. Setting $\alpha = \psi - 2\sqrt{\kappa\lambda}$, we may rewrite (4.5) as

$$\begin{aligned} p(u, v; \chi, \psi, \kappa, \lambda) = &\sqrt{\chi\kappa} \exp\{\sqrt{\chi}\sqrt{(\alpha + 2\sqrt{\kappa}\sqrt{\lambda})}\}b(u, v) \\ &\times \exp\{-\tfrac{1}{2}\chi u^{-1} + \alpha u + \kappa u^2 v^{-1} + \lambda v)\}, \end{aligned} \tag{4.7}$$

where $b(u, v) = (2\pi)^{-1}u^{-1/2}v^{-3/2}$. Thus $[N^-, N^-](\chi, \psi, \kappa, \lambda)$ is an exponential family of order 4 with $(\chi, \alpha, \kappa, \lambda)$ as canonical parameter and having cumulant transform

$$-\sqrt{\chi}\sqrt{(\alpha + 2\sqrt{\kappa}\sqrt{\lambda})} - \tfrac{1}{2} \log(\chi\kappa). \tag{4.8}$$

Now, let $(u_1, v_1), \ldots, (u_n, v_n)$ be a sample from (4.5). It follows immediately from the form of (4.8) that

$$(\bar{u}, \bar{v}) \sim [N^-, N^-](n\chi, n\psi, n\kappa, n\lambda), \tag{4.9}$$

where $\bar{u} = n^{-1}u. = n^{-1} \sum u_i.\bar{v} = n^{-1}v. = n^{-1} \sum v_i$. In particular, therefore, if $\psi = \lambda = 0$ the distribution of $n^{-2}u., n^{-4}v.)$ is the same whatever the value of n, so that the distribution $[N^-, N^-](\chi, 0, \kappa, 0)$ could be said to be bivariate stable of index $(\frac{1}{2}, \frac{1}{4})$. For recent work on multivariate stable distributions, see Hudson and Mason (1981) and references therein.

As a corollary we obtain that the marginal distribution of v under $[N^-, N^-](\chi, 0, \kappa, 0)$, whose probability density function is

$$p(v; \chi, \kappa) = (2\pi)^{-1}\sqrt{\chi\kappa}\, v^{-3/2} \int_0^\infty u^{-1/2} \exp\{-\tfrac{1}{2}(\chi u^{-1} + \kappa v^{-1}u^2)\}\, du,$$

is a strictly stable distribution of index $\frac{1}{4}$. This latter result may also be seen as a consequence of the proposition, from subordination of processes, that a stable α-process directed by a stable β-process leads to a stable $\alpha\beta$-process (Feller, 1971, p. 452).

Suppose $n > 1$ and let $w = \sum(u_i^{-1} - u^{-1})$ and $z = \sum(u_i^2 v_i^{-1} - \bar{u}^2\bar{v}^{-1})$. Then

$$\hat{\chi} = n/w, \qquad \hat{\psi} = n/(w\bar{u}^2), \qquad \hat{\kappa} = n/z, \qquad \hat{\lambda} = (n\bar{u}^2)/(z\bar{v}^2).$$

Using $(\bar{u}, \bar{v}) \sim [N^-, N^-](n\chi, n\psi, n\kappa, n\lambda)$ we can prove that w, z and $(\bar{u}, \bar{v})$ are independent, and that $w \sim \Gamma(\frac{1}{2}n - \frac{1}{2}, \frac{1}{2}\chi)$ and $z \sim \Gamma(\frac{1}{2}n - \frac{1}{2}, \frac{1}{2}\kappa)$.

By (4.8) and (2.4) the conditional Laplace transform of the canonical statistic $t = -\frac{1}{2}(\sum u_i^{-1}, u., \sum u_i^2 v_i^{-1}, v.)$, given $(\bar{u}, \bar{v})$ and under $(\chi, \psi, \kappa, \lambda) = (1, 0, 1, 0)$, is expressible as

$$E_{(1,0,1,0)}[\exp\{(\theta - \theta_0) \cdot t\} | (\bar{u}, \bar{v})]$$

$$= (\chi\kappa)^{-(1/2)n} \exp\{-n\sqrt{\chi}\sqrt{(\alpha + 2\sqrt{\kappa}\sqrt{\lambda})}\} \frac{p(\bar{u}, \bar{v}; \theta)}{p(\bar{u}, \bar{v}; \theta_0)},$$

where $\theta = (\chi, \alpha, \kappa, \lambda)$ and $\theta_0 = (1, 0, 1, 0)$. Hence, by (4.9),

$$E_{(1,0,1,0)}[\exp\{(\theta - \theta_0) \cdot t\} | (\bar{u}, \bar{v})]$$

$$= (\chi\kappa)^{-(1/2)n+1/2} \exp[-\tfrac{1}{2}n\{(\chi - 1)\bar{u}^{-1} + \alpha\bar{u} + (\kappa - 1)\bar{u}^2\bar{v}^{-1} + \lambda\bar{v}\}]$$

or

$$E_{(1,0,1,0)}[\exp\{-\tfrac{1}{2}(\chi w + \kappa z)\} | (\bar{u}, \bar{v})] = (1 + \chi)^{-(1/2)n+1/2}(1 + \kappa)^{-(1/2)n+1/2}. \tag{4.10}$$

As the conditional distribution of (w, z) given $(\bar{u}, \bar{v})$ is exponential the above assertion is implied by (4.10).

Thus, in particular, the bivariate distribution (4.5) allows of certain analogues of the analysis of variance for normal observations, as is the case also for the inverse Gaussian distribution; see Tweedie (1957).

By means of the derived results on the distributions of w, z and $(\bar{u}, \bar{v})$ it is, moreover, simple to show that (1.1) is exact for the distribution of $(\hat{\chi}, \hat{\psi}, \hat{\kappa}, \hat{\lambda})$.

The model $[N^-, N^-]$ of Example 4.2 was established by a suitable combination of two inverse Gaussian distributions. It is possible to obtain several other models with properties similar to those of $[N^-, N^-]$ from the three elemental distributions N, Γ and N^-. The models in question are discussed, among others, from a quite different point of view by Barndorff-Nielsen and Blæsild (1983). Of some particular interest is the model $[N^-, N]$ which is defined as the family of distributions of a two-dimensional random variable (u, v) such that $u \sim N^-(\chi, \psi)$ and $v|u \sim N(u\xi, u\sigma^2)$. This model has a physical interpretation in terms of two Brownian motions with drifts and it possesses properties that are completely analogous to those discussed for $[N^-, N^-]$. It also allows an exact F test for equality of the drift parameters. The details are given in the paper just referred to. It is also feasible to construct higher-order examples of a similar kind, for instance by attaching a third N^--distribution to $[N^-, N^-]$.

In conclusion we shall briefly indicate the main steps of the proof of the following result, which is true subject only to minimal regularity conditions that are virtually always met in practice.

Theorem 4.1. *Formula* (1.1) *is exact for all transformation models.*

A proof of this result was outlined by Barndorff-Nielsen (1980) for the case where the group G acts freely on the sample space $\mathscr{X}$. The property that the action is free is a basic assumption in Fraser's (1968, 1979) work on transformation models; Fraser speaks of this property as exactness of the group of transformations on the sample space. However, for a complete assessment of when (1.1) is exact, and for establishing formula (2.6), it is essential to drop the assumption that the action of G is free. One is then forced to work with the theory of invariant measures on homogeneous spaces, see, for instance, Barut and Raczka (1980, §4.3) or Santaló (1979, Chapter 10). In its full generality the theorem may be verified by the following train of arguments, if we recall the terminology for transformation models presented at the end of §1.

(i) Let K be the subgroup of G given by $K = \{k \in G : p(k^{-1}.) = p(.)\}$ and let H be a subset of G such that every $g \in G$ is uniquely factorizable as $g = hk$ where $h \in H$, $k \in K$. The set H may then be taken as the parameter domain of the model, and we denote the maximum likelihood estimator of h by $\hat{h}$.

(ii) Choose the orbital decomposition $(\hat{h}, u)$ of x such that $K = \{k \in G\colon ku = u\}$ for every u.

(iii) By Theorem 3.1 of Barndorff-Nielsen et al. (1982) it now follows that the conditional model function for $\hat{h}$, given the maximal invariant u and relative to invariant measure ν on H, is of the form

$$p(\hat{h}; h|u) = c(u)\bar{L}\{h; (\hat{h}, u)\} \quad \langle\nu\rangle. \tag{4.11}$$

(iv) By definition the measure ν is invariant under the action $\dot{\gamma}$ of G on H induced from $\mathscr{X}$ by the mapping $\hat{h}$. It may be shown that, on account of the definition of the maximum likelihood estimator $\hat{h}$, $\dot{\gamma}$ coincides with the natural action of G on the homogeneous space $G/K \equiv H$.

(v) Therefore, viewing H as an open subset of some Euclidean space one has

$$d\nu(\hat{h}) = J_{\dot{\gamma}(\hat{h})}(e)^{-1} d\lambda(\hat{h}), \tag{4.12}$$

where we use $J_f(y)$ to denote the Jacobian determinant of a mapping f of a space $\mathscr{Y}$, evaluated at $y \in \mathscr{Y}$.

(vi) Next, by differentiation of the likelihood equation it is shown that

$$|\hat{j}| = |j(\hat{h}; x)| = |J_{\dot{\gamma}(\hat{h})}(e)|^{-2}|j(e; u)|. \tag{4.13}$$

(vii) The theorem now follows on combining formulae (4.11), (4.12), (4.13) and (2.3).

Details of the proof and exemplifications are available in an unpublished report by the author. As a point of some special interest it may be mentioned that for the hyperboloid model on the unit hyperboloid H^{d-1} the relevant factorization $G = HK$ is the factorization of the special pseudo-orthogonal group $G = \mathrm{so}^+(1, d-1)$ into the special orthogonal group $K = \mathrm{so}(d-1)$ and a subset H of G which, for $d = 4$, is the set of transformations known in relativity theory as "pure Lorentz transformations" or "boosts".

Note that the derivation of the theorem discussed above yields the crucial result that, considering now an arbitrary parametrization of the model, the measure on Ω defined by $|j(\hat{\omega})|^{1/2} d\hat{\omega}$ is invariant under the action of G on Ω induced by $\hat{\omega}$, for any given value of the maximal invariant statistic. Similarly, one sees from (2.3) that $|i(\hat{\omega})|^{1/2} d\hat{\omega}$ is an invariant measure under this action.

5. Marginal Likelihood for Composite Transformation Models

If for a given model $\mathscr{M}$ there exists a group G acting on the sample space $\mathscr{X}$ in such a way that $\mathscr{M}$ is a union of disjoint submodels $\mathscr{M}_\kappa$, $\kappa \in \mathscr{K}$, each $\mathscr{M}_\kappa$ being a transformation model under the action of G, then $\mathscr{M}$ is said to be a composite transformation model with index parameter κ. According to the principle of G-sufficiency (Barnard, 1963; Barndorff-Nielsen, 1978a, p. 52), inference on κ is then to be drawn from the marginal model for the statistic u that is the maximal invariant relative to G's action.

In concrete cases it is generally not possible to obtain a simple expression for the probability function of u relative to Lebesgue measure or some other standard measure. However, it is often possible to find an explicit form for the marginal likelihood function for κ by means of formula (2.6). The derivation of that formula is based on the following theorem, which extends results of Wijsman (1979) and Andersson (1982).

Theorem 5.1. *Consider a composite transformation model with model function $p(x; g, \kappa)$, relative to invariant measure μ on $\mathscr{X}$. Let $G = HK$ be a factorization of G, as considered in the proof of Theorem* 4.1, *such that $p(x; k, \kappa) = p(x; e, \kappa)$ for every x and κ, and for every $k \in K$. Then the likelihood function depends on g through h only and the marginal likelihood function for κ is given by*

$$L(\kappa; u) = \int L(h, \kappa; x) \Delta(h)^{-1} \, d\nu(h), \tag{5.1}$$

where ν is invariant measure on H, and Δ is the modular function of G.

We outline the proof; more details and the requisite regularity conditions are given in the unpublished report by the author, referred to above. Formula (18) of Andersson (1982) may be recast as

$$L(\kappa; u) = \int L(g, \kappa; x) \, d\beta(g),$$

where β is right invariant measure on G. This extends an earlier result of Wijsman (1979, Theorem 3.1.1). It follows from the theory of quotient measures (Andersson, 1982, §3; Barut and Raczka, 1980, §4.3) that there exists a measure ν on H such that, with $g = hk$ as the factorization of g and with β_K denoting right invariant measure on K,

$$L(\kappa; u) = \int L(g, \kappa; x)\, d\beta(g) = \int_H \int_K L(hk\acute{k}^{-1}, \kappa; x)\Delta(h)^{-1}\, d\beta_K(\acute{k})\, d\nu(h)$$
$$= \int_H L(h, \kappa; x)\Delta(h)^{-1}\, d\nu(h),$$

where ν is the same invariant measure on H as occurs in (4.11) and (4.12).

If $\psi = \psi(h)$ is any smooth, one-to-one transformation of h we may rewrite (5.1) as $L(\kappa; u) = \int L(\psi, \kappa; x)\Delta(\psi)^{-1} d\sigma(\psi)$, where the full likelihood function is now expressed in terms of (ψ, κ) and where σ is the transformed measure derived from ν. By construction the measure σ is invariant under the action of G on $\Psi = \psi(H)$ and in view of the last paragraph of §4 we obtain (2.6). A similar expression is available in terms of observed information.

Acknowledgments

I am much indebted to Steen A. Andersson, David R. Cox, Preben Blæsild, Jens L. Jensen and Jørgen G. Pedersen for stimulating and helpful discussions. The paper is based on part of the Forum Lectures "Parametric statistical models and inference", Wrocław, September 1981.

References

Andersson, S.A. (1975). Invariant normal models. *Ann. Statist.*, **3**, 132–154.

Andersson, S.A. (1982). Distributions of maximal invariants using quotient measures. *Ann. Statist.*, **10**, 955–961.

Barnard, G.A. (1963). Some logical aspects of the fiducial argument. *J. Roy. Statist. Soc. Ser. B*, **25**, 111–114.

Barndorff-Nielsen, O. (1978a). *Information and Exponential Families*. Wiley, Chichester.

Barndorff-Nielsen, O. (1978b). Hyperbolic distributions and distributions on hyperbolae. *Scand. J. Statist.*, **5**, 151–157.

Barndorff-Nielsen, O. (1980). Conditionality resolutions. *Biometrika*, **67**, 293–310.

Barndorff-Nielsen, O. and Blæsild, P. (1983). Reproductive exponential models. *Ann. Statist.*, **11**. To appear.

Barndorff-Nielsen, O., Blæsild, P., Jensen, J.L., and Jørgensen, B. (1982). Exponential transformation models. *Proc. Roy. Soc. London Ser. A*, **379**, 41–65.

Barndorff-Nielsen, O. and Cox, D.R. (1979). Edgeworth and saddlepoint approximations with statistical applications (with discussion). *J. Roy. Statist. Soc. Ser. B*, **41**, 279–312.

Bartlett, M.S. (1936). The information available in small samples. *Proc. Cambridge Philos. Soc.*, **32**, 560–566.

Barut, A.O. and Raczka, R. (1980). *Theory of Group Representations and Applications*. Polish Scientific, Warsaw.
Breslow, N. (1981). Odds ratio estimators when the data are sparse. *Biometrika*, **68**, 73–84.
Chaubey, Y.P. and Gabor, G. (1981). Another look at Fisher's solution to the problem of the weighted mean. *Comm. Statist. A*, **10**, 1225–37.
Cox, D.R. (1975). Partial likelihood. *Biometrika*, **62**, 269–276.
Cox, D.R. (1980). Local ancillarity. *Biometrika*, **67**, 273–278.
Daniels, H.E. (1954). Saddlepoint approximations in statistics. *Ann. Math. Statist.*, **25**, 631–650.
Daniels, H.E. (1980). Exact saddlepoint approximations. *Biometrika*, **67**, 53–8.
Durbin, J. (1980a). Approximations for densities of sufficient estimators. *Biometrika*, **67**, 311–333.
Durbin, J. (1980b). The approximate distribution of partial serial correlation coefficients calculated from residuals from regression on Fourier series. *Biometrika*, **67**, 335–349.
Feller, W. (1971). *An Introduction to the Theory of Probability and its Applications*, vol. **2**, 2nd ed. Wiley, New York.
Fisher, R.A. (1934). Two new properties of mathematical likelihood. *Proc. Roy. Soc. London Ser. A*, **144**, 285–307.
Fraser, D.A.S. (1968). *The Structure of Inference*. Krieger, New York.
Fraser, D.A.S. (1979). *Inference and Linear Models*. McGraw-Hill, Toronto.
Hinkley, D.V. (1980). Likelihood as approximate pivotal distribution. *Biometrika*, **67**, 287–292.
Hudson, W.N. and Mason, J.D. (1981). Operator-stable distributions on R^2 with multiple exponents. *Ann. Probab.*, **9**, 482–489.
Jeffreys, H. (1961). *Theory of Probability*, 3rd ed. Clarendon, Oxford.
Jensen, J.L. (1981). On the hyperboloid distribution. *Scand. J. Statist.*, **8**, 193–206.
Kalbfleisch, J.D. and Sprott, D.A. (1970). Applications of likelihood methods to models involving large numbers of parameters (with discussion). *J. Roy. Statist. Soc. Ser. B*, **32**, 175–208.
Muirhead, R.B. (1978). Latent roots and matrix variates: A review of some asymptotic results. *Ann. Statist.*, **6**, 5–33.
Nelder, J.A. and Wedderburn, R.W.M. (1972). Generalized linear models. *J. Roy. Statist. Soc. Ser. A*, **135**, 370–384.
Neyman, J. and Scott. E.L. (1948). Consistent estimates based on partially consistent observations. *Econometrica*, **16**, 1–32.
Santaló. L.A. (1979). *Integral Geometry and Geometric Probability*. Encyclopedia of Mathematics and its Applications, vol. **1**, Addison-Wesley, London.
Tweedie, M.C.K. (1957). Statistical properties of inverse Gaussian distribution. *Ann. Math. Statist.*, **28**, 362–377.
Wedderburn, R.W.M. (1974). Quasi-likelihood functions, generalized linear models, and the Gauss–Newton method. *Biometrika*, **61**, 439–448.
Wijsman, R.A. (1979). Stopping times of invariant sequential probability ratio tests. In *Developments in Statistics*, vol. **2**, (P.R. Krishnaiah, ed.), pp. 235–314. Academic Press, New York.

[illegible]

Introduction to Rousseeuw (1984) Least Median of Squares Regression

D.G. Simpson
University of Illinois

1. Introduction and Historical Background

Following seminal papers by Box (1953) and Tukey (1960), which demonstrated the need for robust statistical procedures, the theory of robust statistics blossomed in the 1960s and 1970s. An early milestone was Huber's (1964) paper which introduced univariate M-estimators and the minimax asymptotic variance criterion. Hampel (1974) proposed the influence function of an estimator as a way to describe the effect of a single outlier. In order to measure the effect of several outliers, he introduced the breakdown value (Hampel, 1971) in a general and asymptotic setting. Donoho and Huber (1983) advocated a finite-sample version of the breakdown value, in line with Hodges's (1967) study in the univariate framework. Heuristically, the *breakdown point* is the largest percentage of ill-fitting data that a method can cope with. For a formal definition, see equation (2.1) of the reprinted Rousseeuw (1984).

The driving force for the Rousseeuw (1984) paper was the search for a practical, equivariant regression estimator with a 50% breakdown point. For univariate data, the sample median and the median absolute deviation are 50% breakdown estimators of location and scale. In other situations, such estimators proved harder to find. Maronna (1976) established an upper bound on the breakdown point for a class of multivariate M-estimates including nearly all of the estimates known at the time. The bound decreased to zero inversely with the dimension. Then, in the context of multivariate location and scatter, Stahel (1981) and Donoho (1982) introduced the first affine equivariant and a 50% breakdown estimator. The methods were not widely used, however, probably because of the lack of software to compute exact or approximate estimates. Only recently, Maronna and Yohai (1995) established the asymptotic properties and finite-sample behavior. For regression, Siegel (1982) proposed a 50% breakdown estimator, but it lacked equivariance with respect to linear transformations on the regressors. The asymptotic properties of this estimator were established only recently (Hossjer et al., 1994). In the late 1970s and early 1980s there was much activity on M- and generalized M-estimators of regression (for a survey, see Chapter 6 of Hampel et al. (1986)), which are equivariant but have an intrinsically lower breakdown value.

Against this backdrop the Rousseeuw (1984) paper was a breakthrough. It was the first practical method for equivariant regression with a 50% breakdown point and it pointed the way to many further developments. The many references to "least median of squares" in the literature since 1984 attest to the importance of this paper. The references range from papers on advanced asymptotic theory, to computational statistics, to application articles in other fields such as environmental research, chemometrics, and computer vision. Moreover, the 1984 paper initiated an important research area in statistics with positive-breakdown points, aiming to be robust against one or several unannounced outliers that may occur anywhere in the data Rousseeuw (1984) credits Hampel (1975) for the germ of the idea behind LMS, but the fully developed theory and methodology of Rousseeuw's paper, as well as the marvelous presentation, have influenced many further developments.

2. Overview of the Paper

The main focus of the Rousseeuw (1984) paper was on the introduction of the least median of squares (LMS) method for multiple linear regression. The simple and elegant idea of replacing the sum in the least squares criterion by a median was easily explained and compelling.

However, the paper contained much more. It laid out an entire methodology for positive-breakdown statistics, consisting of a self-contained theory with theorems and proofs, a novel algorithmic approach with corresponding implementation, and guidance on detecting outliers by their deviation from a positive-breakdown fit.

It is remarkable how many results were already given in this initial paper. The existence of an LMS solution and its breakdown value (Lemma 1 and Theorem 1) have unusual linear algebra proofs that served as models for subsequent work. The *exact fit property* (Corollary 1) is formulated, and proved as a consequence of the breakdown value. Remark 1 formalizes regression equivariance, which is standard terminology nowadays, and gives the maximal breakdown value for all equivariant estimators. (The proof is tucked away in the Appendix, as is the LMS asymptotic behavior.) Theorem 2 gives the geometric interpretation of the LMS fit using the narrowest (hyper)strip covering half the data, a key to its computation in simple regression. For multiple regression, the reader is referred to the program PROGRESS. The relatively large computation times are explained by showing the close connection with (linear) projection pursuit. The example section illustrates the detection of outliers with this approach. Following publication of Rousseeuw (1984), the book of Rousseeuw and Leroy (1987) provided detailed discussion of these and other issues in robust data analysis.

Under the classical assumptions underlying least squares regression, LMS is statistically inefficient. Noting the close connection with the "shorth," the univariate location estimate defined as the center of the shortest interval containing half of the data, Rousseeuw argued that the LMS regression estimate converges only at the rate $n^{-1/3}$, where n is the sample size, Later this was established rigorously by Kim and Pollard (1990) and Davies (1990). The Rousseeuw (1984) paper also suggested ways to improve on the efficiency by using LMS as a starting value and then doing one step of an M-estimate computing algorithm. Jurečkova and Portnoy (1987) followed up on the suggestion to use one Newton–Raphson step, establishing the root-n convergence rigorously. Another appealing suggestion, one step of the reweighted least squares algorithm, was later shown by He and Portnoy (1992) to yield the same slow rate of convergence as LMS.

It should be noted that the 1984 paper also introduced several other estimators that are in widespread use today. Section 4 proposes combining the initial LMS with a one-step M-estimator, and provides the resulting efficiency. Also the one-step reweighted LS approach is described, as implemented in PROGRESS. Another new method is *least trimmed squares* (LTS) regression given in (4.1), with its asymptotic behavior for an arbitrary trimming proportion (not necessarily 50%). Also tests and orthogonal regression are mentioned. For affine equivariant estimation of multivariate location and scatter, Section 4 proposes two new methods: the *minimum volume ellipsoid estimator* (MVE) and the *minimum covariance determinant* estimator (MCD).

3. Impact

It may be argued that the 1984 paper pioneered the positive-breakdown approach not because of a single estimator or theorem, but by developing a compact and coherent framework with a unifying principle and a diverse set of estimators and results, by clarifying connections with existing literature and concepts, by providing a careful implementation and making this software widely available, by presenting nontrivial examples, and by opening up several directions for further research.

The 1984 paper also provided the groundwork for the development of more efficient breakdown resistant estimators, such as S-estimates (Rousseeuw and Yohai, 1984), MM-estimates (Yohai 1987), τ-estimates (Yohai and Zamar 1988), and one-step bounded influence methods (Simpson, Ruppert and Carroll, 1992; Coakley and Hettmansperger, 1993; Simpson and Chang, 1996). Regression diagnostics based on LMS and related methods have been developed by several authors, including Rousseeuw and van Zomeren (1990), Atkinson and Weisberg (1991), and Atkinson (1994). Recently, the LMS idea has been extended to nonlinear and generalized linear models by Stromberg (1993a) and Christmann (1994).

There has been considerable interest in the computational aspects of LMS and related techniques. The original PROGRESS algorithm for LMS regression was an approximate method based on randomly sampling elemental sets from the data. Elemental sets are minimal sets of observations which just determine a regression plane. Since then, there have been a number of computational developments. A number of researchers have worked on the problem of finding the exact LMS solution; these include Souvaine and Steele (1987), Edelsbrunner and Souvaine (1990), Stromberg (1993b), and Hawkins (1993a). Cook, Hawkins, and Weisberg (1993) and Hawkins (1993b) considered the related problem of computing the MVE solution for location and scatter. Hawkins, Simonoff, and Stromberg (1994) showed how to distribute the LMS computation over a network. The "ltsreg" and "cov.mve" functions in S-Plus (Mathsoft, Inc.) currently use a genetic algorithm to compute the LTS and MVE estimates. The computational aspects are by no means resolved. For instance, Woodruff and Rocke (1994) argued that random sampling algorithms fail in higher dimensions, and proposed a partition algorithm suitable for parallel computing.

Several aspects of Rousseeuw (1984) broke with tradition and were sometimes considered controversial, such as the nature of the algorithms and the low statistical efficiency of the raw (initial) estimators. For interesting debate see the papers by Cook, Hawkins, and Weisberg (1992), Hettmansperger and Sheather (1992), Davies (1994), and Rousseeuw (1994). It often happens that one commentator sees a flaw where another sees an advantage. Much of the controversy can be attributed to context. Traditional methods such as least squares linear regression have been generally expected to be applied to "clean" data with a relatively low proportion of "outliers." In such contexts, investigators might feel there is little need for an ability to screen nearly half of the data. A particular area of controversy is the detection of missing variables in the regression and whether a high breakdown procedure is a help or a hindrance in the effort to find and model such variables. Ideally the ability to fit a submodel while ignoring a large fraction of outliers would enhance the ability to detect missing structure in the model. For further exposition of this viewpoint, see Rousseeuw (1994).

Although LMS, LTS, MVE, MCD, and other methods with high breakdown points were developed in the context of regression analysis and multivariate statistics, new application areas have emerged in finance, analytical chemistry, automated connection of optical fiber cables, power networks in electrical engineering, and computer vision; see, e.g., Rousseeuw (1996). In each of these applications, the ability to breakdown resistant procedures to filter and hence detect a substantial portion of "different" data, or a substructure, is essential. The use of breakdown-resistant methods is thus expanding well beyond the original setting in which they were developed, and new application areas are likely to emerge in the future.

4. About the Author

Peter J. Rousseeuw was born on October 13, 1956, in Antwerp, Belgium. His publishing career began at age 13 when he wrote an adventure story and bought his first bike with the royalties. From 1978 to 1980 he visited the ETH, Zurich, to write a Ph.D. thesis with Frank Hampel, where he worked on the influence function for tests and the change-of-variance function. These results were later summarized in Hampel et al. (1986). After the Ph.D. in 1981 he worked on topics in applied probability and statistical mechanics, before performing the research (including 3 months in Berkeley) leading to the 1984 paper reprinted here. He was also active in various fields of application such as chemometrics and engineering, and did substantial work on cluster analysis (see Kaufman and Rousseeuw (1990)) yielding L^1-based dissimilarity methods, the silhouette plot and new fuzzy clustering techniques. Professor Rousseeuw is the author of numerous other research articles and several books. His publications display broad interests, great inventiveness, and wonderful clarity. Currently he is Professor in the Department of Mathematics at the University of Antwerp.

References

Atkinson, A.C. (1994). Fast very robust methods for the detection of multiple outliers. *J. Amer. Stat. Assoc.*, **89**, 1329–1339.

Atkinson, A.C. and Weisberg, S. (1991). Simulated annealing for the detection of multiple outliers using least squares and least median of squares fitting. *Directions in Robust Statistics and Diagnostics.* Part I, pp. 7–20.

Box, G.E.P. (1953). Non-normality and tests on variances. *Biometrika*, **40**, 318–335.

Christmann, A. (1994). Least median of weighted squares in logistic regression with large strata. *Biometrika*, **81**, 413–417.

Coakley, C.W. and Hettmansperger, T.P. (1993). A bounded influence, high breakdown, efficient regression estimator. *J. Amer. Statist. Assoc.*, **88**, 872–880.

Cook, R.D., Hawkins, D.M., and Weisberg, S. (1992). Comparison of model misspecification diagnostics using residuals from least mean of squares and least median of squares fits. *J. Amer. Statist. Assoc.*, **87**, 419–424.

Cook, R.D., Hawkins, D.M., and Weisberg, S. (1993). Exact iterative computation of the robust multivariate minimum volume ellipsoid estimator. *Statist. Probab. Lett.*, **16**, 213–218.

Davies, L. (1990). The asymptotics of *S*-estimators in the linear regression model. *Ann. Statist.*, **18**, 1651–1675.

Davies, L. (1994). Desirable properties, breakdown and efficiency in the linear regression model. *Statist. Probab. Lett.*, **19**, 361–370.

Donoho, D.L. (1982). Breakdown properties of multivariate location estimators. Qualifying paper, Harvard University, Boston.

Donoho, D.L. and Huber, P.J. (1983). The notion of breakdown point. In *A Festschrift for Erich Lehmann* (P. Bickel, K. Doksum, and J.L. Hodges, eds.). Wadsworth, Belmont, CA.

Edelsbrunner, H. and Souvaine, D.L. (1990). Computing least median of squares regression lines and guided topological sweep. *J. Amer. Statist. Assoc.*, **85**, 115–119.

Hampel, F.R. (1971). A general qualitative definition of robustness. *Ann. Math. Statist.*, **42**, 1887–1896.

Hampel, F.R. (1974). The influence curve and its role in robust estimation. *J. Amer. Statist. Assoc.*, **69**, 383–393.

Hampel, F.R. (1975). Beyond location parameters: Robust concepts and methods. *Bull. Internat. Statist. Assoc.*, **46**, 375–382.

Hampel, F.R., Ronchetti, E.M., Rousseeuw, P.J., and Stahel, W.A. (1986). *Robust Statistics: The Approach Based on Influence Functions.* Wiley, New York.

Hawkins, D.M. (1993a). The feasible set algorithm for least median of squares regression. *Comput. Statist. Data Anal.*, **16**, 81–101.

Hawkins, D.M. (1993b). A feasible solution algorithm for the minimum volume ellipsoid estimator in multivariate data. *Comput. Statist. Quart.*, **8**, 95–107.

Hawkins, D.M., Simonoff, J.S., and Stromberg, A. (1994). Distributing a computationally intensive estimator: The case of exact LMS regression. *Comput. Statist. Quart.*, **9**, 83–95.

He, X. and Portnoy, S. (1992). Reweighted LS estimators converge at the same rate as the initial estimator. *Ann. Statist.*, **20**, 2161–2167.

Hettmansperger. T.P. and Sheather, S.J. (1992). A cautionary note on the method of least median squares. *Amer. Statist.*, **46**, 79–83.

Hodges, J.L. (1967). Efficiency in normal samples and tolerance of extreme values for some estimates of location. *Proceedings of the 5th Berkeley Symposium on Mathematical Statistics and Probability*, Vol. 1 pp. 163–168.

Hossjer, O., Rousseeuw, P.J., and Croux, C. (1994). Asymptotics of the repeated median slope estimator. *Ann. Statist.*, **22**, 1478–1501.

Huber, P.J. (1964). Robust estimation of a location parameter. *Ann. Math. Statist.*, **35**, 73–101.

Jurečkova, J. and Portnoy, S. (1987). Asymptotics for one-step *M*-estimators in regression with application to combining efficiency and high breakdown point. *Commun. Statist., Theory and Methods*, **16**(8), 2187–2199.

Kaufman, L. and Rousseeuw, P.J. (1990). *Finding Groups in Data.* Wiley, New York.

Kim, J. and Pollard, D. (1990). Cube root asymptotics. *Ann. Statist.*, **18**, 191–219.

Maronna, R.A. (1976). Robust *M*-estimators of multivariate location and scatter. *Ann. Statist.*, **4**, 51–67.

Maronna, R.A. and Yohai, V.J. (1995). The behavior of the Stahel–Donoho robust multivariate estimator. *J. Amer. Statist. Assoc.*, **90**, 330–341.

Martin, R.D., Yohai, V.J., and Zamar, R.H. (1989). Min-max bias robust regression. *Ann. Statist.*, **17**, 1608–1630.

Rousseeuw, P.J. (1984). Least median of squares regression. *J. Amer. Statist. Assoc.*, **79**, 871–880.

Rousseeuw, P.J. (1985). Multivariate estimation with high breakdown point. In *Mathematical Statistics and Applications*, vol. B (W. Grossmann, G. Pflug, I. Vincze, and W. Wertz, eds.). Reidel, Dordrecht, pp. 283–297.

Rousseeuw, P.J. (1994). Unconventional features of positive-breakdown estimators. *Statist. Probab. Lett.*, **19**, 417–431.

Rousseeuw, P.J. (1996). Robust Regression, Positive Breakdown. *Encyclopedia of Statistical Sciences Update Volume*, to appear.

Rousseeuw, P.J. and Leroy, A.M. (1987). *Robust Regression and Outlier Detection.* Wiley–Interscience, New York.

Rousseeuw, P.J. and van Zomeren, B.C. (1990). Unmasking multivariate outliers and leverage points. *J. Amer. Statist. Assoc.*, **85**, 633–651.

Rousseeuw, P.J. and Yohai, V.J. (1984). Robust regression by means of *S*-estimators. In *Robust and Nonlinear Time Series Analysis.* Lecture Notes in Statistics, No. 26 (J. Franke, W. Haerdle, and R.D. Martin, eds.). Springer-Verlag, New York, pp. 256–272.

Siegel, A.F. (1982). Robust regression using repeated medians. *Biometrika*, **69**, 242–244.

Simpson, D.G. and Chang, Y.-C.I. (1996). Reweighting approximate GM estimators: Asymptotics and residual-based graphics. *J. Statist. Plann. Inference*, **53**, in press.

Simpson, D.G., Ruppert, D., and Carroll, R.J. (1992). On one-step GM-estimates and stability of inferences in linear regression. *J. Amer. Statist. Assoc.*, **87**, 439–450.

Souvaine, D.L. and Steele, J.M. (1987). Time- and space-efficient algorithms for least median of squares regression. *J. Amer. Statist. Assoc.*, **82**, 794–801.

Stahel, W.A. (1981). Robuste Schaetzungen: Infinitesimale Optimalitaet und Schaetzungen von Kovarianzmatrizen. Ph.D. thesis, ETH Zurich.

Stromberg, A. (1993a). Computation of high breakdown nonlinear regression parameters. *J. Amer. Statist. Assoc.*, **88**, 237–244.

Stromberg, A. (1993b). Computing the exact least median of squares estimate and stability diagnostics in multiple linear regression. *SIAM J. Sci. Statist. Comput.*, **14**, 1289–1299.

Tukey, J. (1960). A survey on sampling from contaminated distributions. In *Contributions to Probability and Statistics.* (I. Olkin, ed.). Stanford University Press, Stanford, CA.

Woodruff, D.L. and Rocke, D.M. (1994). Computable robust estimation of multivariate location and shape in high dimension using compound estimators. *J. Amer. Statist. Assoc.*, **89**, 888–896.

Yohai, V.J. (1987). High breakdown-point and high efficiency robust estimates for regression. *Ann. Statist.*, **15**, 642–656.

Yohai, V.J. and Zamar, R.H. (1988). High breakdown-point estimates of regression by means of the minimization of an efficient scale. *J. Amer. Statist. Assoc.*, **83**, 406–413.

Least Median of Squares Regression

Peter J. Rousseeuw*

Classical least squares regression consists of minimizing the sum of the squared residuals. Many authors have produced more robust versions of this estimator by replacing the square by something else, such as the absolute value. In this article a different approach is introduced in which the sum is replaced by the *median* of the squared residuals. The resulting estimator can resist the effect of nearly 50% of contamination in the data. In the special case of simple regression, it corresponds to finding the narrowest strip covering half of the observations. Generalizations are possible to multivariate location, orthogonal regression, and hypothesis testing in linear models.

1. Introduction

The classical linear model is given by $y_i = x_{i1}\theta_1 + \cdots + x_{ip}\theta_p + e_i$ ($i = 1, \ldots, n$), where the error e_i is usually assumed to be normally distributed with mean zero and standard deviation σ. The aim of multiple regression is to estimate $\theta = (\theta_1, \ldots, \theta_p)^t$ from the data $(x_{i1}, \ldots, x_{ip}, y_i)$. The most popular estimate $\hat{\theta}$ goes back to Gauss or Legendre (see Stigler (1981) for a recent

* Peter J. Rousseeuw is Professor, Department of Mathematics and Informatics, Delft University of Technology, Julianalaan 132, 2628 BL Delft, The Netherlands. This work was supported by the Belgian National Science Foundation. The author is grateful to W. Stahel and F. Hampel for bringing the subject of high breakdown regression to his attention and providing interesting discussions. Helpful comments and suggestions were made by P. Bickel, D. Donoho, D. Hoaglin, P. Huber, R.D. Martin, R.W. Oldford, F. Plastria, A. Samarov, A. Siegel, J. Tukey, R. Welsch, V. Yohai, and two referees. Special thanks go to A. Leroy for assistance with the programming.

historical discussion) and corresponds to

$$\underset{\hat{\theta}}{\text{minimize}} \sum_{i=1}^{n} r_i^2, \tag{1.1}$$

where the residuals r_i equal $y_i - x_{i1}\hat{\theta}_1 - \cdots - x_{ip}\hat{\theta}_p$. Legendre called it the *method of least squares* (LS), and it became a cornerstone of statistics. But in spite of its mathematical beauty and computational simplicity, this estimator is now being criticized more and more for its dramatic lack of robustness. Indeed, one single outlier can have an arbitrarily large effect on the estimate. In this connection Hampel (1971) introduced the notion of the *breakdown point* ε^*, extending a definition of Hodges (1967): ε^* is the smallest percentage of contaminated data that can cause the estimator to take on arbitrarily large aberrant values. In the case of least squares, $\varepsilon^* = 0$.

A first step toward a more robust regression estimator came from Edgeworth (1887), improving a proposal of Boscovich. His *least absolute values* or L_1 criterion is

$$\underset{\hat{\theta}}{\text{minimize}} \sum_{i=1}^{n} |r_i|. \tag{1.2}$$

This generalizes the median of a one-dimensional sample and, therefore, has to be made unique (Harter, 1977). But whereas the breakdown point of the sample median is 50%, it can be shown that L_1 regression yields the same value $\varepsilon^* = 0$ as LS. Although L_1 regression protects against outlying y_i, it cannot cope with grossly aberrant values of $x_i = (x_{i1}, \ldots, x_{ip})$, which have a large influence (called *leverage*) on the fit.

The next step in this direction was the *M estimator* (Huber, 1973, p. 800), based on the idea of replacing r_i^2 in (1.1) by $\rho(r_i)$, where ρ is a symmetric function with a unique minimum at zero. Unlike (1.1) or (1.2), however, this is not invariant with respect to a magnification of the error scale. Therefore one often estimates the scale parameter simultaneously:

$$\sum_{i=1}^{n} \psi(r_i/\hat{\sigma})x_i = 0, \tag{1.3}$$

$$\sum_{i=1}^{n} \chi(r_i/\hat{\sigma}) = 0, \tag{1.4}$$

where ψ is the derivative of ρ and χ is a symmetric function. (Finding the simultaneous solution of this system of equations is not trivial, and in practice one uses an iteration scheme based on reweighted least squares or Newton–Raphson.) Motivated by minimax asymptotic variance arguments, Huber proposed to use the function

$$\psi(u) = \min(k, \max(u, -k)),$$

where k is some constant, usually around 1.5. As a consequence, such M estimators are statistically more efficient than L_1 at a central model with Gaussian errors. However, again $\varepsilon^* = 0$ because of the possibility of leverage points.

Because of this vulnerability to leverage points, *generalized M estimators* (GM estimators) were introduced, with the basic purpose of bounding the influence of outlying x_i, making use of some weight function w. Mallows (1975) proposed to replace (1.3) by

$$\sum_{i=1}^{n} w(x_i)\psi(r_i/\hat{\sigma})x_i = 0, \tag{1.5}$$

whereas Schweppe (see Hill (1977)) suggested using

$$\sum_{i=1}^{n} w(x_i)\psi(r_i/(w(x_i)\hat{\sigma}))x_i = 0. \tag{1.6}$$

Making use of influence functions, good choices of ψ and w were made (Hampel, 1978; Krasker, 1980; Krasker and Welsch, 1982). It turns out, however, that the GM estimators now in use have a breakdown point of at most $1/(p+1)$, where p is the dimension of x_i (Maronna, Bustos, and Yohai, 1979; Donoho and Huber, 1983). Various other estimators have been proposed by Theil (1950), Brown and Mood (1951), Sen (1968), Jaeckel (1972), and Andrews (1974); but none of them achieves $\varepsilon^* = 30\%$ in the case of simple regression ($p = 2$).

All of this raises the question whether robust regression with a high breakdown point is at all possible. The affirmative answer was given by Siegel (1982), who proposed the *repeated median* with a 50% breakdown point. Indeed, 50% is the best that can be expected (for larger amounts of contamination, it becomes impossible to distinguish between the "good" and the "bad" parts of the sample). Siegel's estimator is defined as follows: For any p observations $(x_{i1}, y_{i1}), \ldots, (x_{ip}, y_{ip})$, which determine a unique parameter vector, the jth coordinate of this vector is denoted by $\theta_j(i_1, \ldots, i_p)$. The repeated median is then defined coordinatewise as

$$\hat{\theta} = \operatorname*{med}_{i_1}(\cdots(\operatorname*{med}_{i_p-1}(\operatorname*{med}_{i_p} \theta_j(i_1, \ldots, i_p)))\ldots). \tag{1.7}$$

This estimator can be calculated explicitly, but is not equivariant for linear transformations of the x_i. It was applied to a biological problem by Siegel and Benson (1982).

Let us now return to (1.1). A more complete name for the LS method would be *least sum of squares*, but apparently few people have objected to the deletion of the word "sum"—as if the only sensible thing to do with n positive numbers would be to add them. Perhaps as a consequence of this historical name, most people have tried to make this estimator robust by replacing the square by something else, not touching the summation sign.

Why not, however, replace the sum by a median, which is very robust? This yields the *least median of squares* (LMS) estimator, given by

$$\underset{\hat{\theta}}{\text{minimize}} \ \underset{i}{\text{med}} \ r_i^2. \tag{1.8}$$

This proposal is essentially based on an idea of Hampel (1975, p. 380). In the next section it is shown that the LMS satisfies $\varepsilon^* = 50\%$ but has a very low efficiency. In Section 4 some variants with higher efficiency are given.

2. Properties of the Least Median of Squares Method

We shall now investigate the behavior of the LMS technique. The n observations $(x_i, y_i) = (x_{i1}, \ldots, x_{ip}, y_i)$ belong to the linear space of row vectors of dimension $p+1$, and the unknown parameter θ is a p-dimensional column vector $(\theta_1, \ldots, \theta_p)^t$. The unperturbed linear model states that $y_i = x_i\theta + e_i$, where e_i is distributed according to $N(0, \sigma)$. Throughout this section it is assumed that all observations with $x_i = 0$ have been deleted, because they give no information on θ. (This condition is automatically satisfied if the model has an intercept because then the last coordinate of each x_i equals 1.) Moreover, it is assumed that in the $(p+1)$-dimensional space of the (x_i, y_i), there is no vertical hyperplane containing more than $[n/2]$ observations. (Here, a vertical hyperplane is a p-dimensional subspace that contains $(0, \ldots, 0)$ and $(0, \ldots, 0, 1)$. The notation $[r]$ stands for the largest integer less than or equal to r.) The proofs of the following results can be found in the Appendix.

Lemma 1. *There always exists a solution to* (1.8).

In what follows we shall say the observations are in general position when any p of them give a unique determination of θ. For example, in the case in which $p = 2$, this means that any pair of observations (x_{i1}, x_{i2}, y_i), and (x_{j1}, x_{j2}, y_j) determines a unique nonvertical plane through zero, which implies that $(0, 0, 0)$, (x_{i1}, x_{i2}, y_i), and (x_{j1}, x_{j2}, y_j) cannot be collinear. When the observations come from continuous distributions, this event has probability one.

Let us now discuss the breakdown properties of the LMS method. Hampel's (1971) original definition of the breakdown point was asymptotic in nature. In this article, however, I use a version introduced by Donoho and Huber (1983) that is intended for finite samples, like the precursor ideas of Hodges (1967). Take any sample X of n data points (x_i, y_i) and a regression estimator T. Let $\beta(m; T, X)$ be the supremum of $\|T(X') - T(X)\|$ for all corrupted samples X', where any m of the original data points are replaced by arbitrary values. Then the breakdown point of T at X is

$$\varepsilon^*(T, X) = \min\{m/n; \beta(m; T, X) \text{ is infinite}\}. \tag{2.1}$$

In other words, it is the smallest amount of contamination that can cause the estimator to take on values arbitrarily far from $T(X)$. Note that this definition contains no probability distributions! For least squares, $\varepsilon^*(T, X) = 1/n$ because one bad observation can already cause breakdown. For least median of squares, however, this is no longer the case.

Theorem 1. *If $p > 1$ and the observations are in general position, then the breakdown point of the LMS method is $([n/2] - p + 2)/n$.*

Note that the breakdown point depends only slightly on n. To have only a single value, one often considers the limit for $n \to \infty$ (with p fixed); so it can be said that LS has a breakdown point of 0%, whereas the breakdown point of the LMS technique is as high as 50%, the best that can be expected. The following corollary gives a special case that shows the large resistance of the LMS method.

Corollary 1. *If $p > 1$ and there exists some θ such that at least $n - [n/2] + p - 1$ of the observations satisfy $y_i = x_i\theta$ exactly and are in general position, then the LMS solution equals θ whatever the other observations are.*

Remark 1. The breakdown point in Theorem 1 is slightly smaller than that of the repeated median, although they are both 50% breakdown estimators. I am indebted to A. Siegel (personal communication) for a way to overcome this. Instead of taking the median of the ordered squared residuals, consider the kth order statistic $(r^2)_{k:n}$, where $k = [n/2] + [(p+1)/2]$, and minimize $(r^2)_{k:n}$. It turns out (analogous to the proof of Theorem 1) that this variant of the LMS has breakdown point $([(n-p)/2] + 1)/n$, which is exactly the same value as for Siegel's repeated median. In the Appendix, it is shown that this is the maximal value for all regression-equivariant estimators. (By regression equivariance, I mean the property

$$T(\{(x_i, y_i + x_i v); \quad i = 1, \ldots, n\}) = T(\{(x_i, y_i); \quad i = 1, \ldots, n\}) + v$$

for any vector v.) For this variant of the LMS, Corollary 1 holds whenever strictly more than $\frac{1}{2}(n + p - 1)$ of the observations are in an exact fit situation, which also corresponds to the repeated median.

It is well known that the LS estimator reduces to the arithmetic mean in the special case of one-dimensional estimation of location, obtained by putting $p = 1$ and $x_i = 1$ for all i. Interestingly, in that special case, the LMS estimator also corresponds to something we know.

Theorem 2. *Let $p = 1$ and all $x_i = 1$, so the sample reduces to $(y_i)_{i=1,\ldots,n}$. If*

$$m_T^2 : \operatorname*{med}_i r_i^2 = \operatorname*{med}_i (y_i - T)^2$$

$$\text{equals } \min_\theta \operatorname*{med}_i (y_i - \theta)^2,$$

then both $T - m_T$ and $T + m_T$ are observations in the sample.

Theorem 2 makes it easy to determine T in the location case because one has to determine only the shortest half of the sample. (This is done by finding the smallest of the values

$$y_{h:n} - y_{1:n}, y_{h+1:n} - y_{2:n}, \ldots, y_{n:n} - y_{n-h+1:n},$$

where $h = [n/2] + 1$ and $y_{1:n} \leq y_{2:n} \leq \cdots \leq y_{n:n}$ are the ordered observations.) By Theorem 2 T simply equals the midpoint of this shortest interval. (In case there are several shortest halfs, which happens with probability zero when the distribution is continuous, one could take the average of their midpoints.) This is reminiscent of the estimator called *shorth* in the Princeton Monte Carlo study (Andrews et al., 1972), where the mean of all of the observations in the shortest half is taken. The shorth converges like $n^{-1/3}$; therefore its influence function is not well defined, but its finite-sample breakdown behavior is extremely good (Andrews et al., 1972, pp. 50, 33, 103). All of these properties are shared by the LMS (details on its asymptotic behavior can be found in the Appendix). The fact that the LMS converges like $n^{-1/3}$ does not trouble me very much, because I consider the LMS mainly as a data analytic tool, for which statistical efficiency is not the most important criterion. In Section 4, I will construct variants with higher efficiency. Although the LMS has no well-defined influence function, it is possible to get some idea of its local robustness properties by constructing stylized sensitivity curves, as was done by Andrews et al. (1972, p. 101) for the shorth. For the LMS this yields Figure 1 for $n = 10$; for large sample sizes the upward and downward peaks become thinner and higher.

Theorem 2 can also be used in the more general case of regression with a constant, obtained by putting $x_{i,p} = 1$ for all i. From Theorem 2, it follows that for an LMS solution, both hyperplanes $y = x\hat{\theta} - m_T$ and $y = x\hat{\theta} + m_T$ contain at least one observation.

In the special case of simple regression, there are only a single independent variable and a single dependent variable to be fitted to the model $y_i = ax_i + b + e_i$. One therefore has to find the slope and the intercept of a line determined by n points in the plane. By Theorem 2 the LMS solution corresponds to finding the narrowest strip covering half of the observations. (To be exact the thickness of the strip is measured in the vertical direction, and we want at least $h = [n/2] + 1$ points on it.) It is easy to write a computer program to determine the LMS line, because for each value of a, the

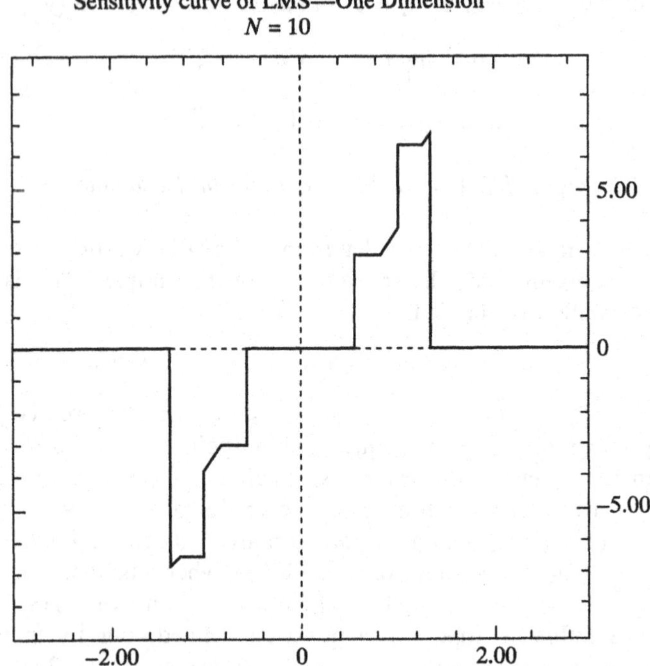

Figure 1. Stylized sensitivity curve of the LMS in one dimension, given by $SC(x) = n(T_n(y_1, \ldots, y_{n-1}, x) - T_{n-1}(y_1, \ldots, y_{n-1}))$, where $y_i = \Phi^{-1}(i/n)$ for $n = 10$.

above algorithm for the location case can be used to calculate

$$m_a^2 = \min_b \operatorname*{med}_i ((y_i - ax_i) - b)^2$$

immediately. Therefore, one only has to minimize the continuous function m_a^2 of the single variable a. This technique will be used in Section 3. For data analytic purposes, D. Hoaglin (personal communication) proposed to compute not only the global minimum of the objective function but also the second best local minimum (if it exists), because this might reflect a possible ambiguity in the data.

For problems in higher dimensions, another program has been developed (Leroy and Rousseeuw, 1984), making use of brute force minimization of the objective function $\text{med}_i\, r_i^2$. Table 1 lists some computation times on the CDC computer of the University of Brussels for different values of n and p. (For larger values of n, the time is roughly proportional to n for fixed p.) These times are large, which is not so surprising in view of the relationship of the LMS to the projection pursuit technique (Friedman and Tukey, 1974). Indeed, consider the $(p+1)$-dimensional space of observations (x_i, y_i). We want to find a direction, given by some vector $(\theta, -1)$, such that the pro-

Table 1. Computation Times* for the LMS Multiple Regression Program With Intercept for Different n and p.

	p								
n	2	3	4	5	6	7	8	9	10
25	.23	1.28	1.95	2.89	4.13	4.91	5.74	6.76	7.79
50	1.52	2.08	3.16	4.50	5.98	6.89	7.82	8.87	10.02
100	2.34	3.82	5.55	7.65	10.10	11.02	12.11	13.45	16.83
150	3.39	5.51	8.05	10.89	14.00	15.15	16.50	17.98	19.56
200	4.47	7.16	10.28	13.84	17.82	19.33	20.85	22.39	24.42

* In CP seconds on a CDC 750 computer.

jection of the data on the y axis, in the direction orthogonal to $(\theta, -1)$, possesses the smallest dispersion (measured by the median of squares).

In addition to the regression coefficients $\theta_1, \ldots, \theta_p$, the scale parameter σ has to be estimated in a robust way. Once the LMS solution T has been found, with $m_T^2 : \min_\theta \operatorname{med}_i r_i^2$, a natural estimator for σ is

$$S = 1.483c(n,p)m_T, \tag{2.2}$$

where $1/\Phi^{-1}(0.75) \simeq 1.483$ is an asymptotic correction factor for the case of normal errors, because then

$$\operatorname*{med}_i r_i^2 \to \sigma^2 \operatorname{med}(\chi_1^2) = \sigma^2(\Phi^{-1}(.75))^2,$$

where Φ denotes the standard normal cumulative. The constant $c(n,p)$ is a finite-sample correction factor larger than 1, which is necessary to make S approximately unbiased when simulating samples with normal errors. Work is in progress to determine $c(n,p)$ empirically. For n tending to infinity, $c(n,p)$ converges to 1.

3. Examples

In the first example, 30 "good" observations were generated according to the linear relation $y_i = ax_i + b + e_i$, where $a = 1, b = 2$, x_i is uniformly distributed on (1, 4), and e_i is normally distributed with mean zero and standard deviation .2. The number p of coefficients to be estimated therefore equals 2. Then a cluster of 20 "bad" observations were added, possessing a spherical bivariate normal distribution with mean (7, 2) and standard deviation $\frac{1}{2}$. This yielded 40% of contamination in the pooled sample, which is high. This amount was actually chosen to demonstrate what happens if one goes above the upper bound $1/(p+1) \simeq 33.3\%$ on the breakdown point of the GM estimators now in use.

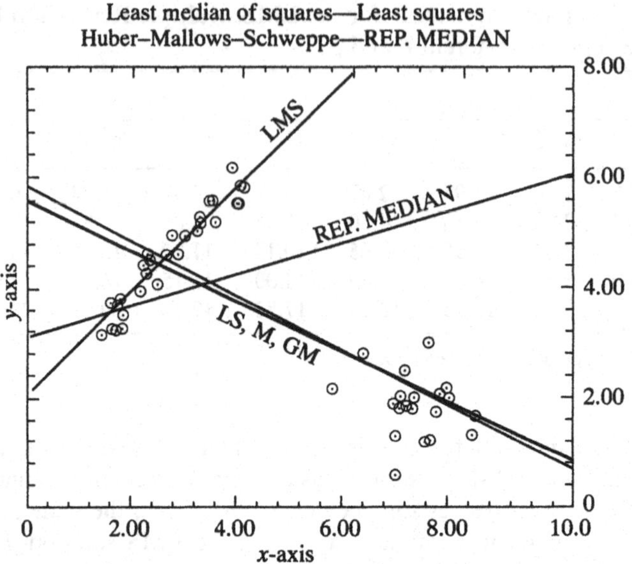

Figure 2. Regression lines for the simulated data of the first example, using six methods. (LMS = least median of squares; LS = least squares; M = Huber's M estimator; GM = Mallows's and Schweppe's G-M estimator; REP. MEDIAN = repeated median; ⊙ = 30 "good" points generated according to a linear relation $y_i = x_i + 2 + e_i$ and 20 "bad" points in a spherical cluster around (7, 2).

Let us now see which estimator succeeds best in describing the pattern of the majority of the data. The classical least squares method yields $\hat{a} = -.47$ and $\hat{b} = 5.62$: it clearly fails because it tries to suit both the good and the bad data points. Making use of the ROBETH library of subroutines (Marazzi, 1980), three robust estimators were applied: Huber's M estimator (1.3) – (1.4) with $\psi(x) = \min(1.5, \max(-1.5, x))$, Mallows's GM estimator (1.5) with Hampel weights, and Schweppe's GM estimator (1.6) with Hampel-Krasker weights (both Mallows's and Schweppe's using the same Huber function ψ). All three methods, however, gave results virtually indistinguishable from the LS solution: the four lines almost coincide in Figure 2. The repeated median estimator (1.7) yields $\hat{a} = .30$ and $\hat{b} = 3.11$. If the cluster of "bad" points is moved further down, the repeated median line follows it a little more and then stops. Therefore this method does not break down. Finally, the LMS (1.8), calculated by means of the algorithm for simple regression described in Section 2, yields $\hat{a} = .97$ and $\hat{b} = 2.09$, which comes close to the original values of a and b. When the cluster of bad points is moved further away, this solution does not change. Moreover, the LMS method does not break down even when only 26 "good" points and 24 outliers are used.

It may seem unfair to consider such large amounts of contamination (although they sometimes occur, e.g., in the case of ancient astronomical observations (Huber, 1974) or in certain sloppy medical data sets). The breakdown point of the currently used GM estimators, however, is less than $1/(p+1)$, which is small in problems with several independent variables; so very common amounts of contamination already necessitate the use of a more robust regression estimator. Moreover, it still has to be investigated empirically whether the upper bound $1/(p+1)$ on the asymptotic breakdown point can actually be reached in finite sample situations.

Note that looking at the least squares residuals (possibly followed by a rejection of outlying ones) is not sufficient. In fact the least squares fit often masks bad data points: in Figure 2, the largest LS residuals correspond to good data! In problems with several variables, a very robust estimator like the LMS can be used for finding the outlying observations, as shall be seen in the next example.

When faced with a practical application, it seems like a good idea to run both an LMS and an LS regression. If they agree closely, the LS result can be trusted. If, on the other hand, there is a significant difference, then we know which observations are responsible by looking at the LMS residuals.

Let us now look at a second example, containing multidimensional real data. It seems that an entirely real example with "messy" data might not be completely convincing, because we would end up with different results for LS and LMS without a conclusive way to decide which analysis is best, possibly causing some debates. Therefore, we start with a real data set that is rather well behaved and contaminate it by replacing a few observations. It would be easy to illustrate the resistance of the LMS by throwing in some very bad outliers, but I would like to put the LMS to a harder test by considering a more delicate situation. To show that the LMS also works in small samples, I selected a data set containing 20 points with six parameters to be estimated. The raw data came from Draper and Smith (1966, p. 227) and were used to determine the influence of anatomical factors on wood specific gravity, with five independent variables and an intercept (Draper and Smith conclude that x_{i2} could be deleted from the model, but this matter is not considered for the present purpose). Table 2 lists a contaminated version of these data, in which a few observations have been replaced by outliers. Applying least squares yields

$$\hat{y}_i = .44069x_{i1} - 1.47501x_{i2} - .26118x_{i3} + .02079x_{i4} + .17082x_{i5} + .42178.$$

Table 2 lists the LS residuals, divided by the LS scale estimate $\hat{\sigma}_{LS} = .02412$. It is not easy to spot the outliers just by looking at the observations, and the LS result (without a more detailed analysis) is of little help. Indeed, the standardized residuals look very inconspicuous, except for obser-

Table 2. Modified Data on Wood Specific Gravity With Standardized Residuals From Least Squares and Least Median of Squares.

								Residual/Scale	
i	x_{i1}	x_{i2}	x_{i3}	x_{i4}	x_{i5}	x_{i6}	y_i	LS	LMS
1	.5730	.1059	.4650	.5380	.8410	1.000	.5340	−.7250	−.0827
2	.6510	.1356	.5270	.5450	.8870	1.000	.5350	.0472	.0013
3	.6060	.1273	.4940	.5210	.9200	1.000	.5700	1.2427	.2836
4	.4370	.1591	.4460	.4230	.9920	1.000	.4500	.3547	−7.6137
5	.5470	.1135	.5310	.5190	.9150	1.000	.5480	1.0024	.9020
6	.4440	.1628	.4290	.4110	.9840	1.000	.4310	−.4518	−9.1023
7	.4890	.1231	.5620	.4550	.8240	1.000	.4810	.9067	.3746
8	.4130	.1673	.4180	.4300	.9780	1.000	.4230	−.0349	−8.9077
9	.5360	.1182	.5920	.4640	.8540	1.000	.4750	−.3959	−.3746
10	.6850	.1564	.6310	.5640	.9140	1.000	.4860	−.4150	−.2071
11	.6640	.1588	.5060	.4810	.8670	1.000	.5540	1.9859	.0013
12	.7030	.1335	.5190	.4840	.8120	1.000	.5190	−1.1977	−.9656
13	.6530	.1395	.6250	.5190	.8920	1.000	.4920	−.4854	.0013
14	.5860	.1114	.5050	.5650	.8890	1.000	.5170	−1.2612	−.6709
15	.5340	.1143	.5210	.5700	.8890	1.000	.5020	−.5886	−.1733
16	.5230	.1320	.5050	.6120	.9190	1.000	.5080	.5237	.0013
17	.5800	.1249	.5460	.6080	.9540	1.000	.5200	−.2548	.0013
18	.4480	.1028	.5220	.5340	.9180	1.000	.5060	.2838	−.1090
19	.4170	.1687	.4050	.4150	.9810	1.000	.4010	−1.0837	−10.7265
20	.5280	.1057	.4240	.5660	.9090	1.000	.5680	.5450	.0013

vation 11 (and this is a false trail). Because of this, many people would probably be satisifed with the LS fit (especially when not expecting trouble). By means of the subroutines that are presently at my disposal, I was unable to obtain a GM estimate essentially different from the LS one, although it is possible that a more refined GM program would do the job. The LMS estimate can also be computed, however, as described in Remark 1 of Section 2, taking $c(20, 6) = 1.8$ in (2.2). This yields

$$\begin{aligned}\hat{y}_i = &.26870x_{i1} - .23806x_{i2} - .53572x_{i3} \\ &- .29373x_{i4} + .45096x_{i5} + .43474.\end{aligned}$$

Now look at the LMS residuals divided by the LMS scale estimate .0195, which are given in the last column of Table 2. These standardized residuals make it easy to spot the four outliers. This example illustrates the use of the LMS as a data analytic tool: as a next step in the analysis, LS could be computed again without these four observations.

4. Related Approaches

A disadvantage of the LMS method is its lack of efficiency because of its $n^{-1/3}$ convergence. Of course it is possible to take an extreme point of view, wanting to stay on the safe side, even if it costs a lot. However, it is not so difficult to improve the efficiency of the LMS estimator. One first has to calculate the LMS estimate T and the corresponding scale estimate S given by (2.2). With these starting values, one can compute a *one-step M estimator* (Bickel, 1975). If one uses a redescending ψ function, like the one of the hyperbolic tangent estimator (Hampel, Rousseeuw, and Ronchetti, 1981) or the biweight (Beaton and Tukey, 1974), the large outliers will not enter into the computation. Such a one-step M estimator converges like $n^{-1/2}$ and possesses the same asymptotic efficiency (for normal errors) as a fully iterated M estimator. This was proven by Bickel (1975) when the starting value was $n^{1/2}$ consistent, but in general it even holds when the starting value is better than $n^{1/4}$ consistent (Bickel, personal communication, 1983). In particular, the combined procedure (LMS + one-step M) achieves the asymptotic efficiency $e = (\int \psi' \, d\Phi)^2/(\int \psi^2 \, d\Phi)$. For instance, the choice $c = 4.0$ and $k = 4.5$ in Table 2 of Hampel, Rousseeuw, and Ronchetti (1981) already yields an efficiency of more than 95%.

Another possibility is to use reweighted least squares. To each observation (x_i, y_i), one assigns a weight w_i that is a nonincreasing function of $|r_i/S|$ and that becomes zero starting from, say, $|r_i/S|$ equal to three or four. Then one replaces all observations (x_i, y_i) by $(w_i^{1/2}x_i, w_i^{1/2}y_i)$, which means that points with large LMS residuals disappear entirely. On these weighted observations, a standard least squares program is used to obtain the final estimate. Actually, applying this estimator to the data in Section 3 amounts to deleting the 20 "bad" points from the first example and deleting the four outliers from the second.

The main idea of the LMS is to minimize the scatter of the residuals. From this point of view, there already exist other estimators with a similar objective, using other measures of scatter. For instance, LS (1.1) minimizes the mean square of the residuals, and L_1 (1.2) minimizes their mean deviation (from zero). Jaeckel's (1972) estimator is also defined by minimizing a dispersion measure of the residuals. For this purpose, Jaeckel uses linear combinations of the ordered residuals. Since these estimators of scale are translation invariant, it is not possible to estimate a constant term in the regression model. Although these scale estimates can be quite robust, their largest breakdown point (with regard to both explosion and implosion of $\hat{\sigma}$) is $\varepsilon^* = 25\%$, which is achieved by the interquartile range. Therefore, the breakdown point of a Jaeckel estimator is at most 25%. A way to improve this would be to replace the scale estimator by another one having a larger breakdown point. For this reason Hampel (1975) suggested using the median absolute deviation, which essentially yields the LMS. The LMS, indeed, possesses a 50% breakdown point but unfortunately only converges like $n^{-1/3}$. This slow rate of convergence can be improved by computing a one-step M estimator afterwards, as seen eariler in this section. There is also another way to achieve this, by using a different objective function. The *least trimmed squares* (LTS) esimator (Rousseeuw, in press) is given by

$$\underset{\hat{\theta}}{\text{minimize}} \sum_{i=1}^{h} (r^2)_{i:n}, \tag{4.1}$$

where $(r^2)_{1:n} \leq \cdots \leq (r^2)_{n:n}$ are the ordered squared residuals. If $h = [n/2] + 1$ is chosen, the breakdown point of Theorem 1 is obtained, and for $h = [n/2] + [(p+1)/2]$, the result of Remark 1 holds. In general, h may depend on some trimming proportion α, for instance by means of $h = [n(1-\alpha)] + 1$. The LTS converges like $n^{-1/2}$ (Rousseeuw, in press), with the same asymptotic efficiency at the normal distribution as the M estimator defined by $\psi(x) = x$ for $|x| \leq \Phi^{-1}(1 - (\alpha/2))$ and $\psi(x) = 0$ otherwise, which is called the *Huber skipped mean*. The main disadvantage of the LTS is that its objective function requires sorting of the squared residuals, which takes $O(n \log n)$ operations compared with only $O(n)$ operations for the median, hereby blowing up the already large computation times in Table 1. Another possiblity is to use the so-called *S estimators* defined by

$$\underset{\hat{\theta}}{\text{minimize}}\ S(\theta), \tag{4.2}$$

where $S(\theta)$ is a certain type of M estimator of scale on the residuals $r_1(\theta), \ldots, r_n(\theta)$. (These estimators are now being investigated in collaboration with V. Yohai, following similar suggestions by J. Tukey and R. Martin (personal communications).) It appears that S estimators have essentially the same asymptotic behavior as regression M estimators, but they can also achieve a high breakdown point.

In the case of simple regression, we saw that the LMS method corresponds to finding the narrowest strip covering half of the points. Taking the word "narrowest" literally amounts to replacing the usual squared residuals by the squared distances from the observations to the fitted line. This way the LMS can be generalized to orthogonal regression. Another generalization can be made to hypothesis testing in linear models, replacing the usual sums of squared residuals by their medians (or, better still, by the objective function of (4.1) or the square of (4.2)). The ratio of two such medians would then be compared to critical values, possibly obtained by simulation.

The LMS technique can also be used to estimate the location of a spherical multivariate distribution. If the sample $x_1, \ldots, x_n$ consists of p-dimensional vectors, then the LMS estimator T is defined by

$$\underset{T}{\text{minimize}} \ \underset{i}{\text{med}} \ \|T - x_i\|^2, \tag{4.3}$$

which corresponds to finding the center of the smallest disk (or ball) covering half of the points. The LTS analog would be to minimize the sum of the first h ordered values of $\|T - x_i\|^2$. Both procedures have a breakdown point of 50% and are equivariant with respect to magnification factors, translations, and orthogonal transformations.

Recently a more difficult problem has received attention, namely to construct a high breakdown estimator of multivariate location that is equivariant for affine transformations, which means that $T(Ax_1 + b, \ldots, Ax_n + b) = AT(x_1, \ldots, x_n) + b$ for any vector b and any nonsingular matrix A. The first solution to this problem was obtained independently by Stahel (1981) and Donoho (1982). For each observation x_i, one looks for a one-dimensional projection in which x_i is most outlying and then downweights x_i according to this worst result. The Stahel–Donoho estimator is then defined as the weighted mean of the observations, which is affine equivariant and possesses a breakdown point of 50%. Making use of the same weights, they also compute a covariance estimator. We shall now see that it is also possible to generalize the LMS to meet the joint objectives of affine equivariance and 50% breakdown point: Define T as the center of the minimal volume ellipsoid covering at least h observations (Rousseeuw, in press). The corresponding generalization of the LTS, which also yields $\varepsilon^* = 50\%$, takes the mean of the h points for which the determinant of the covariance matrix is minimal. Both the minimal volume ellipsoid estimator and the minimal covariance determinant estimator yield robust covariance estimators at the same time (for instance, by computing the covariance of the selected h observations, multiplied with a correction factor to obtain consistency in the case of multivariate normality).

It is possible to use the LMS idea in multivariate location, but one can also extend the Stahel–Donoho estimator to regression. For each observation (x_i, y_i), one could define u_i as the supremum of

$$|r_i(\theta)| / \underset{j}{\text{med}} \, |r_j(\theta)|$$

over all possible vectors θ. Then one could assign a weight $w(u_i)$ to each point and compute a weighted least squares estimator. Unfortunately this technique involves an optimization for each data point, so it requires about n times the computational cost of the LMS.

Finally, a word of caution. Some people have objected to the notion of the breakdown point, on the ground that it is very crude, and have pointed out the possibility of constructing high breakdown procedures that are not good on other grounds (Oldford, 1983). It is indeed true that the breakdown point is only one out of several criteria; so a high breakdown point alone is not a sufficient condition for a good method. I consider a good breakdown point to be more like a necessary condition: If it is not satisfied because the procedure is vulnerable to a certain type of contamination (such as leverage points in regression), one cannot guarantee that such contamination will never occur in practice.

Appendix

Proof of Lemma 1

For this proof, we work in the $(p+1)$-dimensional space E of the observations (x_i, y_i). The space of the x_i is the horizontal hyperplane through the origin, which is denoted by $(y = 0)$ because the y coordinates of all points in this plane are zero (in a space E of dimension $p + 1$, a hyperplane is a p-dimensional subspace). There are two cases.

Case A. This is really a special case for which there exists a $(p-1)$-dimensional subspace of V of $(y = 0)$ containing (at least) $[n/2]+1$ of the x_i. The observations (x_i, y_i) corresponding to these x_i now generate a subspace S of E (in the sense of linear algebra), which is at most p dimensional. Because it was assumed that E has no vertical hyperplane containing $[n/2]+1$ observations, it follows that S does not contain (0, 1); hence the dimension of S is at most $p-1$. This means that there exists a nonvertical hyperplane H given by some equation $y = x\theta$, which includes S. For this value of θ, clearly $\text{med}_i\, r_i^2 = 0$, which is the minimal value.

Case B. Let us now assume that we are in the general situation in which case A does not hold. The rest of the proof will be devoted to showing that there exists a ball around the origin, which may be very large, to which attention can be restricted for finding a minimum of $\text{med}_i\, r_i^2(\theta)$. Because $\text{med}_i\, r_i^2(\theta)$ is continuous in θ, this is sufficient for the existence of a minimum. Put $\delta = \frac{1}{2}\inf\{\eta > 0;$ there exists a $(p-1)$-dimensional subspace V of $(y = 0)$ such that V^η covers (at least) $[n/2]+1$ of the $x_i\}$. Here V^η is the set of all x with distance to V not larger than η. Because we are not in case A, $\delta > 0$. Also $M : \max_i |y_i|$. Now attention may be restricted to the ball around the origin with radius $(\sqrt{2}+1)M/\delta$. Indeed, for any θ with $\|\theta\| > (\sqrt{2}+1)M/\delta$, it will be shown that

$$\operatorname*{med}_i r_i^2(\theta) > \operatorname*{med}_i y_i^2 = \operatorname*{med}_i r_i^2(0),$$

so smaller objective functions cannot be found outside the ball. A geometrical construction is needed to prove this. Such a θ determines a nonvertical hyperplane H given by $y = x\theta$. By the dimension theorem of linear algebra, the intersection $H \cap (y = 0)$ has dimension $p - 1$. Therefore $(H \cap (y = 0))^\delta$ contains at most $[n/2]$ of the x_i. For each of the remaining observations (x_i, y_i), we construct the vertical two-dimensional plane through (x_i, y_i) and orthogonal to $(H \cap (y = 0))$. (This plane does not pass through zero, so to be called vertical, it has to go through both (x_i, y_i) and $(x_i, y_i + 1)$.) We see that

$$|r_i| = |x_i\theta - y_i| \geq \| x_i\theta| - |y_i\|$$

with $|x_i\theta| > \delta|tg(\alpha)|$, where α is the angle (between $-\pi/2$ and $\pi/2$) formed by H and the horizontal line in P_i. Therefore $|\alpha|$ is the angle between the line orthogonal to H and (0, 1); hence

$$\begin{aligned}|\alpha| &= \arccos\{|(-\theta, 1)(0, 1)^t|/(\|(-\theta, 1)\| \cdot \|(0, 1)\|)\}\\ &= \arccos(1/(1 + \|\theta\|^2)^{1/2}),\end{aligned}$$

and finally, $|tg(\alpha)| = \|\theta\|$. Because $\|\theta\| > (\sqrt{2}+1)M/\delta$, it follows that $|x_i\theta| > \delta\|\theta\| > M \geq |y_i|$, so $|r_i| > (\delta\|\theta\| - |y_i|)$. But then $r_i^2 > ((\sqrt{2}+1)M - |y_i|)^2 > 2M^2$ for at least $n - [n/2]$ observations, hence $\operatorname{med}_i(r_i^2) > M^2 \geq \operatorname{med}_i(y_i^2)$. Such θ outside the ball yield an objective function larger than the one for $\theta = 0$; hence such θ can be disregarded.

Proof of Theorem 1

1. We first show that

$$\varepsilon^*(T, X) \geq ([n/2] - p + 2)/n$$

for any sample $X = \{(x_i, y_i); i = 1, \ldots, n\}$ consisting of n points in general position. By Lemma 1 the sample X yields a solution θ of (1.8). We now have to show that the LMS remains bounded when $n - ([n/2] - p + 2) + 1$ points are unchanged. For this purpose construct any corrupted sample $X' = \{(x_i', y_i'); i = 1, \ldots, n\}$ by retaining $n - [n/2] + p - 1$ observations of X—which will be called the "good" observations—and by replacing the others by arbitrary values. It suffices to prove that $\|\theta - \theta'\|$ is bounded, where θ' corresponds to X'. For this purpose some geometry is needed. We work in the $(p+1)$-dimensional space E of the observations (x_i, y_i). The space of the x_i is the horizontal hyperplane through the origin, denoted by $(y=0)$ because the y coordinates of all points in this plane are zero. (We call this subspace a hyperplane because its dimension is p, which is one less than the dimension of the total space E.) Put $\rho: \frac{1}{2}\inf\{\eta > 0$; there exists a $(p-1)$-dimensional subspace V of $(y=0)$ such that V^η covers at least p of the $x_i\}$. Here V^η is the set of all x with distance to V not larger than η. Because X is in general positon, it holds that $\rho > 0$. Also put $M: \max_i |r_i|$, where r_i are the residuals $y_i - x_i\theta$.

The rest of the proof of part 1 will be devoted to showing that $\|\theta - \theta'\| < 2(\|\theta\| + M/\rho)$, which is sufficient because the right member is a finite constant. Denote by H the nonvertical hyperplane given by the equation $y = x\theta$, and let H' correspond in the same way to θ'. Without loss of generality assume that $\theta' \neq \theta$, hence $H' \neq H$. By the dimension theorem of linear algebra, the intersection $H \cap H'$ has dimension $p-1$. If $\text{pr}(H \cap H')$ denotes the vertical projection of $H \cap H'$ on $(y=0)$, it follows that at most $p-1$ of the good x_i can lie on $(\text{pr}(H \cap H'))^\rho$. A is defined as the set of remaining good observations, containing at least $n - [n/2] + p - 1 - (p-1) = n - [n/2]$ points. Now consider any (x_a, y_a) belonging to A, and put $r_a = y_a - x_a\theta$ and $r'_a = y_a - x_a\theta'$. Construct the vertical two-dimensional plane P_a through (x_a, y_a) and orthogonal to $\text{pr}(H \cap H')$. It follows, as in the proof of Lemma 1, that

$$|r'_a - r_a| = |x_a\theta' - x_a\theta| > \rho|tg(\alpha') - tg(\alpha)|$$

$$\geq \rho\big||tg(\alpha')| - |tg(\alpha)|\big| = \rho\big|\|\theta'\| - \|\theta\|\big|,$$

where α is the angle formed by H and some horizontal line in P_a and α' corresponds in the same way to H'. Since

$$\|\theta' - \theta\| \leq \|\theta\| + \|\theta'\| = 2\|\theta\| + (\|\theta'\| - \|\theta\|)$$

$$\leq \big|\|\theta'\| - \|\theta\|\big| + 2\|\theta\|,$$

it follows that $|r'_a - r_a| > \rho(\|\theta' - \theta\| - 2\|\theta\|)$. Now the median of squared residuals of the new sample X' with respect to the old θ, with at least $n - [n/2] + p - 1$ of these residuals being the same as before, is less than or equal to M^2. Because θ' is a solution of (1.8) for X', it also follows that $\text{med}_i(y'_i - x'_i\theta')^2 \leq M^2$. If it is now assumed that $\|\theta' - \theta\| \geq 2(\|\theta\| + M/\rho)$, then for all a in A, it holds that

$$|r'_a - r_a| > \rho(\|\theta' - \theta\| - 2\|\theta\|) \geq 2M,$$

so $|r'_a| \geq |r'_a - r_a| - |r_a| > 2M - M = M$, and finally, $\text{med}_i(y'_i - x'_i\theta')^2 > M^2$, a contradiction. Therefore, $\|\theta' - \theta\| < 2(\|\theta\| + M/\rho)$ for any X'.

2. Let us now show that the breakdown point can be no larger than the announced value. For this purpose, consider corrupted samples in which only $n - [n/2] + p - 2$ of the good observations are retained. Start by taking

$p-1$ of the good observations, which determine a $(p-1)$-dimensional subspace L of E. Now construct any nonvertical hyperplane H' through L, which determines some θ' through the equation $y = x\theta'$. If all of the "bad" observations are put on H', then X' has a total of $([n/2]-p+2)+(p-1) = [n/2]+1$ points that satisfy $y_i' = x_i'\theta'$ exactly; so the median squared residual of X' with respect to θ' is zero, hence θ' satisfies (1.8) for X'. By choosing H' steeper and steeper, one can make $\|\theta' - \theta\|$ as large as one wants.

Proof of Corollary 1

There exists some θ such that at least $n-[n/2]+p-1$ of the observations lie on the hyperplane H given by the equation $y = x\theta$. Then θ is a solution of (1.8), because $\text{med}_i r_i^2(\theta) = 0$. Suppose that there is another solution $\theta' \neq \theta$, corresponding to a hyperplane $H' \neq H$ and yielding residuals $r_i(\theta')$. As in the proof of Theorem 1, $(H \cap H')$ has dimension $p-1$ and thus contains at most $p-1$ observations. For all remaining observations in H, it holds that $r_i^2(\theta') > 0$ and there are at least $n-[n/2]$ of them. Therefore $\text{med}_i\, r_i^2(\theta') > 0$, so θ' cannot be a solution.

Remark 1 of Section 2

Let us now show that any regression equivariant estimator T satisfies

$$\varepsilon^*(T, X) \leq ([(n-p)/2]+1)/n$$

at all samples $X = \{(x_i, y_i); i = 1, \ldots, n\}$. Suppose that the breakdown point is strictly larger than this value. This would mean that there exists finite constant b such that $T(X')$ lies in the ball $B(T(X), b)$ for all samples X' containing at least $m{:}\, n-[(n-p)/2]-1$ points of X. Here $B(T(X), b)$ is defined as the set of all θ for which $\|T(X) - \theta\| \leq b$. Now construct a p-dimensional column vector $v \neq 0$ such that $x_1 v = 0, \ldots, x_{p-1} v = 0$. Inspection shows that $2m-(p-1) \leq n$. Therefore the first $2m-(p-1)$ points of X can be replaced by

$$(x_1, y_1), \ldots, (x_{p-1}, y_{p-1}), (x_p, y_p), \ldots, (x_m, y_m),$$
$$(x_p, y_p + x_p \lambda v), \ldots, (x_m, y_m + x_m \lambda v)$$

for any $\lambda > 0$. For this new sample X', the estimate $T(X')$ belongs to $B(T(X), b)$. But looking at X' in another way reveals that $T(X')$ can also be written as $T(X'') + \lambda v$, where $T(X'')$ is in $B(T(X), b)$, hence $T(X')$ also belongs to $B(T(X) + \lambda v, b)$. This is a contradiction, however, because the intersection of $B(T(X), b)$ and $B(T(X) + \lambda v, b)$ is empty for large enough values of λ.

Proof of Theorem 2

Case A. First suppose that n is odd, with $n = 2k - 1$. Then $\text{med}(r_i^2)$ is reached by the kth square. Therefore at least one of the points $T + m_T$ or $T - m_T$ is an observation; without loss of generality, suppose that $T + m_T$ is and $T - m_T$ is not. There is a partition of the r_i^2 into $k - 1$ squares $\geq m_T^2$, 1 square $= m_T^2$, and $k - 1$ squares $\leq m_T^2$. Now take the smallest observation y_j that is larger that $T - m_T$ (choose one if there are several), and define

$$T' = \tfrac{1}{2}((T + m_T) + y_j)$$

and

$$m^2 = (\tfrac{1}{2}|(T + m_T) - y_j|)^2 < m_T^2.$$

Then there is a partition of the $(r_i')^2 = (y_i - T')^2$ into $k - 1$ squares $\geq m^2$ (corresponding to the same points as before), $k - 2$ squares $\leq m^2$ (corresponding to the same points as before, except y_j), and 2 squares $= m^2$. Finally $\text{med}(r_i')^2 = m^2 < m_T^2$, a contradiction.

Case B. Suppose that n is even, with $n = 2k$. If the ordered squares are denoted by $r_{(1)}^2 \leq \cdots \leq r_{(n)}^2$, then

$$m_T^2 = \tfrac{1}{2}(r_{(k)}^2 + r_{(k+1)}^2).$$

There is a partition of the squared residuals into $k - 1$ squares $\leq r_{(k)}^2$, $r_{(k)}^2$ itself, $r_{(k+1)}^2$ itself, and $k - 1$ squares $\geq r_{k+1}^2$. If $T + m_T$ is an observation and $T - m_T$ is not (or conversely), we can repeat the reasoning of case A. Now suppose that neither $T + m_T$ nor $T - m_T$ is an observation, which implies that $r_{(k)}^2 < r_{(k+1)}^2$ because otherwise $r_{(k)}^2 = m_T^2 = r_{(k+1)}^2$. Therefore, at least $r_{(k+1)}^2 > 0$.

Case B.1. Assume that $r_{(k)}^2 = 0$. In that case T coincides with exactly k observations, the nearest other observation (call it y_d) being at a distance $|r_{(k+1)}|$. Putting $T' = \frac{1}{2}(T + y_d)$, however, we find

$$\begin{aligned} \text{med}(y_i - T')^2 &= \tfrac{1}{2}((\tfrac{1}{2}r_{(k+1)})^2 + (\tfrac{1}{2}r_{(k+1)})^2) \\ &= \tfrac{1}{4}r_{(k+1)}^2 < \tfrac{1}{2}r_{(k+1)}^2 = m_T^2, \end{aligned}$$

a contradiction.

Case B.2. Assume that $r_{(k)}^2 > 0$. Denote by y_j some observation corresponding to $r_{(k)}^2$ and by y_d some observation corresponding to $r_{(k+1)}^2$. If the observations leading to $r_{(k)}^2$ and $r_{(k+1)}^2$ are all larger than T or all smaller than T, one can again repeat the reasoning of case A. Therefore, one may assume without loss of generality that $y_j < T < y_d$. Putting $T' = \frac{1}{2}(y_j + y_d)$, we find

$$\operatorname{med}(y_i - T')^2 = \tfrac{1}{2}((y_j - T')^2 + (y_d - T')^2)$$
$$< \tfrac{1}{2}((y_j - T)^2 + (y_d - T)^2)$$

because the function $g(t) = (a - t)^2 + (b - t)^2$ attains its unique minimum at $t = \frac{1}{2}(a + b)$.

Asymptotic Behavior of the LMS Estimator

Suppose that the observations $y_1, \ldots, y_n$ are iid according to $F(y - \theta)$, where F is a symmetric and strongly unimodal distribution with density f. Then the distribution of the LMS estimator T_n converges weakly as follows:

$$\mathscr{L}(n^{1/3}(T_n - \theta)) \to \mathscr{L}(A\tau/f(F^{-1}(.75))).$$

Here $A = (\frac{1}{2}\Lambda^2(F^{-1}(.75)))^{-1/3}$, where $\Lambda = -f'/f$ corresponds to the maximum likelihood scores, and τ is the random time s for which $s^2 + Z(s)$ attains its minimum, where $Z(s)$ is a standard Brownian motion. This result is obtained by repeating parts 1, 2, and 3 of the heuristic reasoning of Andrews et al. (1972, p. 51), putting $\alpha = .25$, which yields the constant A. The remaining part of the calculation is slightly adapted, using the same notation. If $\theta = 0$ and $\hat{t}$ is the minimizing value of t, then the main asymptotic variability of T_n is given by

$$\tfrac{1}{2}(F_n^{-1}((\tfrac{1}{2} + \hat{t}) + .25) + F_n^{-1}((\tfrac{1}{2} + \hat{t} - .25))$$
$$\simeq \hat{t}(F^{-1})'(\tfrac{1}{2} + .25) = \hat{t}/f(F^{-1}(.75)),$$

where $n^{1/3}\hat{t}$ behaves asymptotically like $A\tau$.

References

Andrews, D.F. (1974). A robust method for multiple linear regression. *Technometrics*, **16**, 523–531.

Andrews, D.F., Bickel, P.J., Hampel, F.R., Huber, P.J., Rogers, W.H., and Tukey, J.W. (1972). *Robust Estimates of Location: Survey and Advances*. Princeton University Press, Princeton, N.J.

Beaton, A.E. and Tukey, J.W. (1974). The fitting of power series, meaning polynomials, illustrated on band-spectroscopic data. *Technometrics*, **16**, 147–158.

Bickel, P.J. (1975). One-step Huber estimates in the linear model. *J. Amer. Statist. Assoc.*, **70**, 428–434.

Brown, G.W. and Mood, A.M. (1951). On median tests for linear hypotheses. *Proceedings of the 2nd Berkeley Symposium on Mathematical Statistics and Probability*, pp. 159–166.

Donoho, D.L. (1982). Breakdown properties of multivariate location estimators. Qualifying paper, Harvard University, Statistics Department.

Donoho, D.L. and Huber, P.J. (1983). The notion of breakdown point. In *A Festschrift for Erich L. Lehmann* (P. Bickel, K. Doksum, and J.L. Hodges, Jr., eds.). Wadsworth, Belmont, CA.

Draper, N.R. and Smith, H. (1966). *Applied Regression Analysis*. Wiley, New York.

Edgeworth, F.Y. (1887). On observations relating to several quantities. *Hermathena*, **6**, 279–285.

Friedman, J.H. and Tukey, J.W. (1974). A projection pursuit algorithm for exploratory data analysis *IEEE. Transactions Comput.* **C-23**, 881–889.

Hampel, F.R. (1971). A general qualitative definition of robustness. *Ann. Math. Statist.*, **42**, 1887–1896.

Hampel, F.R. (1975). Beyond location parameters: Robust concepts and methods. *Bull. Internat. Statist. Instit.* **46**, 375–382.

Hampel, F.R. (1978). Optimally bounding the gross-error-sensitivity and the influence of position in factor space. *Proceedings of the Statistical Computing Section, American Statistical Association*, pp. 59–64.

Hampel, F.R., Rousseeuw, P.J., and Ronchetti, E. (1981). The change-of-variance curve and optimal redescending M-estimators. *J. Amer. Statist. Assoc.*, **76**, 643–648.

Harter, H.L. (1977). Nonuniqueness of least absolute values regression. *Commun. Statist.*, **A6**, 829–838.

Hill, R.W. (1977). Robust regression when there are outliers in the carriers. Unpublished Ph.D. dissertation, Harvard University.

Hodges, J.L., Jr. (1967). Efficiency in normal samples and tolerance of extreme values for some estimates of location. *Proceedings of the 5th Berkeley Symposium on Mathematical Statistics and Probability*, Vol. 1, pp. 163–168.

Huber, P.J. (1973). Robust regression: Asymptotics, conjectures and Monte Carlo. *Ann. Statist.* **1**, 799–821.

Huber, P.J. (1974). Early cuneiform evidence for the planet Venus. Paper presented at the Annual Meeting of the American Association for the Advancement of Science, Feb. 25, San Francisco.

Huber, P.J. (1981). *Robust Statistics*, Wiley, New York.

Jaeckel, L.A. (1972). Estimating regression coefficients by minimizing the dispersion of residuals. *Ann. Math. Statist.*, **5**, 1449–1458.

Krasker, W.S. (1980). Estimation in linear regression models with disparate data points. *Econometrica*, **48**, 1333–1346.

Krasker, W.S. and Welsch, R.E. (1982). Efficient bounded-influence regression estimation. *J. Amer. Statist. Assoc.* **77**, 595–604.

Leroy, A. and Rousseeuw, P. (1984). PROGRES: A program for robust regression. Research Report No. 201, University of Brussels, Centrum Voor Statistiek en Operationeel Onderzoek.

Mallows, C.L. (1975). On some topics in robustness. Unpublished memorandum, Bell Telephone Laboratories, Murray Hill, NJ.

Marazzi, A. (1980). Robust linear regression programs in ROBETH. Research Report No. 23, Fachgruppe für Statistik, ETH Zürich.

Maronna, R.A., Bustos, O., and Yohai, V. (1979). Bias- and efficiency-robustness of general M-estimators for regression with random carriers. In *Smoothing Techniques for Curve Estimation* (T. Gasser and M. Rosenblatt, eds.). Springer-Verlag, New York, pp. 91–116.

Oldford, R.W. (1983). A note on high breakdown regression estimators. Technical Report, Massachusetts Institute of Technology, Center for Computational Research in Economics and Management Science.

Rousseeuw, P.J. (in press). Multivariate estimation with high breakdown point. *Proceedings of the Fourth Pannonian Symposium on Mathematical Statistics and Probability*. Bad Tatzmannsdorf, Austria, September 4–9, 1983.

Sen, P.K. (1968). Estimates of the regression coefficient based on Kendall's tau. *J. Amer. Statist. Assoc.*, **63**, 1379–1389.

Siegel, A.F. (1982). Robust regression using repeated medians. *Biometrika*, **69**, 242–244.

Siegel, A.F. and Benson, R.H. (1982). A robust comparison of biological shapes. *Biometrics*, **38**, 341–350.

Stahel, W.A. (1981). Robuste Schätzungen: Infinitesimale Optimalität und Schätzungen von Kovarianzmatrizen. Unpublished Ph.D. dissertation, ETH Zürich, Statistics Deptartment.

Stigler, S.M. (1981). Gauss and the invention of least squares. *Ann. Statist.*, **9**, 465–474.

Theil, H. (1950). A rank-invariant method of linear and polynomial regression analysis (Parts 1–3). *Nederl. Akad. Wetensch. Proc. Ser. A*, **53**, 386–392; 521–525; 1397–1412.

Introduction to Liang and Zeger (1986) Longitudinal Data Analysis Using Generalized Linear Models

Peter J. Diggle
Department of Mathematics and Statistics,
Lancaster University

1. About the Authors

Kung-Yee Liang and Scott Zeger have been colleagues in the Department of Biostatistics, Johns Hopkins University, since 1982. They received the Snedecor Award in 1987 for their methodological work on longitudinal data analysis.

Kung-Yee Liang received his first degree, in mathematics, from the Tsin Hua University, Taiwan, in 1973. He subsequently studied in the United States, receiving his MS in Statistics from the University of South Carolina in 1979, and his Ph.D. in biostatistics at the University of Washington, Seattle, in 1982. In Seattle he worked on the asymptotic equivalence of conditional and unconditional inference procedures under the guidance of Norman Breslow.

Scott Zeger first graduated in biology, at the University of Pennsylvania in 1974. He later studied for an M.S. in mathematics at Drexel University, awarded in 1978, and for a Ph.D. in statistics at Princeton University, awarded in 1982, where his Ph.D. adviser was Peter Bloomfield. His thesis topic concerned time series analysis with applications in atmospheric science.

As members of the Biostatistics Department at Johns Hopkins, Liang and Zeger were exposed to a wide range of substantive problems in public health sciences, many of which hinged on the analysis of longitudinal data. It became clear that their previously disparate research backgrounds, in inference and in time series, respectively, were both highly relevant to the development of appropriate methodology for dealing with such data. This led to a fruitful, and continuing, period of collaboration betwen them. Their 1986 *Biometrika* paper exemplifies the dual role which theory and practice can play in the development of good statistical methodology.

2. The Scientific Background to the Paper

The phrase *longitudinal data* is usually taken to mean a set of data in which a time-ordered sequence of measurements has been taken on each of a number of experimental units, or *subjects*. Formally, a dataset of this kind can be described by a pair of arrays, $(Y_{ij}, t_{ij}): j = 1, \ldots, n_i; i = 1, \ldots, m$, say, in which Y_{ij} denotes the jth of n_i measurements of a stochastic response variable taken on the ith of m subjects, and t_{ij} the time at which Y_{ij} is taken. In this setting, different measurements on the same subject are typically dependent, but measurements on different subjects can often be assumed to be independent. In many applications, a primary objective is to make inferences about the set of mean responses, $\mu_{ij} = \mathrm{E}(Y_{ij})$, which may depend on time, on treatment allocation or on a number of other covariates. In general, explanatory variables in this context may be attached either to individual subjects, to individual times, or to individual responses. Examples of these three kinds of explanatory variable are, respectively: allocation to one or other treatment arm in a clinical trial; time since start of treatment; alcohol intake on the preceding day, when the response variable is a measurement of blood pressure.

Depending on one's point of view, it is possible to approach longitudinal data from a range of perspectives: hierarchical analysis of variance (with variation entering both between and within subjects); multivariate analysis (the sequence of measurements on a single subject defining a random vector); time series (with replication between subjects).

The simplest expression of the hierarchical analysis of variance perspective is to make a direct analogy with the split-plot experiment in the agricultural sciences. A model-based description on this is that

$$Y_{ij} = \mu_{ij} + U_i + Z_{ij}, \tag{1}$$

where the $U_i \sim \mathrm{N}(0, v^2)$ represent independent random variation between subjects (plots) and the $Z_{ij} \sim \mathrm{N}(0, \tau^2)$ independent random variation between times (subplots) within subjects. The form of the underlying model is important because, in contrast to the agricultural context in which the split-plot analysis was first proposed, there is no randomization justification for the analysis—time cannot be randomized. Despite this potentially serious defect, several standard textbooks on applied statistics refer to the analysis based on (1) as "*the* repeated measures analysis of variance" (e.g., Winer (1977), but my italics).

The multivariate perspective led to the development of multivariate Gaussian models for the vector of observations on a subject. Given sufficiently many subjects, and sufficiently few observations per subject, a classical multivariate analysis of variance can be used, as in Cole and Grizzle (1966). However, efficiency gains can be achieved by making structural assumptions about the covariance matrix, for example, by incorporating random subject-effects of a more general kind than the simple scalar in (1). Perhaps the culmination of this perspective lies in a series of papers by C.R. Rao on what has come to be called the "growth curve model." See, for example, Rao (1965).

The time series perspective suggests different ways of modeling the covariance matrix; for example, by assuming stationarity, the number of parameters in an n by n covariance matrix is immediately reduced from $\frac{1}{2}n(n+1)$ to n, and further reductions to perhaps two or three parameters can be achieved at the expense of parametric assumptions about the serial correlation in the data.

In practice, a very important distinction between longitudinal data analysis and classical time series analysis is that the latter concentrates almost entirely on the analysis of long, single series, whereas typical longitudinal datasets consist of many short series. Furthermore, in many longitudinal studies the covariance structure of the data is not of direct interest, but only indirectly relevant because of its impact on inferences about the mean response, whereas in classical time-series analysis the covariance structure is usually the primary focus. This second consideration makes it less important in the longitudinal setting to model the covariance structure for its own sake, than is typically the case in classical time series application. Also, the replication in a longitudinal study allows the statistician to devise methods for making inferences about the mean response which are both reasonably efficient and somewhat robust to mis-specification of the covariance structure.

An important paper which represents something of a synthesis of these three perspectives within the framework of the Gaussian linear model was Laird and Ware (1982). They proposed a structured covariance matrix which combined random effects to describe the variation between subjects with possibly serially correlated variation within subjects. Their model could therefore be written as

$$Y_{ij} = d'_{ij}U_i + Z_{ij}, \tag{2}$$

where now the U_i are mutually independent multivariate Gaussian vectors of zero-mean random effects, the d_i are corresponding vectors of explanatory variables, and the $Z_{ij}: j = 1, \dots, n$, are possibly correlated sequences of Gaussian random variables, realized independently between subjects. Most users of the Laird–Ware model (2) have concentrated on its random effects aspect, specializing to mutually independent Z_{ij}. In particular, a very common use of the model is in growth studies, using $d_i = (1, t_{ij})$ to specify random intercepts and slopes for the different subjects, together with mutually independent Z_{ij}. Later authors proposed particular classes of models with serially correlated Z-sequences. For example, Diggle (1988) proposed the model

$$Y_{ij} = \mu_{ij} + U_i + W_i(t_{ij}) + Z_{ij}, \tag{3}$$

where $U_i \sim N(0, \nu^2)$, $Z_{ij} \sim N(0, \tau^2)$, and $W_i(t)$ is a stationary Gaussian process with variance σ^2 and correlation function $\rho(u) = \exp(-\alpha u^\kappa)$.

In all of these approaches, the overwhelming majority of applications assume a general linear model for the specification of the mean responses, μ_{ij}. Many also assume a common set of equally spaced observation times $t_{ij} = j$ for all subjects, although Diggle explicitly defined his model (3) in continuous time to accommodate irregularly spaced time-sequences for which standard time-series models formulated in discrete time (e.g., the ARMA class developed systematically in Box and Jenkins (1970) are not appropriate.

At the time that Liang and Zeger did the work for their 1986 paper, the Laird–Ware methodology had therefore provided essentially all the basic requirements for the analysis of longitudinal data under Gaussian assumptions. However, Liang and Zeger were concerned with longitudinal studies in which the response measurement was a count, or even a binary presence/absence of some primary health indicator. For datasets of this kind, no comparably general methodology existed although papers covering specific cases were beginning to appear [e.g., Stiratelli, Laird, and Ware (1984)].

3. The Contribution of the Paper

In the context of mutually independent responses, the analysis of non-Gaussian data had been profoundly influenced by Nelder and Wedderburn's (1972) embedding of the Gaussian linear model within the wider framework of generalized linear models. As Liang and Zeger make clear from their own summary, the primary motivation of their paper was to achieve a similar embedding of linear Gaussian regression models for correlated data within a comparably general framework. They recognized, but perhaps did not emphasize sufficiently, that in this wider setting it would be necessary to distinguish between population-averaged and subject-specific regression parameters. Suppose, for example, that in the model (1), the mean response is a linear function of t, so that

$$Y_{ij} = \alpha + \beta t_{ij} + U_i + Z_{ij}. \tag{4}$$

Then, the regression parameter β measures the increase in mean response per unit time, both for the population at large and for any individual subject, whatever the value of $\nu^2 = \mathrm{Var}(U_i)$. However, in a random-intercept logistic model, say

$$\log\{p_{ij}/(1 - p_{ij})\} = \alpha + \beta t_{ij} + U_i, \tag{5}$$

where $p_{ij} = P(Y_{ij} = 1 | U_i)$ and $U_i \sim N(0, \nu^2)$ and before, then the population averaged mean response is determined not only by α and β but also by the random effect variance, ν^2.

Liang and Zeger argue that in many longitudinal studies, a primary aim is to understand how explanatory variables affect the mean response. They therefore take this as their primary objective, adopt a generalized linear model for the mean response, and then address the question of how to make inferences about this mean response. Acknowledging that no sufficiently rich class of non-Gaussian distributions is available, they look for ways of obtaining approximate inferences under minimal assumptions. This leads them to propose a form of quasi-likelihood inference (Wedderburn, 1974; McCullagh, 1983) but for correlated data. They derive estimating equations (now widely known as GEE, an acronym for *generalized estimating equations*) for their regression parameters under working assumptions about the correlation structure, establish asymptotic Normality for the resulting estimators, and show how the asymptotic covariance matrix of the estimators can be estimated consistently, using the empirical correlation structure of the Pearson residuals from the fitted model. Liang and Zeger also suggest that the working correlation matrix may also include one or more parameters which can be estimated from the data. Their method of calculating the asymptotic covariance matrix of the regression parameters is usually called the *information sandwich*, as it takes the form $\mathrm{Var}(\hat{\beta}) = I_0^{-1} I_1 I_0^{-1}$, where I_0 and I_1 correspond to "true" and "working" information matrices, respectively. The information sandwich earlier appeared as a corollary to Theorem 3 in Huber (1967), and was discussed in some detail in White (1982).

A simple special case of GEE arises when the working correlation matrix is the identity matrix. The analysis then consists of ignoring the longitudinal nature of the data altogether to obtain point estimates of the parameters of interest, but then estimating the (unstructured) true correlation matrix in order to obtain approximately correct inferences. Liang and Zeger emphasize that whilst the resulting estimators ofen have good efficiency, the notional standard errors implied by the identity working correlation matrix can be seriously wrong—and usually are!

The GEE methodology has since become very widely used, not only for the analysis of longitudinal data but also for other forms of "clustered" data in which response measurements occur in correlated sets, or *clusters*. The attractive features of the GEE methodology include: generality—continuous, discrete, and binary responses are handled in a unified way, as with classical generalized linear models; simplicity—much of the analysis can be carried out using standard regression packages, with only the information sandwich requiring special-purpose software, and the approach feels reassuringly familiar to nonspecialists in longitudinal methods; robustness—validation of an assumed covariance structure is notoriously difficult with small-to-moderate datasets, and the availability of valid, albeit approximate, inferences are obtained under mis-specification of the correlation structure is reassuring.

4. Limitations of the GEE Methodology

As Liang and Zeger took great care to emphasize, the robustness properties of GEE are obtained in the limit as the number of subjects tends to infinity, and in practice reliability of the estimated standard errors requires that the data consist of many, short series. Not all subsequent users of GEE have followed this advice!

The GEE methodology assumes that mean response profiles, or population-averaged parameters, are the focus. Whether or not this is so should depend on the practical context. As a simple example, in an investigation into the effects of air quality on asthma incidence, a public health officer's primary concern might be to know by how much the proportion of asthmatic children in the population might be reduced by a given reduction in air pollution, whereas a parent may be more concerned to know the reduction in the probability that their child would suffer from asthma. Chapter 7 of Diggle, Liang, and Zeger (1994) discusses this issue in more detail.

The GEE methodology is unable to deal with missing values in the data unless these are completely random in the sense of Little and Rubin (1987). Recent work by Robins, Rotnitzky, and Zhao (1995) develops an extension of GEE which is valid under the weaker assumption of data missing at random.

Finally, and depending on one's philosophical stance with regard to statistical modeling and inference, the absence of an explicit underlying probability model may be discomforting. In this connection, Crowder (1995) notes that when estimating a parameter, α say, in a working correlation matrix, difficulties can arise because of the uncertain status of α when the adopted form for the working correlation matrix is different from the true correlation structure.

Acknowledgment

I thank Mary Joy Argo for helpful background information, and the Department of Statistics, University of Adelaide, for their support through an Australian Research Council grant.

References

Box, G.P. and Jenkins, G.M. (1970). *Time Series Analysis —Forecasting and Control*, rev. ed. Holden-Day, San Francisco, CA.

Cole, J.W.L. and Grizzle, J.E. (1966). Applications of multivariate analysis of variance to repeated measurements experiments. *Biometrics*, **22**, 810–828.

Crowder, M. (1995). On the use of a working correlation matrix in using generalised linear models for repeated measures. *Biometrika*, **82**, 407–410.

Diggle, P.J. (1988). An approach to the analysis of repeated measures. *Biometrics*, **44**, 959–971.

Diggle, P.J., Liang K.Y., and Zeger, S.L. (1994). *Analysis of Longitudinal Data*. Oxford University Press, Oxford.

Huber, P.J. (1967). The behaviour of maximum likelihood estimates under nonstandard conditions. *Proceedings of the Fifth Berkeley Symposium on Mathematical Statistics and Probability*, **1**, 221–233.

Laird, N.M. and Ware, J.H. (1982). Random-effects models for longitudinal data. *Biometrics*, **38**, 963–974.

Little, R.J. and Rubin, D.B. (1987). *Statistical Analysis with Missing Data*. Wiley, New York.

McCullagh, P. (1983). Quasi-likelihood functions. *Ann. Statist.*, **11**, 59–67.

Nelder, J.A. and Wedderburn, R.W.M. (1972). Generalized linear models. *J. Roy. Statist. Soc. Ser. A*, **135**, 370–384.

Rao, C.R. (1965). The theory of least squares when the parameters are stochastic and its application to the analysis of growth curves. *Biometrika*, **52**, 447–458.

Robins, J.M., Rotnitzky, A., and Zhao, L.P. (1995). Analysis of semiparametric regression models for repeated outcomes in the presence of missing data. *J. Amer. Statist. Assoc.*, **90**, 106–121.

Stiratelli, R., Laird, N., and Ware, J.H. (1984). Random effects models for serial observations with binary responses. *Biometrics*, **40**, 961–971.

Wedderburn, R.W.M. (1974). Quasi-likelihood functions, generalized linear models and the Gaussian method. *Biometrika*, **61**, 439–447.

White, H. (1982). Maximum likelihood estimation of misspecified models. *Econometrica*, **50**, 1–25.

Winer, B.J. (1977). *Statistical Principles in Experimental Design*, 2nd edn. McGraw-Hill, New York.

Longitudinal Data Analysis Using Generalized Linear Models

Kung-Yee Liang and Scott L. Zeger
Johns Hopkins University

Summary

This paper proposes an extension of generalized linear models to the analysis of longitudinal data. We introduce a class of estimating equations that give consistent estimates of the regression parameters and of their variance under mild assumptions about the time dependence. The estimating equations are derived without specifying the joint distribution of a subject's observations yet they reduce to the score equations for multivariate Gaussian outcomes. Asymptotic theory is presented for the general class of estimators. Specific cases in which we assume independence, m-dependence and exchangeable correlation structures from each subject are discussed. Efficiency of the proposed estimators in two simple situations is considered. The approach is closely related to quasi-likelihood.

1. Introduction

Longitudinal data sets, comprised of an outcome variable, y_{it}, and a $p \times 1$ vector of covariates, x_{it}, observed at times $t = 1, \ldots, n_i$ for subjects $i = 1, \ldots, K$ arise often in applied sciences. Typically the scientific interest is either in the pattern of change over time, e.g. growth, of the outcome measures or more simply in the dependence of the outcome on the covariates. In the latter case, the time dependence among repeated measurements for a subject is a nuisance. For example, the severity of respiratory disease along with the nutritional status, age, sex and family income of children might be observed once every three months for an 18 month period. The dependence of the outcome variable, severity of disease, on the covariates is of interest.

With a single observation for each subject ($n_i = 1$), a generalized linear model (McCullagh and Nelder, 1983) can be applied to obtain such a description for a variety of continuous or discrete outcome variables. With repeated observations, however, the correlation among values for a given subject must be taken into account. This paper presents an extension of generalized linear models to the analysis of longitudinal data when regression is the primary focus.

When the outcome variable is approximately Gaussian, statistical methods for longitudinal data are well developed, e.g. Laird and Ware (1982) and Ware (1985). For non-Gaussian outcomes, however, less development has taken place. For binary data, repeated measures models in which observations for a subject are assumed to have exchangeable correlations have been proposed by Ochi and Prentice (1984) using a probit link, by Stiratelli, Laird and Ware (1984) using a logit link and by Koch et al. (1977) using log linear models. Only the model proposed by Stiratelli, Laird and Ware allows for time-dependent covariates. Zeger, Liang and Self (1985) have proposed a first-order Markov chain model for binary longitudinal data which, also, however, requires time independent covariates. One difficulty with the analysis of non-Gaussian longitudinal data is the lack of a rich class of models such as the multivariate Gaussian for the joint distribution of y_{it} $(t = 1, \dots, n_i)$. Hence likelihood methods have not been available except in the few cases mentioned above.

The approach in this paper is to use a working generalized linear model for the marginal distribution of y_{it}. We do not specify a form for the joint distribution of the repeated measurements. Instead, we introduce estimating equations that given consistent estimates of the regression parameters and of their variances under weak assumptions about the joint distribution. We model the marginal rather than the conditional distribution given previous observations although the conditional approach may be more appropriate for some problems. The methods we propose reduce to maximum likelihood when the y_{it} are multivariate Gaussian.

The estimating equations introduced here are similar to those described by Jorgensen (1983) and by Morton (1981). However our problem differs from the one considered by Jorgensen in that the correlation parameters do not appear in the estimating equations in an additive way: it is different than the problem considered by Morton in that pivots cannot be used to remove the nuisance correlation parameters.

To establish notation, we let $Y_i = (y_{i1}, \dots, y_{in_i})^{\mathrm{T}}$ be the $n_i \times 1$ vector of outcome values and $X_i = (x_{i1}, \dots, x_{in_i})^{\mathrm{T}}$ be the $n_i \times p$ matrix of covariate values for the ith subject $(i = 1, \dots, K)$. We assume that the marginal density of y_{it} is

$$f(y_{it}) = \exp[\{y_{it}\theta_{it} - a(\theta_{it}) + b(y_{it})\}\phi], \tag{1}$$

where $\theta_{it} = h(\eta_{it})$, $\eta_{it} = x_{it}\beta$. By this formulation, the first two moments of y_{it} are given by

$$E(y_{it}) = a'(\theta_{it}), \qquad \text{var}(y_{it}) = a''(\theta_{it})/\phi. \tag{2}$$

When convenient to simplify notation, we let $n_i = n$ without loss of generality.

Section 2 presents the "independence" estimating equation which arises by adopting the working assumption that repeated observations for a subject are independent. It leads to consistent estimates of β and of its variance given only that the regression model for $E(y)$ is correctly specified. Section 3 introduces and presents asymptotic theory for the "generalized" estimating equation in which we borrow strength across subjects to estimate a "working" correlation matrix and hence explicitly account for the time dependence to achieve greater asymptotic efficiency. In §4, examples of specific models to be used in the analysis of longitudinal data are given. Section 5 considers questions of efficiency. The final section discusses several issues concerning the use of these estimating procedures.

2. Independence Estimating Equations

In this section, we present an estimator, $\hat{\beta}_I$, of β which arises under the working assumption that repeated observations from a subject are independent of one another. Under the independence working assumption, the score equations from a likelihood analysis have the form

$$U_I(\beta) = \sum_{i=1}^{K} X_i^T \Delta_i S_i = 0, \tag{3}$$

where $\Delta_i = \text{diag}(d\theta_{it}/d\eta_{it})$ is an $n \times n$ matrix and $S_i = Y_i - a_i'(\theta)$ is of order $n \times 1$ for the ith subject. The estimator $\hat{\beta}_I$ is defined as the solution of equation (3).

Define for each i the $n \times n$ diagonal matrix $A_i = \text{diag}\{a''(\theta_{it})\}$. Under mild regularity conditions we have the following theorem.

Theorem 1. *The estimator $\hat{\beta}_I$ of β is consistent and $K^{1/2}(\hat{\beta}_I - \beta)$ is asymptotically multivariate Gaussian as $K \to \infty$ with zero mean and covariance matrix V_1 given by*

$$V_I = \lim_{K\to\infty} K\left(\sum_{i=1}^{K} X_i^{\mathrm{T}}\Delta_i A_i \Delta_i X_i\right)^{-1}\left(\sum_{i=1}^{K} X_i^{\mathrm{T}}\Delta_i \,\text{cov}(Y_i)\Delta_i X_i\right)$$

$$\times \left(\sum_{i=1}^{K} X_i^{\mathrm{T}} \Delta_i A_i \Delta_i X_i \right)^{-1}$$

$$= \lim_{K \to \infty} K\{H_1(\beta)\}^{-1} H_2(\beta) \{H_1(\beta)\}^{-1}, \quad (4)$$

where the moment calculations for the Y_i's are taken with respect to the true underlying model.

The proof of the theorem is straightforward and is omitted. The variance of $\hat{\beta}_I$ given in Theorem 1 can be consistently estimated by

$$\{H_1(\hat{\beta}_I)\}^{-1} \left(\left[\sum_{i=1}^{K} X_i^{\mathrm{T}} \Delta_i S_i S_i^{\mathrm{T}} \Delta_i X_i \right]_{\hat{\beta}_I} \right) \{H_1(\hat{\beta}_I)\}^{-1}.$$

Note that the estimation of ϕ is unnecessary for estimating V_I even though the latter is a function of ϕ.

The estimator $\hat{\beta}_I$ has several advantages. It is easy to compute with existing software, e.g. GLIM (Baker and Nelder, 1978). Both $\hat{\beta}_I$ and $\mathrm{var}(\hat{\beta}_I)$ are consistent given only a correct specification of the regression which is the principal interest. Note that this requires missing data to be missing completely at random in the sense of Rubin (1976). As discussed in §5. $\hat{\beta}_I$ can be shown to be reasonably efficient for a few simple designs. The principal disadvantage of $\hat{\beta}_I$ is that it may not have high efficiency in cases where the autocorrelation is large. The next section proposes a "generalized" estimating equation that leads to estimators with higher efficiency.

3. Generalized Estimating Equations

3.1. General

In this section, we present a class of estimating equations which take the correlation into account to increase efficiency. The resulting estimators of β remain consistent. In addition, consistent variance estimates are available under the weak assumption that a weighted average of the estimated correlation matrices converges to a fixed matrix.

To begin, let $R(\alpha)$ be a $n \times n$ symmetric matrix which fulfills the requirement of being a correlation matrix, and let α be an $s \times 1$ vector which fully characterizes $R(\alpha)$. We refer to $R(\alpha)$ as a "working" correlation matrix.

Define

$$V_i = A_i^{1/2} R(\alpha) A_i^{1/2} / \phi, \quad (5)$$

which will be equal to cov (Y_i) if $R(\alpha)$ is indeed the true correlation matrix for the Y_i's.

We define the general estimating equations to be

$$\sum_{i=1}^{K} D_i^{\mathrm{T}} V_i^{-1} S_i = 0, \tag{6}$$

where $D_i = d\{a_i'(\theta)\}/d\beta = A_i \delta_i X_i$. Two remarks are worth mentioning. First, equation (6) reduces to the independence equations in §2 if we specify $R(\alpha)$ as the identity matrix. Second, for each i, $U_i(\beta, \alpha) = D_i^{\mathrm{T}} V_i^{-1} S_i$ is similar to the function derived from the quasilikelihood approach advocated by Wedderburn (1974) and McCullagh (1983) except that the V_i's here are not only a function of β but of α as well. Equation (6) can be reexpressed as a function of β alone by first replacing α in (5) and (6) by $\hat{\alpha}(y, \beta, \phi)$, a $K^{1/2}$-consistent estimator of α when β and ϕ are known, that is $\hat{\alpha}$ for which $K^{1/2}(\hat{\alpha} - \alpha) = O_p(1)$. Except for particular choices of R and $\hat{\alpha}$, the scale parameter ϕ will generally remain in (6). To complete the process, we replace ϕ by $\hat{\phi}(Y, \beta)$, a $K^{1/2}$-consistent estimator when β is known. Consequently, (6) has the form

$$\sum_{i=1}^{K} U_i[\beta, \hat{\alpha}\{\beta, \hat{\phi}(\beta)\}] = 0. \tag{7}$$

and $\hat{\beta}_G$ is defined to be the solution of equation (7). The next theorem states the large-sample property for $\hat{\beta}_G$.

Theorem 2. *Under mild regularity conditions and given that:*

(i) *$\hat{\alpha}$ is $K^{1/2}$-consistent given β and ϕ;*
(ii) *$\hat{\phi}$ is $K^{1/2}$-consistent given β; and*
(iii) *$|\partial \hat{\alpha}(\beta, \phi)/\partial \phi| \leq H(Y, \beta)$ which is $O_p(1)$;*

then $K^{1/2}(\hat{\beta}_G - \beta)$ is asymptotically multivariate Gaussian with zero mean and covariance matrix V_G given by

$$V_G = \lim_{K \to \infty} K \left(\sum_{i=1}^{K} D_i^{\mathrm{T}} V_i^{-1} D_i \right)^{-1} \left\{ \sum_{i=1}^{K} D_i^{\mathrm{T}} V_i^{-1} \operatorname{cov}(Y_i) V_i^{-1} D_i \right\} \times \left(\sum_{i=1}^{K} D_i^{\mathrm{T}} V_i^{-1} D_i \right)^{-1}.$$

A sketch of the proof is given in the Appendix. The variance estimate $\hat{V}_G$ of $\hat{\beta}_G$ can be obtained by replacing cov (Y_i) by $S_i S_i^{\mathrm{T}}$ and β, ϕ, α by their estimates in the expression V_G. As in the independence case, the consistency of $\hat{\beta}_G$ and $\hat{V}_G$ depends only on the correct specification of the mean, not on the correct choice of R. This again requires that missing observations be missing completely at random (Rubin, 1976). Note that the asymptotic variance of $\hat{\beta}_G$ does not depend on choice of estimator for α and ϕ among those that are $K^{1/2}$-consistent. Analogous results are known for the Gaussian data case and in quasi-likelihood where the variance of the regression parameters does not depend on the choice of estimator of ϕ. In our problem, where the likelihood is not fully specified, the result follows from choosing estimating equations for β in which an individual's contribution, U_i, is a product of two terms: the first involving α but not the data, and the second independent of α and with expectation zero. Then $\sum E(\hat{c}U_{ij}\partial\alpha)$ is $o_p(K)$ and var $(\hat{\beta}_G)$ does not depend on $\hat{\alpha}$ or $\hat{\phi}$ as can be seen from the discussion in the Appendix.

3.2 Connection with the Gauss–Newton Method

To compute $\hat{\beta}_G$, we iterate between a modified Fisher scoring for β and moment estimation of α and ϕ. Given current estimates $\hat{\alpha}$ and $\hat{\phi}$ of the nuisance parameters, we suggest the following modified iterative procedure for β:

$$\hat{\beta}_{j+1} = \hat{\beta}_j - \left\{\sum_{i=1}^{K} D_i^{\mathrm{T}}(\hat{\beta}_j)\tilde{V}_i^{-1}(\hat{\beta}_j)D_j(\hat{\beta}_j)\right\}^{-1}\left\{\sum_{i=1}^{K} D_i^{\mathrm{T}}(\hat{\beta}_j)\tilde{V}_i^{-1}(\hat{\beta}_j)S_i(\hat{\beta}_j)\right\}. \tag{8}$$

where $\tilde{V}_i(\beta) = V_i[\beta, \hat{\alpha}\{\beta, \hat{\phi}(\beta)\}]$. This procedure can be viewed as a modification of Fisher's scoring method in that the limiting value of the expectation of the derivative of $\sum U_i[\beta, \hat{\alpha}\{\beta, \hat{\phi}(\beta)\}]$ is used for correction.

Now, define $D = (D_1^{\mathrm{T}}, \dots, D_K^{\mathrm{T}})^{\mathrm{T}}$, $S = (S_1^{\mathrm{T}}, \dots, S_K^{\mathrm{T}})$ and let $\tilde{V}$ be a $nK \times nK$ block diagonal matrix with $\tilde{V}_i$'s as the diagonal elements. Define the modified dependent variable

$$Z = D\beta - S,$$

and then the iterative procedure (8) for calculating $\hat{\beta}_G$ is equivalent to performing an iteratively reweighted linear regression of Z on D with weight $\tilde{V}^{-1}$.

3.3 Estimators of α and ϕ

At a given iteration the correlation parameters α and scale parameter ϕ can be estimated from the current Pearson residuals defined by

$$\hat{r}_{it} = \{y_{it} - a'(\hat{\theta}_{it})\}/\{a''(\hat{\theta}_{it})\}^{1/2},$$

where $\hat{\theta}_{it}$ depends upon the current value for β. We can estimate ϕ by

$$\hat{\phi}^{-1} = \sum_{i=1}^{K} \sum_{t=1}^{n_i} \hat{r}_{it}^2/(N-p).$$

where $N = \sum n_i$. This is the longitudinal analogue of the familiar Pearson statistic (Wedderburn, 1974; McCullagh, 1983). It is easily shown to be $K^{1/2}$-consistent given that the fourth moments of the y_{it}'s are finite. To estimate α consistently, we borrow strength over the K subjects. The specific estimator depends upon the choice of $R(\alpha)$. The general approach is to estimate α by a simple function of

$$\hat{R}_{uv} = \sum_{i=1}^{K} \hat{r}_{iu}\hat{r}_{iv}/(N-p).$$

Specific estimators are given in the next section.

Alternative estimators of ϕ such as one based upon the log likelihood described by McCullagh and Nelder (1983, p. 83) are available. Because we do not specify the entire joint distribution of Y_i, the analogous estimators for α are not available. Note, however, that the asymptotic distribution of $\hat{\beta}_G$ does not depend on the specific choice of α and ϕ among those that are $K^{1/2}$-consistent. The finite sample performance of $\hat{\beta}_G$ for a variety of α, ϕ estimators requires further study.

4. Examples

In this section several specific choices of $R(\alpha)$ are discussed. Each leads to a distinct analysis. The number of nuisance parameters and the estimator of α vary from case to case.

Example 1. Let $R(\alpha)$ be R_0, any given correlation matrix. When $R_0 = I$, the identity matrix, we obtain the independence estimating equation. However for any R_0, $\hat{\beta}_G$ and $\hat{V}_G$ will be consistent. Obviously, choosing R_0 closer to the true correlation gives increased efficiency. Note that for any specified R_0, no knowledge on ϕ is required in estimating β and var $(\hat{\beta}_G)$.

Example 2. Let $\alpha = (\alpha_1, \ldots, \alpha_{n-1})^{\mathrm{T}}$, where $\alpha_t = \mathrm{corr}(Y_{it}, Y_{i,t+1})$ for $t = 1, \ldots, n-1$. A natural estimator of α_t, given β and ϕ, is

$$\hat{\alpha}_t = \phi \sum_{i=1}^{K} \hat{r}_{it}\hat{r}_{i,t+1}/(K-p).$$

Now let $R(\alpha)$ be tridiagonal with $R_{t,t+1} = \alpha_t$. This is equivalent to the one-dependent model. An estimator of ϕ is unnecessary for calculating $\hat{\beta}_G$ and

$\hat{V}_G$ when the α_t's above are used since the ϕ which appears in the formula for $\hat{\alpha}_t$ cancels in the calculation of V_i. As a special case, we can let $s = 1$ and $\alpha_t = \alpha (t = 1, \ldots, n-1)$. Then the common α can be estimated by

$$\hat{\alpha} = \sum_{t=1}^{n-1} \hat{\alpha}_t/(n-1).$$

An extension to m-dependence is straightforward.

EXAMPLE 3. Let $s = 1$ and assume that $\mathrm{corr}(y_{it}, y_{it'}) = \alpha$ for all $t \neq t'$. This is the exchangeable correlation structure obtained from a random effects model with a random level for each subject, e.g. Laird and Ware (1982). Given ϕ, α can be estimated by

$$\hat{\alpha} = \phi \sum_{i=1}^{K} \sum_{t>t'} \hat{r}_{it}\hat{r}_{it'} \Bigg/ \left\{ \sum_{i=1}^{K} \tfrac{1}{2} n_i(n_i - 1) - p \right\}.$$

As in Examples 1 and 2, ϕ need not be estimated to obtain $\hat{\beta}_G$ and V_G. Note that an arbitrary number of observations and observation times for each subject are possible with this assumption.

EXAMPLE 4. Let $\mathrm{corr}(y_{it}, y_{it'}) = \alpha^{|t-t'|}$. For y_{it} Gaussian, this is the correlation structure of the continuous time analogue of the first-order autoregressive process. AR-1 (Feller, 1971, p. 89). Since under this model, $E(\hat{r}_{it}\hat{r}_{it'}) \simeq \alpha^{|t-t'|}$, we can estimate α by the slope from the regression of $\log(\hat{r}_{it}\hat{r}_{it'})$ on $\log(|t-t'|)$. Note that an arbitrary number and spacing of observations can be accommodated with this working model. But $\hat{\phi}$ must be calculated in the determination of $\hat{\beta}_G$ and $\hat{V}_G$.

EXAMPLE 5. Let $R(\alpha)$ be totally unspecified, that is $s = \frac{1}{2}n(n-1)$. Now R can be estimated by

$$\frac{\phi}{K} \sum_{i=1}^{K} A_i^{-1/2} S_i S_i^{\mathrm{T}} A_i^{-1/2}. \tag{9}$$

Note that for this case, equations (6) and (9) together give the actual likelihood equations if the Y_i's follow a multivariate Gaussian distribution. Further, the asymptotic covariance, V_G, reduces to

$$\lim_{K\to\infty} \left\{ \sum_{i=1}^{K} D_i^{\mathrm{T}} \mathrm{cov}^{-1}(Y_i) D_i/K \right\}^{-1}$$

since R is the true correlation matrix. Again, no estimation of ϕ is required to obtain $\hat{\beta}_G$. However, this assumption is useful only with a small number of observation times.

5. Efficiency Considerations

In this section, we consider two very simple data configurations and ask the following questions: (i) how much more efficient is $\hat{\beta}_G$ than $\hat{\beta}_I$; and (ii) how do $\hat{\beta}_G$ and $\hat{\beta}_I$ compare to the maximum likelihood estimator when further distributional assumptions on the Y_i's are made? To address the first question, consider the generalized linear model with natural link so that

$$\theta_{it} = x_{it}\beta \qquad (t = 1, \ldots, 10).$$

We assume that each $X_i = (x_{i1}, \ldots, x_{i,10})'$ is generated from a distribution with mean $(0.1, 0.2, \ldots, 1.0)'$ and finite covariance. Table 1 then gives the asymptotic relative efficiency of $\hat{\beta}_I$ and $\hat{\beta}_G$'s for three distinct correlation assumptions to the generalized estimator in which the correlation matrix is correctly specified. The correlation structures are one-dependent, exchangeable and first-order autoregressive, Examples 2, 3 and 4. The upper and lower entries are for $\alpha = 0.3$ and 0.7 respectively.

There is little difference between $\hat{\beta}_I$ and the $\hat{\beta}_G$'s when the true correlation is moderate, 0.3 say. However, lower entries of Column 1 indicate that substantial improvement can be made by correctly specifying the correlation matrix when α is large. The efficiency of $\hat{\beta}_I$ relative to the $\hat{\beta}_G$ using the correct correlation matrix is lowest, 0.74, when R has the one-dependent form and highest, 0.99, when R has the exchangeable pattern. That $\hat{\beta}_I$ is efficient relative to $\hat{\beta}_G$ in the latter case is because $n_i = 10$ for all i so that the extra-binomial variation introduced by the exchangeable correlation is the same for all subjects and no misweighting occurs by ignoring it. If instead, we assume that n_i takes values from 1 to 8 with equal probability, the relative efficiency of $\hat{\beta}_I$ drops to 0.82. Note that the results in Table 1 hold regardless of the underlying marginal distribution.

To address the second question, we consider a two-sample configuration with binary outcomes. Subjects are in two groups, with marginal expectations satisfying logit $\{E(y_{it})\} = \beta_0 + \beta_1 x_i$, where $x_i = 0$ for Group 0,

Table 1. Asymptotic Relative Efficiency of $\hat{\beta}_I$ and $\hat{\beta}_G$ to Generalized Estimator with Correlation Matrix Correctly Specified for $\eta_{it} = \beta_0 + \beta_1 t/10$. Here, $\beta_0 = \beta_1 = 1$, $n_i = 10$. For Upper Entry $\alpha = 0.3$; Lower entry $\alpha = 0.7$.

	Working R			
True R	Independence	1-Dependence	Exchangeable	AR-1
1-Dependence	0.97	1.0	0.97	0.99
	0.74	1.0	0.74	0.81
Exchangeable	0.99	0.95	1.0	0.95
	0.99	0.23	1.0	0.72
AR-1	0.97	0.99	0.97	1.0
	0.88	0.75	0.88	1.0

Table 2. Asymptotic Relative Efficiency of $\hat{\beta}_I$ and $\hat{\beta}_G$ Assuming AR1 Correlation Structure to the Maximum Likelihood Estimate for First-Order Markov Chain with $\theta_{it} = \beta_0 + \beta_1 x_i$. $x_i = 0$ for Group 0, $x_i = 1$ for Group 1. Here $\beta_0 = 0, \beta = 1$, and for Upper Entry $n_i = 10$, Lower Entry $n_i = 1, \ldots, 8$ with Equal Probabilities.

	Correlation α						
	0.0	0.1	0.2	0.3	0.5	0.7	0.9
$\hat{\beta}_I$	1.0	1.0	0.99	0.97	0.94	0.91	0.92
	1.0	1.0	0.98	0.96	0.92	0.86	0.81
$\hat{\beta}_G$ (AR 1)	1.0	1.0	0.99	0.99	0.98	0.97	0.98
	1.0	1.0	0.99	0.99	0.98	0.98	0.99

and 1 for Group 1. The repeated observations are assumed to come from a Markov chain of order 1 with first lag autocorrelation α. In Table 2, we compare the asymptotic relative efficiencies of $\hat{\beta}_I$ and $\hat{\beta}_G$ using the AR-1 correlation structure. Example 4, to the maximum likelihood estimator. For the upper entry, $n_i = 10$ for all i; for the lower, $n_i = 1$ to 8 with equal probability. The results indicate that both $\hat{\beta}_I$ and $\hat{\beta}_G$ are highly efficient for smaller α. As α increases, $\hat{\beta}_G$ retains nearly full efficiency while $\hat{\beta}_I$ does not. The contrast between $\hat{\beta}_I$ and $\hat{\beta}_G$ is strongest for the unequal sample size case.

6. Discussion

The analysis of non-Gaussian longitudinal data is difficult partly because few models for the joint distribution of the repeated observations for a subject are available. On the other hand, longitudinal data offer the advantage that data from distinct subjects are independent. The methods we propose avoid the need for multivariate distributions by only assuming a functional form for the marginal distribution at each time. The covariance structure across time is treated as a nuisance. We rely, however, on the independence across subjects to estimate consistently the variance of the proposed estimators even when the assumed correlation is incorrect, as we expect it often will be.

Modelling the marginal expectation and treating the correlation as a nuisance may be less appropriate when the time course of the outcome for each subject, e.g. growth, is of primary interest or when the correlation itself has scientific relevance. The random effects model for binary data discussed by Stiratelli, Laird and Ware (1984) can be extended to the generalized linear model family and is more appropriate for the study of growth. When the time dependence is central, models for the conditional distribution of y_t

given $y_{t-1}, y_{t-2}, \ldots, y_1$ may be more appropriate. Cox (1970, p. 72) has proposed such a model for binary outcomes. Korn and Whittemore (1979) have applied this model to air pollution data.

The examples in §4 provide several alternative methods for analysing longitudinal data sets. The method in Example 1, which includes the independence estimating equation as a special case, requires the fewest assumptions. Only the regression specification must be correct to obtain consistent estimates of β and $\text{var}(\hat{\beta})$. In §5, the independence estimator was shown to have high efficiency when the correlation is moderate in a simple situation with binary outcomes, Table 2. We believe that it may be less efficient in more realistic situations with more heterogeneity among both the X_i's and n_i's. Further study is needed.

Among the remaining methods implied by the generalized estimating equation, allowing R to have $\frac{1}{2}n(n-1)$ parameters. Example 5, gives the most efficient estimator. This approach, however, is only useful when there are few observation times. The remaining estimators will be as efficient only if the true correlation matrix can be expressed in terms of the chosen $R(\alpha)$ for some α. In particular, all generalized estimating equation estimators will be efficient if observations for a subject are independent. Note that each estimator and its variance will be consistent as long as α and ϕ can be estimated consistently for any correlation.

Missing data are common in some longitudinal studies. For $\hat{\beta}_G$ and $\hat{V}_G$ to be consistent even when R is misspecified, we require that data be missing completely at random (Rubin, 1976). That is, whether an observation in missing cannot depend on previous outcomes. Intuitively, we should not expect to handle complicated missing value patterns unless our working model is correct. When R is the true correlation, the missing completely at random assumption can be unnecessary. For Gaussian outcomes, the missing data pattern can depend arbitrarily on past observations and consistency is retained. For binary outcomes, the pattern can depend on any single previous outcome.

If the elements of R are proportional to those of α, then the scale parameter, ϕ, does not have to be determined as a step in solving the general estimating equation. This was the case for all examples above except 4. Note that ϕ also is eliminated from the estimation of β in quasi-likelihood methods (Wedderburn, 1974). In addition, the variance of $\hat{\beta}_G$ does not depend on the choice of estimator of the nuisance parameters, α and ϕ among those that are $K^{1/2}$-consistent. This is also the case in quasi-likelihood where the only nuisance parameter is ϕ. The estimating equations described in this paper can be thought of as an extension of quasi-likelihood to the case where the second moment cannot be fully specified in terms of the expectation but rather additional correlation parameters must be estimated. It is the independence across subjects that allows us to consistently estimate these nuisance parameters where this could not be done otherwise.

Acknowledgements

We thank the referee and Professor Nan Laird for helpful comments.

Appendix

Proof of Theorem 2

Write $\alpha^*(\beta) = \hat{\alpha}\{\beta, \hat{\phi}(\beta)\}$ and under some regularity conditions $K^{1/2}(\hat{\beta}_G - \beta)$ can be approximated by

$$\left[\sum_{i=1}^{K} -\frac{\delta}{\delta\beta} U_i\{\beta, \alpha^*(\beta)\}/K\right]^{-1} \left[\sum_{i=1}^{K} U_i\{\beta, \alpha^*(\beta)\}/K^{1/2}\right],$$

where

$$\begin{aligned}\delta U_i\{\beta, \alpha^*(\beta)\}/\delta\beta &= \partial U_i\{\beta, \alpha^*(\beta)\}/\partial\beta + [\partial U_i\{\beta, \alpha^*(\beta)\}/\partial\alpha^*]\{\partial\alpha^*(\beta)/\partial\beta\} \\ &= A_i + B_i C. \qquad \text{(A1)}\end{aligned}$$

Let β be fixed and Taylor expansion gives

$$\begin{aligned}\frac{\sum U_i\{\beta, \alpha^*(\beta)\}}{K^{1/2}} &= \frac{\sum U_i(\beta, \alpha)}{K^{1/2}} + \frac{\sum \partial/\partial\alpha U_i(\beta, \alpha)}{K} K^{1/2}(\alpha^* - \alpha) + o_p(1) \\ &= A^* + B^* C^* + o_p(1), \qquad \text{(A2)}\end{aligned}$$

where the sums are over $i = 1, \ldots, K$. Now, $B^* = o_p(1)$, since $\partial U_i(\beta, \alpha)/\partial\alpha$ are linear functions of S_i's whose means are zero, and conditions (i) to (iii) give

$$\begin{aligned}C^* &= K^{1/2}[\hat{\alpha}\{\beta, \hat{\phi}(\beta)\} - \hat{\alpha}(\beta, \phi) + \hat{\alpha}(\beta, \phi) - \alpha] \\ &= K^{1/2}\left\{\frac{\partial\hat{\alpha}}{\partial\phi}(\beta, \phi^*)(\hat{\phi} - \phi) + \hat{\alpha}(\beta, \phi) - \alpha\right\} = O_p(1).\end{aligned}$$

Consequently, $\sum U_i\{\beta, \alpha^*(\beta)\}/K^{1/2}$ is asymptotically equivalent to A^* whose asymptotic distribution is multivariate Gaussian with zero mean and covariance matrix

$$\lim_{K\to\infty}\left\{\sum_{i=1}^{K} D_i^{\mathrm{T}} V_i^{-1} \operatorname{cov}(Y_i) V_i^{-1} D_i/K\right\}.$$

Finally, it is easy to see that $\sum B_i = o_p(K)$, $C = O_p(1)$ and that $\sum A_i/K$ converges as $K \to \infty$ to $-\sum D_i^{\mathrm{T}} V_i^{-1} D_i/K$. This completes the proof. □

References

Baker, R.J. and Nelder, J.A. (1978). *The GLIM System, Release 3. Generalized Linear Interactive Modelling*. Numerical Algorithms Group, Oxford.

Cox, D.R. (1970). *The Analysis of Binary Data*. Methuen, London.

Feller, W. (1971). *An Introduction to Probability Theory*, vol. **2**, 2nd ed. Wiley, New York.

Jorgensen, B. (1983). Maximum likelihood estimation and large-sample inference for generalized linear and nonlinear regression models. *Biometrika*, **70**, 19–28.

Koch, G.C., Landis, J.R., Freeman, J.L., Freeman, D.H., and Lehman, R.G. (1977). A general methodology for the analysis of repeated measurements of categorical data. *Biometrics*, **33**, 133–158.

Korn, E.L. and Whittemore, A.S. (1979). Methods for analyzing panel studies of acute health effects of air pollution. *Biometrics*, **35**, 795–802.

Laird, N.M. and Ware, J.H. (1982). Random-effects models for longitudinal data. *Biometrics*, **38**, 963–974.

McCullagh, P. (1983). Quasi-likelihood functions. *Ann. Statist.*, **11**, 59–67.

McCullagh, P. and Nelder, J.A. (1983). *Generalized Linear Models*. Chapman and Hall, London.

Morton, R. (1981). Efficiency of estimating equations and the use of pivots. *Biometrika*, **68**, 227–233.

Ochi, Y. and Prentice, R.L. (1984). Likelihood inference in correlated probit regression. *Biometrika*, **71**, 531–543.

Rubin, D.B. (1976). Inference and missing data. *Biometrika*, **63**, 81–92.

Stiratelli, R., Laird, N., and Ware, J. (1984). Random effects models for serial observations with binary responses. *Biometrics*, **40**, 961–971.

Ware, J.H. (1985). Linear models for the analysis of longitudinal studies. *Amer. Statist.*, **39**, 95–101.

Wedderburn, R.W.M. (1974). Quasi-likelihood functions, generalized linear models, and the Gauss–Newton method. *Biometrika*, **61**, 439–447.

Zeger, S.L., Liang, K.Y., and Self, S.G. (1985). The analysis of binary longitudinal data with time independent covariates. *Biometrika*, **72**, 31–38.

Introduction to Hall (1988) Theoretical Comparison of Bootstrap Confidence Intervals

E. Mammen

Bootstrap was introduced in Efron (1979) [for a reprint with discussion see also Volume II of Kotz and Johnson (1992)]. For a long time it was one of the most active fields in mathematical and applied statistics. It has been discussed for a variety of statistical applications. Now we have a clear and detailed knowledge on the advantages and restrictions of the bootstrap approach. Bootstrap has become a standard method in the tool kit of statisticians. Some books on bootstrap [Beran and Ducharme (1991), Hall (1992), Mammen (1992), Efron and Tibshirani (1993), and Shao and Tu (1995)] have appeared already. Bootstrap has found its way into introductory text books on statistics.

In the first days of bootstrap it was argued that bootstrap is a magic approach that is automatic and allows us to "let the data speak for themselves." In particular, it would not require a deep understanding of the assumed statistical model, as is needed for more traditional approaches. No theoretical reasoning and technical mathematical calculus would be required for its application. Meanwhile it has become clear that this view was wrong. In many applications it is not clear how bootstrap has to be used and it is a difficult problem to decide if a specific implementation of bootstrap works well. These questions have motivated many deep mathematical considerations and they have stimulated the revival of classical asymptotic mathematical approaches like the Edgeworth expansions. The following reprinted paper by Peter Hall is one of the most famous examples of these mathematical contributions to the theory of bootstrap. At the time when this paper was written a debate was going on about how bootstrap should be used to construct confidence sets. Different proposals had been made. In this paper, Hall compares these different bootstrap confidence intervals. The analysis by Hall is done without prejudice. All proposals are considered using the same rigorous mathematical analysis. The only criterion is the asymptotic performance of the bootstrap confidence intervals. The material presented in this paper is one of the main ingredients of the book by Hall (1992). In the book a detailed analysis based on Edgeworth expansions is also given for other applications of the bootstrap.

The aim of bootstrap is the estimation of the operational characteristics of a statistical procedure. A simple example is the variance $\text{var}_p(\hat{\theta})$ of a statistic $\hat{\theta}$. The variance depends on the distribution P of the observed data sample X. Formally we can write $L(P) = \text{var}(\hat{\theta})$ where L is a map from the class $\mathscr{P}$ of possible distributions of the sample X to $\mathbb{R}$. The bootstrap estimate of $L(P)$ is given by $L(\hat{P}_n)$ where $\hat{P}_n$ is an estimate of P. Variance estimation was one of the motivations for the introduction of bootstrap in Efron (1979). In another more general setting, $L(P)$ may be defined as the distribution of a root $R = R(X, P)$. A root is a random variable depending on the sample X and on the distribution P. A simple example of a root is the unStudentized statistic $\hat{\theta} - \theta(P)$ or the Studentized statistic $(\hat{\theta} - \theta(P))/\hat{\sigma}$ where $\hat{\theta}$ is an estimate of a functional $\theta(P)$ and $\hat{\sigma}^2$ is an estimate of the variance σ^2 of $\hat{\theta}$.

For a good bootstrap performance it is required that $\hat{P}_n$ is an accurate estimate of P and that $L(Q)$ depends continuously on Q for Q in a neighborhood of P. An extreme example where this is fulfilled is when $L(Q)$ does not depend on Q. In this case roots are called *pivotal.* In an asymptotic setting, roots with the property that the distribution of (properly rescaled) roots R converges to a fixed distribution L (for Q in a neighborhood of P) are called *asymptotic pivots*. For the accuracy of the bootstrap estimate $L(\hat{P}_n)$ the smoothness of $L(Q)$ in a neighborhood of P is crucial. The smoothness of $L(Q)$ can be increased by a good choice of the root R. Asymptotic theory suggests that bootstrap for Studentized statistics is more accurate than for unStudentized statistics. This point has been made in Babu and Singh (1984) and subsequently in a lot of papers (e.g., in the following paper) for different setups. It has been explained by the fact that for Studentized statistics the limiting standard normal distribution does not depend on Q, i.e., the Studentized statistics are asymptotic pivots [see also Hartigan (1986) and the discussion in the Introduction to the following paper]. The relation between the accuracy of bootstrap and the pivotal nature of the bootstrapped root was clarified and explained in detail in the work of Beran [see, e.g., Beran (1987)]. He had also proposed iterative applications of bootstrap to construct pivotal roots with the property that $L(Q)$ becomes even less dependent on the distribution Q.

In different settings, statistical procedures have been proposed that make explicit use of bootstrap estimates. This has been done, e.g., for confidence sets, for testing and for model choice. Bootstrap confidence sets can be based on the bootstrap estimate $\hat{k}_{1-\alpha}$ of the $1 - \alpha$ quantile of a root R. A bootstrap $1 - \alpha$ confidence set is then given by $\{Q : R(Q, X) \leq \hat{k}_{1-\alpha}\}$. A discussion of this general setup of bootstrap confidence sets can be found in

Beran (1987). Bootstrap confidence sets where the root is an unStudentized statistic $\hat{\theta} - \theta(P)$ or a Studentized statistic $(\hat{\theta} - \theta(P))/\hat{\sigma}$ were first discussed in Efron (1981). In the following paper they are called *hybrid method* (with critical points $\hat{\theta}_{\mathrm{HYB}}$) and *percentile t-method* (with critical points $\hat{\theta}_{\mathrm{STUD}}$), respectively. Another approach to bootstrap confidence intervals dates back to Efron (1979). Here (two-sided) confidence intervals are given by the $\alpha/2$ and the $1 - \alpha/2$ quantile of the bootstrap estimate of the distribution of $\hat{\theta}$. [More precisely, B (say) bootstrap resamples are generated and B values of the estimate $\hat{\theta}$, based on the resamples, are recalculated. The boundary points of the bootstrap confidence interval are given by the $B\alpha/2$ and the $B(1 - \alpha/2)$ smallest value.] These bootstrap confidence intervals are difficult to motivate. In the following paper this method is called *percentile method* (with critical points $\hat{\theta}_{\mathrm{BACK}}$). There is a simple symmetry between the percentile bootstrap intervals and the hybrid bootstrap intervals: $\hat{\theta}_{\mathrm{BACK}}(1 - \alpha/2) - \hat{\theta} = \hat{\theta} - \hat{\theta}_{\mathrm{HYB}}(\alpha/2)$ and $\hat{\theta}_{\mathrm{HYB}}(1 - \alpha/2) - \hat{\theta} = \hat{\theta} - \hat{\theta}_{\mathrm{BACK}}(\alpha/2)$. In the following paper Hall argues that the percentile method is based on "an erroneous choice, since it is the bootstrap version of looking up the ... tables, backwards." Another point of view is given in Efron and Tibshirani (1993). There it is argued that the percentile t-method "can give somewhat erratic results, and can be heavily influenced by a few outlying data points" (in particular for samples with small sample sizes). Furthermore, it is pointed out that the percentile confidence interval for a transformed parameter is given by the transformed interval. This is not true for the percentile t-method. For a better level of accuracy, Efron has proposed a modification of the percentile method: the *bias-corrected method*. Details of its definition and for its motivation can be found in Efron and Tibshirani (1993) and in the following article by Hall. In Schenker (1985) it was shown that the percentile and the bias-corrected methods "perform poorly in the relatively simple problem of setting a confidence interval for the variance of a normal distribution." For this example, the assumptions underlying the use of these methods were discussed and it was concluded that both bootstrap methods "should be used with caution in complex models." Motivated by Schenker's example, in Efron (1985) another modification of the percentile method "the *accelerated bias-corrected method*" was proposed, which was aimed to result "in second-order correctness in a wide variety of problems." The mathematical comparison of these different bootstrap confidence intervals and their higher-order performance is the content of this paper by Hall. As a result of the analysis, the percentile t-method is favored. However, it is mentioned that the "criticism of the percentile method" and the "preference for percentile t over accelerated bias correction lose much of their force when a stable estimate of σ^2 is not available." (See also Hall (1992, pp. 128–129).)

Peter Hall's mathematical approach is based on higher-order Edgeworth expansions. Edgeworth expansions have been used in statistics for two purposes: for the mathematical study of statistical procedures and for the construction of new procedures with higher-order performance. Examples for the second application are higher-order analytic corrections of critical points (see the references given by Hall in his Introduction). The usefulness of the second application has been questioned by some statisticians: Trivially, higher-order expansions require more restrictive conditions than first-order expansions. So a good performance of higher-order expansions can only be expected if the accuracy of the first-order expansion is already reasonable. If first-order asymptotics break down there is no hope that higher-order asymptotics will do better. Often, an approximate accuracy (e.g., for the coverage probability of a confidence set) suffices and reliability may be more important. Then higher-order modifications of a statistical procedure may do just the wrong thing: they may make the procedure less reliable and the only advantage may be that they improve the accuracy of the statistical procedure for cases when it is not really needed. A further discussion of these points can be found in Hall (1992, p. 121). The distrust of a good performance of such methods (based on explicit higher-order Edgeworth expansions) is the main reason why they are not widespread in statistics. Psychological aversions against tedious algebraic calculations (required by higher-order corrections) may be another reason. This point has often been stressed when analytic higher-order procedures are compared with bootstrap procedures. From a scientific point of view this point should be of less importance.

Edgeworth expansions have been and are a basic tool in analyzing the performance of bootstrap. Similiar arguments to those in the last paragraph apply to the interpretation of the results in such studies. We should always be cautious before making too general conclusions. Some remarks in this direction are made by Hall, too. However, often we are only interested in qualitative statements, e.g., a certain bootstrap approach is more accurate than another bootstrap method. These statements may generalize to a much more general setting than quantitative statements (that are required for the analysis and development of analytic higher-order corrections discussed in the last paragraph). For instance, often a second-order Edgeworth correction overcorrects errors of a first-order approximation. Then an explicit use of a second-order Edgeworth expansion may result in lower accuracy. Nevertheless, the second-order terms can help to understand the performance of the first-order approximation. To give an example: that bootstrap of Studentized statistics performs better than bootstrap of unStudentized statistics seems to be a rather general phenomenon [although it is not always true, see the above remark by Peter Hall and the simulation example of a high-dimensional linear model in Mammen (1992)]. This is related to the approximate pivotal nature of Studentized statistics, see the discussion at the beginning of these comments. There seems to be no reason to conjecture that it requires that the higher-order Edgeworth expansion is accurate. On the other hand, the good asymptotic performance of accelerated bias-corrected bootstrap confidence intervals (shown by Hall) should be interpreted with more caution because an analytic correction (motivated by

Edgeworth expansions) is involved in the construction of these confidence intervals. The good asymptotic behavior of these confidence sets relies strongly on the accuracy of the Edgeworth expansions. More generally, we may argue, like Hall, that "... computer-intensive methods such as the bootstrap, which are designed to avoid tedious analytic corrections, should not have to appeal to such corrections."

There seems to be no general alternative to Edgeworth expansions. For a few models it has been shown that bootstrap works under weaker conditions than other classical methods (see Bickel and Freedman (1983) and Mammen (1992)). We may argue that results of this type have a much higher practical relevance. They imply that, in these models, bootstrap is more reliable. However, no general mathematical approach of this type is available.

In the following paper higher-order expansions are developed under the "smooth function" model [Bhattacharya and Ghosh (1978), see Section 1.3]. In this model, it is assumed that a smooth functional of a vector mean is estimated by its sample analogues. In particular, this includes the usual t-statistic because a sample $X = (X_1, \ldots, X_n)$ can be rewritten as a sample of duples $Z = ((X_1, X_1^2), \ldots, (X_n, X_n^2))$. Now, the sample mean and variance of X can be written as smooth functions of the two-dimensional sample mean of Z. Although this model has a very simple structure it is very general. It offers an attractive framework for an asymptotic case study on the performance of bootstrap. Mathematical details are not given in this paper. Some were published in Hall (1986). Rigorous and detailed proofs are given in the book by Hall (1992). The mathematics are not trivial. It is complicated by the fact that, for bootstrap confidence sets $\{Q\colon R(Q, X) \leq \hat{k}_{1-\alpha}\}$, both quantities $R(Q, X)$ and $\hat{k}_{1-\alpha}$ are random.

Peter Hall is one of the most active and influential current researchers in statistics and probability theory. He was born in 1951. In 1976 he received an M.Sc. at the Australian National University. In the same year he obtained his D.Phil at Oxford University with a thesis on a topic in probability theory ("Problems in limit theory"). Since 1978, he has been with the Australian National University (since 1988: Personal Chair in Statistics). Peter Hall has worked, and still works, in a wide range of different research areas such as: discrete time martingales, limit theory for sums of independent random variables, convergence rates, extreme value theory, curve estimation, coverage processes and random set theory, bootstrap methods, and fractal analysis; see also Hall and Heyde (1980) and Hall (1982, 1988, 1992). In his statistical research he has shown how advanced mathematical tools can successfully be used for questions of direct practical importance.

References

Babu, G.J. and Singh, K. (1985). Edgeworth expansions for sampling without replacement for finite populations. *J. Multiv-Anal.*, **17**, 261–278.

Beran, R. (1987). Prepivoting to reduce level error of confidence sets. *Biometrika*, **74**, 457–468.

Beran, R. and Ducharme, G.R. (1991). *Asymptotic Theory for Bootstrap Methods in Statistics*. Université de Montréal, Montréal.

Bhattachary, R.N. and Ghosh, J.K. (1978). On the validity of the formal Edgeworth expansion, *Ann Statist*, **6**, 435–451.

Bickel, P.J. and Freedman, D.A. (1983). Bootstrapping regression models with many parameters. In *A Festschrift for Erich L. Lehmann* (P.J. Bickel, K.A. Doksum, and J.C. Hodges, Jr., eds.). Wadsworth, Belmont, CA, pp. 28–48.

Efron, B. (1979). Bootstrap methods: Another look at the jackknife. *Ann. Statist.*, **7**, 1–26.

Efron, B. (1981). Nonparametric standard errors and confidence intervals, *Canad. J. Statist.*, **9**, 139–158 (Discussion 158–172).

Efron, B. (1985). Bootstrap confidence intervals for a class of parametric problems. *Biometrika*, **72**, 45–58.

Efron, B. and Tibshirani, R.J. (1986). Bootstrap methods for standard errors, confidence intervals, and other measures of statistical accuracy. *Statist. Sci.*, **1**, 50–75 (Discussion 75–77).

Efron, B. and Tibsharani, R.J. (1993). *An Introduction to the Bootstrap*. Chapman and Hall, London.

Hall, P. (1982). *Rates of Convergence in the Central Limit Theorem*. Pitman, London.

Hall, P. (1986). On the bootstrap and confidence intervals. *Ann. Statist.* **14**, 1431–1452.

Hall, P. (1988). *Introduction to the Theory of Coverage Processes*. Wiley, New York.

Hall, P. (1992). *The Bootstrap and Edgeworth Expansion*, Springer Verlag, New York.

Hall, P. and Heyde, C.C. (1980). *Martingale Limit Theory and its Application*. Academic Press, New York.

Hartigan, J. (1986). In discussion of Efron and Tibshirani (1986), pp. 75–77.

Mammen, E. (1992). *When Does Bootstrap Work? Asymptotic Results and Simulations*. Lecture Notes in Statistics, vol. 77. Springer-Verlag, New York.

Schenker, N. (1985). Qualms about bootstrap confidence intervals. *J. Amer. Statist. Assoc.*, **80**, 360–361.

Shao, J. and Tu, T. (1995). *The Jackknife and Bootstrap*, Springer-Verlag, New York.

Theoretical Comparison of Bootstrap Confidence Intervals

Peter Hall
Australian National University

Abstract

We develop a unified framework within which many commonly used bootstrap critical points and confidence intervals may be discussed and compared. In all, seven different bootstrap methods are examined, each being usable in both parametric and nonparametric contexts. Emphasis is on the way in which the methods cope with first- and second-order departures from normality. Percentile-t and accelerated bias-correction emerge as the most promising of existing techniques. Certain other methods are shown to lead to serious errors in coverage and position of critical point. An alternative approach, based on "shortest" bootstrap confidence intervals, is developed. We also make several more technical contributions. In particular, we confirm Efron's conjecture that accelerated bias-correction is second-order correct in a variety of multivariate circumstances, and give a simple interpretation of the acceleration constant.

1. Introduction and Summary

1.1 Introduction

There exists in the literature an almost bewildering array of bootstrap methods for constructing confidence intervals for a univariate parameter θ. We can identify at least five which are in common use, and others which have been proposed. They include the so-called "percentile method" (resulting in critical points designated by $\hat{\theta}_{\mathrm{BACK}}$ in this paper), the "percentile-t method" (resulting in $\hat{\theta}_{\mathrm{STUD}}$), a hybrid method (resulting in $\hat{\theta}_{\mathrm{HYB}}$), a bias-corrected method (resulting in $\hat{\theta}_{\mathrm{BC}}$) and an accelerated bias-corrected method (resulting in $\hat{\theta}_{\mathrm{ABC}}$). See Efron (1981, 1982, 1987). [The great majority of nontechnical statistical work using bootstrap methods to construct confidence intervals does not make it clear which of these five techniques is employed. Our enquiries of users indicate that the percentile method (not percentile-t) is used in more than half of cases and that the hybrid method is used in almost all the rest. Some users are not aware that there is a difference between hybrid and percentile methods.] Our aim in this paper is to develop a unifying theoretical framework within which different bootstrap critical points may be discussed, compared and evaluated. We draw a variety of conclusions and challenge some preconceptions about ways in which bootstrap critical points should be assessed.

Let $\hat{\theta}$ be our estimate of θ, based on a sample of size n and with asymptotic variance $n^{-1}\sigma^2$. Let $\hat{\sigma}^2$ be an estimate of σ^2. There is a variety of "theoretical critical points" which could be used in the "ideal" circumstance where the distributions of $n^{1/2}(\hat{\theta}-\theta)/\sigma$ and $n^{1/2}(\hat{\theta}-\theta)/\hat{\sigma}$ were known. If we knew σ, then we could look up "ordinary" tables of the distribution of $n^{1/2}(\hat{\theta}-\theta)/\sigma$, and if σ were unknown, we could consult "Studentized" tables of the distribution of $n^{1/2}(\hat{\theta}-\theta)/\hat{\sigma}$. Obviously we would commit errors if we were to get those tables mixed up or to confuse upper quantiles with lower quantiles. Nevertheless, if we insisted on using the wrong tables, we could perhaps make amends for some of our errors by looking up a slightly different probability level. For example, if we were bent on using standard normal tables when we should be employing Student's t-tables, and if we sought the upper 5% critical point, then for a sample of size 5 we could reduce our error by looking up the $2\frac{1}{2}\%$ point instead of the 5% point.

We argue that most bootstrap critical points are just elementary bootstrap estimates of theoretical critical points, often obtained by "looking up the wrong tables." *Bootstrap approximations are so good that if we use bootstrap estimates of erroneous theoretical critical points, we commit noticeable errors*. This observation will recur throughout our paper and will be the source of many of our conclusions about bootstrap critical points. Using the common "hybrid" bootstrap critical points is tantamount to looking up the wrong tables, and using the percentile method critical point amounts to looking up the wrong tables backwards. Bias-corrected methods use adjusted probability levels to correct some of the errors incurred by looking up wrong tables backwards.

There are other ways of viewing bootstrap critical points, although they do not lend themselves to the development of a unifying framework. The distinction between looking up "ordinary" and "Studentized" tables is sometimes expressed by arguing that $n^{1/2}(\hat{\theta}-\theta)/\sigma$ is pivotal if σ is known, whereas $n^{1/2}(\hat{\theta}-\theta)/\hat{\sigma}$ is pivotal if σ is unknown [e.g., Hartigan (1986)]. However, it is often the case that neither of these quantities is strictly pivotal in the sense in which that term is commonly used in inference [e.g., Cox and Hinkley (1974), page 211].

Much of our discussion ranges around the notion of second-order correctness, defined in Section 2.3. Our accelerated bias-corrected bootstrap critical point is *deliberately designed* to be second-order correct and in that sense it is superficially a little different from Efron's (1987) accelerated bias-corrected point, which is motivated via transformation theory. However, Efron conjectures that his accelerated bias-corrected critical point is second-order correct and in each circumstance where that conjecture is valid, his critical point and ours coincide *exactly*. Indeed, one of the technical contributions of our paper is to verify Efron's conjecture in important cases—for the parametric bootstrap in multivariate exponential family models and for the nonparametric bootstrap in cases where estimators can be expressed as functions of multivariate vector means. Previously, verification of the conjecture was confined to univariate, parametric models. We also provide a very simple interpretation of the acceleration constant (see Section 2.4).

We argue that second-order correctness is of major importance to one-sided confidence intervals, but that its impact is reduced for two-sided intervals, even though it is most often discussed in that context. There, interval length is influenced by third-order rather than second-order properties, although second-order characteristics do have an effect on coverage. Note particularly that the difference between standard normal tables and Student's t-tables is a third-order effect—it results in a term of size $(n^{-1/2})^3$ in the formula for a critical point for the mean. We argue that percentile-t does a better job than accelerated bias-correction of getting third-order terms right, provided the variance estimate $\hat{\sigma}^2$ is chosen correctly.

It is shown that as a rule, coverage is not directly related to interval length, since the majority of bootstrap intervals are not designed to have minimum length for given coverage. Nevertheless, it is possible to construct "shortest" bootstrap confidence intervals, of shorter length than percentile-t intervals. Curiously, these short intervals can have much improved coverage accuracy as well as shorter length, in important cases; see Section 4.6.

We should stress that our theoretical comparisons of critical points comprise only part of the information needed for complete evaluation of bootstrap methods. Simulation studies [e.g., Efron (1982), Hinkley and Wei (1984) and Wu (1986)] and applications to real data provide valuable additional information. Nevertheless, we suggest that the theoretical arguments in this paper amount to a strong case *against* several bootstrap methods which currently enjoy popularity: the percentile method (distinct from the percentile-t method), the hybrid method, and the bias-corrected method (distinct from the accelerated bias-corrected method). It will be clear from our analysis that of the remaining established techniques, we favour percentile-t over accelerated bias-correction, although our choice is not unequivocal. Our decision is based on third-order properties of two-sided confidence intervals (see Section 4.4), on a philosophical aversion to looking up "ordinary" tables when we should be consulting "Studentized" tables (see particularly the example in the first paragraph of Section 4), and on a prejudice that computer-intensive methods such as the bootstrap, which are designed to avoid tedious analytic correction, should not have to appeal to such correction. There exist *many* devices for achieving second-order and even third-order correct critical points via analytic corrections, without resampling [e.g., Johnson (1978), Pfanzagl (1979), Cox (1980), Hall (1983, 1985, 1986), Withers (1983, 1984) and McCullagh (1984)], and it does seem cumbersome to have to resample as well as analytically correct. On the other hand, accelerated bias-correction enjoys useful properties of invariance under transformations, not shared by percentile-t. See for example Lemma 1 of Efron (1987).

Just as theoretical arguments are indecisive when attempting a choice between percentile-t and accelerated bias-correction, so too are simulation studies. Efron [(1981), page 154] reports a case where percentile-t intervals fluctuate erratically, and this can be shown to happen in other circumstances unless the variance estimate $\hat{\sigma}^2$ is chosen carefully. Conversely, simulations of equal-tailed accelerated bias-corrected intervals for small samples and large nominal coverage levels can produce abnormally short intervals, due to the fact that those intervals shrink to a point as coverage increases, for any given sample.

We should also point out that in some situations there are practical reasons for using "suboptimal" procedures. In complex circumstances it can be quite awkward to estimate σ^2; "utilitarian" estimates such as the jackknife may fluctuate erratically. A "suboptimal" confidence interval can be better than no interval at all. Our criticism of the percentile method and our preference for percentile-t over accelerated bias correction, lose much of their force when a stable estimate of σ^2 is not available.

Later in this section we define what we mean by parametric and nonparametric forms of the bootstrap, discuss a general model for the estimators $\hat{\theta}$ and $\hat{\sigma}$, and review elements of the theories of Edgeworth expansion and Cornish–Fisher expansion. Much of our discussion is based on inverse Cornish–Fisher expansions of bootstrap critical points and on Edgeworth expansions of coverage errors. Related work appears in Bickel and Freedman (1981), Singh (1981) and Hall (1986), although in each of those cases attention is focussed on particular versions of the bootstrap. Our *comparison* of bootstrap critical points, using asymptotic expansion methods, indicates among other things that there is often not much to choose between computationally expensive critical points such as $\hat{\theta}_{\mathrm{HYB}}$ and $\hat{\theta}_{\mathrm{BACK}}$ and the simple normal-theory critical point; see Section 4.5.

Section 2 introduces theoretical critical points, and derives their main properties. Bootstrap estimates of those points are defined in Section 3 and their properties are discussed in Section 4. Section 5 gives brief notes on some rigorous technical arguments which are omitted from our work.

1.2. Parametric and Nonparametric Bootstraps

Assume that $\hat{\theta}$ and $\hat{\sigma}$ are constructed from a random n-sample $\mathscr{X}$. In the parametric case, suppose the density h_λ of the sampling distribution is completely determined except for a vector λ of unknown parameters. Use $\mathscr{X}$ to estimate λ (e.g., by maximum likelihood) and write $\mathscr{X}^*$ for a random n-sample drawn from the population with density $h_{\hat{\lambda}}$. We call $\mathscr{X}^*$ a "resample." In the nonparametric case, $\mathscr{X}^*$ is simply drawn at random (with replacement) from $\mathscr{X}$. In either case, let $\hat{\theta}^*$ and $\hat{\sigma}^*$ be version of $\hat{\theta}$ and $\hat{\sigma}$ computed in the same manner as before, but with the resample $\mathscr{X}^*$ replacing the sample $\mathscr{X}$.

Two examples are helpful in explaining parametric and nonparametric versions of the bootstrap. Suppose first that we are in a parametric context and that $\hat{\theta}$ and $\hat{\sigma}$ are "bootstrap estimates" (that is, obtained by replacing functionals of a distribution function by functionals of the empiric). Assume that the unknown parameters λ are functions of location and scale of $\hat{\theta}$, and that the statistics $n^{1/2}(\hat{\theta}-\theta)/\sigma$ and $n^{1/2}(\hat{\theta}-\theta)/\hat{\sigma}$ are location and scale invariant. Cases in point include inference about θ in an $N(\theta, \sigma^2)$ population and about the mean θ of an exponential distribution. Then the distributions of $n^{1/2}(\hat{\theta}^*-\hat{\theta})/\hat{\sigma}$ and $n^{1/2}(\hat{\theta}^*-\hat{\theta})/\hat{\sigma}^*$ (either conditional on $\mathscr{X}$ or unconditionally) are *identical* to those of $n^{1/2}(\hat{\theta}-\theta)/\sigma$ and $n^{1/2}(\hat{\theta}-\theta)/\hat{\sigma}$, respectively.

Next, suppose we wish to estimate the mean θ of a continuous distribution, without making parametric assumptions. Let $\hat{\theta}$ and $\hat{\sigma}^2$ denote, respectively, sample mean and sample variance, the latter having divisor n rather than $n-1$. Then the distributions of $n^{1/2}(\hat{\theta}^*-\hat{\theta})/\hat{\sigma}$ and $n^{1/2}(\hat{\theta}^*-\hat{\theta})/\hat{\sigma}^*$, conditional on $\mathscr{X}$, approximate the unconditional distributions of $n^{1/2}(\hat{\theta}-\theta)/\sigma$ and $n^{1/2}(\hat{\theta}-\theta)/\hat{\sigma}$, respectively.

During our discussion of bootstrap critical points we shall use these examples to illustrate arguments and the conclusion.

We should point out that "bootstrap population moments," on which depend coefficients of bootstrap polynomials such as $\hat{p}_i$ and $\hat{q}_i$ (see Section 4), have different interpretations in parametric and nonparametric circumstances. In the parametric case, bootstrap population moments are moments with respect to density $h_{\hat{\lambda}}$; in the nonparametric case, they are moments of the sample $\mathscr{X}$. In the parametric case we assume that $\int xh_{\hat{\lambda}}(x)\,dx$ equals the mean $\bar{X}$ of $\mathscr{X}$. For example, this is true if h_λ is from an exponential family and $\hat{\lambda}$ is the maximum likelihood estimator.

1.3. The "Smooth Function" Model

All of our explicit calculation of Edgeworth expansions will be in the context of the following model. Assume that the data comprising the sample $\mathscr{X}$

are in the form of n independent and identically distributed d-vectors $X_1, \ldots, X_n$. Let X have the distribution of the X_i's and put $\mu \equiv E(X)$ and $\bar{X} \equiv n^{-1} \sum X_i$. We suppose that for known real-valued smooth functions f and g, $\theta = f(\mu)$ and $\sigma^2 = g(\mu)$. Estimates of θ and σ^2 are taken to be $\hat{\theta} \equiv f(\bar{X})$ and $\hat{\sigma}^2 \equiv g(\bar{X})$, respectively. Examples include parametric inference in exponential families and nonparametric estimation of means, of ratios or products of means, of variances, of ratios or products of variances, of correlation coefficients, etc. Rigorous Edgeworth expansion theory developed by Bhattacharya and Ghosh (1978) was tailored to this type of model.

Vector components will be denoted by bracketed superscripts. For example, $X_i = (X_i^{(1)}, \ldots, X_i^{(d)})$. We write $f_{(i_1 \ldots i_p)}(x)$ for $(\partial^p / \partial x^{(i_1)} \ldots \partial x^{(i_p)}) f(x)$, $a_{i_1 \ldots i_p}$ for $f_{(i_1 \ldots i_p)}(\mu)$, $\mu_{i_1 \ldots i_p}$ for $E\{(X-\mu)^{(i_1)} \ldots (X-\mu)^{(i_p)}\}$, c_i for $g_{(i)}(\mu)$ and $A(x)$ for $f(x) - f(\mu)$.

1.4. Edgeworth Expansion and Cornish–Fisher Inversion

Let $A: \mathbb{R}^d \to \mathbb{R}$ be a smooth function satisfying $A(\mu) = 0$. Then the cumulants of $U \equiv n^{1/2} A(\bar{X})$ are

$$k_1(U) \equiv E(U) = n^{-1/2} A_1 + O(n^{-3/2}),$$

$$k_2(U) \equiv E(U^2) - (EU)^2 = \sigma^2 + O(n^{-1}),$$

and

$$k_3(U) \equiv E(U^3) - 3E(U^2)E(U) + 2(EU)^3 = n^{-1/2} A_2 + O(n^{-3/2}),$$

where if $a_{i_1 \ldots i_p} \equiv A_{(i_1, \ldots i_p)}(\mu)$, then $\sigma^2 \equiv \sum\sum a_i a_j \mu_{ij}$, $A_1 \equiv \frac{1}{2} \sum\sum a_{ij} \mu_{ij}$ and

$$A_2 \equiv \sum\sum\sum a_i a_j a_k \mu_{ijk} + 3 \sum\sum\sum\sum a_i a_j a_{kl} \mu_{ik} \mu_{jl}.$$

In consequence,

$$P(U/\sigma \le x) = \Phi(x) + n^{-1/2} p_1(x) \phi(x) + O(n^{-1}), \tag{1.1}$$

where $-p_1(x) \equiv \sigma^{-1} A_1 + \frac{1}{6} \sigma^{-3} A_2 (x^2 - 1)$ and ϕ and Φ are the standard normal density and distribution functions, respectively [e.g., Wallace (1958)].

If we estimate σ^2 using $\hat{\sigma}^2 \equiv g(\bar{X})$, then (1.1) becomes

$$P(U/\hat{\sigma} \le x) = \Phi(x) + n^{-1/2} q_1(x) \phi(x) + O(n^{-1}), \tag{1.2}$$

where with $B \equiv A/g^{1/2}$, $b_{i_1 \ldots i_p} \equiv B_{(i_1 \ldots i_p)}(\mu)$, $B_1 \equiv \frac{1}{2} \sum\sum b_{ij} \mu_{ij}$ and

$$B_2 \equiv \sum\sum\sum b_i b_j b_k \mu_{ijk} + 3 \sum\sum\sum\sum b_i b_j b_{kl} \mu_{ik} \mu_{jl},$$

we have $-q_1(x) \equiv B_1 + \frac{1}{6} B_2 (x^2 - 1)$. Let $c_i \equiv g_{(i)}(\mu)$. It may be shown that $b_i = a_i \sigma^{-1}$ and $b_{ij} = a_{ij} \sigma^{-1} - \frac{1}{2}(a_i c_j + a_j c_i) \sigma^{-3}$ and thence that

$$p_1(x) - q_1(x) \equiv -\tfrac{1}{2} \sigma^{-3} (\textstyle\sum\sum a_i c_j \mu_{ij} x^2). \tag{1.3}$$

Clearly, if x_α, y_α and z_α are defined by $P(U/\sigma \le x_\alpha) = P(U/\hat{\sigma} \le y_\alpha) = \Phi(z_\alpha) = \alpha$, then

$$\begin{aligned} x_\alpha &= z_\alpha - n^{-1/2} p_1(z_\alpha) + O(n^{-1}), \\ y_\alpha &= z_\alpha - n^{-1/2} q_1(z_\alpha) + O(n^{-1}). \end{aligned} \tag{1.4}$$

Results (1.1) and (1.2) are Edgeworth expansions; results (1.4) are (inverse) Cornish–Fisher expansions. The definition $z_\alpha \equiv \Phi^{-1}(\alpha)$ will be used throughout this paper.

More generally, suppose that for some $\nu \ge 1$,

$$P(U/\sigma \le x) = \Phi(x) + \sum_{i=1}^{\nu} n^{-i/2} p_i(x)\phi(x) + O(n^{-(\nu+1)/2}),$$

$$P(U/\hat{\sigma} \le x) = \Phi(x) + \sum_{i=1}^{\nu} n^{-i/2} q_i(x)\phi(x) + O(n^{-(\nu+1)/2}).$$

Then p_i and q_i are polynomials of degree $3i-1$ and odd/even indexed polynomials are even/odd functions, respectively. The quantiles x_α and y_α defined earlier admit the expansions

$$x_\alpha = z_\alpha + \sum_{i=1}^{\nu} n^{-i/2} p_{i1}(z_\alpha) + O(n^{-(\nu+1)/2}),$$

$$y_\alpha = z_\alpha + \sum_{i=1}^{\nu} n^{-i/2} q_{i1}(z_\alpha) + O(n^{-(\nu+1)/2}),$$

where p_{i1} and q_{i1} may be defined in terms of p_j and q_j for $j \le i$. In particular,

$$\begin{aligned} p_{11}(x) &= -p_1(x), \\ p_{21}(x) &= p_1(x)p_1'(x) - \tfrac{1}{2} x p_1(x)^2 - p_2(x), \end{aligned} \tag{1.5}$$

with similar relations for the q's. The polynomials p_{i1} and q_{i1} are of degree $i+1$ and odd/even indices correspond to even/odd functions.

2. Theoretical Critical Points

2.1. Introduction

In this section we work under the assumption that the distribution functions

$$\begin{aligned} H(x) &\equiv P\{n^{1/2}(\hat{\theta} - \theta)/\sigma \le x\} \\ \text{and} \quad & \\ K(x) &\equiv P\{n^{1/2}(\hat{\theta} - \theta)/\hat{\sigma} \le x\} \end{aligned} \tag{2.1}$$

are known. We discuss critical points which could be used in that ideal circumstance. Such points will be called *theoretical critical points*. Some of those points will be clearly inadvisable, and that fact does a lot to explain difficulties inherent in bootstrap estimates of the points (see Section 3). None of the critical points introduced in the present section involves the bootstrap in any way.

2.2. Ordinary, Studentized, Hybrid and Backwards Critical Points

Let $x_\alpha \equiv H^{-1}(\alpha)$ and $y_\alpha \equiv K^{-1}(\alpha)$ denote α-level quantiles of H and K, respectively. Suppose we seek a critical point $\hat{\theta}(\alpha)$ with the property, $P\{\theta \le \hat{\theta}(\alpha)\} \simeq \alpha$. If σ were known, we could use the "ordinary" critical point

$$\hat{\theta}_{\text{ord}}(\alpha) \equiv \hat{\theta} - n^{-1/2}\sigma x_{1-\alpha}.$$

If σ were unknown, the "Studentized" point

$$\hat{\theta}_{\text{Stud}}(\alpha) \equiv \hat{\theta} - n^{-1/2}\hat{\sigma} y_{1-\alpha}$$

would be an appropriate choice. These points are both "exact" in the sense that

$$P\{\theta \le \hat{\theta}_{\text{ord}}(\alpha)\} = P\{\theta \le \hat{\theta}_{\text{Stud}}(\alpha)\} = \alpha.$$

Should we get the quantiles $x_{1-\alpha}$ and $y_{1-\alpha}$ muddled, we might use the "hybrid" point

$$\hat{\theta}_{\text{hyb}}(\alpha) \equiv \hat{\theta} - n^{-1/2}\hat{\sigma} x_{1-\alpha}$$

instead of $\hat{\theta}_{\text{Stud}}$. This is analogous to mistakenly looking up normal tables instead of Student's t tables in problems of inference about a normal mean. Should we hold those tables upside down, we might confuse $y_{1-\alpha}$ with $-x_\alpha$ and obtain the "backwards" critical point

$$\hat{\theta}_{\text{back}}(\alpha) \equiv \hat{\theta} + n^{-1/2}\hat{\sigma} x_\alpha.$$

Thus, $\hat{\theta}_{\text{back}}$ is the result of *two* errors—looking up the wrong tables, backwards.

2.3. Bias-Corrected Critical Points

Bias corrections attempt to remedy the errors in $\hat{\theta}_{\text{back}}$. They might be promoted as follows. Clearly $\hat{\theta}_{\text{back}}(\alpha)$ is an inappropriate choice. But if we are bent on looking up the wrong tables backwards, we might reduce some of our errors by using something else instead of α. Perhaps if we choose β correctly, $\hat{\theta}_{\text{back}}$ might not be too bad. For example, choosing β such that

$-x_\beta = y_{1-\alpha}$ will improve matters, for in that case $\hat{\theta}_{\text{back}}(\beta)$ is just the exact critical point $\hat{\theta}_{\text{Stud}}(\alpha)$. More generally, if

$$-x_\beta = y_{1-\alpha} + O(n^{-1}),$$

then $\hat{\theta}_{\text{back}}(\beta)$ agrees with $\hat{\theta}_{\text{Stud}}(\alpha)$ to order $n^{-1} = (n^{-1/2})^2$, that is, to second order. In that case we say that $\hat{\theta}_{\text{back}}(\beta)$ is *second-order correct*. [This discussion has ignored properties of translation invariance which bias-corrected critical points enjoy. See Efron (1987).]

We may look at this problem from the point of view of coverage error rather than position of critical point. Suppose H and K admit Edgeworth expansions

$$H(x) = \Phi(x) + n^{-1/2}p_1(x)\phi(x) + O(n^{-1}),$$

$$K(x) = \Phi(x) + n^{-1/2}q_1(x)\phi(x) + O(n^{-1}).$$

Then $x_\alpha \equiv z_\alpha - n^{-1/2}p_1(z_\alpha) + O(n^{-1})$ and so the interval $(-\infty, \hat{\theta}_{\text{back}}(\alpha)]$ has coverage

$$\begin{aligned} P\{\theta \le \hat{\theta}_{\text{back}}(\alpha)\} &= P\{n^{1/2}(\hat{\theta} - \theta)/\hat{\sigma} \ge -z_\alpha + n^{-1/2}p_1(z_\alpha) + O(n^{-1})\} \\ &= \alpha - n^{-1/2}\{p_1(z_\alpha) + q_1(z_\alpha)\}\phi(z_\alpha) + O(n^{-1}). \end{aligned} \tag{2.2}$$

Therefore, coverage error is proportional to $p_1(z_\alpha) + q_1(z_\alpha)$ in large samples. This function is an even quadratic polynomial in z_α. *Bias correction eliminates the constant term in* $p_1(z_\alpha) + q_1(z_\alpha)$; *accelerated bias correction eliminates all of* $p_1(z_\alpha) + q_1(z_\alpha)$ and so reduces coverage error of the one-sided interval from $O(n^{-1/2})$ to $O(n^{-1})$. This is equivalent to second-order correctness. We shall show in Section 4 that bootstrap versions of bias-correction and accelerated bias-correction operate in precisely the same manner.

We deal first with ordinary bias-correction. Let $G(x) \equiv P(\hat{\theta} \le x)$ and put

$$\begin{aligned} m \equiv \Phi^{-1}\{G(\theta)\} = \Phi^{-1}\{H(0)\} &= \Phi^{-1}\{\tfrac{1}{2} + n^{-1/2}p_1(0)\phi(0) + O(n^{-1})\} \\ &= n^{-1/2}p_1(0) + O(n^{-1}). \end{aligned}$$

Take $\beta \equiv \Phi(z_\alpha + 2m)$. Then $z_\beta = z_\alpha + 2m$ and so

$$\begin{aligned} x_\beta &= z_\beta - n^{-1/2}p_1(z_\beta) + O(n^{-1}) \\ &= z_\alpha + n^{-1/2}\{2p_1(0) - p_1(z_\alpha)\} + O(n^{-1}). \end{aligned} \tag{2.3}$$

The (theoretical) *bias-corrected critical point* is

$$\hat{\theta}_{\text{bc}}(\alpha) \equiv \hat{\theta}_{\text{back}}(\beta) = \hat{\theta} + n^{-1/2}\hat{\sigma}[z_\alpha + n^{-1/2}\{2p_1(0) - p_1(z_\alpha)\} + O(n^{-1})].$$

The argument leading to (2.2) shows that the interval $(-\infty, \hat{\theta}_{\text{bc}}(\alpha)]$ has coverage

$$P\{\theta \le \hat{\theta}_{\text{bc}}(\alpha)\} = \alpha + n^{-1/2}\{2p_1(0) - p_1(z_\alpha) - q_1(z_\alpha)\}\phi(z_\alpha) + O(n^{-1}). \tag{2.4}$$

This is the same as (2.2) except that the term $2p_1(0)$ cancels out the constant component of the even quadratic polynomial $-\{p_1(z_\alpha) + q_1(z_\alpha)\}$. [Note that $p_1(0) = q_1(0)$, since $H(0) = K(0)$.]

The quantity $n^{-1/2}\{2p_1(0) - p_1(z_\alpha) - q_1(z_\alpha)\}$ appearing in (2.4) may be written as $-az_\alpha^2$, where a does not depend on z_α. To completely remove this term from (2.4), replace β by any number β_a satisfying

$$\beta_a = \Phi\{z_\alpha + 2m + az_\alpha^2 + O(n^{-1})\}. \tag{2.5}$$

The argument leading to (2.3) shows that $x_{\beta_a} = z_\alpha + n^{-1/2}q_1(z_\alpha) + O(n^{-1})$. The (theoretical) *accelerated bias-corrected critical point* is

$$\hat{\theta}_{\text{abc}}(\alpha) \equiv \hat{\theta}_{\text{back}}(\beta_a) = \hat{\theta} + n^{-1/2}\hat{\sigma}\{z_\alpha + n^{-1/2}q_1(z_\alpha) + O(n^{-1})\}$$

and the corresponding one-sided interval $(-\infty, \hat{\theta}_{\text{abc}}(\alpha)]$ has coverage equal to $\alpha + O(n^{-1})$. Notice that $\hat{\theta}_{\text{abc}}(\alpha) = \hat{\theta}_{\text{Stud}}(\alpha) + O_p(n^{-3/2})$, since

$$\hat{\theta}_{\text{Stud}}(\alpha) = \hat{\theta} + n^{-1/2}\hat{\sigma}\{z_\alpha + n^{-1/2}q_1(z_\alpha) + O(n^{-1})\}.$$

Therefore, $\hat{\theta}_{\text{abc}}$ is second-order correct.

2.4. The Acceleration Constant

We call a the *acceleration constant*. The preceding argument explains why accelerated bias-correction works, but provides little insight into the nature of the constant. We claim that a is simply one-sixth of the third moment (skewness) of the first-order approximation to $n^{1/2}(\hat{\theta} - \theta)/\sigma$, at least in many important cases. For the "smooth function" model introduced in Section 1.3,

$$n^{1/2}(\hat{\theta} - \theta)/\sigma = (n^{1/2}/\sigma)\sum_{i=1}^{d}(\bar{X} - \mu)^{(i)}a_i + O_p(n^{-1/2}),$$

and so our claim is that

$$\begin{aligned} a &\equiv \tfrac{1}{6}E\left\{(n^{1/2}/\sigma)\sum_{i=1}^{d}(\bar{X} - \mu)^{(i)}a_i\right\}^3 \\ &= n^{-1/2}\tfrac{1}{6}\sigma^{-3}\sum\sum\sum a_i a_j a_k \mu_{ijk}. \end{aligned} \tag{2.6}$$

To check this, recall that

$$\begin{aligned} b \equiv n^{1/2}6\sigma^3 a &= 6\sigma^3 z_\alpha^{-2}\{p_1(z_\alpha) + q_1(z_\alpha) - 2p_1(0)\} \\ &= 3\sum\sum a_i c_j \mu_{ij} - 2\sum\sum\sum a_i a_j a_k \mu_{ijk} \\ &\quad - 6\sum\sum\sum\sum a_i a_j a_{kl} \mu_{ik} \mu_{jl}, \end{aligned} \tag{2.7}$$

the last equality following from results in Section 1.4. [Note particularly (1.3) and remember that $c_i \equiv g_i(\mu)$.] We treat parametric and nonparametric cases separately.

CASE (i) (Exponential Family Model). Assume X has density

$$h_\lambda(x) \equiv \exp\{\lambda^T x - \psi(\lambda)\} h_0(x),$$

where ψ and h_0 are known functions and λ is a d-vector of unknown parameters. Then $\mu^{(i)} = \psi_{(i)}(\lambda)$, $\mu_{ij} = \psi_{(ij)}(\lambda)$ and $\mu_{ijk} = \psi_{(ijk)}(\lambda)$. Write $M \equiv (\mu_{ij})$ and $N = (\nu_{ij}) \equiv M^{-1}$, both $d \times d$ matrices. Inverting the matrix of equations $\partial\mu^{(i)}/\partial\lambda^{(j)} = \mu_{ij}$, we conclude that $\partial\lambda^{(i)}/\partial\mu^{(j)} = \nu_{ij}$, whence

$$\frac{\partial}{\partial\mu^{(k)}} \psi_{(ij)}(\lambda) = \sum_l \frac{\partial\lambda^{(l)}}{\partial\mu^{(k)}} \frac{\partial}{\partial\lambda^{(l)}} \psi_{(ij)}(\lambda) = \sum_l \nu_{kl}\mu_{ijl}.$$

Remembering that $g(\mu) = \sigma^2 = \sum\sum f_{(i)}(\mu) f_{(j)}(\mu) \psi_{(ij)}(\lambda)$ (see Section 1.4), we obtain

$$\begin{aligned} c_k = g_{(k)}(\mu) &= \sum_i \sum_j \Big[\{f_{(ik)}(\mu) f_{(j)}(\mu) + f_{(i)}(\mu) f_{(jk)}(\mu)\} \psi_{(ij)}(\lambda) \\ &\qquad + f_{(i)}(\mu) f_{(j)}(\mu) \frac{\partial}{\partial\mu^{(k)}} \psi_{(ij)}(\lambda) \Big] \\ &= 2 \sum_i \sum_j a_i a_{jk} \mu_{ij} + \sum_i \sum_j \sum_l a_i a_j \nu_{kl} \mu_{ijl}. \end{aligned}$$

From this formula for c_k and the fact that

$$\sum_p \sum_k a_p \left(\sum_i \sum_j \sum_l a_i a_j \nu_{kl} \mu_{ijl} \right) \mu_{pk} = \sum_i \sum_j \sum_l a_i a_j a_l \mu_{ijl}$$

[since $(\nu_{ij}) = (\mu_{ij})^{-1}$], we conclude that

$$\sum \sum a_i c_j \mu_{ij} = 2 \sum_i \sum_j \sum_k \sum_l a_i a_j a_{kl} \mu_{ik} \mu_{jl} + \sum_i \sum_j \sum_k a_i a_j a_k \mu_{ijk}. \tag{2.8}$$

Substituting into (2.7), we find that $b = \sum\sum\sum a_i a_j a_k \mu_{ijk}$, which is equivalent to (2.6).

CASE (ii) (Nonparametric Inference). Recall from Section 1.4 that

$$\sigma^2 = \sum \sum f_{(i)}(\mu) f_{(j)}(\mu) \{E(X^{(i)} X^{(j)}) - \mu^{(i)} \mu^{(j)}\}.$$

If the products $X^{(i)} X^{(j)}$ are not components of the vector X, we may always adjoin them to X. Let $\langle i,j\rangle$ denote that index k such that $X^{(k)} \equiv X^{(i)}X^{(j)}$. Then

$$g(\mu) = \sigma^2 = \sum_i \sum_j f_{(i)}(\mu) f_{(j)}(\mu)(\mu^{(\langle i,j\rangle)} - \mu^{(i)}\mu^{(j)}).$$

On this occasion, a little algebra gives us the relation

$$c_k = g_{(k)}(\mu) = 2\sum_i \sum_j a_i a_{jk} \mu_{ij} - 2a_k \sum_i a_i \mu^{(i)} + \sum_i \sum_j{}_{(k)} a_i a_j,$$

where $\sum_i \sum_{j(k)}$ denotes summation over values (i,j) such that $\langle i,j\rangle = k$, From this formula and the fact that $\mu_{ijl} = \mu_{kl} - \mu^{(i)}\mu_{jl} - \mu^{(j)}\mu_{il}$ if $\langle i,j\rangle = k$, we conclude that (2.8) holds. As before, that leads to (2.6).

2.5. "Shortest" Intervals

Let $0 < \alpha < \frac{1}{2}$. Since we are assuming that we know that distribution of $n^{1/2}(\hat{\theta} - \theta)/\hat{\sigma}$, we may choose v, w to minimize $v + w$ subject to

$$P\{-w \le n^{1/2}(\hat{\theta} - \theta)/\hat{\sigma} \le v\} = 1 - 2\alpha. \tag{2.9}$$

We call $I_0 \equiv [\hat{\theta} - n^{-1/2}\hat{\sigma}v, \hat{\theta} + n^{-1/2}\hat{\sigma}w]$ the "shortest" confidence interval. It has the same coverage as the equal-tailed interval $[\hat{\theta}_{\text{Stud}}(\alpha), \hat{\theta}_{\text{Stud}}(1-\alpha)]$, but usually (except in cases of near-symmetry) has strictly shorter length. If the distribution of $n^{1/2}(\hat{\theta} - \theta)/\hat{\sigma}$ is unimodal, then the shortest confidence interval is equivalent to a likelihood-based confidence interval [Cox and Hinkley (1974), page 236].

Suppose the distribution function K admits the expansion

$$K(x) = \Phi(x) + n^{-1/2}q_1(x)\phi(x) + n^{-1}q_2(x)\phi(x) + O(n^{-3/2}).$$

Put $\phi_i(x) \equiv q_i(x)\phi(x)$ for $i \ge 1$, $\phi_0(x) \equiv \Phi(x)$, $\phi_{ik}(x) \equiv (\partial/\partial x)^k \phi_i(x)$ and $\psi_{ik} \equiv \phi_{ik}(z_{1-\alpha})$. A little calculus shows that the numbers v, w which minimize $v + w$ subject to (2.9), satisfy

$$v = z_{1-\alpha} + \sum_{i=1}^{\nu} n^{-i/2} v_i + O(n^{-(\nu+1)/2}),$$

$$w = z_{1-\alpha} + \sum_{i=1}^{\nu} (-n^{-1/2})^i v_i + O(n^{-(\nu+1)/2}),$$

where

$$v_1 \equiv -\psi_{11}\psi_{02}^{-1}, \qquad v_2 \equiv (\tfrac{1}{2}\psi_{11}^2\psi_{02}^{-1} - \psi_{20})\psi_{01}^{-1} \tag{2.10}$$

and higher-order v_i's admit more complex formulae.

See Pratt (1961, 1963), Harter (1964), Wilson and Tonascia (1971) and Kendall and Stuart [(1979), pages 125–129] for discussions of "short" confidence intervals.

3. Bootstrap Critical Points

3.1. Introduction

In this section we suggest that commonly used bootstrap critical points are elementary estimates of theoretical critical points introduced in Section 2. We argue that the bootstrap approximation is so good that bootstrap versions of erroneous theoretical critical points are also erroneous.

Bootstrap versions of distribution functions H and K [see (2.1)] are

$$\hat{H}(x) \equiv P\{n^{1/2}(\hat{\theta}^* - \hat{\theta})/\hat{\sigma} \leq x|\mathscr{X}\} \quad \text{and} \quad \hat{K}(x) \equiv P\{n^{1/2}(\hat{\theta}^* - \hat{\theta})/\hat{\sigma}^* \leq x|\mathscr{X}\},$$

respectively. For any distribution function F, define $F^{-1}(x) \equiv \sup\{x: F(x) \leq \alpha\}$.

3.2. Ordinary, Studentized, Hybrid and Backwards Critical Points

Bootstrap estimates of x_α and y_α are $\hat{x}_\alpha = \hat{H}^{-1}(\alpha)$ and $\hat{y}_\alpha = \hat{K}^{-1}(\alpha)$, respectively. Bootstrap versions of $\hat{\theta}_{\text{ord}}$, $\hat{\theta}_{\text{Stud}}$, $\hat{\theta}_{\text{hyb}}$ and $\hat{\theta}_{\text{back}}$ are obtained by replacing true quantiles x_α and y_α by these estimates:

$$\hat{\theta}_{\text{ORD}}(\alpha) \equiv \hat{\theta} - n^{-1/2}\sigma\hat{x}_{1-\alpha}, \qquad \hat{\theta}_{\text{STUD}}(\alpha) \equiv \hat{\theta} - n^{-1/2}\hat{\sigma}\hat{y}_{1-\alpha},$$

$$\hat{\theta}_{\text{HYB}}(\alpha) \equiv \hat{\theta} - n^{-1/2}\hat{\sigma}\hat{x}_{1-\alpha}, \qquad \hat{\theta}_{\text{BACK}}(\alpha) \equiv \hat{\theta} + n^{-1/2}\hat{\sigma}\hat{x}_{\alpha}.$$

In the existing literature, $\hat{\theta}_{\text{HYB}}$ and $\hat{\theta}_{\text{BACK}}$ are usually motivated using other arguments. For example, some statisticians employ $\hat{G}^{-1}(\alpha)$ as a critical point, where

$$\hat{G}(x) \equiv P(\hat{\theta}^* \leq x|\mathscr{X})$$

is the conditional distribution function of $\hat{\theta}^*$. This is often referred to as percentile-method critical point, although any one of $\hat{\theta}_{\text{ORD}}$, $\hat{\theta}_{\text{STUD}}$, $\hat{\theta}_{\text{HYB}}$ and $\hat{\theta}_{\text{BACK}}$ could be called percentile-method points. Since $\hat{G}(x) \equiv \hat{H}\{n^{1/2}(x - \hat{\theta})$ then $\hat{G}^{-1}(\alpha)$ is none other than $\hat{\theta}_{\text{BACK}}(\alpha)$. Our argument views this as an erroneous choice, since it is the bootstrap version of "looking up the wrong tables, backwards"; see Section 2.2.

Sometimes statisticians argue that the appropriate quantile is $\hat{\theta} - \xi_{1-\alpha}$, where $\xi_{1-\alpha}$ is the $(1-\alpha)$-level quantile of $\hat{\theta}^* - \hat{\theta}$:

$$\xi_{1-\alpha} \equiv \sup\{x: P(\hat{\theta}^* - \hat{\theta} \leq x|\mathscr{X}) \leq 1 - \alpha\}.$$

This is tantamount to saying that the conditional distribution of $\hat{\theta}^* - \hat{\theta}$ is a good approximation to the distribution of $\hat{\theta} - \theta$. Since $P(\hat{\theta}^* - \hat{\theta} \leq x|\mathscr{X}) \equiv \hat{H}(n^{1/2}x/\hat{\sigma})$, then $\hat{\theta} - \xi_{1-\alpha}$ is none other than $\hat{\theta}_{\text{HYB}}(\alpha)$. We view this as an incorrect choice, since it is the bootstrap version of "looking up the wrong tables." On the other hand, our argument suggests that $\hat{\theta}_{\text{STUD}}$ is a reasonable choice when σ is unknown and $\hat{\theta}_{\text{ORD}}$ a good choice when σ is known.

3.3. Bias-Corrected Critical Points

Recall that theoretical versions of bias-corrected and accelerated bias-corrected critical points were just $\hat{\theta}_{\text{back}}(\beta)$ and $\hat{\theta}_{\text{back}}(\beta_a)$, respectively. To obtain bootstrap analogues, we simply replace β and β_a by their bootstrap estimates $\hat{\beta}$ and $\hat{\beta}_a$ and use $\hat{\theta}_{\text{BACK}}$ instead of $\hat{\theta}_{\text{back}}$.

To define $\hat{\beta}$, remember that $\beta \equiv \Phi(z_\alpha + 2m)$, where $m \equiv \Phi^{-1}\{G(\theta)\}$. The bootstrap estimate of G is of course $\hat{G}$ and so we take $\hat{m} \equiv \Phi^{-1}\{\hat{G}(\hat{\theta})\}$ and $\hat{\beta} \equiv \Phi(z_\alpha + 2\hat{m})$. [Efron (1982, 1985, 1987) uses the notation z_0 instead of $\hat{m}$.]

To estimate the acceleration constant a, remember that

$$a \equiv n^{-1/2} z_\alpha^{-2}\{p_1(z_\alpha) + q_1(z_\alpha) - 2p_1(0)\},$$

where p_1 and q_1 are even quadratic polynomials appearing in Edgeworth expansions of H and K. Some coefficients of these polynomials may be functions of unknown characteristics of the distribution. Replace those quantities by their bootstrap estimates and call the resulting polynomials $\hat{p}_1$ and $\hat{q}_1$, respectively. As we shall see in Section 4, the polynomials $\hat{p}_1$ and $\hat{q}_1$ appear in Edgeworth expansions of $\hat{H}$ and $\hat{K}$. Put

$$\begin{aligned} \hat{a} &\equiv n^{-1/2} z_\alpha^{-2}\{\hat{p}_1(z_\alpha) + \hat{q}_1(z_\alpha) - 2\hat{p}_1(0)\}, \\ \hat{\beta}_\alpha &\equiv \Phi[\hat{m} + (\hat{m} + z_\alpha)\{1 - \hat{a}(\hat{m} + z_\alpha)\}^{-1}]. \end{aligned} \tag{3.1}$$

Of course,

$$\hat{m} + (\hat{m} + z_\alpha)\{1 - \hat{a}(\hat{m} + z_\alpha)\}^{-1} = z_\alpha + 2\hat{m} + \hat{a}z_\alpha^2 + O_p(n^{-1})$$

and so (3.1) compares directly with the definition (2.5) of β_a. The argument of Φ in (3.1) could be replaced by any one of many quantities satisfying $z_\alpha + 2\hat{m} + \hat{a}z_\alpha^2 + O_p(n^{-1})$, without upsetting the main conclusions we shall reach about properties of accelerated bias-correction. The particular choice (3.1) was motivated by Efron (1987) via considerations of transformation theory and is eminently reasonable.

Bootstrap versions of bias-corrected and accelerated bias-corrected critical points are

$$\hat{\theta}_{\text{BC}}(\alpha) \equiv \hat{\theta}_{\text{BACK}}(\hat{\beta}) \qquad \text{and} \qquad \hat{\theta}_{\text{ABC}}(\alpha) \equiv \hat{\theta}_{\text{BACK}}(\hat{\beta}_a),$$

respectively. It is readily seen that $\hat{\theta}_{\text{BC}}$ is identical to the bias-corrected point proposed by Efron (1982); work in the next section shows that $\hat{\theta}_{\text{ABC}}$ is identical to Efron's accelerated bias-corrected point, at least in many important cases.

3.4. The Acceleration Constant

Recall from Section 2.4 that in the cases studied there, $a \equiv n^{-1/2}\frac{1}{6}\sigma^{-3}\sum\sum\sum a_i a_j a_k \mu_{ijk}$. Our estimate of a is of course

$$\hat{a} \equiv n^{-1/2}\,\tfrac{1}{6}\,\hat{\sigma}^{-3}\sum\sum\sum \hat{a}_i\hat{a}_j\hat{a}_k\hat{\mu}_{ijk},$$

where the "hats" denote bootstrap estimates. We shall prove that this estimate of a coincides with that given by Efron (1987). Section 4 will show that $\hat{\theta}_{\mathrm{ABC}}$ is second-order correct and together these results confirm Efron's conjecture about second-order correctness of his accelerated bias-corrected critical points, at least in the cases studied here.

The reader is referred to Section 2.4 for necessary notation.

Case (i) (Exponential Family Model). Efron's estimate is

$$\hat{a}_{\mathrm{Ef}} \equiv n^{-1/2}\tfrac{1}{6}\hat{\psi}^{(3)}(0)\{\hat{\psi}^{(2)}(0)\}^{-3/2},$$

where $\hat{\psi}^{(j)}(0) \equiv (\partial/\partial t)^j\psi(\hat{\lambda}+t\hat{\tau})|_{t=0}$, $\hat{\lambda}$ is an estimate of λ (e.g., maximum likelihood estimator, although it could be something else) and $\hat{\tau}$ is obtained from the d-vector $\tau = (\tau^{(i)})$ defined in the following, on replacing λ by $\hat{\lambda}$:

$$\tau^{(i)}(\lambda) \equiv \sum_j \nu_{ij}(\lambda)\frac{\partial}{\partial\lambda^{(j)}}\,\theta(\lambda).$$

Now

$$\frac{\partial\theta}{\partial\lambda^{(j)}} = \sum_k \frac{\partial\theta}{\partial\mu^{(k)}}\frac{\partial\mu^{(k)}}{\partial\lambda^{(j)}} = \sum_k a_k\mu_{kj} = \sum_k \mu_{jk}a_k,$$

whence, since $(\nu_{ij}) = (\mu_{ij})^{-1}$,

$$\tau^{(i)}(\lambda) = \sum_j\sum_k \nu_{ij}\mu_{jk}a_k = a_i.$$

It is now relatively easy to prove that

$$(\partial/\partial t)^l\psi(\lambda+t\tau)|_{t=0} = \begin{cases} \sigma^2 & \text{if } l=2,\\ \sum_i\sum_j\sum_k a_i a_j a_k\mu_{ijk} & \text{if } l=3,\end{cases}$$

and so the theoretical version of $\hat{a}_{\mathrm{Ef}}$ is just our a. In consequence, $\hat{a}_{\mathrm{Ef}} = \hat{a}$.

Case (ii) (Nonparametric Inference). Efron's estimate is

$$\hat{a}_{\mathrm{Ef}} \equiv \tfrac{1}{6}\left(\sum_{i=1}^n U_i^3\right)\left(\sum_{i=1}^n U_i^2\right)^{-3/2},$$

where

$$U_i \equiv \lim_{\Delta\to 0} [f\{(1-\Delta)\bar{X} + \Delta X_i\} - f(\bar{X})]\Delta^{-1} = \sum_{j=1}^{d} (X_i - \bar{X})^{(j)} f_{(j)}(\bar{X}).$$

Notice that

$$n^{-1}\sum_{k=1}^{n} U_k^2 = \sum_{i=1}^{d}\sum_{j=1}^{d} f_{(i)}(\bar{X}) f_{(j)}(\bar{X}) n^{-1} \sum_{k=1}^{n} (X_k - \bar{X})^{(i)} (X_k - \bar{X})^{(j)},$$

which is simply the bootstrap estimate $\hat{\sigma}^2$ of $\sigma^2 = \sum\sum f_{(i)}(\mu) f_{(j)}(\mu)\mu_{ij}$, obtained by replacing all population moments by sample moments. Similarly, $n^{-1}\sum U_k^3$ is just the bootstrap estimate of $\sum\sum\sum a_i a_j a_k \mu_{ijk}$. Therefore, $\hat{a}_{\mathrm{Ef}} = \hat{a}$.

3.5. "Shortest" Intervals

Recall that the numbers v and w used to construct the "ideal" shortest interval $I_0 \equiv [\hat{\theta} - n^{-1/2}\hat{\sigma}v, \hat{\theta} + n^{-1/2}\hat{\sigma}w]$ in Section 2.5 were defined to minimize $v + w$ subject to $K(v) - K(-w) = 1 - 2\alpha$. Their bootstrap estimates are defined as follows. For each x such that $\hat{K}(x) \geq 1 - 2\alpha$, choose $y = y(x)$ such that $\hat{K}(x) - \hat{K}(-y)$ is as close as possible to $1 - 2\alpha$. Take $(\hat{v}, \hat{w})$ to be that pair (x, y) which minimizes $x + y$. The shortest bootstrap confidence interval is then

$$I_1 \equiv [\hat{\theta} - n^{-1/2}\hat{\sigma}\hat{v}, \hat{\theta} + n^{-1/2}\hat{\sigma}\hat{w}].$$

Buckland (1980) has given an informal treatment of shortest bootstrap confidence intervals, although of a different type from those here. See also Buckland (1983).

4. Properties of Bootstrap Critical Points

4.1. Introduction

Throughout this paper we have stressed difficulties which we have with critical points that are based on "looking up the wrong tables." To delineate our argument, it is convenient to go back to one of the simple examples mentioned in Section 1.2. Suppose our sample is drawn from an $N(\theta, \sigma^2)$ population and we estimate θ and σ^2 via maximum likelihood. As we pointed out in Section 1.2, the distribution functions $\hat{H}$ and H are identical in this case [both being $N(0,1)$], and the distribution functions $\hat{K}$ and K

are identical (both being scale-changed Student's t with $n-1$ degrees of freedom). Therefore, $\hat{G}(\hat{\theta}) = \hat{H}(0) = H(0) = \frac{1}{2}$, whence $\hat{m} = \Phi^{-1}\{\hat{G}(\hat{\theta})\} = 0$, and $\hat{p}_1 \equiv p_1 \equiv \hat{q}_1 \equiv q_1 \equiv 0$, whence

$$\hat{a} \equiv n^{-1/2} z_\alpha^{-2} \{\hat{p}_1(z_\alpha) + \hat{q}_1(z_\alpha) - 2\hat{p}_1(0)\} = 0.$$

In consequence, $\hat{\beta} = \beta = \hat{\beta}_a = \beta_a = \alpha$, $\hat{\theta}_{\mathrm{Stud}}(\alpha) = \hat{\theta}_{\mathrm{STUD}}(\alpha) = \hat{\theta} - \hat{\sigma} y_{1-\alpha}$ and

$$\hat{\theta}_{\mathrm{ord}}(\alpha) = \hat{\theta}_{\mathrm{ORD}}(\alpha) = \hat{\theta}_{\mathrm{hyb}}(\alpha) = \hat{\theta}_{\mathrm{HYB}}(\alpha) = \hat{\theta}_{\mathrm{back}}(\alpha) = \hat{\theta}_{\mathrm{BACK}}(\alpha)$$

$$= \hat{\theta}_{\mathrm{bc}}(\alpha) = \hat{\theta}_{\mathrm{BC}}(\alpha) = \hat{\theta}_{\mathrm{abc}}(\alpha) = \hat{\theta}_{\mathrm{ABC}}(\alpha) = \hat{\theta} - \hat{\sigma} x_{1-\alpha}.$$

Thus, each of the bootstrap critical points $\hat{\theta}_{\mathrm{HYB}}$, $\hat{\theta}_{\mathrm{BACK}}$, $\hat{\theta}_{\mathrm{BC}}$ and $\hat{\theta}_{\mathrm{ABC}}$ is tantamount to looking up standard normal tables, when we should be consulting Student's t-tables. Only $\hat{\theta}_{\mathrm{STUD}}$ is equivalent to looking up the right tables. See also Beran [(1987), Section 3.4].

The situation is not so clear-cut in other circumstances, although the philosophical attractions of $\hat{\theta}_{\mathrm{STUD}}$ are just as strong. In this section we use Edgeworth expansion theory to elucidate and compare properties of bootstrap critical points. We show that $\hat{\theta}_{\mathrm{STUD}}$ and $\hat{\theta}_{\mathrm{ABC}}$ are both second-order correct, but argue that while second-order correctness has a major role to play in the theory of one-sided confidence intervals, its importance for two-sided intervals is diminished. There, third-order properties assume a significant role in determining confidence interval length, although second-order properties do have an influence on coverage. The difference between Student's t-tables and standard normal tables is a third-order effect. We argue that third-order properties of $\hat{\theta}_{\mathrm{STUD}}$ are closer to those of $\hat{\theta}_{\mathrm{Stud}}$ than are those of $\hat{\theta}_{\mathrm{ABC}}$. For example, the expected length of the two-sided interval $[\hat{\theta}_{\mathrm{STUD}}(\alpha), \hat{\theta}_{\mathrm{STUD}}(1-\alpha)]$ is closer to the expected length of $[\hat{\theta}_{\mathrm{Stud}}(\alpha), \hat{\theta}_{\mathrm{Stud}}(1-\alpha)]$ than is the expected length of $[\hat{\theta}_{\mathrm{ABC}}(\alpha), \hat{\theta}_{\mathrm{ABC}}(1-\alpha)]$. In the example at the beginning of this section, $\hat{\theta}_{\mathrm{STUD}}$ got third-order properties exactly right; $\hat{\theta}_{\mathrm{ABC}}$ got them wrong.

We show that the mean length of bootstrap confidence intervals is often not directly related to coverage. However, our examples demonstrate that in the case of equal-tailed two-sided 95% intervals for a population mean based on bootstrap critical points, $\hat{\theta}_{\mathrm{STUD}}$ leads to intervals which tend to be conservative in the sense that they have longer length and greater coverage than their competitors. (This generalization can fail in the case of distributions with exceptionally large positive kurtosis.) Oddly, the shortest bootstrap intervals introduced in Section 3.5 have both shorter length and smaller coverage error than equal-tailed intervals based on $\hat{\theta}_{\mathrm{STUD}}$, in the case of our examples. For example, shortest 95% bootstrap confidence intervals for a population mean have almost 50% smaller coverage error, in large samples, than their equal-tailed competitors based on $\hat{\theta}_{\mathrm{STUD}}$.

4.2. Edgeworth Expansions and Cornish–Fisher Inversions

Edgeworth expansions of the form

$$\begin{aligned} H(x) &= \Phi(x) + \sum_{i=1}^{\nu} n^{-i/2} p_i(x)\phi(x) + O(n^{-(\nu+1)/2}), \\ K(x) &= \Phi(x) + \sum_{i=1}^{\nu} n^{-i/2} q_i(x)\phi(x) + O(n^{-(\nu+1)/2}), \end{aligned} \tag{4.1}$$

have bootstrap analogues

$$\hat{H}(x) = \Phi(x) + \sum_{i=1}^{\nu} n^{-i/2} \hat{p}_i(x)\phi(x) + O_p(n^{-(\nu+1)/2}), \tag{4.2}$$

$$\hat{K}(x) = \Phi(x) + \sum_{i=1}^{\nu} n^{-i/2} \hat{q}_i(x)\phi(x) + O_p(n^{-(\nu+1)/2}), \tag{4.3}$$

in which $\hat{p}_i$ and $\hat{q}_i$ are identical to p_i and q_i except that unknown quantities in coefficients are replaced by bootstrap estimates. [See Hall (1986); technical arguments are outlined in Section 5.] Likewise, Cornish–Fisher inversions of theoretical quantiles, such as

$$\begin{aligned} x_\alpha &\equiv H^{-1}(\alpha) = z_\alpha + \sum_{i=1}^{\nu} n^{-i/2} p_{i1}(z_\alpha) + O(n^{-(\nu+1)/2}), \\ y_\alpha &\equiv K^{-1}(\alpha) = z_\alpha + \sum_{i=1}^{\nu} n^{-i/2} q_{i1}(z_\alpha) + O(n^{-(\nu+1)/2}), \end{aligned}$$

have analogues

$$\hat{x}_\alpha \equiv \hat{H}^{-1}(\alpha) = z_\alpha + \sum_{i=1}^{\nu} n^{-i/2} \hat{p}_{i1}(z_\alpha) + O_p(n^{-(\nu+1)/2}), \tag{4.4}$$

$$\hat{y}_\alpha \equiv \hat{K}^{-1}(\alpha) = z_\alpha + \sum_{i=1}^{\nu} n^{-i/2} \hat{q}_{i1}(z_\alpha) + O_p(n^{-(\nu+1)/2}). \tag{4.5}$$

Polynomials $\hat{p}_{i1}$ are related to $\hat{p}_j$ and $\hat{q}_{i1}$ are related to $\hat{q}_j$ in the usual manner. For example, the bootstrap analogue of (1.5) holds; that suffices for our purposes.

4.3. Expansions of Bootstrap Critical Points

We begin with bias-corrected points. By (4.2), noting that $\hat{p}_2(0) = 0$ since $\hat{p}_2$ is odd, we have

$$\begin{aligned} z_{\hat{\beta}} &= z_\alpha + 2\hat{m} = z_\alpha + 2\Phi^{-1}\{\hat{H}(0)\} \\ &= z_\alpha + 2\Phi^{-1}\{\tfrac{1}{2} + n^{-1/2}\hat{p}_1(0)\phi(0) + O_p(n^{-3/2})\} \\ &= z_\alpha + n^{-1/2} 2\hat{p}_1(0) + O_p(n^{-3/2}). \end{aligned}$$

Therefore, by (4.4),

$$\hat{x}_{\hat{\beta}} = z_{\hat{\beta}} + \sum_{i=1}^{2} n^{-i/2}\hat{p}_{i1}(z_{\hat{\beta}}) + O_p(n^{-3/2})$$
$$= z_\alpha + n^{-1/2}\{\hat{p}_{11}(z_\alpha) + 2\hat{p}_1(0)\} + n^{-1}\{\hat{p}_{21}(z_\alpha) + 2\hat{p}'_{11}(z_\alpha)\hat{p}_1(0)\} + O_p(n^{-3/2}). \tag{4.6}$$

Also, $\hat{a} \equiv n^{-1/2}z_\alpha^{-2}\{\hat{p}_1(z_\alpha) + \hat{q}_1(z_\alpha) - 2\hat{p}_1(0)\}$ and

$$z_{\hat{\beta}_a} = \hat{m} + (\hat{m} + z_\alpha)\{1 - \hat{a}(\hat{m} + z_\alpha)\}^{-1}$$
$$= z_\alpha + 2\hat{m} + \hat{a}(z_\alpha^2 + 2z_\alpha\hat{m}) + \hat{a}^2 z_\alpha^3 + O_p(n^{-3/2})$$
$$= z_\alpha + n^{-1/2}\{\hat{p}_1(z_\alpha) + \hat{q}_1(z_\alpha)\}$$
$$+ n^{-1}\{\hat{p}_1(z_\alpha) + \hat{q}_1(z_\alpha)\}\{\hat{p}_1(z_\alpha) + \hat{q}_1(z_\alpha) - 2\hat{p}_1(0)\}z_\alpha^{-1} + O_p(n^{-3/2}).$$

Therefore, by (4.4),

$$\hat{x}_{\hat{\beta}_a} = z_{\hat{\beta}_a} + \sum_{i=1}^{2} n^{-i/2}\hat{p}_{i1}(z_{\hat{\beta}_a}) + O_p(n^{-3/2})$$
$$= z_\alpha + n^{-1/2}\{\hat{p}_1(z_\alpha) + \hat{q}_1(z_\alpha) + \hat{p}_{11}(z_\alpha)\} + n^{-1}(\{\hat{p}_1(z_\alpha) + \hat{q}_1(z_\alpha)\}$$
$$\times [\{\hat{p}_1(z_\alpha) + \hat{q}_1(z_\alpha) - 2\hat{p}_1(0)\}z_\alpha^{-1} + \hat{p}'_{11}(z_\alpha)] + \hat{p}_{21}(z_\alpha)) + O_p(n^{-3/2}). \tag{4.7}$$

Together, results (4.4)–(4.7) give expansions of all the quantile estimates used to construct the six bootstrap critical points. Using those formulae and noting that $\hat{p}_{11} = -\hat{p}_1$ and $\hat{q}_{11} = -\hat{q}_1$ [see (1.5)], we obtain the expansions

$$\hat{\theta}_{\mathrm{ORD}}(\alpha) = \hat{\theta} + n^{-1/2}\sigma\{z_\alpha + n^{-1/2}\hat{p}_1(z_\alpha) + n^{-1}\hat{p}_{21}(z_\alpha)\} + O_p(n^{-2}),$$
$$\hat{\theta}_{\mathrm{STUD}}(\alpha) = \hat{\theta} + n^{-1/2}\hat{\sigma}\{z_\alpha + n^{-1/2}\hat{q}_1(z_\alpha) + n^{-1}\hat{q}_{21}(z_\alpha)\} + O_p(n^{-2}),$$
$$\hat{\theta}_{\mathrm{HYB}}(\alpha) = \hat{\theta} + n^{-1/2}\hat{\sigma}\{z_\alpha + n^{-1/2}\hat{p}_1(z_\alpha) + n^{-1}\hat{p}_{21}(z_\alpha)\} + O_p(n^{-2}),$$
$$\hat{\theta}_{\mathrm{BACK}}(\alpha) = \hat{\theta} + n^{-1/2}\hat{\sigma}\{z_\alpha - n^{-1/2}\hat{p}_1(z_\alpha) + n^{-1}\hat{p}_{21}(z_\alpha)\} + O_p(n^{-2}),$$
$$\hat{\theta}_{\mathrm{BC}}(\alpha) = \hat{\theta} + n^{-1/2}\hat{\sigma}[z_\alpha + n^{-1/2}\{2\hat{p}_1(0) - \hat{p}_1(z_\alpha)\}$$
$$+ n^{-1}\hat{p}_{21}(z_\alpha) - 2\hat{p}'_1(z_\alpha)\hat{p}_1(0)\}] + O_p(n^{-2}),$$
$$\hat{\theta}_{\mathrm{ABC}}(\alpha) = \hat{\theta} + n^{-1/2}\hat{\sigma}\{z_\alpha + n^{-1/2}\hat{q}_1(z_\alpha) + n^{-1}(\{\hat{p}_1(z_\alpha) + \hat{q}_1(z_\alpha)\}$$
$$\times [\{\hat{p}_1(z_\alpha) + \hat{q}_1(z_\alpha) - 2\hat{p}_1(0)\}z_\alpha^{-1} - \hat{p}'_1(z_\alpha)]$$
$$+ \hat{p}_{21}(z_\alpha))\} + O_p(n^{-2}).$$

Of course, $\hat{p}_{21}(x) = \hat{p}_1(x)\hat{p}'_1(x) - \frac{1}{2}x\hat{p}_1(x)^2 - \hat{p}_2(x)$, with a similar formula for $\hat{q}_{21}$; see (1.5).

Expansions of the "ideal" critical points $\hat{\theta}_{\text{ord}}$ and $\hat{\theta}_{\text{Stud}}$ may be derived similarly but more simply; they are

$$\hat{\theta}_{\text{ord}}(\alpha) = \hat{\theta} + n^{-1/2}\sigma\{z_\alpha + n^{-1/2}p_1(z_\alpha) + n^{-1}p_{21}(z_\alpha)\} + O(n^{-2}),$$

$$\hat{\theta}_{\text{Stud}}(\alpha) = \hat{\theta} + n^{-1/2}\hat{\sigma}\{z_\alpha + n^{-1/2}q_1(z_\alpha) + n^{-1}q_{21}(z_\alpha)\} + O_p(n^{-2}).$$

Comparing all these expansions and noting that $\hat{p}_1 = p_1 + O_p(n^{-1/2})$ and $\hat{q}_1 = q_1 + O_p(n^{-1/2})$, we conclude that $|\hat{\theta}_{\text{STUD}} - \hat{\theta}_{\text{Stud}}|$ and $|\hat{\theta}_{\text{ABC}} - \hat{\theta}_{\text{Stud}}|$ are both $O_p(n^{-3/2})$. Therefore, $\hat{\theta}_{\text{STUD}}$ and $\hat{\theta}_{\text{ABC}}$ are second-order correct, while $\hat{\theta}_{\text{HYB}}$, $\hat{\theta}_{\text{BACK}}$ and $\hat{\theta}_{\text{BC}}$ are usually only first-order correct. This is exactly the behaviour noted in Section 2 for the theoretical versions of these critical points. Bootstrap approximations to theoretical critical points are so good that they reflect the inferior properties of points such as $\hat{\theta}_{\text{hyb}}$, $\hat{\theta}_{\text{back}}$ and $\hat{\theta}_{\text{bc}}$.

If it so happens that the polynomials p_1 and q_1 are identical, then of course the hybrid critical point *is* second-order correct. Indeed, in that case *the hybrid and accelerated bias-corrected critical points are third-order equivalent*. To see this, observe from the preceding expansions that

$$\hat{\theta}_{\text{HYB}}(\alpha) - \hat{\theta}_{\text{ABC}}(\alpha) = n^{-3/2}\hat{\sigma}\{\hat{p}_1(z_\alpha) + \hat{q}_1(z_\alpha)\} \times [\{\hat{p}_1(z_\alpha) + \hat{q}_1(z_\alpha) - 2\hat{p}_1(0)\}z_\alpha^{-1} - \hat{p}_1'(z_\alpha)] + O_p(n^{-2}).$$

When $p_1 \equiv q_1$, we have $\hat{p}_1(x) = \hat{q}_1(x) = \hat{C}_1 + \hat{C}_2 x^2$ for random variables $\hat{C}_1$ and $\hat{C}_2$ and for all x. Therefore,

$$\{\hat{p}_1(z_\alpha) + \hat{q}_1(z_\alpha) - 2\hat{p}_1(0)\}z_\alpha^{-1} - \hat{p}_1'(z_\alpha) = 2\hat{C}_2 z_\alpha^2 \cdot z_\alpha^{-1} - 2\hat{C}_2 z_\alpha = 0.$$

In consequence, $\hat{\theta}_{\text{HYB}}(\alpha) - \hat{\theta}_{\text{ABC}}(\alpha) = O_p(n^{-2})$, implying that $\hat{\theta}_{\text{HYB}}$ and $\hat{\theta}_{\text{ABC}}$ are third-order equivalent. This circumstance arises when θ is a slope parameter in general regression problems, such as multivariate linear or polynomial regression. Although regression problems do not fit easily into the discussion in this paper, it is nevertheless true that hybrid and accelerated bias-corrected critical points for slope parameters are third-order equivalent.

Unlike the other bootstrap critical points, $\hat{\theta}_{\text{ORD}}$ is designed for use when σ is known, and so should be compared with $\hat{\theta}_{\text{ord}}$ rather than $\hat{\theta}_{\text{Stud}}$. When viewed in these terms $\hat{\theta}_{\text{ORD}}$ is second-order correct, since $|\hat{\theta}_{\text{ORD}} - \hat{\theta}_{\text{ord}}| = O_p(n^{-3/2})$.

EXAMPLE 1 (Nonparametric Estimation of Mean). Let $Y_1, \ldots, Y_n$ be independent and identically distributed observations from a continuous univariate population with mean $\theta \equiv E(Y_1)$, variances $\sigma^2 \equiv E(Y_1 - \theta)^2$, standardized skewness $\gamma \equiv \sigma^{-3}E(Y_1 - \theta)^3$ and standardized kurtosis $\kappa \equiv \sigma^{-4}E(Y_1 - \theta)^4 - 3$. Sample versions of these quantities are $\hat{\theta} \equiv n^{-1}\sum Y_i$, $\hat{\sigma}^2 \equiv n^{-1}\sum(Y_i - \bar{Y})^2$, $\hat{\gamma} \equiv \hat{\sigma}^{-3}n^{-1}\sum(Y_i - \bar{Y})^3$ and $\hat{\kappa} \equiv$

$\hat{\sigma}^{-4}n^{-1}\sum(Y_i - \bar{Y})^4 - 3$, respectively. The polynomials which interest us are on this occasion

$$p_1(x) \equiv -\tfrac{1}{6}\gamma(x^2-1),$$
$$q_1(x) \equiv \tfrac{1}{6}\gamma(2x^2+1),$$
$$p_2(x) \equiv -x\{\tfrac{1}{24}\kappa(x^2-3) + \tfrac{1}{72}\gamma^2(x^4-10x^2+15)\},$$
$$q_2(x) \equiv x\{\tfrac{1}{12}\kappa(x^2-3) - \tfrac{1}{18}\gamma^2(x^4+2x^2-3) - \tfrac{1}{4}(x^2+3)\},$$

[see, e.g., Geary (1947), Petrov (1975), page 138]. Polynomials $\hat{p}_1, \hat{p}_2, \hat{q}_1$ and $\hat{q}_2$ are identical to their theoretical counterparts, except that γ and κ are replaced by $\hat{\gamma}$ and $\hat{\kappa}$, respectively. Noting that $\hat{\gamma} = \gamma + O_p(n^{-1/2})$ and $\hat{\kappa} = \kappa + O_p(n^{-1/2})$, we may derive the following expansions of critical points:

$$\begin{aligned}
\hat{\theta}_{\text{STUD}}(\alpha) &= \hat{\theta} + n^{-1/2}\hat{\sigma}[z_\alpha + n^{-1/2}\tfrac{1}{6}\hat{\gamma}(2z_\alpha^2+1) \\
&\quad + n^{-1}z_\alpha\{-\tfrac{1}{12}\kappa(z_\alpha^2-3) + \tfrac{5}{72}\gamma^2(4z_\alpha^2-1) + \tfrac{1}{4}(z_\alpha^2+3)\}] + O_p(n^{-2}), \\
\hat{\theta}_{\text{HYB}}(\alpha) &= \hat{\theta} + n^{-1/2}\hat{\sigma}[z_\alpha - n^{-1/2}\tfrac{1}{6}\hat{\gamma}(z_\alpha^2-1) \\
&\quad + n^{-1}z_\alpha\{\tfrac{1}{24}\kappa(z_\alpha^2-3) - \tfrac{1}{36}\gamma^2(2z_\alpha^2-5)\}] + O_p(n^{-2}), \\
\hat{\theta}_{\text{BACK}}(\alpha) &= \hat{\theta} + n^{-1/2}\hat{\sigma}[z_\alpha + n^{-1/2}\tfrac{1}{6}\hat{\gamma}(z_\alpha^2-1) \\
&\quad + n^{-1}z_\alpha\{\tfrac{1}{24}\kappa(z_\alpha^2-3) - \tfrac{1}{36}\gamma^2(2z_\alpha^2-5)\}] + O_p(n^{-2}), \\
\hat{\theta}_{\text{BC}}(\alpha) &= \hat{\theta} + n^{-1/2}\hat{\sigma}[z_\alpha + n^{-1/2}\tfrac{1}{6}\hat{\gamma}(2z_\alpha^2+1) \\
&\quad + n^{-1}z_\alpha\{\tfrac{1}{24}\kappa(z_\alpha^2-3) - \tfrac{1}{36}\gamma^2(2z_\alpha^2-9)\}] + O_p(n^{-2}), \\
\hat{\theta}_{\text{ABC}}(\alpha) &= \hat{\theta} + n^{-1/2}\hat{\sigma}[z_\alpha + n^{-1/2}\tfrac{1}{6}\hat{\gamma}(2z_\alpha^2+1) \\
&\quad + n^{-1}z_\alpha\{\tfrac{1}{24}\kappa(z_\alpha^2-3) + \tfrac{1}{36}\gamma^2(2z_\alpha^2+11)\}] + O_p(n^{-2}),
\end{aligned} \tag{4.8}$$

$$\begin{aligned}
\hat{\theta}_{\text{Stud}}(\alpha) &= \hat{\theta} + n^{-1/2}\hat{\sigma}[z_\alpha + n^{-1/2}\tfrac{1}{6}\gamma(2z_\alpha^2+1) \\
&\quad + n^{-1}z_\alpha\{-\tfrac{1}{12}\kappa(z_\alpha^2-3) + \tfrac{5}{72}\gamma^2(4z_\alpha^2-1) + \tfrac{1}{4}(z_\alpha^2+3)\}] + O_p(n^{-2}),
\end{aligned} \tag{4.9}$$

Similar expansions may be derived for $\hat{\theta}_{\text{ORD}}$ and $\hat{\theta}_{\text{ord}}$.

Example 2 (Estimation of Exponential Mean). Let $Y_1, \ldots, Y_n$ be independent and identically distributed observations from the distribution with density $h_\theta(y) \equiv \theta^{-1}\exp(-\theta^{-1}y)$, $y > 0$. The maximum likelihood estimate of θ is the sample mean $\hat{\theta} \equiv n^{-1}\sum Y_i$ and is also the maximum likelihood estimator of σ $(= \theta)$. As noted in Section 1.2, the distribution functions H and $\hat{H}$ and identical, and the distribution functions K and $\hat{K}$ are identical, in this case. Therefore, bootstrap critical points are identical to their theoretical counterparts. The polynomials are $p_1(x) \equiv -(1/3)(x^2-1)$, $p_2(x) \equiv -(1/36)x(2x^4-11x^2+3)$, $q_1(x) \equiv (1/3)(2x^2+1)$ and $q_2(x) = -(1/36)x(8x^4-11x^2+3)$. In consequence,

$$\begin{aligned}
\hat{\theta}_{\mathrm{STUD}}(\alpha) &= \hat{\theta}_{\mathrm{Stud}}(\alpha) \\
&= \hat{\theta}+n^{-1/2}\hat{\sigma}\{z_\alpha+n^{-1/2}\tfrac{1}{3}(2z_\alpha^2+1)+n^{-1}\tfrac{1}{36}\,z_\alpha(13z_\alpha^2+17)\}+O_p(n^{-2}), \\
\hat{\theta}_{\mathrm{HYB}}(\alpha) &= \hat{\theta}+n^{-1/2}\hat{\sigma}\{z_\alpha-n^{-1/2}\tfrac{1}{3}(z_\alpha^2-1)+n^{-1}\tfrac{1}{36}\,z_\alpha(z_\alpha^2-7)\}+O_p(n^{-2}), \\
\hat{\theta}_{\mathrm{BACK}}(\alpha) &= \hat{\theta}+n^{-1/2}\hat{\sigma}\{z_\alpha+n^{-1/2}\tfrac{1}{3}(z_\alpha^2-1)+n^{-1}\tfrac{1}{36}\,z_\alpha(z_\alpha^2-7)\}+O_p(n^{-2}), \\
\hat{\theta}_{\mathrm{BC}}(\alpha) &= \hat{\theta}+n^{-1/2}\hat{\sigma}\{z_\alpha+n^{-1/2}\tfrac{1}{3}(z_\alpha^2+1)+n^{-1}\tfrac{1}{36}\,z_\alpha(z_\alpha^2+9)\}+O_p(n^{-2}), \\
\hat{\theta}_{\mathrm{ABC}}(\alpha) &= \hat{\theta}+n^{-1/2}\hat{\sigma}\{z_\alpha+n^{-1/2}\tfrac{1}{3}(2z_\alpha^2+1)+n^{-1}\tfrac{1}{36}\,z_\alpha(13z_\alpha^2+17)\}+O_p(n^{-2}).
\end{aligned}$$

Therefore, $\hat{\theta}_{\mathrm{STUD}}$ and $\hat{\theta}_{\mathrm{ABC}}$ are both *third-order* correct. This contrasts with the parametric example which we treated in Section 4.1, where we showed that $\hat{\theta}_{\mathrm{ABC}}$ failed to be third-order correct.

4.4. Lengths of Two-Sided Equal-Tailed Intervals

Each of the critical points $\hat{\theta}_{\mathrm{STUD}}$, $\hat{\theta}_{\mathrm{HYB}}$, $\hat{\theta}_{\mathrm{BACK}}$, $\hat{\theta}_{\mathrm{ABC}}$ and $\hat{\theta}_{\mathrm{Stud}}$ admits an expansion of the form

$$\hat{\theta}(\alpha) = \hat{\theta}+n^{-1/2}\hat{\sigma}\left\{z_\alpha+\sum_{i=1}^{3} n^{-i/2}\hat{s}_i(z_\alpha)\right\}+O_p(n^{-5/2}), \tag{4.10}$$

where $\hat{s}_1$ and $\hat{s}_3$ are even polynomials and $\hat{s}_2$ is an odd polynomial. The two-sided, equal-tailed confidence interval $I(1-2\alpha) \equiv [\hat{\theta}(\alpha), \hat{\theta}(1-\alpha)]$ therefore has length

$$\begin{aligned}
l(1-2\alpha) &\equiv \hat{\theta}(1-\alpha)-\hat{\theta}(\alpha) \\
&= 2n^{-1/2}\hat{\sigma}\{z_{1-\alpha}+n^{-1}\hat{s}_2(z_{1-\alpha})\}+O_p(n^{-5/2}).
\end{aligned} \tag{4.11}$$

Note particularly that second-order terms have cancelled entirely. Equal-tailed intervals based on $\hat{\theta}_{\mathrm{HYB}}$ and $\hat{\theta}_{\mathrm{BACK}}$ always have exactly the same length, but usually have different centres.

In the case of $\hat{\theta}_{\mathrm{Stud}}$, the polynomial $\hat{s}_2$ is of course deterministic; we write it as $s_{2,\mathrm{Stud}}$. The version $\hat{s}_{2,\mathrm{STUD}}$ of $\hat{s}_2$ in the case of $\hat{\theta}_{\mathrm{STUD}}$ is derived by replacing unknowns in the coefficients of $s_{2,\mathrm{Stud}}$ by their bootstrap estimates. This means that the lengths $l_{\mathrm{Stud}}(1-2\alpha)$ and $l_{\mathrm{STUD}}(1-2\alpha)$ of the intervals $[\hat{\theta}_{\mathrm{Stud}}(\alpha), \hat{\theta}_{\mathrm{Stud}}(1-\alpha)]$ and $[\hat{\theta}_{\mathrm{STUD}}(\alpha), \hat{\theta}_{\mathrm{STUD}}(1-\alpha)]$ differ only by a term of $O_p(n^{-2})$. In general, none of the other bootstrap intervals track the "ideal" equal-tailed interval $[\hat{\theta}_{\mathrm{Stud}}(\alpha), \hat{\theta}_{\mathrm{Stud}}(1-\alpha)]$ as closely as this; the error in length is usually $O_p(n^{-3/2})$. In the case of accelerated bias-correction and nonparametric estimation of a mean, this is clear from comparison of expansions (4.8) and (4.9).

The closeness with which the interval $I_{\mathrm{STUD}}(1-2\alpha)$ tracks $I_{\mathrm{Stud}}(1-2\alpha)$ is even plainer if we base our comparison on *mean* interval length. Since $E(\hat{\sigma}\hat{s}_{2,\mathrm{STUD}}) = \sigma s_{2,\mathrm{Stud}} + O(n^{-1}) = E(\hat{\sigma})s_{2,\mathrm{Stud}} + O(n^{-1})$, then $E\{l_{\mathrm{STUD}}(1-2\alpha)\} = E\{l_{\mathrm{Stud}}(1-2\alpha)\} + O(n^{-5/2})$, whereas in general, $E\{l_{\mathrm{ABC}}(1-2\alpha)\} = E\{l_{\mathrm{Stud}}(1-2\alpha)\} + O(n^{-3/2})$.

4.5. Coverage

Let $\hat{\theta}(\alpha)$ be a critical point admitting expansion (4.10) and let s_1 and s_2 denote the theoretical versions of $\hat{s}_1$ and $\hat{s}_2$. Put $U(\alpha) \equiv n^{1/2}\{\hat{s}_1(z_\alpha) - s_1(z_\alpha)\}$, $S \equiv n^{1/2}(\hat{\theta}-\theta)/\hat{\sigma}$ and $T \equiv S + n^{-1}U(\alpha)$. The confidence interval $(-\infty, \hat{\theta}(\alpha)]$ has coverage

$$\begin{aligned}\pi(\alpha) &\equiv P\{\theta \le \hat{\theta}(\alpha)\}\\ &= P\left\{0 \le T + z_\alpha + \sum_{i=1}^{2} n^{-i/2}s_i(z_\alpha) + O_p(n^{-3/2})\right\}\\ &= P\left\{T \ge -z_\alpha - \sum_{i=1}^{2} n^{-i/2}s_i(z_\alpha)\right\} + O(n^{-3/2}),\end{aligned} \tag{4.12}$$

assuming that the $O_p(n^{-3/2})$ term makes a $O(n^{-3/2})$ contribution to be probability. (See Section 5.)

We may deduce a more concise formula for the coverage π by developing an Edgeworth expansion of the distribution of T. That expansion is very close to the one we already know for S [see 4.1)]; indeed,

$$P(T \le x) = P(S \le x) - n^{-1}ux\phi(x) + O(n^{-3/2}) \tag{4.13}$$

uniformly in x, where $u = u(\alpha)$ is a constant satisfying $E\{SU(\alpha)\} = u + O(n^{-1})$ as $n \to \infty$. (See Section 5.) It may now be shown after some algebra that

$$\begin{aligned}\pi(\alpha) &= \alpha + n^{-1/2}\{s_1(z_\alpha) - q_1(z_\alpha)\}\phi(z_\alpha)\\ &\quad - n^{-1}[\tfrac{1}{2}s_1(z_\alpha)^2 z_\alpha + s_1(z_\alpha)\{q_1'(z_\alpha) - q_1(z_\alpha)z_\alpha\}\\ &\qquad - q_1(z_\alpha) - s_2(z_\alpha) - uz_\alpha]\phi(z_\alpha) + O(n^{-3/2}).\end{aligned} \tag{4.14}$$

We should stress that the polynomial $s_1 - q_1$ appearing in the coefficient of the $n^{-1/2}$ term is *even*. This observation is important when calculating the coverage of equal-tailed two-sided confidence intervals.

The simple "normal theory" critical point $\hat{\theta}_{\text{Norm}}(\alpha) \equiv \hat{\theta} + n^{-1/2}\hat{\sigma}z_\alpha$ is based on the fact that $n^{1/2}(\hat{\theta} - \theta)/\hat{\sigma}$ is approximately normally distributed. It has coverage

$$\pi_{\text{Norm}}(\alpha) = \alpha - n^{-1/2}q_1(z_\alpha)\phi(z_\alpha) + n^{-1}q_2(z_\alpha)\phi(z_\alpha) + O(n^{-3/2}).$$

The most important point to notice from (4.14) is that the coverage error $\pi(\alpha) - \alpha$ is of order n^{-1} for all α if and only if $s_1 \equiv q_1$; that is, if and only if the critical point $\hat{\theta}(\alpha)$ is second-order correct. Critical points which fail to be second-order correct lead to coverage errors of order $n^{-1/2}$, rather than n^{-1}, in the case of one-sided confidence intervals.

Note too that the term of order $n^{-1/2}$ in (4.14) is exactly as it would be if the bootstrap critical point $\hat{\theta}(\alpha)$ were replaced by its theoretical version. Indeed, the theoretical version of $\hat{\theta}(\alpha)$ satisfies

$$\hat{\theta}_{\text{theor}}(\alpha) = \hat{\theta} + n^{-1/2}\hat{\sigma}\{z_\alpha + n^{-1/2}s_1(z_\alpha) + O(n^{-1})\}$$

and

$$\begin{aligned} P\{\theta \le \hat{\theta}_{\text{theor}}(\alpha)\} &= P\{S \ge -z_\alpha - n^{-1/2}s_1(z_\alpha) + O(n^{-1})\} \\ &= \alpha + n^{-1/2}\{s_1(z_\alpha) - q_1(z_\alpha)\} + O(n^{-1}), \end{aligned}$$

by (4.1). This reinforces the theme which underlies our paper: bootstrap approximations are so good that bootstrap versions of erroneous theoretical critical points are themselves erroneous.

The situation is quite different in the case of two-sided intervals. Notice that the polynomial $s_1 - q_1$ appearing in (4.14) is even and that the order $n^{-3/2}$ remainder in (4.14) may be written as $n^{-3/2}r(z_\alpha)\phi(z_\alpha) + O(n^{-2})$, where r is an even polynomial. Therefore, the equal-tailed interval $I(1 - 2\alpha) \equiv [\hat{\theta}(\alpha), \hat{\theta}(1 - \alpha)]$ has coverage

$$\begin{aligned} \pi(1 - \alpha) - \pi(\alpha) = 1 - 2\alpha - 2n^{-1}[\tfrac{1}{2}s_1(z_{1-\alpha})^2 z_{1-\alpha} &+ s_1(z_{1-\alpha}) \\ &\times \{q_1'(z_{1-\alpha}) - q_1(z_{1-\alpha})z_{1-\alpha}\} - q_2(z_{1-\alpha}) \\ &- s_2(z_{1-\alpha}) + uz_{1-\alpha}]\phi(z_{1-\alpha}) + O(n^{-2}). \end{aligned} \tag{4.15}$$

The issue of second-order correctness has relatively little influence on coverage in this circumstance. Of course, the precise form of $\hat{s}_1$ does have some bearing on the coefficient of order n^{-1} in (4.15), but it does not affect the order of magnitude of the coverage error.

Formulae (4.14) and (4.15) may be used to develop approximations to coverage error of bootstrap confidence intervals in a wide variety of circumstances. We shall treat only the examples discussed in Section 4.3.

EXAMPLE 1 (Nonparametric Estimation of Mean). (See Section 4.3 for notation and other details.) Here the value of u is $(\kappa - \frac{3}{2}\gamma^2)\gamma^{-1}s_1(z_\alpha)$ and in consequence the versions of $\pi(\alpha)$ in (4.14) reduce to

$$\pi_{\rm STUD}(\alpha) = \alpha - n^{-1}(\kappa - \tfrac{3}{2}\gamma^2)\tfrac{1}{6} z_\alpha(2z_\alpha^2+1)\phi(z_\alpha) + O(n^{-3/2}),$$

$$\pi_{\rm HYB}(\alpha) = \alpha - n^{-1/2}\tfrac{1}{2}\gamma z_\alpha^2\phi(z_\alpha) - n^{-1}z_\alpha\{-\tfrac{1}{24}\kappa(7z_\alpha^2-13) + \tfrac{1}{24}\gamma^2(3z_\alpha^4+6z_\alpha^2-11) + \tfrac{1}{4}(z_\alpha^2+3)\}\phi(z_\alpha) + O(n^{-3/2}),$$

$$\pi_{\rm BACK}(\alpha) = \alpha - n^{-1/2}\tfrac{1}{6}\gamma(z_\alpha^2+2)\phi(z_\alpha) - n^{-1}z_\alpha\{\tfrac{1}{24}\kappa(z_\alpha^2+5) + \tfrac{1}{72}\gamma^2(z_\alpha^4+2z_\alpha^2-9) + \tfrac{1}{4}(z_\alpha^2+3)\}\phi(z_\alpha) + O(n^{-3/2}),$$

$$\pi_{\rm BC}(\alpha) = \alpha - n^{-1/2}\tfrac{1}{6}\gamma z_\alpha^2\phi(z_\alpha) - n^{-1}z_\alpha\{\tfrac{1}{24}\kappa(z_\alpha^2+13) + \tfrac{1}{72}\gamma^2(z_\alpha^4-2z_\alpha^2-41) + \tfrac{1}{4}(z_\alpha^2+3)\}\phi(z_\alpha) + O(n^{-3/2}),$$

$$\pi_{\rm ABC}(\alpha) = \alpha - n^{-1}z_\alpha\{\tfrac{1}{24}\kappa(5z_\alpha^2+13) - \tfrac{1}{8}\gamma^2(2z_\alpha^2+5) + \tfrac{1}{4}(z_\alpha^2+3)\}\phi(z_\alpha) + O(n^{-3/2}).$$

Of course, $\pi_{\rm Stud}(\alpha) = \alpha$.

Coverage probabilities of two-sided bootstrap confidence intervals are more meaningful when they are compared with interval length. We shall do this in the case of two-sided 95% intervals. Observe from (4.11) and (4.14) that interval length $l(1-2\alpha)$ and coverage $\pi(1-\alpha)-\pi(\alpha)$ of an interval $[\hat\theta(\alpha), \hat\theta(1-\alpha)]$ may be written in the form

$$\begin{aligned} l(1-2\alpha) &= 2n^{-1/2}\hat\sigma\{z_{1-\alpha} + n^{-1}s(z_{1-\alpha})\} + O_p(n^{-2}), \\ \pi(1-\alpha)-\pi(\alpha) &= 1-2\alpha + n^{-1}2t(z_{1-\alpha})\phi(z_{1-\alpha}) + O(n^{-2}), \end{aligned} \tag{4.16}$$

for polynomials s and t. For the case of 95% intervals, $\alpha = 0.025$ and $z_{1-\alpha} = 1.95996$. Table 1 relates interval length and coverage error in

Table 1. Length and Coverage of Two-Sided 95% Intervals in Nonparametric Case. The Column Head $s(z_{1-\alpha})$ Is Proportional to the Amount by which Interval Length Exceeds $2z_{1-\alpha}n^{-1/2}\hat\sigma$; The Column Headed $t(z_{1-\alpha})$ Is Proportional To Coverage Error $\pi(1-\alpha)-\pi(\alpha)-(1-2\alpha)$. Standardized Skewness and Kurtosis Are Denoted by γ and κ, Respectively.

Type of critical point	$s(z_{1-\alpha})$ (length)	$t(z_{1-\alpha})$ (coverage error)
STUD	$-0.14\kappa + 1.96\gamma^2 + 3.35$	$-2.84\kappa + 4.25\gamma^2$
HYB	$0.069\kappa - 0.15\gamma^2$	$1.13\kappa - 4.60\gamma^2 - 3.35$
BACK	$0.069\kappa - 0.15\gamma^2$	$-0.72\kappa - 0.37\gamma^2 - 3.35$
BC	$0.069\kappa + 0.072\gamma^2$	$-1.38\kappa + 0.92\gamma^2 - 3.35$
ABC	$0.069\kappa + 0.81\gamma^2$	$-2.63\kappa + 3.11\gamma^2 - 3.35$
Norm	0	$0.14\kappa - 2.12\gamma^2 - 3.35$
Stud	$-0.14\kappa + 1.96\gamma^2 + 3.35$	0

this circumstance. The simple "normal theory" confidence interval $I_{\text{Norm}}(1-2\alpha) \equiv [\hat\theta - n^{-1/2}\hat\sigma z_{1-\alpha}, \hat\theta + n^{-1/2}\hat\sigma z_{1-\alpha}]$ is included for the sake of comparison. The coverage of the interval $(-\infty, \hat\theta + n^{-1/2}\hat\sigma z_\alpha]$ equals

$$\begin{aligned}\pi_{\text{Norm}}(\alpha) = \alpha - n^{-1/2}\tfrac{1}{6}\gamma(2z_\alpha^2+1)\phi(z_\alpha)\\ + n^{-1}z_\alpha\{\tfrac{1}{12}\kappa(z_\alpha^2-3) - \tfrac{1}{18}\gamma^2(z_\alpha^4+2z_\alpha^2-3) - \tfrac{1}{4}(z_\alpha^2+3)\}\\ \times\phi(z_\alpha) + O(n^{-3/2}).\end{aligned}$$

If skewness γ and kurtosis κ are both zero, then $\hat\theta_{\text{STUD}}$ gives rise to two-sided confidence intervals with coverage errors $O(n^{-2})$, not just $O(n^{-1})$. All the other equal-tailed bootstrap confidence intervals have coverage errors $O(n^{-1})$. Indeed when $\gamma = \kappa = 0$, other bootstrap intervals *undercover* by an amount $3.35n^{-1}\phi(1.96)$; see Table 1. The term -3.35 appearing in the third column of Table 1 arises from the difference between the standard normal distribution function and an expansion of Student's t distribution function. In the case of distributions with nonzero skewness or kurtosis, we see from Table 1 that serious undercoverage can occur if we use $\hat\theta_{\text{HYB}}$ when $\kappa \le 0$ and if we use $\hat\theta_{\text{BACK}}$ when $\kappa \ge 0$.

It is clear from Table 1 that in large samples, intervals based on $\hat\theta_{\text{STUD}}$ usually tend to be longer and have greater coverage than intervals based on any of the other equal-tailed bootstrap intervals, and than the normal-theory interval. This generalization only fails in cases of large positive kurtosis and indicates that the interval $[\hat\theta_{\text{STUD}}(0.025), \hat\theta_{\text{STUD}}(0.975)]$ is conservative in many circumstances. Nevertheless, there is no general relationship between coverage error and interval length. For example, in the case of distributions with nonzero skewness the ordinary bias-corrected interval tends to be shorter than the accelerated bias-corrected interval, but has smaller coverage only when $\kappa < 1.74\gamma^2$.

Example 2 (Estimation of Exponential Mean). (See Section 4.3 for notation.) We shall content ourselves here with an analogue of Table 1; see Table 2. This shows that equal-tailed two-sided intervals based on $\hat\theta_{\text{HYB}}$,

Table 2. Length and Coverage of Two-Sided 95% Confidence Intervals in Exponential Case. Notation as for Table 1. See (4.16) for Definitions of s and t.

Type of critical point	$s(z_{1-\alpha})$ (length)	$t(z_{1-\alpha})$ (coverage error)
STUD, ABC, Stud	3.64	0
HYB	−0.17	−8.24
BACK	−0.17	−2.44
BC	0.70	−1.21
Norm	0	−4.29

$\hat{\theta}_{\text{BACK}}$, $\hat{\theta}_{\text{BC}}$ and $\hat{\theta}_{\text{Norm}}$ tend to have shorter length and lower coverage than intervals based on $\hat{\theta}_{\text{Stud}}$, $\hat{\theta}_{\text{STUD}}$ and $\hat{\theta}_{\text{ABC}}$. The latter three critical points are all third-order equivalent, but although $\hat{\theta}_{\text{Stud}} \equiv \hat{\theta}_{\text{STUD}}$, these points are not exactly the same as $\hat{\theta}_{\text{ABC}}$.

Of particular interest is the fact that equal-tailed intervals based on the normal theory critical point $\hat{\theta}_{\text{Norm}}$ appear to have better coverage properties than intervals based on the computationally expensive bootstrap point $\hat{\theta}_{\text{HYB}}$.

4.6. "Shortest" Interval

Some of the properties of theoretical "shortest" confidence intervals were discussed in Section 2.5. Bootstrap analogues of those intervals were introduced in Section 3.5 and have similar properties. In particular, the bootstrap estimates $\hat{v}$ and $\hat{w}$ of v and w satisfy

$$\hat{v} = z_{1-\alpha} + \sum_{j=1}^{3} n^{-j/2}\hat{v}_j + O_p(n^{-2})$$

and

$$\hat{w} = z_{1-\alpha} + \sum_{j=1}^{3}(-n^{-1/2})^j\hat{v}_j + O_p(n^{-2}),$$

where $\hat{v}_j$ is the bootstrap estimate of v_j. (See Section 2.5 for a definition of v_j.) Interval length is

$$\begin{aligned} n^{-1/2}\hat{\sigma}(\hat{v}+\hat{w}) &= 2n^{-1/2}\hat{\sigma}(z_{1-\alpha}+n^{-1}\hat{v}_2) + O_p(n^{-5/2}) \\ &= 2n^{-1/2}\hat{\sigma}(z_{1-\alpha}+n^{-1}v_2) + O_p(n^{-2}), \end{aligned} \tag{4.17}$$

and mean interval length is

$$\begin{aligned} E\{n^{-1/2}\hat{\sigma}(\hat{v}+\hat{w})\} &= E\{2n^{-1/2}\hat{\sigma}(z_{1-\alpha}+n^{-1}\hat{v}_2)\} + O(n^{-5/2}) \\ &= E\{2n^{-1/2}\hat{\sigma}(z_{1-\alpha}+n^{-1}v_2)\} + O(n^{-5/2}) \\ &= E\{n^{-1/2}\hat{\sigma}(v+w)\} + O(n^{-5/2}). \end{aligned}$$

Coverage probability may be found by an argument similar to that leading to (4.15) and is $\beta_1 \equiv 1 - 2\alpha + n^{-1}2uz_{1-\alpha}\phi(z_{1-\alpha}) + O(n^{-3/2})$, where on the present occasion u is given by

$$E\{Sn^{1/2}(\hat{v}_1 - v_1)\} = u + O(n^{-1}).$$

Length of the shortest bootstrap confidence interval is less than that of the equal-tailed interval I_{STUD} by an amount of order $n^{-3/2}$. This is the same order as the difference between standard normal and Student's t-critical points for the mean θ of an $N(\theta, \sigma^2)$ population.

Example 1 (Nonparametric Estimation of Mean). (See Section 4.3 for notation.) Here $v_1 \equiv -\frac{1}{6}\gamma(2z_{1-\alpha}^2 - 3)$,

$$v_2 \equiv z_{1-\alpha}\{-\tfrac{1}{12}\kappa(z_{1-\alpha}^2 - 3) + \tfrac{1}{72}\gamma^2(20z_{1-\alpha}^2 - 21) + \tfrac{1}{4}(z_{1-\alpha}^2 + 3)\} \qquad (4.18)$$

[see (2.10)] and $u \equiv -\frac{1}{6}(\kappa - \frac{3}{2}\gamma^2)(2z_{1-\alpha}^2 - 3)$. Therefore, coverage is

$$\beta_1 = 1 - 2\alpha - n^{-1}\tfrac{1}{3}(\kappa - \tfrac{3}{2}\gamma^2)z_{1-\alpha}(2z_{1-\alpha}^2 - 3)\phi(z_{1-\alpha}) + O(n^{-3/2}).$$

If

$$\beta_2 \equiv 1 - 2\alpha - n^{-1}\tfrac{1}{3}(\kappa - \tfrac{3}{2}\gamma^2)z_{1-\alpha}(2z_{1-\alpha}^2 + 1)\phi(z_{1-\alpha}) + O(n^{-3/2})$$

denotes coverage of the equal-tailed interval $I_{\mathrm{STUD}}(1-2\alpha) \equiv [\hat{\theta}_{\mathrm{STUD}}(\alpha), \hat{\theta}_{\mathrm{STUD}}(1-\alpha)]$, and $\beta_0 \equiv 1 - 2\alpha$ denotes nominal coverage, then the ratio of coverage errors $(\beta_1 - \beta_0)/(\beta_2 - \beta_0)$ converges to $(2z_{1-\alpha}^2 - 3)/(2z_{1-\alpha}^2 + 1)$ as $n \to \infty$. This quantity is always positive for $\beta_0 > 0.78$, and equals 0.38, 0.54 and 0.72 in the important cases $\beta_0 = 0.90, \beta_0 = 0.95$ and $\beta_0 = 0.99$, respectively. Therefore, the "shortest" confidence interval not only results in a reduction in length compared with the equal-tailed interval $I_{\mathrm{STUD}}(1-2\alpha)$, but also a reduction in coverage error, at least in large samples.

Substituting formula (4.18) for v_2 into formula (4.17) for interval length and comparing with formula (4.11) for length of equal-tailed intervals, we see that interval length has been reduced by an amount $n^{-3/2}(4/9)\sigma\gamma^2 z_{1-\alpha} + O_p(n^{-2})$, compared with the equal-tailed interval $I_{\mathrm{STUD}}(1-2\alpha)$.

Example 2 (Estimation of Exponential Mean). Here the shortest bootstrap interval and $I_{\mathrm{STUD}}(1-2\alpha)$ both have zero coverage error. The former has length shorter by an amount $n^{-3/2}(16/9)\theta z_{1-\alpha} + O_p(n^{-2})$.

5. Technical Arguments

Technical arguments are distinctly different in parametric and nonparametric cases. A detailed account will be published elsewhere. In the nonparametric case, many technical arguments are expanded versions of proofs from Hall (1986). For example, result (4.3) for $\nu = 1$ appears in Proposition 5.1 of Hall (1986); inverse Cornish–Fisher expansions such as (4.5) are given in Section 3 and in step (iii) of the proof of Theorem 2.1 of Hall (1986); coverage expansions such as (4.12) and (4.14) appear in step (iv) of the proof of Theorem 2.1 of Hall (1986). In some respects, the parametric case is simpler than the nonparametric one treated in Hall (1986), since the population from which the bootstrap resample $\mathscr{X}^*$ is drawn is continuous.

Result (4.13) follows from the fact that all but the second of the first four cumulants of S and T are identical up to (but not including) terms of order $n^{-3/2}$ and that $k_2(T) = k_2(S) + n^{-1}2u + O(n^{-2})$. To understand why the fourth cumulants agree, it is helpful to notice that $E\{S^3U(\alpha)\} = 3u + O(n^{-1})$ if S and $U(\alpha)$ may be approximated by sums of independent random variables (which is the case under the "smooth function model" introduced in Section 1.3, for example). Note that $E(S^2) = 1 + O(n^{-1})$.

References

Beran, R. (1987). Prepivoting to reduce level error of confidence sets. *Biometrika*, **74**, 457–468.

Bhattacharya, R.N. and Ghosh, J.K. (1978). On the validity of the formal Edgeworth expansion. *Ann. Statist.*, **6**, 435–451.

Bickel, P.J. and Freedman, D. (1981). Some asymptotics on the bootstrap. *Ann. Statist.*, **9**, 1196–1217.

Buckland, S.T. (1980). A modified analysis of the Jolly–Seber capture–recapture model. *Biometrics*, **36**, 419–435.

Buckland, S.T. (1983). Monte Carlo methods for confidence interval construction using the bootstrap technique. *Bias*, **10**, 194–212.

Cox, D.R. (1980). Local ancillarity. *Biometrika*, **67**, 279–286.

Cox, D.R. and Hinkley, D.V. (1974). *Theoretical Statistics*. Chapman and Hall, London.

Efron, B. (1981). Nonparametric standard errors and confidence intervals (with discussion). *Canad. J. Statist.*, **9**, 139–172.

Efron, B. (1982). *The Jackknife, the Bootstrap and Other Resampling Plans*. SIAM. Philadelphia, PA.

Efron, B. (1985). Bootstrap confidence intervals for a class of parametric problems. *Biometrika*, **72**, 45–58.

Efron, B. (1987). Better bootstrap confidence intervals (with discussion). *J. Amer. Statist. Assoc.*, **82**, 171–200.

Efron, B. and Tibshirani, R. (1986). Bootstrap methods for standard errors, confidence intervals, and other measures of statistical accuracy (with discussion). *Statist. Sci.*, **1**, 54–77.

Geary, R.C. (1947). Testing for normality. *Biometrika*, **34**, 209–242.

Hall, P. (1983). Inverting an Edgeworth expansion. *Ann. Statist.*, **11**, 569–576.

Hall, P. (1985). A tabular method for correcting skewness. *Math. Proc. Cambridge Philos. Soc.*, **97**, 525–540.

Hall, P. (1986). On the bootstrap and confidence intervals. *Ann. Statist.*, **14**, 1431–1452.

Harter, H.L. (1964). Criteria for best substitute interval estimator, with an application to the normal distribution. *J. Amer. Statist. Assoc.*, **59**, 1133–1140.

Hartigan, J.A. (1986). Discussion of "Bootstrap methods for standard errors, confidence intervals, and other measures of statistical accuracy" by Efron and Tibshirani. *Statist. Sci.*, **1**, 75–77.

Hinkley, D. and Wei, B.-C. (1984). Improvement of jackknife confidence limit methods. *Biometrika*, **71**, 331–339.

Johnson, N.J. (1978). Modified *t*-tests and confidence intervals for asymmetrical populations. *J. Amer. Statist. Assoc.*, **73**, 536–544.

Kendall, M.G. and Stuart, A. (1979). *The Advanced Theory of Statistics*, vol. **2**, Griffin, London.

McCullagh, P. (1984). Local sufficiency. *Biometrika*, **71**, 233–244.

Petrov, V.V. (1975). *Sums of Independent Random Variables*. Springer-Verlay, Berlin.

Pfanzagl, J. (1979). Nonparametric minimum contrast estimators. In *Selecta Statistica Canadiana*, vol. **5**, pp. 105–140.

Pratt, J.W. (1961). Length of confidence intervals. *J. Amer. Statist. Assoc.*, **56**, 549–567.

Pratt, J.W. (1963). Shorter confidence intervals for the mean of a normal distribution with known variance. *Ann. Math. Statist.*, **34**, 574–586.

Singh, K. (1981). One the asymptotic accuracy of Efron's bootstrap. *Ann. Statist.*, **9**, 1187–1195.

Wallace, D.L. (1958). Asymptotic approximations to distributions. *Ann. Math. Statist.*, **29**, 635–654.

Wilson, P.D. and Tonascia, J. (1971). Tables for shortest confidence intervals on the standard deviation and variance ratio from normal distribution. *J. Amer. Statist. Assoc.*, **66**, 909–912.

Withers, C.S. (1983). Expansions for the distribution and quantiles of a regular functions of the empirical distribution with applications to nonparametric confidence intervals. *Ann. Statist.*, **11**, 577–587.

Withers, C.S. (1984). Asymptotic expansions for distributions and quantiles with power series cumulants. *J. Roy. Statist. Soc. Ser. B*, **46**, 389–396.

Wu, C.F. (1986). Jackknife, bootstrap and other resampling methods in regression analysis (with discussion). *Ann. Statist.*, **14**, 1261–1350.

Introduction to Gelfand and Smith (1990) Sampling-Based Approaches to Calculating Marginal Densities

D.M. Titterington
University of Glasgow
[with some remarks on Gelfand, Hills, Racine-Poon, and Smith (1990)]

1. Introduction

The Bayesian approach to statistical inference has long had its devoted adherents, as well as many dedicated opponents. Between these two companies there have been very many others who have been more "moderate," because of either pragmatism, indifference, lack of a dogmatic outlook, or whatever. Many find the coherence of the use of Bayes's theorem undeniably appealing but find it difficult to accept the concept of combining the different types of probability function represented by the prior and the likelihood, or they baulk at the purported realism of proposed prior distributions. Many others have been discouraged by the fact that, unless the problem exhibits simple conjugacy, it is conceptually trivial but operationally prohibitive to turn the Bayesian handle with a view to obtaining reliable numerical answers to practical questions, in the form of posterior probabilities of prescribed events or of functionals associated with marginal posterior distributions, for instance.

One such context of particular interest to this commentator is the analysis of finite mixture distributions. Suppose it is assumed that the likelihood function is generated by a sample of size n from a mixture of k distributions, and that neat conjugate priors exist for the individual mixture components. Then the joint posterior density for the mixing weights and other parameters does have an explicitly known form, but it is expressible only as a combination of up to k^n terms. As a result, it is typically unusable for practical purposes, especially if the calculation of interest involves integration, as will almost always be the case. As is mentioned at the beginning of Gelfand and Smith (1990), much effort has been expended in the development of numerical analytic tools for Bayesian analysis and in the use of approximations such as the Laplace method (Tierney and Kadane, 1986), but the former require specialist skills and the latter may not be reliable in many practical problems. It is for the publicization of simple-to-use computational procedures for Bayesian calculations that the paper by Gelfand and Smith earns its place in this volume. As we shall mention, the paper describes a number of general methods, but it promotes in particular the method known as Gibbs sampling; in the companion paper by Gelfand et al. (1990), where substantial practical scenarios are worked through in detail, only Gibbs sampling is illustrated.

2. Gibbs Sampling

A thumbnail description of Gibbs sampling is as follows. Suppose we are interested in a joint distribution of a vector v of m variables, $v_1, \ldots, v_m$, represented by the density function or probability mass function, $p(v) = p(v_1, \ldots, v_m)$. Suppose also that we have available to us all the full conditional densities $\{p(v_i|v_j, j \neq i), i = 1, \ldots, m\}$. Then Gibbs sampling generates a realization from $p(v)$ by initializing a vector v and then recursively sampling from the individual full conditionals, updating the components of v as we go along. If it can be established that the joint density is completely determined by the set of m full conditionals, then the stationary distribution of the Markov chain created by the sampling procedure is indeed the joint density of interest, so that, provided the sampling procedure goes on long enough, it does generate a realization from the required distribution. Repetition of the procedure, or extension of the single chain, with due care about serial correlation, generates a sample of arbitrary size from $p(v)$. The resulting sample can then be used as the basis for density estimation, for instance, using the kernel method, either of $p(v)$ itself or of any marginal or conditional distribution of interest. Similarly, functionals of the distribution can be approximated by natural sample averages.

The reader may well have noticed that there is no explicit mention in the previous paragraph of Bayesian statistics, and indeed Gibbs sampling is not intrinsically a Bayesian concept. It is, instead, just one of the class of so-called Markov Chain Monte Carlo (MCMC) techniques that use a Markov chain to generate realizations from probability distributions that are hard to simulate from directly. Nevertheless, the impact of Gibbs sampling on Bayesian statistics has been enormous, thanks largely to the pointers provided in Gelfand and Smith (1990). In the Bayesian context, $p(v)$ is usually some distribution conditional on the observed data; it might be a posterior distribution, in which case v denotes the set of parameters; it might be the joint distribution of the parameters along with a future observation, a marginal distribution of which gives the predictive distribution; it might be the joint distribution of the parameters along with some missing values in the data, a marginal distribution of which gives the posterior distribution of the parameters. The last case is relevant to the mixture problem, in which the missing values are the indicators of the mixture component to which each observation belongs. Whichever case obtains, it is arranged, and usually automatically occurs, that all the full conditionals associated with $p(v)$ are explicitly known, simple, and easily simulated. In examples like the mixture problem, Bayesian analysis is often quite straightforward if the missing indicator variables are known, and this combines with the simplicity of the conditional distributions of the indicators to make the method easy. The importance of the simplicity of complete-data Bayesian analysis is reminiscent of what facilitates the M-step within the EM algorithm, as used for the computation of maximum likelihood estimates from incomplete data. [The EM algorithm consists of a sequence of iterations, each of which consists of two steps, the E (Expectation)-step and the M (Maximization)-step. If there are explicit formulas for maximum likelihood estimates of parameters from complete data, then typically the M-step can be done explicitly.]

3. Overview of the Papers

At this point it is appropriate to give a brief overview of the two papers. We recall first our remark that Gelfand and Smith (1990) is not dedicated totally to the Gibbs sampler, but also reviews other sampling approaches, in particular, adaptations of the data-augmentation algorithm of Tanner and Wong (1987). They point out the relationships among the methods, indicating that the data-augmentation approach is potentially more efficient than the Gibbs sampler, but that it requires the existence of more conditional distributions than simply the set of full conditional distributions of the individual variables. In the mixture context, for instance, we might use an iterative procedure consisting of two substeps, in the former of which one simulates the *set* of all missing items (indicator vectors), from their joint conditional distribution, and the second in which we simulate the *set* of all parameters in a similar way. This would correspond to data-augmentation, whereas in the Gibbs sampling the different missing items and the different parameters are generated successively from their own full conditionals. Of course, in the mixture context, the different indicator vectors are conditionally independent, so that the two approaches are operationally the same in the first substep. If, however, the unknown parameters include the set of mixing weights, then the two methods will differ in the second phase; for more details of this example see Diebolt and Robert (1994). Gelfand and Smith (1990) also relate the Gibbs sampling to the SIR method of Rubin (1988), which includes an importance-sampling aspect. The examples in Section 3 of the paper illustrate some of the cases for which one or more of the sampling approaches offer straightforward Monte Carlo alternatives to what can be very difficult Bayesian analysis. They include *manifestly* incomplete-data problems (multivariate Normal data with missing values), problems that can be *interpreted* as missing-data problems (variance-component models), and problems where conjugacy is the key to the feasibility of the sampling methods (hierarchical models). Section 4 reports numerical results on two illustrations. In one of these the Gibbs sampling and substitution sampling (a version of data augmentation) methods do very well, converging quickly and out-perfoming Rubin's algorithm.

The bulk of Gelfand et al. (1990) is devoted to a deeper discussion of a number of examples, and concentrates on Gibbs sampling. The authors freely admit that they make no claim that Gibbs sampling is the most efficient way to deal with particular problems, but they point out its generality, its frequent simplicity, its ability to treat otherwise inaccessible problems in Bayesian analysis, and its frequently remarkable empirical performance in terms of speedy convergence, particularly in the examples described in the paper. The examples include those involving variance components, the comparison of a set of Normal means (this time with order restrictions among the means), and hierarchical modeling, that were touched on in Gelfand and Smith (1990), as well as a problem involving a cross-over trial with missing data. Various difficulties, such as the need to impose order restrictions or to cope with missing data, are easily dealt with; see their summary for a highlighted list of these and other benefits.

Although Gelfand and Smith (1990) is the more fundamental of the two papers and appears to be the more highly cited, the practical illustrative material presented in Gelfand et al. (1990) has surely been extremely influential in impressing on practitioners the potential and simplicity of the Gibbs sampler.

4. Antecedents and Alternatives

In this section we say a little about the origins of the Gibbs sampling scheme. Like many Markov Chain Monte Carlo techniques, it is familiar in the literature of statistical physics, known variously as the heat-bath method (Hertz, Krogh, and Palmer, 1991) and Glauber dynamics (Amit, 1989). Besag and Green (1993) highlight the review by Sokal (1989), and point out that Grenander (1983) used Gibbs sampling in Bayesian analysis. The method became well known in the statistical community, however, through the paper by Geman and Geman (1984), and that paper's stimulation of a huge amount of work in statistical image analysis.

The main idea was to base inferences, about a true scene, on its (posterior) distribution given a distorted image of the scene. A prior Markovian structure was assumed for the true scene such that, although the posterior distribution is not directly accessible for practical purposes, yet all full conditional distributions are comparatively simple. Thus, in principle, realizations from the posterior distributions can be generated by the Gibbs sampling procedure. In fact, Geman and Geman (1984) did not use this as a means of generating, say, general summaries of the posterior distribution. Instead, they combined the MCMC procedure with an *annealing* step, in which they based the Gibbs sampler on the (T^{-1})th power of the posterior density, and let the value of the parameter T tend to zero from above, during the algorithm. The idea is that the sampling procedure should then home in on the mode of the original posterior distribution, leading to the Maximum a Posteriori estimate of the true scene. Concentration on this apparently unambitious form of analysis of the posterior distribution is perhaps not surprising in the context of the analysis of pixellated images, given the massive dimensionsality of the state space of the unknown quantities. The contribution of Gelfand and Smith (1990) is to apply the method to statistical problems of more conventional (i.e., much smaller) dimension, which are, nevertheless, difficult to deal with exactly.

As hinted by some of the references in the previous paragraph, the Gibbs sampler has been used in the neural-computing literature, and Titterington and Anderson (1994) emphasize its application by Aarts and Korst (1989) to the so-called Boltzmann machines, which are characterized by somewhat unwieldy multivariate binary distributions. Aarts and Korst (1989) apply other MCMC techniques to the problem of simulating realizations from these distributions, and these include the Metropolis method of Metropolis et al. (1953). The method used is, in turn, a version of what is now called the Metropolis–Hastings algorithm in the statistical literature, in deference to another hitherto unsung, but now seminal, contribution by Hastings (1970). Indeed, both the original Metropolis method and the Gibbs sampler arise as special cases of the general Metropolis–Hastings algorithm. In its turn, the Metropolis–Hastings algorithm is itself subsumed within the even wider class of Markov Chain Monte Carlo procedures, any one of which simulates realizations from a (usually high dimensional) ergodic, irreducible Markov chain whose stationary distribution is the probability distribution of prime interest; see, for instance, Besag and Green (1993) for discussion of other members of this wider class.

5. Current Related Trends and Influence

It could be said that the period of direct influence of these papers is now passing, in that much subsequent work has been done in the development of the basic Gibbs sampler and other MCMC methods; see Tierney (1994) and the associated discussion for many points of theoretical and practical interest. There is considerable evidence that there are advantages in using non-Gibbs MCMC algorithms, especially those with a Metropolis flavor; see, for instance, Besag (1994). Furthermore, many important articles on MCMC find it legitimately unnecessary to cite these two papers. However, it is hard to envisage the current level of activity in applying MCMC methods to practical problems and, in particular, to those involving a Bayesian approach, without the stimulus brought about by the movement that was spearheaded by Gelfand and Smith (1990) and Gelfand et al. (1990). The recent compendium edited by Gilks et al. (1996), and all the important ideas and methods it contains, would surely not have come to pass without them.

6. About the Authors

Alan Gelfand is Professor of Statistics at the University of Connecticut. Adrian Smith is Professor of Statistics in Imperial College, London, having previously been Professor at the University of Nottingham. They both enjoy distinguished reputations in Bayesian statistics, and began to collaborate in the late 1980s, when Gelfand was in Nottingham for one year as a UK Science and Engineering Research Council (SERC) Visiting Research Fellow to work with Smith on comparisons of Bayes and Empirical Bayes approaches to hierarchical models. At that time, Smith's group in Nottingham were shifting their emphasis in Bayesian computation from numerical quadrature to methods based on stochastic simulation, and this became the focus of the collaborative programme. A major practical stimulus for the work was provided by important practical problems thrown up by Smith's collaboration with Amy Racine-Poon of the Ciba–Geigy Pharmaceutical Company. Susan Hills, who had recently completed a Ph.D. with Smith, was the SERC-funded Research Assistant on the project.

References

Aarts, E.H.L. and Korst, J.H.M. (1989). *Simulated Annealing and Boltzmann Machines*. Wiley, Chichester.

Amit, D. (1989). *Modeling Brain Function*. Cambridge, Cambridge University Press.

Besag, J. (1994). Discussion of a paper by L. Tierney. *Ann. Statist.*, **22**, 1734–1741.

Besag, J. and Green, P.J. (1993). Spatial statistics and Bayesian computation. *J. Roy. Statist. Soc. Ser. B*, **55**, 25–37.

Diebolt, J. and Robert, C.P. (1994). Estimation of finite mixture distributions by Bayesian sampling. *J. Roy. Statist. Soc. Ser. B*, **56**, 363–375.

Gelfand, A.E., Hills, S.E., Racine-Poon, A., and Smith, A.F.M. (1990). Illustration of Bayesian inference in Normal data models using Gibbs sampling. *J. Amer. Statist. Assoc.*, **85**, 972–985.

Gelfand, A.E. and Smith, A.F.M. (1990). Sampling-based approaches to calculating marginal densities. *J. Amer. Statist. Assoc.*, **85**, 398–409.

Geman, S. and Geman, D. (1984). Stochastic relaxation, Gibbs distributions and the Bayesian restoration of images. *IEEE Trans. Pattern Anal. Machine Intell.*, **6**, 721–741.

Gilks, W., Richardson, S., and Spiegelhalter, D.J. (eds.) (1996). *Practical Markov Chain Monte Carlo.* Chapman & Hall, London.

Grenander, U. (1983). Tutorial in pattern theory. Report. Division of Applied Mathematics, Brown University, Providence, RI.

Hastings, W.K. (1970). Monte Carlo sampling methods using Markov chains, and their applications. *Biometrika*, **57**, 97–109.

Hertz, J., Krogh, A., and Palmer, R.G. (1991) *Introduction to the Theory of Neural Computation*. Addison-Wesley, Redwood City, CA.

Metropolis, N., Rosenbluth, A.W., Rosenbluth, M.N., Teller, A.H., and Teller, E. (1953). Equations of state calculations by fast computing machine. *J. Chem. Phys.*, **21**, 1087–1091.

Rubin, D.B. (1988). Using the SIR algorithm to simulate posterior distributions. In *Bayesian Statistics, vol. 3* (J.M. Bernardo, D.V. Lindley, M.H. DeGroot and A.F.M. Smith, eds.). New York, Oxford University Press, pp. 395–402.

Sokal, A.D. (1989). *Monte Carlo Methods in Statistical Mechanics: Foundations and New Algorithms.* Cours de Troisième Cycle de la Physique en Suisse Romande (Lausanne, June 1989).

Tanner, M. and Wong, W. (1987). The calculation of posterior distributions by data augmentation (with discussion). *J. Amer. Statist. Assoc.*, **82**, 528–550.

Tierney, L. (1994). Markov chains for exploring posterior distributions (with discussion). *Ann. Statist.*, **22**, 1701–1762.

Tierney, L. and Kadane, J. (1986). Accurate approximations for posterior moments and marginal densities. *J. Amer. Statist. Assoc.*, **81**, 82–86.

Titterington, D.M. and Anderson, N.H. (1994). Boltzmann machines. In *Probability, Statistics and Optimization: A Tribute to Peter Whittle* (F.P. Kelly, ed.). Wiley, New York, pp. 255–279.

Sampling-Based Approaches to Calculating Marginal Densities

Alan E. Gelfand and Adrian F.M. Smith

Abstract

Stochastic substitution, the Gibbs sampler, and the sampling–importance–resampling algorithm can be viewed as three alternative sampling- (or Monte Carlo-) based approaches to the calculation of numerical estimates of marginal probability distributions. The three approaches will be reviewed, compared, and contrasted in relation to various joint probability structures frequently encountered in applications. In particular, the relevance of the approaches to calculating Bayesian posterior densities for a variety of structured models will be discussed and illustrated.

1. Introduction

In relation to a collection of random variables, $U_1, U_2, \ldots, U_k$, suppose that either (a) for $i = 1, \ldots, k$, the conditional distributions $U_i|U_j$ $(j \neq i)$ are available, perhaps having for some i reduced forms $U_i|U_j$ $(j \in S_i \subset \{1, \ldots, k\})$, or (b) the functional form of the joint density of $U_1, U_2, \ldots, U_k$ is known, perhaps modulo the normalizing constant, and at least one $U_i|U_j$ $(j \neq i)$ is available, where *available* means that samples of U_i can be straightforwardly and efficiently generated, given specified values of the appropriate conditioning variables.

The problem addressed in this article is the exploitation of the kind of structural information given by either (a) or (b), to obtain numerical estimates of nonanalytically available marginal densities of some or all of the

U_i (when possible) simply by means of simulated samples from available conditional distributions, and without recourse to sophisticated numerical analytic methods. We do not claim that the sampling methods to be described are necessarily computationally efficient compared with expert use of the latter. Instead, the attraction of the sampling-based methods is their conceptual simplicity and ease of implementation for users with available computing resources but without numerical analytic expertise. All that the user requires is insight into the relevant conditional probability structure and techniques for the efficient generation of appropriate random variates (e.g., as described by Devroye (1986) and Ripley (1987)).

In Section 2, we discuss and extend three alternative approaches put forward in the literature for calculating marginal densities via sampling algorithms. These are (variants of) the data-augmentation algorithm described by Tanner and Wong (1987), the Gibbs sampler algorithm introduced by Geman and Geman (1984), and the form of importance-sampling algorithm proposed by Rubin (1987, 1988). We note that the Gibbs sampler has been widely taken up in the image-processing literature and in other large-scale models such as neural networks and expert systems, but that its general potential for more conventional statistical problems seems to have been overlooked. As we show, there is a close relationship between the Gibbs sampler and the substitution or data-augmentation algorithm proposed by Tanner and Wong (1987). We generalize the latter and show that it is as least as efficient as the Gibbs sampler, and potentially more efficient, given the availability of distinct conditional distributions in addition to those in (a). We note that as a consequence of the relationship between the two algorithms, the convergence results established by Geman and Geman (1984) are applicable to the generalized substitution algorithm. The stronger convergence results established by Tanner and Wong (1987) require the availability of a particular set of conditional distributions, including those in (a). Both the substitution and Gibbs sampler algorithms are iterative Monte Carlo procedures, applicable when the kind of structural information given by (a) is available. When the structural information is of the kind described by (b), we see that an importance-sampling algorithm based on that of Rubin (1987, 1988) provides a noniterative Monte Carlo integration approach to calculating marginal densities.

In Section 3, we illustrate various model structures occurring frequently in applications, where one or more of these three approaches offers an easily implemented solution. In particular, we consider the calculation of Bayesian posterior distributions in incomplete-data problems, conjugate hierarchical models, and normal data models. In Section 4, we briefly summarize the results of some preliminary computational experience in two simple cases. (Detailed applications to complex, real-data problems will be presented in a subsequent paper.) Finally, in Section 5 we provide a summary discussion.

2. Sampling Approaches

In the sequel, we assume that we are dealing with real, possibly vector-valued random variables having a joint distribution whose density function is strictly positive over the (product) sample space. This ensures that knowledge of all full conditional specifications [such as in (a) of Sec. 1] uniquely defines the full joint density (e.g., see Besag, 1974). Throughout, we assume the existence of densities with respect to either Lebesgue or counting measure, as appropriate, for all marginal and conditional distributions. The terms *distribution density* and *density* are therefore used interchangeably.

Densities are denoted generically by brackets, so joint, conditional, and marginal forms, for example, appear as $[X, Y]$, $[X|Y]$, and $[X]$. Multiplication of densities is denoted by $*$; for example, $[X, Y] = [X|Y] * [Y]$. The process of marginalization (i.e., integration) is denoted by forms such as $[X|Y] = \int [X|Y, Z, W] * [Z|W, Y] * [W|Y]$, with the convention that all variables appearing in the integrand but not in the resulting density have been integrated out. Thus the integration is with respect to Z and W. More generally, we use notation such as $\int h(Z, W) * [W]$ to denote, for given Z, the expectation of the function $h(Z, W)$ with respect to the marginal distribution for W.

2.1. Substitution or Data-Augmentation Algorithm

The substitution algorithm for finding fixed-point solutions to certain classes of integral equations is a standard mathematical tool that has received considerable attention in the literature (e.g., see Rall, 1969). Its potential utility in statistical problems of the kind we are concerned with was observed by Tanner and Wong (1987) (who called it a data-augmentation algorithm) and the associated discussion. Briefly reviewing the essence of their development using the notation introduced previously, we have

$$[X] = \int [X|Y] * [Y] \tag{1}$$

and

$$[Y] = \int [Y|X] * [X], \tag{2}$$

so substituting (2) into (1) gives

$$\begin{aligned} [X] &= \int [X|Y] * \int [Y|X'] * [X'] \\ &= \int h(X, X') * [X'], \end{aligned} \tag{3}$$

where $h(X, X') = \int [X|Y] * [Y|X']$, with X' appearing as a dummy argument in (3), and of course $[X] \equiv [X']$. Now, suppose that on the right side of (3), $[X']$ were replaced by $[X]_i$, to be thought of as an estimate of $[X] \equiv [X']$ arising at the ith stage of an iterative process. Then, (3) implies that for some $[X]_{i+1}$, $[X]_{i+1} = \int h(X, X') * [X']_i = I_h[X]_i$, in a notation making explicit that I_h is the integral operator associated with h. Exploiting standard theory of such integral operators, Tanner and Wong (1987) showed that under mild regularity conditions this iterative process has the following properties (with obviously analogous results for $[Y]$).

TW1 (Uniqueness). The true marginal density, $[X]$, is the unique solution to (3).

TW2 (Convergence). For almost any $[X]_0$, the sequence $[X]_1, [X]_2, \ldots$ defined by $[X]_{i+1} = I_h[X]_i (i = 0, 1, \ldots)$ converges monotonically in L_1 to $[X]$.

TW3 (Rate). $\int |[X]_i - [X]| \to 0$ geometrically in i.

Extending the substitution algorithm to three random variables X, Y, and Z, we may write [analogous to (1) and (2)]

$$[X] = \int [X, Z|Y] * [Y], \tag{4}$$

$$[Y] = \int [Y, X|Z] * [Z], \tag{5}$$

and

$$[Z] = \int [Z, Y|X] * [X]. \tag{6}$$

Substitution of (6) into (5) and then (5) into (4) produces a fixed-point equation analogous to (3). A new h function arises with associated integral operator I_h, and hence TW1, TW2, and TW3 continue to hold in this extended setting. Extension to k variables is straightforward. A noteworthy by-product, using TW1, is a simple proof that under weak conditions specification of the conditional distributions $[U_{r,r\neq s}|U_s]$ $(s = 1, 2, \ldots, k)$ uniquely determines the joint density.

2.2. Substitution Sampling

Returning to (1) and (2), suppose that $[X|Y]$ and $[Y|X]$ are available in the sense defined at the beginning of Section 1. For an arbitrary (possibly degenerate) initial density $[X]_0$ draw a single $X^{(0)}$ from $[X]_0$. Given $X^{(0)}$, since $[Y|X]$ is available draw $Y^{(1)} \sim [Y|X^{(0)}]$, and hence from (2) the marginal distribution of $Y^{(1)}$ is $[Y]_1 = \int [Y|X] * [X]_0$. Now, complete a cycle by drawing $X^{(1)} \sim [X|Y^{(1)}]$. Using (1), we then have $X^{(1)} \sim [X]_1 = \int [X|Y] * [Y]_1 = \int h(X, X') * [X']_0 = I_h[X]_0$. Repetition of this cycle produces $Y^{(2)}$ and $X^{(2)}$, and eventually, after i iterations, the pair $(X^{(i)}, Y^{(i)})$ such that $X^{(i)} \xrightarrow{d} X \sim [X]$, and $Y^{(i)} \xrightarrow{d} Y \sim [Y]$, by virtue of TW2. Repetition of this sequence m times each to the ith iteration generates m iid pairs $(X_j^{(i)}, Y_j^{(i)})(j = 1, \ldots, m)$. We call this generation scheme *substitution sampling*. Note that though we have independence across j, we have dependence within a given j.

If we terminate all repetitions at the ith iteration, the proposed density estimate of $[X]$ (with an analogous expression for $[Y]$) is the Monte Carlo integration

$$[X]_j = \frac{1}{m} \sum_{j=1}^{m} [X|Y_j^{(i)}]. \tag{7}$$

Note that the X are not used in (7) (see Sec. 2.6).

We note that this version of the substitution-sampling algorithm differs slightly from the imputation–posterior algorithm of Tanner and Wong (1987). At each iteration l $(l = 1, 2, \ldots, i)$, they proposed creation of the mixture density estimate. $[X]_j$, of the form in (7), with subsequent sampling from $[X]$ to begin the next iteration. This mechanism introduces the additional randomness of equally likely selection from the $Y_j^{(l)}$ before obtaining an $X^{(l)}$. We suspect this sampling with replacement of the $Y^{(l)}$ was introduced to allow m to vary across iterations, which may be useful in reducing computational effort.

The L_i convergence of $[\hat{X}]_i$ to $[X]$ is most easily studied by writing $\int |[\hat{X}] - [X]| \leq \int |[\hat{X}]_i - [X]_i| + \int |[X]_i - [X]|$. The second term on the right side can be made arbitrarily small as $i \to \infty$, as a consequence of TW2. The first term on the right can be made arbitrarily small as $m \to \infty$, since $[\hat{X}] \xrightarrow{d} [X]_i$ for almost all X (Glick 1974).

Extension of the substitution-sampling algorithm to more than two random variables is straightforward. We illustrate using the three-variable case, assuming the three conditional distributions in (4)–(6) are available. Taking an arbitrary starting marginal density for X, say $[X]_0$, we draw $X^{(0)} \sim [X]_0$, $(Z^{(0)'}, Y^{(0)'}) \sim [Z, Y|X^{(0)}]$, $(Y^{(1)}, X^{(0)'}) \sim [Y, X, Z^{(0)}]$, and finally $(X^{(1)}, Z^{(1)}) \sim [X, Z|Y^{(1)}]$. A full cycle of the algorithm (i.e., to generate $X^{(1)}$ starting from $X^{(0)}$) thus requires six generated variates, rather than the two we saw earlier. Repeating such a cycle i times produces $(X_j^{(i)}, Y^{(i)}, Z^{(i)})$. The aforementioned theory ensures that $X^{(i)} \xrightarrow{d} X \sim [X]$, $Y^{(i)} \xrightarrow{d} Y \sim [Y]$, and $Z^{(i)} \xrightarrow{d} Z \sim [Z]$. If we repeat the entire process m times we obtain iid $(X_j^{(i)}, Y^{(i)}, Z_j^{(i)})(j = 1, \ldots, m)$ (independent between, but not within, j's). Note that implementation of the substitution-sampling algorithm does not require specification of the full joint distribution. Rather, what is needed is the availability of $[X, Z|Y]$, $[Y, X|Z]$, and $[Z, Y|Z]$. Of course, in many cases sampling from, say, $[X, Z|Y]$ requires, for example, $[X|Y, Z]$ and $[Y|Z]$, that is, the availability of a full conditional and a reduced conditional distribution. Paralleling (7), the density estimator of $[X]$ becomes

$$[\hat{X}]_i = \frac{1}{m}\sum_{j=1}^{m}[X \mid Y_j^{(i)}, Z_j^{(i)}], \tag{8}$$

with analogous expressions for estimating $[Y]$ and $[Z]$. L_1 convergence of (8) to $[X]$ again follows.

For k variables, $U_1, \ldots, U_k$, the substitution-sampling algorithm requires $k(k-1)$ random variate generations to complete a cycle. If we run m sequences out to the ith iteration $[mik(k-1)$ random generations] we obtain m iid k tuples $(U_{tj}^{(i)}, \ldots, U_{kj}^{(i)})$ $(j = 1, \ldots, m)$, with the density estimator for $[U_s]$ $(s = 1, \ldots, k)$ being

$$[\hat{U}_s] = \frac{1}{m}\sum_{j=1}^{m}[U_s \mid U_t = U_{tj}^{(i)}; t \neq s]. \tag{9}$$

2.3. Gibbs Sampling

Suppose that we write (4)–(6) in the form

$$[X] = \int [X \mid Z, Y] * [Z \mid Y] * [Y],$$

$$[Y] = \int [Y \mid X, Z] * [X \mid Z] * [Z],$$

$$[Z] = \int [Z \mid Y, X] * [Y \mid X] * [X]. \tag{10}$$

Implementation of substitution sampling requires the availability of all six conditional distributions on the right side of (10), rarely the case in our applications. As noted at the beginning of Section 2, the full conditional distributions alone, $[X \mid Y, Z]$, $[Y \mid Z, X]$, and $[Z \mid X, Y]$, uniquely determine the joint distribution (and hence the marginal distributions) in the situations under study. An algorithm for extracting the marginal distributions from these full conditional distributions was formally introduced by Geman and Geman (1984) and is known as the Gibbs sampler. An earlier article by Hastings (1970) developed essentially the same idea and suggested its potential for numerical problems arising in statistics.

The Gibbs sampler was developed and has mainly been applied in the context of complex stochastic models involving very large numbers of variables, such as image reconstruction, neural networks, and expert systems. In these cases, direct specification of a joint distribution is typically not feasible. Instead, the set of full conditionals is specified, usually by assuming that an individual full conditional distribution only depends on some "neighborhood" subset of the variables [a reduced form, in the terminology of (a) in Sec. 1]. More precisely, for the set of variables $U_1, U_2, \ldots, U_k$,

$$[U_i | U_j; \neq i] \equiv [U_i | U_j; j \in S_i], \qquad i = 1, \ldots, k, \tag{11}$$

where S_i is a small neighborhood subset of $\{1, 2, \ldots, k\}$. A crucial question is under what circumstances the specification (11) uniquely determines the joint distribution. The answer is taken up in great detail by Geman and Geman (1984), involving concepts such as graphs, neighborhood systems, cliques, Markov random fields, and Gibbs distributions. In all of the examples we consider, the joint distribution is uniquely defined. Our k's will be small to moderate, and the available set of full conditional distributions will, in fact, be calculated from specification of the joint density.

Gibbs sampling is a Markovian updating scheme that proceeds as follows. Given an arbitrary starting set of values $U_1^{(0)}, U_2^{(0)}, \ldots, U_k^{(0)}$, we draw $U_1^{(1)} \sim [U_1 | U_2^{(0)}, \ldots, U_k^{(0)}]$, $U_2^{(1)} \sim [U_2 | U_1^{(1)}, U_3^{(0)}, \ldots, U_k^{(0)}]$, $U_3^{(1)} \sim [U_3 | U_1^{(1)} U_2^{(1)}, U_4^{(0)}, \ldots, U_k^{(0)}]$, and so on, up to $U_k^{(1)} \sim [U_k | U_1^{(1)}, \ldots, U_{k-1}^{(1)}]$. Thus each variable is visited in the natural order and a cycle in this scheme requires K random variate generations. After i such iterations we would arrive at $(U_1^{(i)}, \ldots, U_k^{(i)})$. Under mild conditions, Geman and Geman showed that the following results hold.

GG1 (Convergence). $(U_1^{(i)}, \ldots, U_k^{(i)}) \xrightarrow{d} [U_1, \ldots, U_k]$ and hence for each s, $U_2^{(i)} \xrightarrow{d} U_s \sim [U_s]$ as $i \to \infty$. In fact, a slightly stronger result is proven. Rather than requiring that each variable be visited in repetitions of the natural order, convergence still follows under any visiting scheme, provided that each variable is visited infinitely often (io).

GG2 (Rate). Using the sup norm, rather than the L_1 norm, the joint density of $(U_1^{(i)}, \ldots, U_k^{(i)})$ converges to the true joint density at a geometric rate in i, under visiting in the natural order. A minor adjustment to the rate is required for an arbitrary io visiting scheme.

GGS (Ergodic Theorem). For any measurable function T of $U_1, \ldots, U_k$ whose expectation exists,

$$\lim_{i \to \infty} \frac{1}{i} \sum_{l=1}^{i} T(U_1^{(l)}, \ldots, U_k^{(l)}) \xrightarrow{\text{a.s.}} E(T(U_1, \ldots, U_k)).$$

As in Section 2.3, Gibbs sampling through m replications of the aforementioned i iterations (mik random variate generations) produces m iid k tuples $(U_{1j}^{(i)}, \ldots, U_{kj}^{(i)})$ ($j = 1, \ldots, m$), with the proposed density estimate for $[U_s]$ having form (9).

2.4. Relationship Between Gibbs Sampling and Substitution Sampling

It is apparent that in the case of two random variables Gibbs sampling and substitution sampling are identical. For more than two variables, using (10) and its obvious generalization to k variables, we see that Gibbs sampling assumes the availability of the set of k full conditional distributions (the minimal set needed to determine the joint density uniquely). The substitution-sampling algorithm requires the availability of $k(k-1)$ conditional distributions, including all of the full conditionals.

Gibbs sampling is known to converge slowly in applications with k very large. Regardless, fair comparison with substitution sampling, in the sense of the total amount of random variate generation, requires that we allow the Gibbs sampling algorithm $i(k-1)$ iterations if the substitution-sampling algorithm is allowed i. Even so, there is clearly scope for accelerated convergence from the substitution-sampling algorithm, since it samples from the correct distribution each time, whereas Gibbs sampling only samples from the full conditional distributions. To amplify, we describe how the substitution-sampling algorithm might be carried out under availability of just the set of full conditional distributions. We see that it can be viewed as the Gibbs sampler, but under an io visiting scheme different from the natural one. We present the argument in the three-variable case for simplicity. Returning to (10), if $[Y|X]$ is unavailable we can create a sub-substitution loop to obtain it by means of

$$[Y|X] = \int [Y|X, Z] * [Z|X],$$

$$[Z|X] = \int [Z|X, Y] * [Y|X]. \tag{12}$$

Similar subloops are clearly available to create $[X|Z]$ and $[Z|Y]$. In fact, for k variables this idea can be straightforwardly extended to the estimation of an arbitrary reduced conditional distribution, given the full conditionals. We omit the details.

The previous analysis suggests that we could view the reduced conditional densities such as $[Y|X]$ as available, and that we could thus carry out the substitution algorithm as if all needed conditional distributions were available; however, $[Y|X]$ is not available in our earlier sense. Under the subloop in (12), we can always obtain a density estimate for $[Y|X]$, given any specified X, say $X^{(0)}$. At the next cycle of the iteration, however, we would need a brand-new density estimate for $[Y|X]$ at $X = X^{(1)}$. Nonetheless, suppose we persevered in this manner, making our way through one cycle of (10). The reader may verify that the only distributions actually sampled from are, of course, the available full conditionals, that at the end of the cycle each full conditional will have been sampled from at least once, and thus that under repeated iterations each variable will be visited io. Therefore, this version of the substitution-sampling algorithm is merely Gibbs sampling with a different but still io visiting order. As a result, GG1, GG2, and GG3 still hold (TW1, TW2, and TW3 apply directly only when all required conditional distributions are available). Moreover, there is no gain in implementing the Gibbs sampler in this complicated order; the natural order is simpler and equally good.

This discussion may be readily extended to the case of k variables. As a result, we conclude that when only the set of k full conditionals is available the substitution-sampling algorithm and the Gibbs sampler are equivalent. Furthermore, we can now see when substitution sampling offers the possibility of acceleration relative to Gibbs sampling. This occurs when some reduced conditional distributions, distinct from the full conditional distributions, are available. Suppose that we write the substitution algorithm with appropriate conditioning to capture these available reduced conditionals. As we traverse a cycle, we would sample from these distributions as we come to them, otherwise sampling from the full conditional distributions.

An example will help clarify this idea. One way to carry out the Gibbs sampler in (10) is to follow the substitution order rather than the natural order. That is, given an initial $X^{(0)}$, $Y^{(0)}$, and $Z^{(0)}$, we start at the bottom line of (10), for example, drawing (a) $Y^{(0)}$ from $[Y|X^{(0)}, Z^{(0)}]$, (b) $Z^{(0)'}$ from $[Z|Y^{(0)'}, X^{(0)'}]$, (c) $X^{(0)'}$ from $[X|Z^{(0)'}, Y^{(0)'}]$, (d) $Y^{(1)}$ from $[Y|X^{(0)'}, Z^{(0)'}]$, (e) $Z^{(1)}$ from $[Z|Y^{(1)}, X^{(0)'}]$, and (f) $X^{(1)}$ from $[Z|Y^{(1)}, Z^{(1)}]$. Thus, in this case, one cycle using the substitution order corresponds to two cycles using the natural order. Suppose, however, that in addition to the full conditional distributions, $[Z|Y]$, say, is available and distinct from $[Z|X, Y]$. Following the substitution order, at step (e) we would instead draw $Z^{(1)}$ from the correct distribution, $[Z|Y^{(1)}]$.

In Section 3, we provide classes of examples where distinct reduced conditional distributions are available and classes where they generally are not. Our computational experience shows that the acceleration in convergence that arises from having available distributions in addition to the full conditionals is inconsequential (see Sec. 4).

2.5. The Rubin Importance-Sampling Algorithm

Rubin (1987) suggested a noniterative Monte Carlo method for generating marginal distributions using importance-sampling ideas. We first present the basic idea in the two-variable case. Suppose that we seek the marginal distribution of X, given only the functional form (modulo the normalizing constant) of the joint density $[X, Y]$ and the availability of the conditional distribution $[X|Y]$ [a special case of the conditions described in (b) of Sec. 1].

Suppose further (as is typically the case in applications) that the marginal distribution of Y is not known. Choose an importance-sampling distribution for Y that has positive support wherever $[Y]$ does and that has density $[Y]_s$, say. Then, $[X|Y]*[Y]_s$ provides an importance-sampling distribution for (X, Y). Suppose that we draw iid pairs (X_l, Y_l) $(l = 1, \ldots, N)$ from this joint distribution, for example, by drawing Y_l from $[Y]_s$ and X_l from $[X|Y_l]$. Rubin's idea is to calculate $r_l = [X_l, Y_l]/[X_l|Y_l]*[Y_l]_s$ $(l = 1, \ldots, N)$ and then estimate the marginal density for $[X]$ by

$$[\hat{X}] = \sum_{l=1}^{N} [X|Y_l]r_l \Big/ \sum_{l=1}^{N} r_l. \tag{13}$$

Note the important fact that $[X, Y]$ need only be specified up to a constant, since the latter cancels in (13). In other words, we do not need to evaluate the normalizing constant for $[X, Y]$. This feature is exploited in the examples of Section 3. By dividing the numerator and denominator of (13) by N and using the law of large numbers, we immediately have the following.

R1 (Convergence). $[\hat{X}] \to [X]$ with probability 1 as $N \to \infty$ for almost every X.

In addition, if $[Y|X]$ is available we immediately have an estimate for the marginal distribution of Y: $[\hat{Y}] = \sum_{l=1}^{N}[Y|X_l]r_l / \sum_{l=1}^{N} r_l$.

The successful performance of (13) typically depends strongly on the choice of $[Y]_s$ and its closeness to $[Y]$. Thus the suggestion of Tanner and Wong (1987) in their rejoinder to Rubin, to perhaps use for $[Y]_s$ the density estimate created after i iterations of the substitution algorithm, merits further investigation. In fact, the whole problem of general strategies for synthesizing both the iterative and noniterative approaches under a fixed-budget (total number of random generations) criterion needs considerable further study.

The extension of the Rubin importance-sampling idea to the case of k variables is clear. For instance, when $k = 3$, suppose that we seek the marginal distribution of X, given the functional form of $[X, Y, Z]$ up to a constant and the availability of the full conditional $[X|Y, Z]$. In this case, the pair (Y, Z) plays the role of Y in the two-variable case discussed before, and in general we need to specify an importance-sampling distribution $[Y, Z]_s$. Nevertheless, if $[Y|Z]$ is available, for example, we only need to specify $[Z]_s$. In any case, we draw iid triples $(X_l, Y_l, Z_l)(l = 1, \ldots, N)$ and calculate $r_l = [X_l, Y_l, Z_l]/([X_l|Y_l, Z_l]*[Y_l, Z_l]_s)$. The marginal density estimate for $[X]$ then becomes [analogous to (13)]

$$[\hat{X}] = \sum_{l=1}^{N} [X|Y_l, Z_l]r_l \Big/ \sum_{l=1}^{N} r_l. \tag{14}$$

We note that in the k-variable case the Rubin importance-sampling algorithm requires Nk random variate generations, whereas Gibbs sampling stopped at iteration i requires mik generations. For fair comparison of the two algorithms, we should therefore set $N = mi$. The relationship between the estimators (7) and (13) may be clarified if we resample $Y_1^*, Y_2^*, \ldots, Y_m^*$ from the distribution that places mass $r_l / \sum r_l$ at Y_l $(l = 1, \ldots, N)$. We could then replace (13) with

$$[\hat{X}] = \frac{1}{m} \sum_{j=1}^{m} [X \mid Y_j^*], \tag{15}$$

so (7) and (15) are of the same form. Relative performance on average depends on whether the distribution of $Y^{(i)}$ or Y^* is closer to $[Y]$. Empirical work described in Section 4 suggests that under fair comparison (7) performs better than (14) or (15). It seems preferable to iterate through a learning process with small samples rather than to draw a one-off large sample at the beginning [an idea that underlies much modern work in adaptive Monte Carlo; e.g., see Smith, Skene, Shaw, and Naylor (1987)].

2.6. Density Estimation

In this section, we consider the problem of calculating a final form of marginal density from the final sample produced by either the substitution- or Gibbs sampling algorithms. Since for any estimated marginal the corresponding full conditional has been assumed available, efficient inference about the marginal should clearly be based on using this full conditional distribution. In the simplest case of two variables, this implies that $[X \mid Y]$ and the $Y_j^{(i)}$ $(j = 1, \ldots, m)$ should be used to make inferences about $[X]$, rather than imputing $X_j^{(i)}$ $(j = 1, \ldots, m)$ and basing inference on these $X_j^{(i)}$'s. Intuitively, this follows, because to estimate $[X]$ using the $X_j^{(i)}$ requires a kernel density estimate. Such an estimate ignores the known form $[X \mid Y]$ that is mixed to obtain $[X]$. The formal argument is essentially based on the Rao–Blackwell theorem. We sketch a proof in the context of the density estimator itself. If X is a continuous p-dimensional random variable, consider any kernel density estimator of $[X]$ based on the $X_j^{(i)}$ (e.g., see Devroye and Györfi, 1985) evaluated at X_0: $\Delta_{X_0}^{(i)} = (1/mh_m^p) \sum_{j=1}^{m} \cdot K[(X_0 - X_j^{(i)})/h_m]$, say, where K is a bounded density on R^p and the sequence $\{h_m\}$ is such that as $m \to \infty$, $h_m \to 0$, whereas $mh_m^p \to \infty$. To simplify notation, set $Q_{m,X_0}(X) = (1/h_m^p) K[(X_0 - X)/h_m]$ so that $\Delta_{X_0}^{(i)} = (1/m) \sum_{j=1}^{m} Q_{m,x_0}(X_j^{(i)})$. Define $\gamma_{X_0}^{(i)} = (1/m) \sum_{j=1}^{m} E(Q_{m,X_0}(X) \mid Y_j^{(l)})$. By our earlier theory, both $\Delta_{X_0}^{(i)}$ and $\gamma_{X_0}^{(i)}$ have the same expectation. By the Rao–Blackwell theorem, $\operatorname{var} E(Q_{m,X_0}(X \mid Y)) \leq \operatorname{var} Q_{m,X_0}(X)$, and hence $\mathrm{MSE}(\gamma_{X_0}^{(i)}) \leq \mathrm{MSE}(\Delta_{X_0}^{(i)})$, where MSE denotes the mean squared error of the estimate of $[X_0]$.

Now, for fixed Y, as $m \to \infty$, $E(Q_{m,X_0}(X|Y)) \to [X_0|Y]$ for almost every X_0, by the Lebesgue density theorem (see Devroye and Györfi, 1985, p. 3). Thus in terms of random variables we have $E(Q_{m,X_0}(X|Y)) \xrightarrow{d} [X_0|Y]$, so for large m, $\gamma_{X_0}^{(i)} \sim [\hat{X}_0]_i$ and $\text{MSE}(\gamma_{X_0}^{(i)}) \approx \text{MSE}([\hat{X}_0]_i)$, and hence $[\hat{X}_0]_i$ is preferred to $\Delta_{X_0}^{(i)}$.

The argument is simpler for estimation of $\eta = E(T(X)) = \int T(X) * [X]$, say. Here, $\hat{\eta}_1 = (1/m)\sum_{j=1}^{m} T(X_j^{(i)})$ is immediately seen to be dominated by $\hat{\eta}_2 = (1/m)\sum_{j=1}^{m} E(T(X)|Y_j^{(i)})$.

3. Examples

A major area of potential application of the methodology we have been discussing is in the calculation of marginal posterior densities within a Bayesian inference framework. In recent years, there have been many advances in numerical and analytic approximation techniques for such calculations (e.g., see Geweke, 1988; Naylor and Smith, 1982, 1988; Shaw 1988; Smith et al. 1987; Smith, Skene, Shaw, Naylor, and Dransfield, 1985; Tierney and Kadane 1986), but implementation of these approaches typically requires sophisticated numerical analytic expertise, and possibly specialist software. By contrast, the sampling approaches we have discussed are straightforward to implement. For many practitioners, this feature will more than compensate for any relative computational inefficiency. To provide a flavor of the kinds of areas of application for which the methodology is suited, we present six illustrative examples.

3.1. A Class of Multinomial Models

We extend the one-parameter genetic-linkage example described by Tanner and Wong (1987, p. 530), which in its most general form involves multinomial sampling, where some observations are not assigned to individual cells but to aggregates of cells (see Dempster, Laird, and Rubin 1977; Hartley 1958). We give the model and distribution theory in detail for a two-parameter version, from which the extension to k parameters should be clear. Let the vector $Y = (Y_1, \ldots, Y_5)$ have a multinomial distribution $\text{mult}(n, a_1\theta + b_1, a_2\theta + b_2, a_3\eta + b_3, a_4\eta + b_4, c(1 - \theta - \eta))$, where $a_i, b_i \geq 0$ are known and $0 < c = 1 - \sum_{i=1}^{4} b_i = a_1 + a_2 = a_3 + a_4 < 1$. Thus θ and η range over $\theta \geq 0$, $\eta \geq 0$, and $\theta + \eta \leq 1$, so a three-parameter Dirichlet distribution, Dirichlet$(\alpha_1, \alpha_2, \alpha_3)$, may be a natural choice of prior density for (θ, η). From the form of $[Y|\theta,\eta] * [\theta,\eta]$, note that obtaining the exact marginals $[\theta|Y]$ and $[\eta|Y]$ is somewhat messy (involving a two-dimensional numerical integral). Nevertheless, all three sampling approaches we have described are readily applicable here by considering the unobservable nine-cell multinomial model for $X = (X_1, X_2, \ldots, X_9)$, given by mult$(n, a_1\theta, b_1, a_2\theta, b_2, a_3\eta, b_3, a_4\eta, b_4, c(1-\theta-\eta))$. From the form of $[X|\theta,\eta] * [\theta,\eta]$ we see that $[\theta,\eta|X] \sim$ Dirichlet$(X_1+X_3+\alpha_1, X_5+X_7+\alpha_2, X_9+\alpha_3)$, and hence $[0|X]$ and $[\eta|X]$ are available as beta distributions for sampling. Furthermore, $[\theta|X,\eta]$ and $[\eta|X,\theta]$ are available as scaled beta distributions, scaled to the intervals $[0, 1-\eta]$ and $[0, 1-\theta]$, respectively. If we let $Y_1 = X_1 + X_2$, $Y_2 = X_3 + X_4$, $Y_3 = X_5 + X_6$, $Y_4 = X_7 + X_8$, and $Y_5 = X_9$ and define $Z = (X_1, X_3, X_5, X_7)$, we see that specification of X is equivalent to specification of (Y, Z). In addition, $[Z|Y,\theta,\eta]$ is the product of four independent binomials for X_1, X_3, X_5, and X_7, given by $[X_i|Y,\theta,\eta] =$ binomial$(Y_i, a_i\theta/(a_i\theta+b_i))(i = 1,3,5,7)$, which are therefore readily available for sampling.

In the context of Section 2, we have a three-variable case, (θ, η, Z), with interest in the marginal distributions $[\theta|Y], [\eta|Y]$, and $[Z|Y]$. Gibbs sampling requires $[\theta|Y,Z,\eta]$, $[\eta|Y,Z,\theta]$, and $[Z|Y,\theta,\eta]$, all of which are available. But in this case the reduced distributions $[\theta|Y,Z]$ and $[\eta|Y,Z]$ are available as well, enabling study of accelerated convergence. These reduced distributions substantially simplify the Rubin importance-sampling algorithm in obtaining $[\theta|Y]$ and $[\eta|Y]$; only an importance-sampling distribution $[Z|Y]_s$ need be specified (e.g., a default choice might be binomials with chance equal to $\frac{1}{2}$). Detailed comparison of the performance of the three algorithms for a specific case of this multinomial class is given in Section 4.

3.2. Hierarchical Models Under Conjugacy

Consider a general Bayesian hierarchical model having k stages. In an obvious notation, we write the joint distribution of the data and parameters as

$$[Y|\theta_1] * [\theta_1|\theta_2] * [\theta_2|\theta_3] * \cdots * [\theta_{k-1}|\theta_k] * [\theta_k], \tag{16}$$

where we assume all components of prior specification to be available for sampling. Primary interest is usually in the marginal posterior $[\theta_1|Y]$. The hierarchical structure implies that

$$\begin{aligned}[\theta_i|Y,\theta_{jj} \neq i] &= [\theta_1|Y,\theta_2] && i = 1,\\ &= [\theta_i|\theta_{i-1},\theta_{i+1}], && 1 < i < k-1,\\ &= [\theta_k|\theta_{k-1}], && i = k.\end{aligned} \tag{17}$$

Suppose that we assume proper conjugate distributions at each stage. This is common practice in the formulation of such models, except perhaps for $[\theta_k]$, which is often assumed vague. Nevertheless, conjugate priors can generally be made arbitrarily diffuse by appropriate choices of hyperparameters, so this case is implicitly subsumed within the conjugate framework. In fact, $[\theta_k]$ can be vague, provided $[\theta_k|\theta_{k-1}]$ is still proper and available (see Secs. 3.4 and 3.5). Conjugacy implies that the densities in (17) will be available as updated versions of the respective priors (e.g., see Morris 1983a). Typically, no distinct reduced conditional distributions are available, and Gibbs sampling would be used to estimate the desired marginal posterior densities. To clarify this latter point, consider the case $k = 3$. The six conditional distributions in (10) would be $[\theta_1|y,\theta_2,\theta_3]$, $[\theta_2|y,\theta_1,\theta_3]$, $[\theta_3|y,\theta_1,\theta_2]$, $[\theta_3|y,\theta_2]$, $[\theta_1|y,\theta_3]$, and $[\theta_2|y,\theta_1]$. The first three are available as in (17), the fourth is available but not distinct from the third, and the last two are usually unavailable.

As a concrete illustration, consider an exchangeable Poisson model, which is illustrated further in Section 4 with the reanalysis of a published data set. Suppose that we observe independent counts, s_i, over differing lengths of time, t_i (with resultant rate $\rho_i = s_i/t_i$) $(i = 1,\ldots,p)$. Assume $[s_i|\lambda_i] = p_0(\lambda_i t_i)$ and that the λ_i are iid from $G(\alpha,\beta)$, with density $\lambda_i^{\alpha-1}e^{-\lambda_i/\beta}/\beta^{\alpha}\Gamma(\alpha)$. The parameter α is assumed known (in practice, we might treat α as a tuning parameter, or perhaps, in an empirical Bayes spirit, estimate it from the marginal distribution of the s_i's), and β is assumed to arise from an inverse gamma distribution $\mathrm{IG}(\gamma,\delta)$ with density $\delta^{\gamma}e^{-\delta/\beta}/\beta^{\gamma+1}\Gamma(\gamma)$. (A diffuse version of this final-stage distribution is obtained by taking δ and γ to be very small, perhaps 0.)

Letting $Y = (s_1,\ldots,s_p)$, the conditional distributions $[\lambda_j|Y]$ are sought. The full conditional distribution of λ_j is given by

$$[\lambda_j|Y,\beta,\lambda_{i,i\neq j}] = G(\alpha + s_j,(t_j + 1/\beta)^{-1}), \qquad j = 1,\ldots,p, \tag{18}$$

whereas the full conditional distribution for β is given by

$$[\beta|Y,\lambda_1,\ldots,\lambda_p] = \mathrm{IG}\left(\gamma + p\alpha, \sum \lambda_i + \delta\right). \tag{19}$$

No distinct reduced conditional distributions are available. The conditional distribution of λ_j, given Y and β, is (18), regardless of which or how many λ_i $(i \neq j)$ are given. The conditional distribution of β, given Y and any subset of the λ_j's, is unavailable. Given $(\lambda_1^{(0)},\lambda_2^{(0)},\ldots,\lambda_p^{(0)},\beta^{(0)})$, the Gibbs sampler draws $\lambda_j^{(1)} \sim G(\alpha + s_j,(t_j + 1/\beta^{(0)})^{-1})$ $(j = 1,\ldots,p)$ and $\beta^{(1)} \sim \mathrm{IG}(\gamma + \alpha p, \sum_{j=1}^{p}\lambda_j^{(1)} + \delta)$ to complete one cycle. If we carry out m repetitions each of i iterations, generating $(\lambda_{1l}^{(i)},\ldots,\lambda_{rl}^{(i)},\beta_l^{(i)})$ $(l = 1,\ldots,m)$, the marginal density estimate for λ_j is

$$[\lambda_j \uparrow Y] = \frac{1}{m}\sum_{l=1}^{m} G\left(\alpha + s_j, \left(t_j + \frac{1}{\beta_l^{(1)}}\right)^{-1}\right), \qquad j = 1,\ldots,p, \tag{20}$$

whereas

$$[\beta \uparrow Y] = \frac{1}{m}\sum_{l=1}^{m} \mathrm{IG}\left(\gamma + \alpha p, \sum \lambda_{jl}^{(i)} + \delta\right). \tag{21}$$

Rubin's importance-sampling algorithm is applicable in the setting (16) as well, taking a particularly simple form in the cases $k = 2, 3$. For $k = 3$, suppose that we seek $[\theta_1|y]$. The joint density $[\theta_1, \theta_2, \theta_3|Y] = [Y, \theta_1, \theta_2, \theta_3]/[Y]$, where the functional form of the numerator is given in (16). An importance-sampling density for $[\theta_1, \theta_2, \theta_3|Y]$ could be sampled as $[\theta_1|Y, \theta_2] * [\theta_3|\theta_2] * [\theta_2|Y]_s$ for some $[\theta_2|Y]_s$. As remarked in Section 2.5, a good choice for $[\theta_2|Y]_s$ might be obtained through a few iterations of the substitution-sampling algorithm. In any case, for $l = 1, \ldots, N$ we would generate θ_{2l} from $[\theta_2|Y]_s$, θ_{3l} from $[\theta_3|\theta_{2l}]$, and θ_{1l} from $[\theta_1|Y, \theta_{2l}]$. Calculating

$$r_l = \frac{[Y, \theta_{1l}, \theta_{2l}, \theta_{3l}]}{[\theta_{1l}|Y, \theta_{2l}] * [\theta_{3l}|\theta_{2l}] * [\theta_{2l}|Y]_s},$$

we obtain the density estimator $[\theta_1 \uparrow Y] = \sum[\theta_1|Y, \theta_{2l}]r_l / \sum r_l$. Note that (in the terminology of Rubin) the algorithm in this case can be streamlined by writing the joint density in the numerator of r_l as $[\theta_{1l}|Y, \theta_{2l}] * [Y|\theta_{2l}] * [\theta_{2l}|\theta_{3l}] * [\theta_{31}]$ and noting that r_l does not involve θ_{1l}, so we need not actually generate the θ_{1l}.

Returning to the exchangeable Poisson model, the estimator of the marginal density of λ_j under Rubin's importance-sampling algorithm is

$$[\lambda_j|Y] = \sum_{l=1}^{N} G\left(\alpha + s_j, \left(t_j + \frac{1}{\beta_t}\right)^{-1}\right) r_l \Big/ \sum_{l=1}^{N} r_l.$$

Here $r_l = [Y|\beta_l] * [\beta_l]/[\beta_l|Y]_s$, where $[Y|\beta_l]$ is the product of negative binomial densities; that is,

$$[Y|\beta_l] = \prod_{j=1}^{p} \left(\frac{\Gamma(s_j + \alpha) t_j^{s_j} \beta_l^{\alpha}}{s_j! \Gamma(\alpha)(t_j + \beta_l)^{s_j + \alpha}} \right),$$

and $[\beta_l]$ is the IG prior evaluated at β_l. If $[\beta|Y]_s$ is not obtained from the substitution-sampling algorithm, as in (21), an alternative choice is $\text{IG}(\gamma + \alpha p, \sum_{i=1}^{p} \rho_i + 1)$. This arises because $[\beta|Y] = E_{[\lambda_1, \ldots, \lambda_p|Y]} \cdot [\beta|Y, \lambda_1, \ldots, \lambda_p] \approx [\beta|Y, \hat{\lambda}_1, \ldots, \hat{\lambda}_p]$, using $\hat{\lambda}_j = \rho_j$ in (19).

3.3. Multivariate Normal Sampling

A commonly occurring problem in combining continuous multivariate data is that often not all variables are observed for each experimental unit (e.g., see Dempster et al., 1977). If the data are sampled from multivariate normal populations with conjugate priors for the mean and covariance matrix, we have a general class of models where all full conditional distributions and at least some reduced conditional distributions are available. We illustrate in the simplest case, where we assume that $\binom{U_{1i}}{U_{2i}}$ $(i = 1, \ldots, n_1)$, $\binom{V_{1j}}{V_{2j}}$ $(j = 1, \ldots, n_2)$, and $\binom{W_{1k}}{W_{2k}}$ $(k = 1, \ldots, n_3)$ are all iid $N(\theta, \Delta)$ with $\theta \sim N(\mu, \sum)$, where $\theta = \binom{\theta_1}{\theta_2}$ is not observable but $\mu = \binom{\mu_1}{\mu_2}$, Δ, and $\sum$ are assumed known. Let $U = \binom{U_1}{U_2} = \binom{U_{11} \ldots U_{1n_1}}{U_{21} \ldots U_{2n_1}}$, with similar notation for V and W. Finally, let $X = (U, V, W)$ and $2 \times N$, with $\overline{X} = N^{-1} X 1$, where 1 is a column vector of N 1s and $N = n_1 + n_2 + n_3$. Standard calculations show that $[\theta | X]$ is $N(\eta, \Omega)$, where $\eta = (N\Delta^{-1} + \sum^{-1})^{-1}(N\Delta^{-1}\overline{X} + \sum^{-1}\mu)$ and $\Omega = (N\Delta^{-1} + \sum^{-1})^{-1}$. With the obvious partitioning, $\eta = \binom{\eta_1}{\eta_2}$, $\Omega = \begin{pmatrix} \Omega_{11} & \Omega_{12} \\ \Omega_{21} & \Omega_{22} \end{pmatrix}$, the marginals $[\theta_1 | X] = N(\eta_1, \Omega_{11})$, and $[\theta_2 | X] = N(\eta_2, \Omega_{22})$ are available. Suppose, however, that V_2 and W_1, say, are unobserved. Let $Y = (U, V_1, W_2)$ and $Z = (V_2, W_1)$ so that $X \equiv (Y, Z)$. As in Section 3.1, we have a three-variable problem, here involving θ_1, θ_2, and Z. The full conditional distributions are all normal and hence available. For θ_1 and θ_2, $[\theta_1 | Y, Z, \theta_2] = N(\eta_1 + \Omega_{12}\Omega_{22}^{-1}(\theta_2 - \eta_2), \Omega_{11} - \Omega_{12}\Omega_{22}^{-1}\Omega_{21})$, and $[\theta_2 | Y, Z, \theta_1] = N(\eta_2 + \Omega_{21}\Omega_{11}^{-1}(\theta_1 - \eta_1), \Omega_{22} - \Omega_{21}\Omega_{11}^{-1}\Omega_{12})$. Letting $\overline{U}_1 = n_1^{-1} U_1 1$, with similar notation for $\overline{U}_2, \overline{V}_1, \overline{V}_2, \overline{W}_1$, and $\overline{W}_2$, we note by sufficiency that with regard to Z we only need the full posterior $[\overline{Z} | \overline{Y}, \theta_1, \theta_2]$, where $\overline{Z}^T = (\overline{V}_2, \overline{W}_1)$ and $\overline{Y}^T = (\overline{U}_1, \overline{U}_2, \overline{V}_1, \overline{W}_2)$. Since

$$\overline{X}^T \equiv (\overline{U}_1, \overline{U}_2, \overline{V}_1, \overline{V}_2, \overline{W}_1, \overline{W}_2) | (\theta_1, \theta_2)$$

$$\sim N\left(\begin{pmatrix} \theta \\ \theta \\ \theta \end{pmatrix}, \begin{pmatrix} n_1^{-1}\Delta & 0 & 0 \\ 0 & n_2^{-1}\Delta & 0 \\ 0 & 0 & n_3^{-1}\Delta \end{pmatrix} \right),$$

the conditional distribution $[\overline{Z} | \overline{Y}, \theta_1, \theta_2]$ is clearly normal. With the full conditionals and the reduced conditionals $[\theta_1 | Y, Z]$ and $[\theta_2 | Y, Z]$ available, the accelerated substitution algorithm can be used to obtain $[\theta_1 | Y]$ and $[\theta_2 | Y]$.

The Rubin importance-sampling algorithm is straightforward in this case. Simplifying notation by working with the sufficient statistic $(\overline{Y}, \overline{Z})$, suppose that we seek the density estimator of $[\theta_1 | Y]$ for instance. We have $[\theta_1 | Y] = \sum [\theta_1 | \overline{Y}, \overline{Z}_l, \theta_{2l}] r_l / \sum r_l$, where

$$r_l = \frac{[\overline{X}_l | \theta_{1l}, \theta_{2l}] * [\theta_{1l}, \theta_{2l}]}{[\theta_{1l} | \overline{Y}, \overline{Z}_l, \theta_{2l}] * [\theta_{2l} | \overline{Y}, \overline{Z}_l] * [\overline{Z}_l | \overline{Y}]_s},$$

with $\overline{X}_l \equiv (\overline{Y}, \overline{Z}_l)$ and $[\overline{Z} | \overline{Y}]_s$ a specified importance-sampling density. Thus for $l = 1, \ldots, N$ we generate $\overline{Z}_l \sim [\overline{Z} | \overline{Y}], \theta_{2l} \sim [\theta_2 | \overline{Y}, \overline{Z}_l]$, and $\theta_{1l} \sim [\theta_1 | \overline{Y}, \overline{Z}_l, \theta_{2l}]$. Again, the choice of $[\overline{Z} | \overline{Y}]_s$ could be made using a few iterations of substitution sampling, or perhaps based on the intuitively appealing estimated conditional form, $[\overline{Z} | \overline{Y}, \hat{\theta}_1, \hat{\theta}_2]$, where $\hat{\theta}_1 = (n_1 \overline{U}_1 + n_2 \overline{V}_1)/(n_1 + n_2)$ and $\hat{\theta}_2 = (n_1 \overline{U}_2 + n_3 \overline{W}_3)/(n_1 + n_3)$.

3.4. Variance Component Models

Bayesian inference for variance components has typically required subtle numerical analysis or intricate analytic approximation (e.g., as evidenced by Box and Tiao 1973, Chaps. 5 and 6). In marked contrast to such sophistication, marginal posterior densities for variance components are readily obtained through simple Gibbs sampling.

We illustrate this for the simplest variance components model defined by $Y_{ij} = \theta_i + \varepsilon_{ij}$ $(i = 1, \ldots, K, j = 1, \ldots, J)$, where, assuming conditional independence throughout, $[\theta_i|\mu, \sigma_\theta^2] = N(\mu, \sigma_\theta^2)$ and $[\varepsilon_{ij}|\sigma_e^2] = N(0, \sigma_e^2)$, so $[Y_{ij}|\theta_i, \sigma_e^2] = N(\theta_i, \sigma_e^2)$.

Let $\theta = (\theta_1, \ldots, \theta_K)$ and $Y = (Y_{11}, \ldots, Y_{kJ})$ and assume that μ, σ_θ^2, and σ_e^2 are independent, with priors specified by $[\mu] \sim N(\mu_0, \sigma_0^2)$, $[\sigma_\theta^2] \sim \mathrm{IG}(a_1, b_1)$, and $[\sigma_e^2] \sim \mathrm{IG}[a_2, b_2]$, where $\mu_0, \sigma_0^2, a_1, b_1, a_2$, and b_2 are assumed known (possibly chosen to correspond to diffuse priors).

The joint distribution $[Y, \theta, \mu, \sigma_\theta^2, \sigma_e^2]$ can be written as

$$[Y|\theta, \sigma_e^2] * [\theta|\mu, \sigma_\theta^2] * [\mu] * [\sigma_\theta^2] * [\sigma_e^2], \tag{22}$$

and we follow Box and Tiao (1973, Chap. 5) in focusing interest on $[\sigma_\theta^2|Y]$ and $[\sigma_e^2|Y]$.

From the Gibbs sampling perspective, we have a four-variable system, $(\theta, \mu, \sigma_\theta^2, \sigma_e^2)$, with the following full conditional distributions:

$$\begin{aligned}
[\sigma_\theta^2|Y, \mu, \theta, \sigma_e^2] &= [\sigma_\theta^2|\mu, \theta] \\
&= \mathrm{IG}(a_1 + \tfrac{1}{2}K, b_1 + \tfrac{1}{2}\textstyle\sum(\theta_i - \mu)^2), \\
[\sigma_e^2|Y, \mu, \theta, \sigma_\theta^2] &= [\sigma_e^2|Y, \theta] \\
&= \mathrm{IG}(a_2 + \tfrac{1}{2}KJ, b_2 + \tfrac{1}{2}\textstyle\sum\sum(Y_{ij} - \mu_i)^2), \\
[\mu|Y, \theta, \sigma_\theta^2, \sigma_e^2] &= [\mu|\sigma_\theta^2, \theta] \\
&= N\left(\frac{\sigma_\theta^2\mu_0 + \sigma_0^2\sum\theta_i}{\sigma_\theta^2 + K\sigma_0^2}, \frac{\sigma_\theta^2\sigma_0^2}{\sigma_\theta^2 + K\sigma_0^2}\right),
\end{aligned}$$

and

$$[\theta|Y, \mu, \sigma_\theta^2, \sigma_e^2] = N\left(\frac{J\sigma_\theta^2}{J\sigma_\theta^2 + \sigma_e^2}\,\overline{Y} + \frac{\sigma_e^2}{J\sigma_\theta^2 + \sigma_e^2}\,\mu 1, \frac{\sigma_\theta^2\sigma_e^2}{J\sigma_\theta^2 + \sigma_e^2}\,I\right),$$

where $\overline{Y}^T = (\overline{Y}_1, \ldots, \overline{Y}_{I.})$, $\overline{Y}_{i.} = (1/J)\sum_{j=1}^J Y_{ij}$, 1 is a $K \times 1$ column vector of 1s, and I is a $K \times K$ identity matrix.

Since all of these full conditionals are available, implementation of the Gibbs sampler is straightforward. Moreover, extensions to more elaborate variance component models follow precisely the same pattern, since the full conditional distributions for μ and θ continue to be normal, and those for the variance components continue to be IG.

3.5. Normal Means Model

The exchangeable k-group normal means model with different, unknown measurement variances in each group provides a simple example of an unbalanced class of models that has proved difficult to handle using empirical Bayes approaches to estimating posterior distributions (e.g., see Morris 1983b, 1987). Such models are straightforwardly handled by iterative sampling approaches, as we saw with the Poisson example of Section 3.2 and further illustrate here for this classical normal means example.

Suppose, then, assuming conditional independence throughout, that $Y_{ij} \sim N(\theta_i, \sigma_i^2)$, $\theta_i \sim N(\mu, \tau^2)$, $\sigma_i^2 \sim \mathrm{IG}(a_1, b_1)$ $(i = 1, \ldots, I, j = 1, \ldots, J_i)$, $\mu \sim N(\mu_0, \sigma_0^2)$, and $\tau^2 \sim \mathrm{IG}(a_2, b_2)$, where μ_0, σ_0^2, a_1, b_1, a_2, and b_2 are assumed known (possibly chosen to reflect diffuse prior information). By sufficiency, we can confine attention to $Y = \{(\overline{Y}_i, S_i^2); i = 1, \ldots, I\}$, where $\overline{Y}_i = (1/J_i) \sum Y_{ij}$ and $S_i^2 = (1/J_i) \sum (Y_{ij} - \overline{Y}_{i.})^2$. Then, if we write $\theta = (\theta_1, \ldots, \theta_i)$ and $\sigma^2 = (\sigma_1^2, \ldots, \sigma_I^2)$, the joint distribution of Y, θ, σ^2, μ and τ^2 takes the form

$$[Y|\theta, \sigma^2] * [\theta|\mu, \tau^2] * [\sigma^2] * [\mu] * [\tau^2], \tag{23}$$

where

$$[Y|\theta, \sigma^2] * [\theta|\mu, \tau^2] * [\sigma^2] = \prod_{i=1}^{I} [Y_i|\theta_i, \sigma_i^2] * [S_i^2|\sigma_i^2] * [\theta_i|\mu, \tau^2] * [\sigma_i^2].$$

Of course, there is an obvious similarity between (22) and (23), but here we focus on $[\theta_i|Y]$ $(i = 1, \ldots, I)$. From the Gibbs sampling perspective, this is a $(2I + 2)$-variable problem: (θ_i, σ_i^2) $(i = 1, \ldots, I)$, together with μ and τ^2. To identify the forms of the full conditionals, we first note that

$$[\theta|Y, \sigma^2, \mu, \tau^2] = N(\theta^*, D^*), \tag{24}$$

where $\theta_i^* = (J_i \overline{Y}_{i.} \tau^2 + \mu\sigma_i^2)/(J_i\tau^2 + \sigma_i^2)$, $D_{ii}^* = \sigma_i^2\tau^2/(J_i\tau^2 + \sigma_i^2)$, and $D_{ij}^* = 0$ $(i \neq j)$. Thus the full conditional distributions $[\theta_i|Y, \theta_i j \neq i, \sigma^2, \mu, \tau^2]$ $(i = 1, \ldots, I)$ are just the normal marginals of (24) and therefore available for sampling. From (23), we easily see that $[\sigma^2|Y, \theta, \mu, \tau^2] = [\sigma^2|Y, \theta] = \prod_{i=1}^{I} [\sigma_i^2|\overline{Y}_{i.}, S_i^2, \theta_i]$, where $[\sigma_i^2|\overline{Y}_{i.}, S_i^2, \theta_i] = \mathrm{IG}(a_1 + \frac{1}{2}J_i, b_1 + \frac{1}{2}\sum_j (Y_{ij} - \theta_i)^2)$. Finally, and closely resembling the forms obtained in Section 3.4,

$$[\mu|Y, \theta, \sigma^2, \tau^2] = [\mu|\theta, \tau^2] = N\left(\frac{\tau^2\mu_0 + \sigma_0^2 \sum \theta_i}{\tau^2 + I\sigma_0^2}, \frac{\tau^2\sigma_0^2}{\tau^2 + I\sigma_0^2}\right),$$

and $[\tau^2|Y, \theta, \sigma^2, \mu] = [\tau^2|\theta, \mu] = \mathrm{IG}(a_2 + \frac{1}{2}I, b_2 + \frac{1}{2}\sum(\theta_i - \mu)^2)$.

3.6. An Errors-in-Variables Model

Again, we consider a simple illustrative special case. Consider Y to be a vector of responses assumed related to levels X of a covariate according to the straight-line model

$$Y \sim N\left((1X)\begin{pmatrix}\theta_1\\ \theta_2\end{pmatrix}, \sigma^2 I\right).$$

Responses are obtained at specified levels X_0 of the covariate, but suppose that these are not the actual levels X_a. Rather, given the former, beliefs about the latter are represented by $X_a \sim N(X_0, \tau^2 I)$. Interest centers on $\theta = (\theta_1, \theta_2)$, and to complete the distributional specification, suppose that we place independent conjugate priors on θ, σ^2, and τ^2. The joint distribution on $(Y, X_a, \theta, \sigma^2, \tau^2)$ then has the form

$$[Y|X_a, \theta, \sigma^2] * [X_a|\tau^2] * [\tau^2] * [\theta] * [\sigma^2], \tag{25}$$

where again there is obvious similarity to (22) and (23). The Gibbs sampler requires $[\theta| Y, X_a, \sigma^2, \tau^2] = [\theta| Y, X_a, \sigma^2]$, $[\sigma^2| Y, X_a, \theta, \tau^2] = [\sigma^2| Y, X_a, \theta]$, $[\tau^2| Y, X_a, \theta, \sigma^2] = [\tau^2|X_a]$, and $[X_a| Y, \theta, \sigma^2, \tau^2]$. If we assume a normal prior for θ and IG priors for σ^2 and τ^2, we obtain normal full conditional for θ and X_a and IG full conditionals for σ^2 and τ^2. We omit the details, which are somewhat similar to those in Sections 3.4 and 3.5.

4. Numerical Illustrations

4.1. A Multinomial Model

We provide some preliminary insights into the relative performance and properties of the substitution-, Gibbs, and Rubin importance-sampling approaches by considering an artificial problem based on the class of multinomial models discussed in Section 3.1.

We suppose that data $Y = (Y_1, Y_2, Y_3, Y_4, Y_5) = (14, 1, 1, 1, 5)$ are available as a sample from the multinomial distribution mult(22, $\frac{1}{4}\theta + \frac{1}{8}$, $\frac{1}{4}\eta$, $\frac{1}{4}\eta + \frac{3}{8}, \frac{1}{2}(1 - \theta - \eta)$), and that the prior for (θ, η) is taken to be a Dirichlet (1, 1, 1) distribution. In the general notation of Section 3.1, we therefore have $a_1 = \frac{1}{4}, b_1 = \frac{1}{8}, a_2 = \frac{1}{4}, b_2 = 0, a_3 = \frac{1}{4}, b_3 = 0, a_4 = \frac{1}{4}, b_4 = \frac{3}{8}$, and $\alpha_1 = \alpha_2 = \alpha_3 = 1$, with interest centering on the calculation of the marginal posterior densities $[\theta| Y]$ and $[\eta| Y]$.

By considering instead a split-cell multinomial, which in this case takes the form

$$X = (X_1, X_2, \ldots, X_7) \sim \text{mult}(22, \tfrac{1}{4}\theta, \tfrac{1}{8}, \tfrac{1}{4}\theta, \tfrac{1}{4}\eta, \tfrac{1}{4}\eta, \tfrac{3}{8}, \tfrac{1}{2}(1 - \theta - \eta)),$$

we can use the analysis of Section 3.1 for this special case of a seven-cell multinomial to construct substitution and Gibbs sampling algorithms involving θ, η, and $Z = (X_1, X_5)$.

As noted in Section 3.1, we can compare the two forms of iterative sampling. To do so, we first obtained very accurate numerical estimates of $[\theta|Y]$ and $[\eta|Y]$ using techniques described by Smith et al. (1985, 1987), and from these obtained the true 5, 25, 50, 75, and 95 posterior percentile points for each parameter. Iterative cycles of the two samplers were then run, calibrated so that the total number of random variates generated was the same in both cases (as described in Sec. 2.4). The initialization was defined (for an arbitrary generating seed) in each case by taking independent samples from $\theta \sim U(0,1)$ and $\eta \sim U(0,1)$, subject to $0 \leq \theta + \eta \leq 1$. At each cycle, $m = 10$ drawings of the parameters were then made, and from estimates of the form (9) estimates of the cumulative posterior probabilities corresponding to each of the five true percentile points for each parameter were obtained. This process was replicated 5,000 times, enabling us to study the mean estimates of the cumulative probabilities, together with their standard errors, as well as the percentage of occasions on which each sampler was closest to the true value. A summary of the results following each of the first four cycles is given in Table 1.

Table 1. Comparison of Substitution (S) and Gibbs (G) Samplers.

		Estimate (SE)				S closer than G	
		θ		η			
Cycle	cdf value	G	S	G	S	θ	η
1	.05	.231 (.08)	.217 (.08)	.033 (.01)	.044 (.01)	56%	75%
	.25	.504 (.10)	.492 (.09)	.177 (.04)	.225 (.04)	55%	78%
	.50	.713 (.08)	.706 (.08)	.380 (.06)	.459 (.06)	54%	80%
	.75	.873 (.05)	.871 (.05)	.620 (.06)	.706 (.06)	51%	80%
	.95	.978 (.01)	.978 (.01)	.878 (.04)	.926 (.03)	49%	80%
2	.05	.067 (.04)	.055 (.03)	.047 (.01)	.048 (.01)	56%	51%
	.25	.286 (.07)	.266 (.07)	.236 (.04)	.241 (.04)	56%	52%
	.50	.535 (.08)	.522 (.07)	.478 (.06)	.487 (.06)	53%	52%
	.75	.773 (.06)	.768 (.05)	.728 (.05)	.737 (.05)	51%	52%
	.95	.956 (.02)	.956 (.02)	.940 (.02)	.944 (.02)	51%	53%
3	.05	.052 (.03)	.049 (.03)	.049 (.01)	.049 (.01)	51%	50%
	.25	.254 (.06)	.252 (.06)	.247 (.04)	.247 (.04)	51%	50%
	.50	.505 (.07)	.508 (.07)	.496 (.06)	.496 (.06)	51%	49%
	.75	.754 (.06)	.760 (.05)	.746 (.05)	.747 (.05)	51%	50%
	.95	.951 (.02)	.954 (.02)	.949 (.02)	.949 (.02)	51%	50%
4	.05	.050 (.03)	.047 (.03)	.050 (.01)	.050 (.01)	51%	51%
	.25	.250 (.06)	.249 (.06)	.250 (.04)	.249 (.04)	50%	51%
	.50	.500 (.07)	.505 (.07)	.499 (.06)	.499 (.06)	51%	51%
	.75	.751 (.06)	.757 (.05)	.750 (.05)	.751 (.05)	51%	51%
	.95	.950 (.02)	.953 (.02)	.950 (.02)	.951 (.02)	51%	49%

Note: Standard errors (SE's) are in parentheses.

Table 2. Estimates from the Rubin Importance-Sampling Algorithm.

	Estimates: $m = 40$ (200)	
cdf value	θ	η
.05	.105 (.150)	.049 (.049)
.25	.311 (.351)	.244 (.241)
.50	.521 (.537)	.485 (.477)
.75	.739 (.734)	.729 (.714)
.95	.939 (.932)	.934 (.921)

We note from Table 1 that initially (cycles 1 and 2) the substitution sampler adapts more quickly than the Gibbs sampler, particularly for η. By the time we reach the third and fourth cycles, however, the two approaches are performing indistinguishably. What is astonishing, perhaps, is how remarkably good their performance is. By the fourth cycle, using only $m = 10$ drawings and starting from a default noninformative baseline, the marginal posterior density estimators based on (8) are providing on average extremely accurate estimates of cumulative probabilities. Our experiences with this and other examples (see Sec. 4.2) suggest that satisfactory convergence with iterative sampling requires only a small fraction of the levels of random variate generation reported by Tanner and Wong (1987).

The noniterative Rubin importance-sampling algorithm (Sec. 2.5) requires us to choose a sampling density, $[Z|Y]_s$, and then to proceed as follows, for $l = 1, \ldots, m$: Draw Z_l from $[Z|Y]_s$, η_l from $[\eta|Z, Y]$, and θ_l from $[\theta\eta, Z, Y]$, with the latter two distributions as detailed previously, thus creating a triple (θ_l, η_l, Z_l). Then, calculate

$$r_l = \frac{[Y, Z_l|\theta_l, \eta_l] * [\theta_l, \eta_l]}{[\theta_l|\eta_l, Z_l, Y] * [\eta_l|Z_l, Y] * [Z_l|Y]_s},$$

and form estimates $[\theta \uparrow Y] = \sum_{l=1}^{m} [\theta|\eta_l, Z_l, Y] r_l / \sum_{l=1}^{m} r_l$ and $[\eta \uparrow Y] = \sum_{l=1}^{m} [\eta|\theta_l, Z_l, Y] r_l / \sum_{l=1}^{m} r_l$.

Table 2 shows the average cumulative posterior probability estimates from this approach, based on 2,500 replicates of $m = 40$ and $m = 200$ and taking $[Z|Y]_s$ to be the product of $X_1 \sim \text{binomial}(Y_1, \frac{1}{2})$ and $X_5 \sim \text{binomial}(Y_4, \frac{1}{2})$. Despite the much larger number of drawings compared with the iterative samplers, the estimation is rather poor. In general, experience suggests that the algorithm is highly sensitive to the choice of $[Z|Y]_s$ and that the larger one-off simulation is no match for iterative adaptation via small simulations.

4.2. A Conjugate Hierarchical Model

We apply the exchangeable Poisson model discussed in Section 3.2 to data on pump failures previously analyzed by Gaver and O'Muircheartaigh

Table 3. Pump-Failure Data.

Pump system	s_i	t_i	$\rho_i\ (\times 10^2)$
1	5	94.320	5.3
2	1	15.720	6.4
3	5	62.880	8.0
4	14	125.760	11.1
5	3	5.240	57.3
6	19	31.440	60.4
7	1	1.048	95.4
8	1	1.048	95.4
9	4	2.096	191.0
10	22	10.480	209.9

(1987) (reproduced here in Table 3), where s_i is the number of failures and t_i is the length of time in thousands of hours.

Recalling the model structure of Section 3.2 and the forms of conditional distribution given by (18) and (19), we illustrate the use of the Gibbs sampler for this data set, with $p = 10$, $\delta = 1, \gamma = 0.1$, and, for the purposes of illustration, $\alpha = \bar{\rho}^2/(S_\rho^2 - p^{-1}\bar{\rho}\sum_{t=1}^{p} t_t^{-1})$, with the latter derived by a method-of-moments empirical Bayes argument based on $E(\rho_t) = EE(\rho_t|\lambda_t) = \alpha/\beta \approx \bar{\rho}$:

$$\begin{aligned} V(\rho_t) &= VE(\rho_t|\lambda) + EV(\rho_t|\lambda_t) \\ &= (\alpha/\beta^2) + (\alpha/\beta t_i) \approx S_\rho^2 = p^{-1}\sum(\rho_i - \bar{\rho})^2. \end{aligned}$$

Figure 1 shows a selection of four marginal posterior densities (for $\lambda_2, \lambda_4, \lambda_8, \lambda_9$) calculated from (20) following a run of 10 cycles of the algorithm. In fact, three densities are superposed: One corresponds to $m = 10$, one to $m = 100$, and the third is the exact density calculated using techniques described by Smith et al. (1985, 1987). Even in the cases of λ_8 and λ_9 (chosen as worst cases from $\lambda_1, \ldots, \lambda_{10}$), the densities are hardly distinguishable—a remarkable convergence from such a small number of drawings.

5. Discussion

We have emphasized providing a comparative review and explication of three possible sampling approaches to the calculation of intractable marginal densities. The substitution-, Gibbs, and importance-sampling algorithms are all straightforward to implement in several frequently occurring practical situations, thus avoiding complicated numerical or analytic approximation exercises (often necessitating intricate attention to reparameterization and other subtleties requiring case-by-case consideration). For this

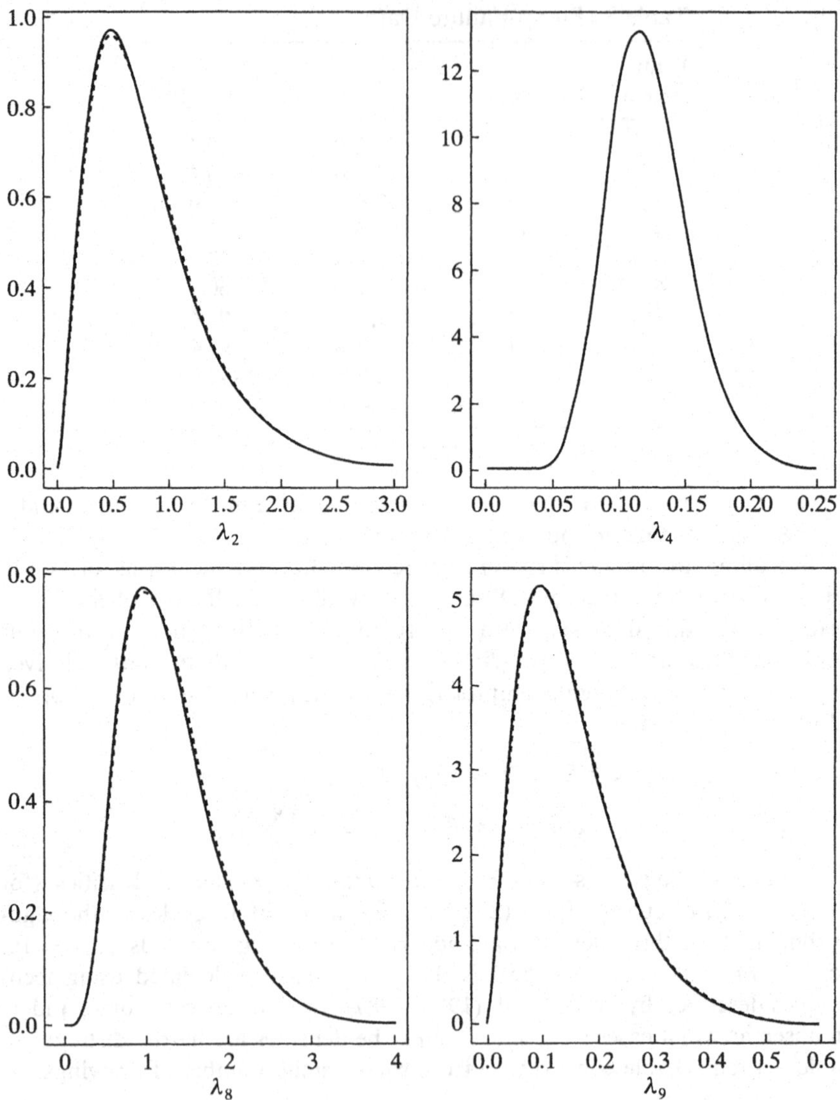

Figure 1. Density estimates for pump-failure data: $\cdots, m = 10$; - - -, $m = 100$; —, exact.

latter reason if for no other the techniques deserve to be better known and experimented with for a wide range of problems. We hope that the unified exposition attempted here will provide a general, clarifying perspective within which to view the work of Geman and Geman (1984), Rubin (1987, 1988), and Tanner and Wong (1987), and to evaluate its potential for other structured problems. For example, in addition to the model structures given in Section 3, the methods find immediate and powerful application to problems involving ordered parameters or change points. In future work we shall provide detailed and extensive numerical illustration of many such problems.

The preliminary computational experience reported here illustrates the following points. Iterative, adaptive sampling (substitution or Gibbs) invariably provides better value, in terms of efficient use of generated variates, than an equivalent sample-size, noniterative, one-off approach (Rubin), provided a suitable structure for iterative sampling exists. In problems where certain reduced conditionals are available, there is scope for accelerating the substitution algorithm so that it becomes more efficient (particularly in early cycles) than the Gibbs algorithm; however, the gain in efficiency is only likely to be of consequence when the number of reduced conditionals is a relatively large fraction of the total number of conditionals involved in a cycle. There are important practical problems in tuning monitoring and stopping-rules procedures for iterative sampling in large-scale complex problems; we shall report on these in future work as well. Finally, we note that even in cases where ultimate convergence of the iterative sampling procedures proves slow, moment or other information provided by a few initial cycles can be used to provide highly effective starting values for more sophisticated numerical or analytic approximation techniques.

References

Besag, J. (1974). Spatial interaction and the statistical analysis of lattice systems (with discussion). *J. Roy. Statist. Soc. Ser. B*, **36**, 192–326.

Box, G.E.P. and Tiao, G.C. (1973). *Bayesian Inference in Statistical Analysis*. Addison-Wesley, Reading, MA.

Dempster, A., Laird, N., and Rubin, D.B. (1977). Maximum likelihood from incomplete data via the EM algorithm (with discussion). *J. Roy. Statist. Soc. Ser. B*, **39**, 1–38.

Devroye, L. (1986). *Non-uniform Random Variate Generation*. Springer-Verlag, New York.

Devroye, L. and Györfi, L. (1985). *Non-parametric Density Estimation: The L_1 View*. Wiley, New York.

Gaver, D. and O'Muircheartaigh, I. (1987). Robust empirical Bayes analysis of event rates. *Technometrics*, **29**, 1–15.

Geman, S. and Geman, D. (1984). Stochastic relaxation, Gibbs distributions and the Bayesian restoration of images. *IEEE Tran. Pattern Anal. Machine Intell.* **6**, 721–741.

Geweke, J. (1988). Antithetic acceleration of Monte Carlo integration in Bayesian inference. *J. Econometrics*, **38**, 73–90.

Glick, N. (1974). Consistency conditions for probability estimators and integrals of density estimators. *Utilitas Math.* **6**, 61–74.

Hartley, H. (1958). Maximum likelihood estimation from incomplete data. *Biometrics*, **14**, 174–194.

Hastings, W.K. (1970). Monte Carlo sampling methods using Markov chains and their applications. *Biometrika*, **87**, 97–109.

Morris, C. (1983a). Natural exponential families with quadratic variance functions: Statistical theory. *Ann. Statist.* **11**, 515–529.

Morris, C. (1983b). Parametric empirical Bayes inference: Theory and applications. *J. Amer. Statist. Assoc.* **78**, 47–59.

Morris, C. (1987). Determining the accuracy of Bayesian empirical Bayes estimates in familiar exponential families. In *Statistical Decision Theory and Related Topics*, vol. 4. (S.S. Gupta and J.O. Berger, eds.). Springer-Verlag, New York, pp. 251–264.

Naylor, J.C. and Smith, A.F.M. (1982). Applications of a method for the efficient computation of posterior distributions. *Appl. Statist.* **31**, 214–225.

Naylor, J.C. and Smith, A.F.M. (1988). Econometric illustrations of novel numerical integration strategies for Bayesian inferences. *J. Econometrics*, **38**, 103–126.

Rall, L. (1969). *Computational Solution of Non-linear Operator Equations*. Wiley, New York.

Ripley, B. (1987). *Stochastic Simulation*. Wiley, New York.

Rubin, D.B. (1987). Comment on "The Calculation of Posterior Distributions by Data Augmentation," by M.A. Tanner and W.H. Wong. *J. Amer. Statist. Assoc.*, **82**, 543–546.

Rubin, D.B. (1988). Using the SIR algorithm to simulate posterior distributions. In *Bayesian Statistics 3* (J.M. Bernardo, M.H. DeGroot, D.V. Lindley, and A.F.M. Smith eds.). Oxford University Press, Oxford, pp. 395–402.

Shaw, J.E.H. (1988). A quasirandom approach to integration in Bayesian statistics. *Ann. Statist.*, **16**, 895–914.

Smith, A.F.M., Skene, A.M., Shaw, J.E.H., and Naylor, J.C. (1987). Progress with numerical and graphical methods for Bayesian statistics. *The Statistician*, **36**, 75–82.

Smith, A.F.M., Skene, A.M., Shaw, J.E.H., Naylor, J.C., and Dransfield, M. (1985). The Implementation of the Bayesian Paradigm. *Communications in Statistics–Theory and Methods*, vol. 14, pp. 1079–1102.

Tanner, M. and Wong, W. (1987). The calculation of posterior distributions by data augmentation (with discussion). *J. Amer. Statist. Assoc.*, **82**, 528–550.

Tierney, L. and Kadane, J. (1986). Accurate approximations for posterior moments and marginal densities. *J. Amer. Statist. Assoc.*, **81**, 82–86.

Index

Springer Series in Statistics

(continued from p. ii)

Pollard: Convergence of Stochastic Processes.
Pratt/Gibbons: Concepts of Nonparametric Theory.
Ramsay/Silverman: Functional Data Analysis.
Read/Cressie: Goodness-of-Fit Statistics for Discrete Multivariate Data.
Reinsel: Elements of Multivariate Time Series Analysis, 2nd edition.
Reiss: A Course on Point Processes.
Reiss: Approximate Distributions of Order Statistics: With Applications to Non-parametric Statistics.
Rieder: Robust Asymptotic Statistics.
Rosenbaum: Observational Studies.
Ross: Nonlinear Estimation.
Sachs: Applied Statistics: A Handbook of Techniques, 2nd edition.
Särndal/Swensson/Wretman: Model Assisted Survey Sampling.
Schervish: Theory of Statistics.
Seneta: Non-Negative Matrices and Markov Chains, 2nd edition.
Shao/Tu: The Jackknife and Bootstrap.
Siegmund: Sequential Analysis: Tests and Confidence Intervals.
Simonoff: Smoothing Methods in Statistics.
Small: The Statistical Theory of Shape.
Tanner: Tools for Statistical Inference: Methods for the Exploration of Posterior Distributions and Likelihood Functions, 3rd edition.
Tong: The Multivariate Normal Distribution.
van der Vaart/Wellner: Weak Convergence and Empirical Processes: With Applications to Statistics.
Vapnik: Estimation of Dependences Based on Empirical Data.
Weerahandi: Exact Statistical Methods for Data Analysis.
West/Harrison: Bayesian Forecasting and Dynamic Models, 2nd edition.
Wolter: Introduction to Variance Estimation.
Yaglom: Correlation Theory of Stationary and Related Random Functions I: Basic Results.
Yaglom: Correlation Theory of Stationary and Related Random Functions II: Supplementary Notes and References.

9 780387 949895